Angiosperm Pollen and Ovules

E. Ottaviano D.L. Mulcahy
M. Sari Gorla G. Bergamini Mulcahy
Editors

Angiosperm Pollen and Ovules

With 208 Illustrations

Springer-Verlag
New York Berlin Heidelberg London Paris
Tokyo Hong Kong Barcelona Budapest

E. Ottaviano (deceased)
Institute of Biology
Genetics Section
University of Milan
via Celoria 26
20133 Milan, Italy

D.L. Mulcahy
Department of Biology
University of Massachusetts
Amherst, MA 01003
USA

M. Sari Gorla
Institute of Biology
Genetics Section
University of Milan
via Celoria 26
20133 Milan, Italy

G. Bergamini Mulcahy
Department of Biology
University of Massachusetts
Amherst, MA 01003
USA

Cover figure adapted from Figure 1 of Chapter 12, "Molecular Genetics of Self-incompatibility in *Nicotiana alata*" by Shaio-Lim Mau et al.

Library of Congress Cataloging-in-Publication Data
Angiosperm pollen and ovules / E. Ottaviano . . . [et al., editors].
p. cm.
Includes bibliographical references and index.
ISBN 0-387-97888-7
1. Angiosperms—Pollen—Congresses. 2. Pollen—Congresses.
3. Ovules—Congresses. 4. Plant biotechnology—Congresses.
I. Ottaviano, Ercole.
QK658.A54 1992
582.13′04463—dc20 92-1605

Printed on acid-free paper.

Production managed by Hal Henglein; manufacturing supervised by Jacqui Ashri.
Camera-ready copy prepared by the contributors.
Printed and bound by Halliday, West Hanover, MA.
Printed in the United States of America.

9 8 7 6 5 4 3 2 1

ISBN 0-387-97888-7 Springer-Verlag New York Berlin Heidelberg
ISBN 3-540-97888-7 Springer-Verlag Berlin Heidelberg New York

Preface

The subject of this volume is the reproductive biology of plants. A steadily growing interest in this field is the result of at least two factors, as pointed out with great foresight by one of the driving forces in the field, H.F. Linskens (Linskens 1964): most of the food consumed by humans takes the form of plant reproductive parts, and molecular biology now provides powerful tools for investigating and manipulating plant reproductive systems. Molecular biology and the allied discipline of biotechnology are solidly represented in the papers in this book. The editors of *Angiosperm Pollen and Ovules* believe that the chapters herein contain some of the most exciting findings of contemporary biology, and hope that the readers of this book will share their enthusiasm.

The editors express sincere and grateful thanks for help from Carla Frova, Enrico Pe, and especially to Giorgio Binelli, all of the University of Milan. Without these three tireless and enthusiastic individuals it would not have been possible to maintain the apparently effortless proceeding of this congress. We thank also the organizing committee and the organizations that generously provided financial support (both listed below). Finally, we extend thanks to M. Cresti, D. Charlesworth, D. Hess, F. Hoekstra, R. Bruce Knox, J.P. Mascarenhas, M.E. Nasrallah, P.L. Pfahler, A. Snow, and M.T.M. Willemse for chairing sessions.

Organizing committee:

CNR—Consiglio Nazionale delle Ricerche

Special project of the Italian Research Council RAISA (Advanced Research for Innovative Agriculture)

Special project of the Italian Ministero Agricoltura e Foreste "Tecnologie Avanzate in Agricoltura" (Advanced Technologies in Agriculture)

Istituto Sperimentale per la Cerealicoltura—Rome

Department of Genetics and Microbiology, University of Milan

SIGA (Italian Society for Agricultural Genetics)

AGI (Italian Society of Genetics)

Financial supporters:

Agrimont

Amity Srl

Kontron Instruments

Perkin-Elmer Cetus

Linskens, H.F. 1964. *Pollen Physiology and Fertilization.* North-Holland Publishing Co., Amsterdam.

Ercole Ottaviano
1937–1991

Dedication

This volume is dedicated, with affection and sadness, to Prof. Ercole Ottaviano, an outstanding investigator, an excellent person, and, most rare of all, a true friend.

Contents

Participants

Abha Agnihotri, Tata Energy Research Institute, 90 Jor Bagh, New Delhi 110003, India

Diego Albani, Paladin Hybrids Inc., 2475 DonReid, Ottawa, Ontario K1H 8P5, Canada

I. Altosaar, Dept. of Biochemistry, University of Ottawa, Ottawa, Ontario K1N 6N5, Canada

Sergio Arcioni, Fac. Agraria, Istituto di Ricerche Miglioramento Genetico-CNR, Borgo XX Giugno 74, 06100 Perugia, Italy

Adela Arnal Olmedilla, C.I.B. (CSIC), Velazquez 144, 28006 Madrid, Spain

Beáta Barnabás, Agricultural Research Institute, Hungarian Academy of Sciences, Martonvasar H-2462, Hungary

Susan Barnes, Dept. of Botany, Natural History Museum, Cromwell Road, London SW7 5BD, United Kingdom

R.J. Bino, Centre for Plant Breeding and Reproduction Research, P.O. Box 32, 6700 AA Wageningen, The Netherlands

Stephen Blackmore, Dept. of Botany, Natural History Museum, Cromwell Road, London SW7 5BD, United Kingdom

Cecilia K. Blomstedt, Botany Dept., University of Melbourne, Grattan Street, Parkville, Victoria 3052, Australia

Robert Bowman, Goldsmith Seeds Inc., P.O. Box 1349, Gilroy, CA 95021, USA

Lone Bruun, Botanical Laboratory, University, Gothersgade 140, DK-1123 Copenhagen, Denmark

Aksel Buchter-Larsen, Maribo Seed, Hojbygaardvej 14, DK-4960 Holeby, Denmark

Bernd Buter, Institute of Plant Science, Swiss Federal Institute of Technology, Station Eschikon 33, 8315 Lindau, Switzerland

Annie Chaboud, RCAP INRA 23879, Université Lyon I-Bât. 741, 43, Blvd. du 11 Novembre, 69622 Villeurbanne Cédex, France

Deborah Charlesworth, Dept. of Ecology and Evolution, University of Chicago, 5630 S. Iglesias Avenue, Chicago, IL 60637, USA

S. V.S. Chauhan, Dept. of Botany, R.B.S. College, Agra, India

Fabrizio Ciampolini, Dip. degli Biologia Ambientale, Università di Studi di Siena, Via P.A. Mattioli 4, 53100 Siena, Italy

Mauro Cresti, Dip. degli Biologia Ambientale, Università di Studi di Siena, Via P.A. Mattioli 4, 53100 Siena, Italy

Mark Crispi, Dept. of Plant Biology, Nesmith Hall, University of New Hampshire, Durham, NH 03824, USA

Isabelle Dajoz, Laboratoire E.S.V., Université Paris-Sud, Bâtiment 362, 91405 Orsay Cédex, France

Rosine De Paepe, Université Paris-Sud, Bâtiment 360, 91405 Orsay Cédex, France

Maria Helena De Souza Goldman, Laboratorium voor Genetica, Rijksuniversiteit Ledeganckstraat 35, B-9000 Gent, Belgium

A.P.M. Den Nijs, Centre for Plant Breeding Research, Postbus 16, 6700 AA Wageningen, The Netherlands

Marcel Devreux, Via Privata, 11, 21020 Ranco, VA, Italy

James Doughty, School of Plant Sciences, University of Reading, Whiteknights Park, Reading RG6 2AS, England

Rivka Dulberger, Dept. of Botany, University of Tel-Aviv, Tel-Aviv, Israel

C. Dumas, RCAP INRA 23879, Université Lyon I-Bât. 741, 43, Blvd. du 11 Novembre, 69622 Villeurbanne Cédex, France

Chen Fang, Biology Dept., Sichuan University, Chengdu, 610064 China

Elisabetta Frascaroli, Istituto di Agronomia, Università di Bologna, Via F. Re, 6, 40126 Bologna, Italy

Luigi Frusciante, Istituto di Agronomia gen. e colt., Via Università 100, 80055 Portici, NA, Italy

L. Gabedava, Digomsky masive, 2 kv. 14 korp. fl. 47, Tbilisi 380059, Georgia

Antonie Gaillard, RCAP INRA 23879, Université Lyon I-Bât. 741, 43, Blvd. du 11 Novembre, 69622 Villeurbanne Cédex, France

Thierry Gaude, RCAP INRA 23879, Université Lyon I-Bât. 741, 43, Blvd. du 11 Novembre, 69622 Villeurbanne Cédex, France

Nina Gorenstein, Institute of Cytology and Genetics, Academy of Sciences, Siberian Branch, 630090 Novosibirsk -90, Russia

Virginie Guyon, Dept. of Botany-School of Plant Sciences, University of Reading, Whiteknights Park, Reading RG6 2AS, England

G.E. Gvaladze, Institute of Botany of Academy of Sciences of the Georgian SSR, Kodjorckoe Shosse 1, 380007 Tbilisi, Georgia

S.K. Hahn, International Institute of Tropical Agriculture, Oyo Road, PMB 5320 Ibadan, Nigeria

David Haig, Dept. of Plant Sciences, South Parks Road, Oxford OX1 3RA, United Kingdom

Cornelia Harte, Institut für Entwicklungsphysiologie, Universität zu Köln, Gyrhofstrasse 17, 5000 Köln 41, Germany

Minami Harufumi, Tokyo Metropolitan Isotope Research Center, 2-11-1 Fukazawa Minato-ku 158 Tokyo, Japan

J.M. Herr, Dept. of Biological Sciences, Universtiy of South Carolina, Columbia, SC 29208, USA

Jack Heslop-Harrison, 137 Bargates, Leominster, Herefordshire HR6 8QS, United Kingdom

Yolande Heslop-Harrison, 137 Bargates, Leominster, Herefordshire HR6 8QS, United Kingdom

Pat Heslop-Harrison, Karyobiology Group, JI Centre for Plants, Colney Lane, Norwich NR4 7UH, United Kingdom

Simon Hiscock, Dept. of Botany-School of Plant Sciences, University of Reading, Whiteknights Park, Reading RG6 2AS, England

Folkert A. Hoekstra, Landbouwuniversiteit, Arboretumlaan 4, 6703 BD Wageningen, The Netherlands

Franz Hoffmann, Developmental & Cellular Biology, University of California, Irvine, CA 92717, USA

M. Ikeda, Graduate School of Science and Technology, Kobe University, Kobe 657, Japan

John F. Jackson, Waite Agricultural Research Institute, Glen Osmond, South Australia 5064, Australia

William A. Jensen, Plant Biology Dept., The Ohio State University, Columbus, OH 43210, USA

Teh-Hui Kao, Molecular & Cell Biology, Pennsylvania State University, University Park, PA 16802, USA

Nadarendra Kaushik, Biochemistry Dept., Royal Holloway, University of London, Egham Hill, Egham, Surrey TW200 EX, United Kingdom

C.J. Keyzer, Dept. of Plant Cytology & Morphology, Arboretumlaan 4, 6703 BD Wageningen, The Netherlands

R.B. Knox, School of Botany, University of Melbourne, Parkville, Victoria 3052, Australia

Géza Kovács, Agricultural Research Institute, Hungarian Academy of Sciences, Martonvasar H-2462, Hungary

Lidia Kovaleva, Institute of Plant Physiology, Botanical str. 35, Moscow 127276, Russia

Erhard Kranz, Institut für Allgemeine Botanik, Universität Hamburg, Ohnhorststrasse 18, D-2000 Hamburg 52, Germany

Udo Kristen, Institut für Allgemeine Botanik, Universität Hamburg, Ohnhorststrasse 18, D-2000 Hamburg 52, Germany

Pierangelo Landi, Istituto di Agronomia, Università di Bologna, Via F. Re, 6, 40126 Bologna, Italy

Knud Larsen, Brostykkevej 83 B, DK-2650 Hvidovre, Denmark

M. Lawrence, School of Biological Sciences, University of Birmingham, Edgbaston, Birmingham B15 2TT, United Kingdom

Erwan Le Deunff, RCAP INRA 23879, Université Lyon I-Bât. 741, 43, Blvd. du 11 Novembre, 69622 Villeurbanne Cédex, France

I. Linskens, Dip. degli Biologia Ambientale, Università di Studi di Siena, Via P.A. Mattioli 4, 53100 Siena, Italy

H.F. Linskens, Dip. degli Biologia Ambientale, Università di Studi di Siena, Via P.A. Mattioli 4, 53100 Siena, Italy

D.M. Lonsdale, Dept. of Molecular Genetics, JI Centre for Plants, Colney Lane, Norwich NR4 7UH, United Kingdom

Elizabeth Lord, 60 Victoria Road, Summertown, Oxford, England

E. Lovites, Institute of Cytology and Genetics, Academy of Sciences, Siberian Branch, 630090 Novosibirsk, Russia

Anna Mariani, Istituto Richerche Miglioramento Genetico-CNR, Borgo XX Giugno 74, 06100 Perugia, Italy

Diane Marshall, Dept. of Biology, University of New Mexico, Albuquerque, NM 87131, USA

J.P. Mascarenhas, Dept. of Biological Sciences, State University of New York at Albany, Albany, NY 12222, USA

E. Matthys-Rochon, RCAP INRA 23879, Université Lyon I-Bât.741, 43, Blvd. du 11 Novembre, 69622 Villeurbanne Cédex, France

Andrew McCubbin, School of Plant Sciences, University of Reading, Whiteknights Park, Reading RG6 2AS, England

N. Melia, Digomsky masive 5 kv. 16 korp. fl. 55 Tbilisi 380059, Georgia

P.L.J. Metz, Centre for Plant Breeding Research, Postbus 16, 6700 AA Wageningen, The Netherlands

Hisako Miki-Hirosige, Kanagawa Dental College, Biological Lab., Inaokacho 82, Yokosuka City 238, Japan

Lloyd Mogensen, Dept. of Biological Sciences, Northern Arizona University, Box 5640, Flagstaff, AZ 86011, USA

J.N.M. Mol, Dept. of Genetics, Free University of Amsterdam, de Bolelaan 1087, 1081 HV Amsterdam, The Netherlands

Michele Morgante, Istituto di Produzione Vegetale, Università di Udine, P.le Kolbe 4, 33100 Udine, Italy

Rosella Motzo, Istituto di Agronomia, Università degli Studi, Via De Nicola 07100 Sassari, Italy

L.A. Mur, Dept. of Genetics, Free University of Amsterdam, de Bolelaan 1087, 1081 HV Amsterdam, The Netherlands

Alain Murigneux, BIOCEM Laboratoire, Campus Université des Cezeaux 24, Av. des Landais, 63170 Aubiere, France

T. Nakanishi, Laboratory of Pomology, Faculty of Agriculture, Kobe University, Kobe 657, Japan

Hyoji Namai, Institute of Agriculture and Forestry, University of Tsukuba, Tsukuba Ibaraki 305, Japan

M.E. Nasrallah, Section of Plant Biology, Division of Biological Sciences, Cornell University, Ithaca, NY 14853, USA

Moshe Negbi, Dept. of Agricultural Botany, Faculty of Agriculture, P.O. Box 12, Rehovot 76100, Israel

Vittoria Nuti Ronchi, Istituto di Mutagenesi e Differenziamento C.N.R., Via Svezia 10, 56124 Pisa, Italy

Margryt Oldenhof, Dept. of Molecular Plant Physiology, University of Nijmegen, Toernooiveld, 6525 ED Nijmegen, The Netherlands

Peter Olesen, Biotechnology Research Division, Langebrogade I, P.O. Box 17, 1001 Copenhagen, Denmark

Angelo Olivieri, Istituto di Produzione Vegetale, Università di Udine, P.le Kolbe 4, 33100 Udine, Italy

Ettore Pacini, Dip. degli Biologia Ambientale, Università degli Studi di Siena, Via P.A. Mattioli 4, 53100 Siena, Italy

Salome Pais, Dept. de Biologia Vegetal, Fac. Ciências de Lisboa, Bloco C2, Campo Grande, 1700 Lisboa, Portugal

Noel Pallais, International Potato Center, P.O. Box 5969, Lima, Peru

Dayakar Pareddy, United Agriseeds, Inc., DowElanco, P.O. Box 4011, Champaign, IL 61824-4011, USA

P.A. Peterson, Dept. of Agronomy, 0407 Agronomy Hall, Iowa State University, Ames, IA 50011, USA

P.L. Pfahler, Agronomy Dept., University of Florida, 304 Newell Hall, Gainesville, FL 32611-0311, USA

Marco Piccirilli, Fac. Agraria, Istituto di Richerche Miglioramento Genetico-CNR, Borgo XX Giugno 74, 06100 Perugia, Italy

Elisabeth Pierson, Dip. degli Biologia Ambientale, Università degli Studi di Siena, Via P.A. Mattioli 4, 53100 Siena, Italy

Vito Polito, Dept. of Pomology, University of California, Davis, CA 95616, USA

Holz Preben Bach, Carlsberg Research Laboratory, 1964 Carlsbergvej, DK-2500 Valby, Denmark

V. Raghavan, Dept. of Plant Biology, The Ohio State University, 1735 Neil Avenue, Columbus, OH 43210, USA

Stephen Read, School of Botany, University of Melbourne, Parkville, Victoria 3052, Australia

M.C. Risueno, Centro de Investigato Biologicas, Civelazguez 144, 28006 Madrid, Spain

T. Robert, Laboratoire E.S.V., Université Paris-Sud, Bâtiment 362, 91405 Orsay Cédex, France

Maria Isabel Rodriguez Garcia, Albareda 1, 18008 Granada, Spain

Mirelle Rougier, RCAP INRA, 23879 Université Lyon I-Bât. 741, 43, Blvd. du 11 Novembre, 69622 Villeurbanne Cédex, France

Marian Ryczkowski, Dept. of Plant Physiology, Jagellonian University, Al. Mickiewicza 3, 31-120 Krakow, Poland

Amelia Sanchez-Pina, C.I.B. (CSIC) Velazquez 144, 28006 Madrid, Spain

Pilar Sanchez-Testillano, C.I.B. (CSIC) Velazquez 144, 28006 Madrid, Spain

Michel Sandmeier, Laboratoire E.S.V., Université Paris-Sud, Bâtiment 362, 91405 Orsay Cédex, France

A. Sarr, Laboratoire E.S.V., Université Paris-Sud, Bâtiment 362, 91405 Orsay Cédex, France

Yo-Ichiro Sato, National Institute of Genetics, Shizuoka-ken 411, Mishima-city, Japan

Jan Schrauwen, A.M. Dept. of Molecular Plant Physiology, University of Nijmegen, Toernooiveld, 6525 ED Nijmegen, The Netherlands

Koike Setsuo, Hokkaido National Agricultural Experiment Station, Hitsujigaoka 1, Toyohira ku, Sapporo 062, Japan

K.R. Shivanna, Dept. of Botany, University of Delhi, Delhi 110007, India

Krishna Shrestha, c/o A.L. Takhtajan Komarov Botanical Institute, Popov street-2, Leningrad 197376, Russia

Bruce Smith, Dept. of Biology, York College of Pennsylvania, York, PA 17403, USA

Allison Snow, Dept. of Botany, Ohio State University, Columbus, OH 43210, USA

Esteban Solorzano Vega, Dept. de Fitotecnia, Universita Autonoma Chapingo, Chapingo Mexicano, C.P. 56230, Mexico

Darlene Southworth, Dept. of Biology, South Oregon State College, Ashland, OR 97520, USA

André Souvré, ENSAT, 145 Avenue de Muret, F-31076 Toulouse Cédex, France

Anna Speranza, Dip. di Biologia E.S., Via Irnerio 42, 40126 Bologna, Italy

Andrew Stephenson, Dept. of Biology, Pennsylvania State University, University Park, PA 16802, USA

Eva Stoger, Institut Mikrobiologie u. Genetik, Althanstrasse 14, 1090 Wien, Austria

M.M.C. Tan, Plant Biotechnology Division, Zaadunie, P.O. Box 26, 1600 AA Enkhuizen, The Netherlands

Stefano Tavoletti, Istituto Ricerche Miglioramento Genetico-CNR, Borgo XX Giugno 74, 06100 Perugia, Italy

Antonio Tiezzi, Dip. degli Biologia Ambientale, Università degli Studi di Siena, Via P.A. Mattioli 4, 53100 Siena, Italy

Irene Till-Bottraud, Laboratoire E.S.V., Universtité Paris-Sud, Bâtiment 362, 91405 Orsay Cédex, France

Veronica E. Franklin-Tong, School of Biological Sciences, University of Birmingham, Edgbaston, Birmingham B15 2TT, United Kingdom

Jaroslav Tupy, Dept. of Genetics, Institute of Experimental Botany, Vltavska 17, CS-150 00 Praha 5, Czechoslovakia

R.W. Van Den Bulk, Centre for Plant Breeding Research, P.O. Box 16, 6700 AA Wageningen, The Netherlands

Marjan Van Grey, C.P.O., Postbus 16, 6700 AA Wageningen, The Netherlands

Marinus Van Herpen, Dept. of Molecular Plant Physiology, University of Nijmegen, Toernooiveld, 6525 ED Nijmegen, The Netherlands

A.J. Van Tunen, Dept. of Genetics, Free University of Amsterdam, de Boelelaan 1087, 1081 HV Amsterdam, The Netherlands

Jaap Van Tuyl, C.P.O., Postbus 16, 6700 AA Wageningen, The Netherlands

G.G. Vendramini,Istituto Miglioramento Genetico Piante Forestali C.N.R., Via San Bartolomeo 13, 50145 Firenze, Italy

Fabio Veronesi, Istituto di Produzione Vegetale, Università di Udine, P.le Kolbe 4, 33100 Udine, Italy

Oscar Vicente, Institut für Mikrobiologie u. Genetik, Althanstrasse 14, A-1090 Wien, Austria

Peter Walton, Dept. of Plant Science, University of Alberta, Edmonton, Alberta T6G 2P5, Canada

E. Wehling, Institut für Angewandte Genetik, Universität Hannover, Herrenhauser Strasse 2, W-3000 Hannover 21, Germany

Koen Weterings, Dept. of Molecular Plant Physiology, University of Nijmegen, Toernooiveld, 6525 ED Nijmegen, The Netherlands

R. Wiermann, Institut für Botanik, Schlossgarten 3, 4400 Münster/Westfalia, Germany

M.T.M. Willemse, Dept. of Plant Cytology & Morphology, Arboretumlaan 4, 6703 BD Wageningen, The Netherlands

G. Wricke, Institut für Angewandte Genetik, Universität Hannover, Herrenhauser Strasse 2, W-3000 Hannover 21, Germany

George Wullems, Dept. of Molecular Plant Physiology, University of Nijmegen, Toernooiveld, 6525 ED Nijmegen, The Netherlands

Minoru Yamada, Staff Sankaido, Bldg. 7F, Akasaka 1-9-13, Minato, Tokyo 107, Japan

L. Lung-Fei Yen, Lab. of Plant Biochemistry, College of Biological Sciences, Beijing Agricultural University, Beijing 100094, China

Helen Young, Biology Dept., Barnard College, 3009 Broadway, New York, NY 10027, USA

Viktor Zarsky, Dept. of Genetics, Institute of Experimental Botany, Vltavska 17, CS-150 00 Praha 5, Czechoslovakia

S.Y. Zee, Botany Department, University of Hong Kong, Hong Kong

Maciej Zenkteler, Faculty of Biology, Adam Mickiewicz University, 61-713 Poznan, Poland

Lali Zhgenti, 4 Gldanskaya Street, Tbilisi 38012, Georgia

Guang Yu Zhou, Shanghai Institute of Biochemistry, Academia Sinica, 320 Yue-Yang Road, Shanghai 200031, China

Molecular Biology

1 Pollen Expressed Genes and Their Regulation

Joseph P. Mascarenhas

Microsporogenesis in flowering plants involves a complex and extensive program of gene expression (Mascarenhas, 1989; Mascarenhas, 1990). Most of the large number of genes that are expressed in the developing pollen grain are also expressed in sporophytic tissues. Only a small fraction of these genes are pollen-specific and are expressed just in pollen and not in other parts of the plant (Tanksley et al., 1981; Willing & Mascarenhas, 1984; Willing et al., 1988). Pollen-expressed genes from several plants have been isolated and preliminary analyses made of their promoter regions (Albani et al., 1990; Guerrero et al., 1990; Mascarenhas 1991; Twell et al., 1990, 1991; van Tunen et al., 1989, 1990).

We have isolated and characterized a genomic clone from maize, designated Zm13, that shows a pollen-specific pattern of expression (Hamilton et al., 1989; Hanson et al., 1989). Zm13 has sequence homology to a pollen-expressed gene (LAT52) from tomato (Twell et al., 1989b). Stably transformed tobacco plants containing either a 1063 bp or a 375 bp Zm13 sequence from -1002 to +61 or -314 to +61 respectively, with respect to the start site of transcription at + 1, expressed a reporter ß-glucuronidase (GUS) gene in a correct temporal and spatial manner (Guerrero et al., 1990). A more detailed dissection of the pollen-specific promoter of Zm13 has now been carried out. This has been done by transient expression of a reporter GUS gene under the control of various sized Zm 13 promoter fragments, using high velocity microprojectile bombardment to transform mature Tradescantia paludosa pollen. These results are discussed with respect to information in the literature on other plants, regarding the cis elements important in directing pollen-specific transcription.

Results and Discussion

An improved helium-driven biolistic device (Ye et al., 1990) was used to introduce DNA coated tungsten particles into mature pollen grains. The particles were coated with plasmids containing Zm13 promoter fragments fused to the GUS reporter gene and with either the nopaline synthase (NOS) or Zm13 3' terminators. Twell et al. (1989a) have previously shown that chimeric genes under the control of a tomato pollen-specific promoter were transiently expressed in tobacco pollen after introduction by high velocity microprojectile bombardment. The system retained the pollen-specificity of expression seen in stably transformed transgenic tobacco plants.

Pollen of Tradescantia paludosa was used in all the experiments because of its experimental advantages compared to maize pollen. Maize pollen is, however, transformable by microprojectile bombardment and transiently expresses a GUS reporter gene when directed by an appropriate pollen promoter. A weighed amount of pollen was spread evenly over the surface of solidified pollen medium in a small petri plate. Within

5 - 10 minutes after spreading the pollen, the plates were bombarded with DNA coated tungsten particles and then assayed for GUS activity after 14 - 20 hours, either by staining with X-glu or by quantitating ß-glucuronidase activity in extracts, using fluorescence measurements (Jefferson, 1987).

The plasmid constructs used in these experiments had various lengths of 5' Zm13 sequence, but all of them terminated at +61 of the untranslated 5' region (Hamilton et al., 1989). The constructs made with the -1001, -260 and -100 promoter fragments all stained blue with the X-glu histochemical stain, although the amount of staining was different. Bombardments of pollen with the -54 and -38 promoter constructs gave no blue staining for GUS activity. All the above constructs contained the NOS 3' terminator. The construct containing a cauliflower mosaic virus 35S (CaMV35S) promoter fused to the GUS gene when introduced into pollen gave practically no blue staining for GUS activity.

When the 35S-GUS construct was bombarded into Tradescantia leaf sections or tobacco suspension culture cells, very active GUS activity was induced. In contrast none of the Zm13 promoter fragment constructs gave blue staining when introduced into leaf tissue or tobacco cells. These results confirm those obtained by Twell et al. (1989a) and indicate that transformation of mature Tradescantia pollen by microprojectile bombardment using a reporter gene driven by a maize pollen promoter, retains the pollen-specificity of expression observed in the maize plant. Since X-glu staining for GUS activity cannot be quantitated, pollen bombarded by the various constructs was analyzed for GUS activity by the fluorometric assay. These results showed that the -260 promoter fragment was the most active. The -1001 promoter construct was about 50% as active as the -260 construct. This suggests that a negative regulatory element is present in the region between -1001 and -260 bp. Promoter activity was dramatically reduced with the -100 construct being only 6% as active as the -260 construct.

When the Zm13 3' terminator was used in place of the NOS terminator with the -100 promoter construct, promoter activity increased from 6% to almost 30%. This 6-fold increase in GUS expression with the Zm13 terminator over that directed by the NOS terminator could be due to any of several reasons. The Zm13 3' fragment could for example, have enhanced activity or it might provide enhanced stability to the messenger RNA. Additional constructs are being made with the Zm13 3' terminator to clarify its functional role in increasing reporter gene activity in transient assays.

The -54-GUS-NOS and the -38-GUS-NOS constructs showed no GUS activity in the fluorometric assay. The -260 promoter fragment showed GUS expression several fold greater than that observed with the -100-GUS-NOS construct. The -100 promoter fragment still correctly directs pollen-specific expression, although at a reduced level. When the promoter region is further deleted to -54 this results in almost total loss of GUS expression which would suggest that the element responsible for pollen-specificity of transcription, is present in the -100 to -54 region.

Twell et al. (1991) have shown that a promoter fragment containing a TGTGG motif acts as an enhanced to a truncated pollen promoter or to a heterologous minimal CaMV35S promoter. There are two such TGTGG motifs, one in each orientation, within the -260 to -100 promoter region of Zm13. The TGTGG motif is similar to the G-box DNA-protein binding region which is associated with many plant promoters such as, light regulated genes (Gilmartin et al., 1990; Green et al., 1988), ABA-regulated genes (Guiltinan et al., 1990) and Adh genes (Ferl & Nick, 1986; McKendree et al., 1990). It is also similar to the Hex sequence located in the promoter regions of the Agrobacterium NOS gene and the CaMV35S gene (Katagiri et al., 1989).

It is proposed that the -260 to -100 region of the Zm13 promoter acts in an enhanced-like fashion to amplify the expression of a pollen-specific core promoter present in the -100 to -54 region. The results presented here and others (Twell et al., 1991; van Tunen et al., 1990) indicate that the elements necessary for pollen-specificity are rather small. A comparison however, of the -100 to -54 region of the Zm13 promoter with the minimal sequence elements of other pollen-specific genes reveals no obvious consensus sequences.

The identity of cis-elements and other factors involved in determining pollen-specific expression of genes as opposed to enhanced functions has not yet been determined. Current work is being directed towards this end.

Acknowledgement

Research in the author's laboratory has been supported by a grant from the National Science Foundation.

References

1. Albani, D.; Robert, L.S.; Donaldson, P.A.; Altosaar, I.; Arnison, P.G.; Fabijanski, S.F. Characterization of a pollen- specific gene family from Brassica napus which is activated during early microspore development. Plant Mol. Biol. 15:605-622; 1990.

2. Ferl, R.J.; Nick, H.S. In vivo detection of regulatory factor binding sites in the 5' flanking region of maize Adh1. J. Biol. Chem. 262:7947-7950; 1986.

3. Gilmartin, P.; Sarokin, L,; Memelink, J.; Chua, N-H. Molecular light switches for plant genes. Plant Cell 2:369-378; 1990.

4. Green, P.J.; Yong, M-H.; Cuozzo, M.; Kano-Murakami, Y.; Silverstein,P.; Chua, N-H. Binding site requirements for pea nuclear protein factor GT-1 correlate with sequences required for light-dependent transcriptional activation of the rbcS-3A gene. EMBO J. 7:4035-4044; 1988.

5. Guiltinan, M.J.; Marcotte, W.R.; Quatrano, R.S. A plant leucine zipper protein that recognizes an abscisic acid response element. Science 250:267-271; 1990.

6. Guerrero, F.D.; Crossland, L.; Smutzer, C.S.; Hamilton, D.A; Mascarenhas, J.P. Promoter sequences from a maize pollen- specific gene direct tissue-specific transcription in tobacco. Mol. Gen. Genet. 224:161-168; 1990.

7. Hamilton, D.A; Bashe, D.M.; Stinson, J.R.; Mascarenhas, J.P. Characterization of a pollen-specific genomic clone from maize. Sex. Plant Reprod. 2:208-212; 1989.

8. Hanson D.D.; Hamilton, D.A; Travis, J.L.; Bashe, D.M.; Mascarenhas, J.P. Characterization of a pollen-specific cDNA clone from Zea mays and its expression. Plant Cell 1:173-179; 1989.

9. Jefferson, R.A. Assaying chimeric genes in plants: the GUS gene fusion system. Plant Mol. Biol. Rep. 5:387-405; 1987.

10. Katagiri, F.; Lam, E.; Chua, N-H. Two tobacco DNA-binding proteins have homology to CREB. Nature 340:727-730; 1989.

11. Mascarenhas, J.P. The male gametophyte of flowering plants. Plant Cell 1:657-664; 1989.

12. Mascarenhas, J.P. Gene activity during pollen development. Ann. Rev. Plant Physiol. Plant Mol. Biol. 41:317-338; 1990.
13. McKendree, W.L.; Paul, A.L.; DeLisle, A.J.; Ferl, R.J. In vivo and in vitro characterization of protein interactions with the dyad G-box of the Arabidopsis Adh gene. Plant Cell 2:207-214; 1990.
14. Tanksley, S.D.; Zamir, D.; Rick, C.M. Evidence for extensive overlap of sporophytic and gametophytic gene expression in Lycopersicon esculentum. Science 213:453-455; 1981.
15. Twell, D.; Klein, T.M.; Fromm, M.E.; McCormick, S. Transient expression of chimeric genes delivered into pollen by microprojectile bombardment. Plant Physiol. 91:1270-1274; 1989a.
16. Twell, D.; Wing, R.A; Yamaguchi, J.; McCormick, S. Isolation and expression of an anther-specific gene from tomato. Mol. Gen. Genet. 217:240-245; 1989b.
17. Twell, D.; Yamaguchi, J.; McCormick, S. Pollen-specific gene expression in transgenic plants: Coordinate regulation of two different tomato gene promoters during microsporogenesis. Develop. 109:705-713; 1990.
18. Twell, D.; Yamaguchi, J.; Wing, R.A; Ushiba, J.; McCormick, S. Promoter analysis of genes that are coordinately expressed during pollen development reveals pollen-specific enhanced sequences and shared regulatory elements. Genes & Develop. 5;496-507; 1991.
19. van Tunen, A.J.; Hartman, S.A.; Mur, L.A.; Mol, J.N.M. Regulation of chalcone flavone isomerase (CHI) gene expression in Petunia hybrida: The use of alternative promoters in corolla, anthers and pollen. Plant Mol. Biol. 12:539-551; 1989.
20. van Tunen, A.J.; Mur, L.A.; Brouns, G.S.; Rienstra, J.D.; Koes, R.E.; Mol, J.N.M. Pollen- and anther-specific promoters from petunia: Tandem promoter regulation of the chiA gene. Plant Cell 2:393-401; 1990.
21. Willing, R.P.; Bashe, D.; Mascarenhas, J.P. An analysis of the quantity and diversity of messenger RNAs from pollen and shoots of Zea mays. Theoret. Appl. Genet. 75:751-753; 1988.
22. Willing, R.P.; Mascarenhas, J.P. Analysis of the complexity and diversity of mRNAs from pollen and shoots of Tradescantia. Plant Physiol. 75:865-868; 1984.
23. Ye, G-N.; Daniell, H.; Sanford, J.C. Optimization of delivery of foreign DNA into higher-plant chloroplasts. Plant Mol. Biol. 15:809-819; 1990.

2

Pollen Allergens: Molecular Cloning and Mechanism for Pollen-induced Asthma

Bruce Knox, Penelope Smith, Cenk Suphioglu, Philip Taylor, Asil Avjioglu, Piyada Theerakulpisut, Terryn Hough, Eng Ong, and Mohan Singh

Grass pollen is the major outdoor environmental factor provoking the immediate hypersensitive response of allergic disease in about one in five of the human population (see review by Howlett and Knox,1984). The symptoms are induced by the production of specific types of human defence molecules of the immune system, especially Immunoglubulin E (IgE). Allergens are environmental antigens which induce the formation of specific IgE, and bind specifically to IgE on the surface of epithelial mast cells. This triggers the mast cells to release the mediators of allergy.

Biochemical techniques were used to isolate the allergens of rye-grass pollen at Cambridge in 1965 (Johnson and Marsh,1965). There are at least six groups of allergens, identified by their different physicochemical and immunological characteristics. The principal allergen is a glycoprotein of 27-35 kD., containing 5% carbohydrate that does not contribute to its allergenicity (Johnson áMarsh,1966). The allergen, known originally as Group I allergen, but now by convention as Lol pI, is the immunodominant component, that is, more than 95% of patients possess specific IgE that binds to this pollen protein. The function of this allergen has only just begun to be understood through molecular analysis.

Genes encoding two important ryegrass allergens have been isolated from a pollen cDNA library and by PCR amplification. One of these clones corresponds to Lol pI (Griffith et al., 1991) while the other encodes a previously unidentified allergen designated Lol pIb (Singh et al., 1991).

Molecular cloning of rye-grass pollen allergens

The deduced amino acid sequence of the full length Lol pI clone contains a characteristic N-terminal sequence [Cottam et al.,1986; Perez et al.,1990; Griffith et al.,1991] : I A K V P ..... In addition, the C-terminus corresponds to the sequence of an internal peptide fragment (Esch and Klapper 1989) previously reported from protein microsequencing. A 23 amino acid signal peptide with similarities to that of barley grain amylase (Jones and Robinson,1989) is also present. This determines secretion of the protein.

The C-terminus of Lol pI shows a high level of homology with the entire 97 amino acid sequences of Lol pII (Mr 11 kD, Ansari et al.,1989a) and Lol pIII (Mr 11 kD, Ansari et al.,1989b). Lol pI and Lol pII have 48% similarity (33% identity). Lol pI and Lol pIII have 44% similarity (26% identity). Lol pII and Lol pIII have 73% similarity (59% identity). However, despite strong homology between amino acids 145 to 240 of Lol pI and Lol pII and III, no serological cross-reactivity has been detected with MAbs raised gainst Lol pI and either Lol pII or III (Ansari et al.,1987).

During the molecular cloning of Lol pI, we have identified a major new allergenic protein, Lol pIb, a 31-kDa basic protein. A cDNA clone encoding Lol pIb (12R) has been isolated, sequenced and characterised (Singh et al.,1991). The sequence of Lol pIb shows no homology with Lol pI, II or III, however it shows strong homology (82% similarity, 71% identity) with the recently identified allergen, Poa pIX from Kentucky bluegrass (Silvanovich et al.,1991) and the deduced N-terminal sequence of Lol pIb: A D A G Y .....is homologous (50% similarity, 20% identity) with that of a major allergen from timothy grass, Phl pV (Matthiesen and Lowenstein, 1991).

Both Lol pI and Lol pIb transcripts are expressed exclusively in pollen of rye-grass. The absence of transcripts in other tissues is consistent with the absence of Lol pI and Lol pIb antigens, as observed by immunoblotting (Singh et al.,1991). These molecules are rare examples of organ-specific proteins that are known allergens. The predicted amino acid sequences of these allergens show no similarities with other proteins of known function in the databases, so that their natural function in pollen remains to be elucidated.

The detection of homologies in amino acid sequence among rye-grass pollen allergens is now possible, since a number of different allergen genes have been isolated and their sequences predicted. These data show that allergens may belong to families of closely related proteins. While their function in pollen remains unknown, the similarity in amino acid sequences suggests that these molecules belong to families of molecules with conserved functions.

The amino acid sequence of a highly conserved antigenic determinant of Fes eI, an allergen from Festuca elatior corresponding to Lol pI, has been obtained from cyanogen bromide cleaved and trypsin generated fragments (Esch and Klapper,1989). This 28 amino acid sequence fragment is common to the group I allergens of several other grasses,including Lolium perenne, Poa pratensis, Anthoxanthum odoratum, Agrostis alba. The possibility exists to use Lol pI cDNA clones as heterologous probes to isolate corresponding genes in other related grasses, and so study the DNA or amino acid sequence changes as a molecular indicator of evolutionary changes in the grasses. The extraordinary similarity in antigenic and allergenic determinants between Lol pI of rye-grass and the corresponding protein in related grasses suggests that regions of the molecules may be highly conserved, while others are not, and the differences in homology may be useful indicators of evolutionary relationships.

Using a newly developed anhydrous fixation technique, we have shown that Lol pI is located in the cytosol (Fig.1) using immuno-gold probes with specific MAbs (Singh et al.,1991). In contrast, using a MAb specific for Lol pIb, the gold marker particles bind mostly to the starch granules. Mature grass pollen is filled with starch granules which are up to 2.5 μm in size, and originate in the amyloplasts. The localization of Lol pIb in the starch granules implies that this protein should be transported from the cytosol to the lumen of the amyloplasts during development.

For transport to chloroplasts, the proteins which are synthesized in the cytosol are synthesized as precursors containing a transit peptide sequence that is cleaved after transport into the organelle. The transit peptide of most chloroplast-targetted proteins

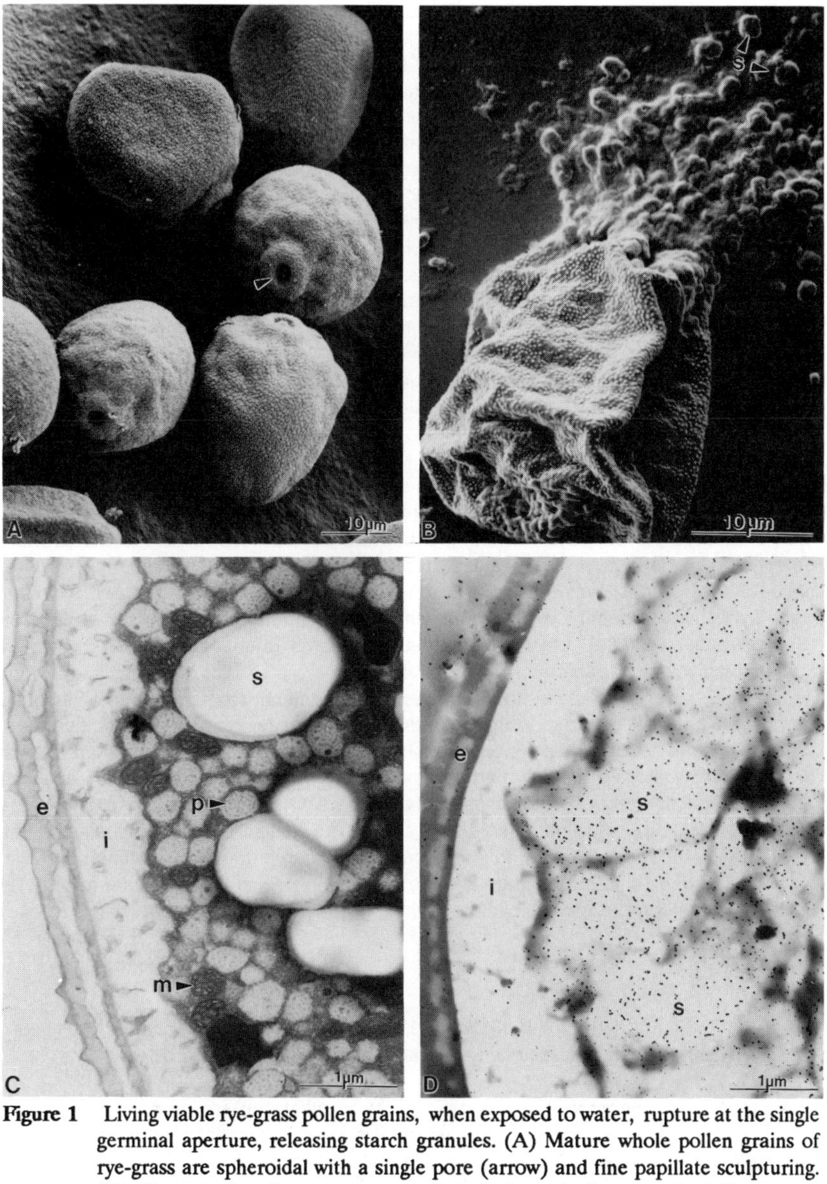

Figure 1 Living viable rye-grass pollen grains, when exposed to water, rupture at the single germinal aperture, releasing starch granules. (A) Mature whole pollen grains of rye-grass are spheroidal with a single pore (arrow) and fine papillate sculpturing. (B) The cytoplasmic contents are released through the pore, including starch granules (s). (C) Thin post-stained section of rye-grass pollen grain showing prominent starch granules (s), polysaccharide particles (p), mitochondria (m), exine (e) and intine (i). (D) Immunogold localisation of two different rye grass pollen allergens, Lol p I and Lol p Ib, by double labeling technique on unstained thin section (Singh et al, 1991). The silver-enhanced gold particles (60nm) show binding of monoclonal antibody specific for Lol pIb mainly to starch granules (s), while smaller 15nm particles show binding of monoclonal antibody specific for Lol pI mainly to cytosol. Mature anthers were processed by anhydrous fixation in a mixture of glutaraldehyde, paraformaldehyde and dimethoxypropane and embedded in LR gold resin. This procedure prevents diffusion of allergens from their cellular sites.

possess a loosely defined cleavage site consensus sequence: (Val/Ile)-X-(Ala/Cys) Ala (Gavel and Von Heijne,1990). Arginine is also found in positions -6 to -10. The corresponding motif in the Lol plb transit peptide sequence is Ser-Tyr-Ala Ala in this position, and there are two arginine residues at positions -4 and -9. We conclude that the Lol plb molecule is synthesized first as a pre-allergen in the cytosol, and is transported to the amyloplast for post-translational modification.

The ecological implications of this finding of a potent allergen in starch granules are profound. Grass pollen (Fig.1A) is known to rupture by osmotic shock, and we showed that this does, in fact, occur in rye-grass pollen (Fig.1B). This rupture ensures the release of the allergen-containing starch granules into the atmospheric aerosol. Furthermore, there are between 500 and 1000 granules in each pollen grain. We first predicted that the granules would be released into the atmosphere during rainfall, and second that these granules constitute the source of the previously unexplained micronic particles (Spieksma et al., 1990). These are particles less than 5|m in diameter that contain grass pollen allergen, and are small enough to enter the airways of susceptible humans to elicit asthma. This is known to be a disease of the lungs, and yet grass pollen is too large to penetrate beyond the oral cavity. Could these starch granules become part of the airspora of the atmosphere?

Aerobiological studies in Melbourne last spring confirmed our hypothesis. Volumetric filter samplers detected the presence of starch granules in the atmosphere, with a 50-fold increase occurring on days following significant rainfall. The starch granules were detected cytochemically by IKI staining patterns on the filters, and using specific antibodies as probes. Previous observations of micronic particles had shown that they were especially frequent on days following rainfall during the grass pollen season (Stewart and Holt,1985). We conclude that the starch granules are likely to be the unexplained micronic particles.

The final evidence would be to determine whether the starch granules, when isolated from whole pollen, can act as allergens in triggering asthma. We have been fortunate to collaborate with Melbourne physicians specialising in allergy and respiratory diseases. Skin prick tests showed that the starch granules are equally effective as pollen extracts, while lung function tests were positive in every case. This indicates that the granules are capable of eliciting an IgE-mediated response, and are the likely elicitors of the asthmatic reaction. The natural function of this molecule in the starch granules remains unknown.

These kind of studies illustrate the power and potential of molecular probes in elucidating the complex biological interactions of grass pollen grains.

References

1. Ansari AA, Shenbagamurthi, P, Marsh, DG. Complete amino acid sequence of a Lolium perenne (perennial rye grass) pollen allergen, Lol p II. J. Biol. Chem. 264: 11181-11185, 1989a.
2. Ansari AA, Shenbagamurthi, P, Marsh, DG. Complete primary structure of a Lolium perenne (perennial rye grass) pollen allergen, Lol p III: Comparison with known Lol p I and Lol p II sequences. Biochem. 28: 8665-8670, 1989b.
3. Ansari AA, Kikhara TK, Marsh DG. Immunochemical studies of Lolium perenne (Rye grass) pollen allergens, Lol p I, II, and III. J. Immunol. 139: 4034-4041, 1987.
4. Cottam GP., Moran DM, Standring R. Physicochemical and immunochemical

characterisation of allergenic proteins from rye grass (Lolium perenne) pollen prepared by a rapid and efficient purification method. Biochem. J. 234: 305-310, 1986.

5. Esch RE, and Klapper DG. Isolation and characterization of a major cross-reactive grass group I allergenic determinant. Mol. Immunol. 26: 557-561, 1989.

6. Gavel Y,and Von Heijne G. Sequence comparisons of transit peptides targeting proteins to chloroplasts. FEBS Letters. 261: 455-458, 1990.

7. Griffith IJ, Smith PM, Pollock J, Theerakulpisut P, Avjioglu A, Davies S, Hough T, Singh MB, Simpson RJ Ward LD, and Knox RB. Cloning and sequencing of Lol pI, the major allergenic protein of rye grass pollen. FEBS Letters 279: 210-215, 1991.

8. Howlett BJ, and Knox RB. Allergic interactions. Enc. Plant Physiol. 17: 656-673, 1984.

9.Johnson P,and Marsh DG. The isolation and characterisation of allergens from the pollen of rye grass (Lolium perenne). Euro. Polymer J. 1: 63-77, 1965.

10. Johnson P and Marsh DG. Allergens from common rye grass pollen, (Lolium perenne). Immunochem. 3: 101-110, 1966.

11. Jones RL, and Robinson DG. Protein secretion in plants. New Phytol. 111: 567-597, 1989.

12. Matthiesen F, and Lowenstein H. Group V allergens in grass pollens- I. Purification and characterisation of the group V allergen from Phleum pratense pollen, Phl p V. and II. Investigation of group V allergens in pollens from ten grasses. Clin. Ex. Allergy. 21: 297-307, 309-320, 1991.

13. Perez M, Ishioka GY, Walker LE, and Chestnut RW. cDNA cloning and immunological characterisation of the rye grass allergen Lol p I. J. Biol.Chem. 265: 16210-16215, 1990.

14. Silvanovich A, Astwood J, Zhang L, Olsen E, Kisil F, Sehon A, Mohapatra S, and Hill R. Nucleotide sequence analysis of three cDNA's coding for Poa pIX isoallergens of Kentucky Bluegrass pollen. J. Biol. Chem. 266: 1204-1210, 1991.

15. Singh MB, Hough T, Theerakulpisut P, Avjioglu A, Davies S, Smith P, Taylor PE, Simpson RJ, Ward L, McCluskey J, Puy R, Knox RB. Isolation of cDNA encoding a newly identified major allergenic protein of rye grass pollen: Intracellular targeting to the amyloplast. Proc Nat Acad Sci, USA . 88: 1384-1388, 1991.

16. Spieksma FTM, Kramps JA, Linden AC. Van der, Nikkels BH, Plomp A, Koerten HK, Dijkman JH. Evidence of grass pollen allergenic activity in the smaller micronic atmospheric aerosol fraction. Clin. Exp. Allergy. 20: 273-280, 1990.

17. Stewart GA, Holt PG. Submicronic airborne allergens. Med J Aust. 143: 426-427,1985.

3

The Analysis of Maize Mutant de-ga MI6 Induced by Transposon Tagging

L. Colombo, M.E. Pe, G. Binelli, and E. Ottaviano

Introduction

The dissection of the complex genetic program that regulates the male gametophytic phase can be accomplished by means of mutant analysis. The defective endosperm with gametophytic function (de-ga) mutants of maize are a category of monogenic viable gametophytic mutants in which the mutation also affects the development of the endosperm, causing a reduction of kernel size more or less severe depending on the mutant. As for the gametophytic effect of the mutation on microsporogenesis and/or on pollen tube growth, it can be easily evaluated by means of simple statistical analysis of the distortion from Mendelian segregation ratios in heterozygous (de-ga/+) plants (Ottaviano et al., 1988).

In order to isolate a few genes belonging to this category, we induced several de-ga mutants by transposon tagging using the Mutator system of transposable elements first discovered by Robertson (1978) and successfully used for the isolation of maize genes (O'Reilly et al., 1985; McCarty et al., 1989; Buckner et al., 1990). In this chapter we describe some characteristics of the mutant de-gaMI6 and the molecular approach followed in the attempt to isolate the mutated gene.

Evaluation of the gametophytic effect

The mutant de-gaMI6 was produced by crossing inbred line W22 with Robertson's mutator line Q60 carrying the Mutator system of transposable elements. Heterozygous (de-ga/+) plants of the M3 generation were selfed, each resulting ear was divided into sectors of equal length from the apex to the base and the proportion of de kernels was determined. The results are summarized in Table 1. The proportion of de/de kernels on the entire ear (PT) is significantly lower than the 0.25 one should expect on the basis of a 3:1 segregation ratio, as tested by the chi-square test (1 d.f.) showing that the de mutation affects pollen development. Furthermore a decrease of the proportion of defective kernels from the apex (P1) to the base (P5), tested by a chi-square test with n-1 d.f., can be interpreted as a reduction of the pollen tube growth rate of de pollen grains.

Table 1. Evaluation of the gametophytic effect of the de-gaMI6 mutant.

Genotype X2(1) kernels	n. of X2(n-1) apex	P1	P2	P3 base	P4	P5	PT
W22/Q60	2766	.224	.225	.198	.176	.125	.197***

** = P < 0.01

Molecular analysis

In order to physically isolate the DNA corresponding to the mutated gene, the hybridization patterns obtained from total genomic DNA of homozygous de-ga plants when probed with the AvaI-BstNI fragment of the Mutator1 (Mu1) clone (Barker et al., 1984) were compared with those of homozygous +/+ plants. A hybridization signal corresponding to a DNA fragment 4.5 kb in size showed cosegregation with the de-gaMI6 mutation when the DNA was digested with the methyl-sensitive enzyme PvuII (Figure 1).

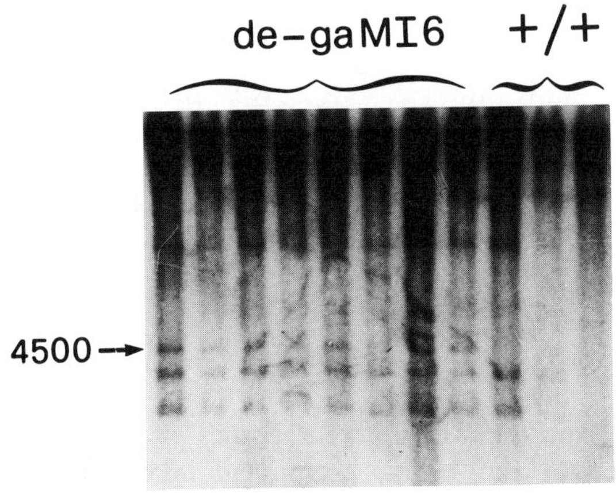

Figure 1. Identification of Mu1-hybridizing PvuII fragment that cosegregates with the de-gaMI6 genotype.Genomic DNA from de-gaMI6 mutants and homozygous +/+ plants digested with PvuII was hybridized to the AvaI-BstNI fragment of Mu1. The arrow indicates the 4.5 kb fragment cosegregating with the mutant genotype.

This DNA fragment was isolated from a lambda-gt10 genomic library and subcloned in the EcoRI site of pUC19. The recombinant clone pDE10 was used to determine the restriction map of the cosegregating fragment. The results are shown in Figure 2 where the position of the region homologous to the Mu1 clone from the Adh1 mutable allele (Barker et al., 1984) and the Mu1 are reported along with the probe used in the isolation.

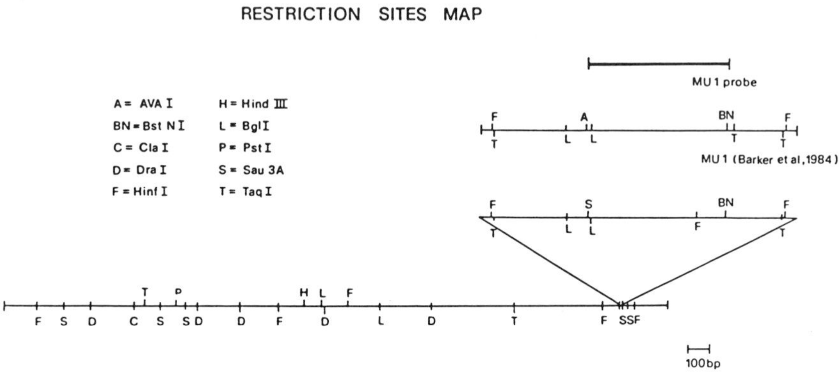

Figure 2. Restriction endonuclease map of the fragment isolated from de-gaMI6 by gene tagging.

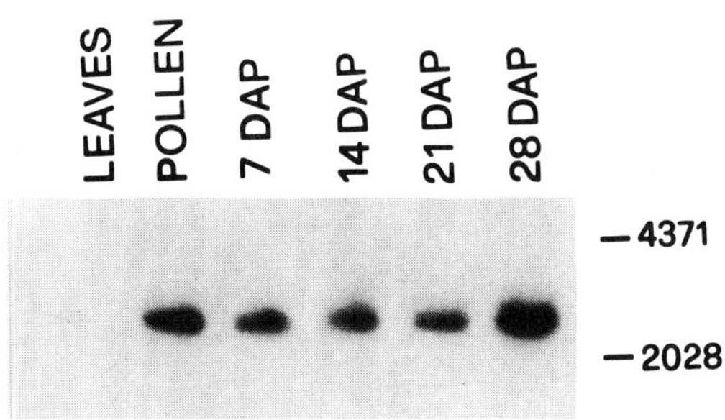

Figure 3. Expression in different tissues of de-gaMI6 fragment. Poly(A)+RNA from leaves, pollen and immature endosperm at different stages of development (DAP=Days After Pollination) was hybridized with the EcoRI-HindIII fragment of pDE10.

The insertion site of the Mu1-like element is shown and compared with the Mu1 map from literature.

The basic structure of the Mu1-like element, deduced by restriction mapping seems to be conserved, although a few differences can be detected.

In the attempt to demonstrate that the cloned DNA fragment contains a portion of the mutated gene, we performed Northern blot experiments using the whole fragment or a portion of it as a probe. Figure 3 shows the results of a Northern blot experiment in which poly(A) + RNA from leaves, mature pollen and developing endosperms at different days after pollination (DAP) of inbred line B37 was probed with the EcoRI-HindIII portion of the fragment, not containing the region homologous to Mu1.

Strong hybridization signals are clearly evident on mature pollen and on developing endosperm RNA; a weaker signal on leaf RNA also appeared when exposure time was increased.

The results of the genetic and molecular analyses and the results on the structure of the clone pDE10, are strong indirect indications that we were able to induce a gametophytic mutant by transposon tagging and to isolate the DNA fragment possibly containing the mutated gene.

Acknowledgments

The Authors wish to thank M. Villa for skillful technical assistance. Research supported by Ministero Agricoltura Foreste, Italy. Special Project "Tecnologie Avanzate in Agricoltura".

Literature Cited

Barker, R.F.; Thompson, D.V.; Talbot, D.R.; Swanson, J.; Bennetzen, J.L. Nucleotide sequence of the maize transposable element Mu1. Nucleic Acid Res. 12: 5955-5967; 1984.

Buckner, B.; Kelson, T.L.; Robertson, D.S. Cloning of the y1 locus of maize, a gene involved in the biosynthesis of carotenoids. The Plant Cell 2: 867-876; 1990.

O'Reilly, C.; Shepherd, N.S.; Pereira, A.; Schwarz-Sommer, Z.; Bertram, I.; Robertson, D.S.; Peterson, P.A.; Saedler, H. Molecular cloning of the a1 locus of Zea mays using the transposable elements. Embo J. 41: 877-882; 1985.

Ottaviano, E.; Petroni, D.; Pè, E. Gametophytic expression of genes controlling endosperm development in maize. Theor. Appl. Genet. 75: 252-258; 1988.

McCarty, D.R.; Carson, C.D.; Stinard, P.S.; Robertson, D.S. Molecular analysis of the viviparous-1 locus in maize. The Plant Cell 1: 523-532; 1989.

Robertson, D.S. Characterization of a mutator system in maize. Mutation Res. 51: 21-28; 1978.

4

Isolation of Genes Preferentially Expressed in Pistils of Tobacco Flowers

Maria Helena S. Goldman, Jef Seurinck, John Van Emmelo, and Celestina Mariani

Flowering is a complex process which begins with the conversion of a vegetative meristem into a floral meristem. Cells divide and differentiate to develop a structure which contains different organ types organized as whorls, occupying precisely determined positions. From outside to inside, the flower is composed of the sterile organs (sepals and petals) and fertile organs (stamens and carpels or pistils).

Flowers are responsible for reproduction in angiosperms. Reproduction involves pollination and, after homologous and compatible pollen-pistil interactions, subsequent fertilization. Consequently, fertilization is a process that requires specific cell-cell recognition events to occur (Goldberg, l988). The processes leading to fertilization are initiated on the stigma. After pollination, the recognition process followed by acceptance or rejection of the male gametophyte takes place in the pistil. As a result of recognition, a chain of sequential, integrated processes in the pistil regulates the post-pollination behavior of the pollen (Shivanna, l982). The vital function of selection of the male gametes in flowering plants is performed by the pistillate tissue, making studies of pollen-pistil interaction of extreme importance in the biology of sexual reproduction (Shivanna, l982). Molecular studies suggest that specific proteins of the pollen wall and of the transmitting tissue of the stigma and style are involved in recognition and the pollen-pistil interactions (Cornish et al., l988; Nasrallah & Nasrallah, l989; Haring et al., l990).

We are interested in the identification and characterization of genes that are preferentially, or exclusively, expressed in pistils and to exploit this information to manipulate the fertilization process in crop plants. An appropriate understanding of pistil biology, of the genes specifically expressed in this organ and of the gene products involved in the recognition process is necessary in order to design experiments entailing the use of hybridization techniques for crop improvement. We have chosen <u>Nicotiana tabacum</u> as a model system because tobacco has large, highly specialized flowers, with all the major organs spatially distinguishable. The pistil of tobacco is comprised of two fused carpels. The style is solid and the stigma is bilobed, papillose, and at maturity is covered with a sticky exudate (Bell & Hicks, l976). Tobacco is easily transformed pistil specific genes manipulated <u>in vitro</u>. In addition, flowers can be obtained directly from superficial tissues of small explants (Tran Than Van l973). In this species, flower mutants with abnormal phenotypes, including apparent shifts in floral organs among the whorls, have been isolated (Malmberg et al., l985) and used in biochemical studies (Evans et al., l988).

Floral organ systems express highly regulated gene sets, and each organ of the flower has a unique mRNA collection that is not detectable in other vegetative and floral organ systems (Goldberg, I988). The organ-specific modulation of gene expression is responsible for the biochemical differentiation of the cells within an organ or tissue (Gasser et al., I989). This biochemical differentiation is determined by the presence of proteins that are the direct or secondary products of floral specific genes and by the absence of proteins specific for other organs. Isolation of the pistil specific genes and determination of the functions of their protein products are important steps in understanding pistil development and its role in the fertilization process.

We have constructed two cDNA libraries from style-stigmas of tobacco flowers (according to Amersham protocols "cDNA Synthesis System Plus RPN1256Y/Z" and "cDNA Cloning System lambda gt10 RPN1257"). One cDNA library represented mRNAs from styles and stigmas of flowers from stage 3 to stage 7 of tobacco flower development (as defined by Goldberg, I988), and it was designated as early stages of flower development (ESFD). A second cDNA library was made with mRNAs from styles of stigmas from stages 8 to 11, and it was designated as advanced stages of flower development (ASFD). Stigmas with adhesion of pollen grains were not used. Differential screening was carried out as desribed by Gasser et al. (I988), using cDNAs probes made from seedlings mRNAs as negative probe. Seedlings were chosen because all the major vegetative organs are represented. The plaques were plated at a low density (about 2,000 plaques onto a 14 cm plate). Replica nylon filters were lifted from each of the plates, one filter was hybridized with the seedling cDNA probe and the other was hybridized with the style-stigma cDNA probe made from either the ESFD or ASFD mRNAs. The plaques showing a positive hybridization signal with the style-stigma cDNA probe and no signal with the seedling cDNA probe were chosen for further analysis and purified by re-screening several times. We have selected nine clones for detailed studies, six clones coming from the ESFD cDNA library and three clones that were isolated from the ADFD cDNA library pistil-specific genes.

Organ Expression of the Selected Clones

On the basis of the Northern hybridization data, it has been possible to divide the isolated clones in four classes. Each class includes independent copies of the same type of (homologous) mRNA as confirmed by sequence analysis of the clones. As there was no selection against mRNAs present in other floral organs, some of the isolated clones are flower-specific instead of being pistil-specific. The expression pattern of each class is shown in Figure 1, in which hybridizations were done with RNA isolated from different organs.

Class I, here presented by the clone pMGo2, corresponds to an mRNA of 0.95kb that accumulates in all floral organs, but at very low level in sepals. Sepals are believed to be very similar to leaves at the mRNA level and the class I gene may represent one example of the biochemical differentiation that occurs in between these two organs. Class II, exemplified by the clone pMG09, hybridizes to an mRNA of 1.8kb that is present at high level in stigmas and styles and at low level in ovaries. This suggests that pMGo8 represents a class of pistil-specific clones. Class III is illustrated by the clone pMG07, which corresponds to a mRNA of 1.9kb. This mRNA is also pistil-specific, accumulating only in stigmas and styles, but at lower levels than class II mRNA. Class IV hybridizes with an mRNA of 2.3kb that is preferentially present in stigmas and styles, but also appears in petals, anthers and ovaries. The genes of classes I and IV are expressed at higher level in stigmas and styles during the early stages of flower development, while the genes of class II and III increase their experession in the advanced stages of development. This increase is notable in the case of class III and only modest for class II genes. These results indicate that the genes represented by the cDNAs under study are developmentally regulated.

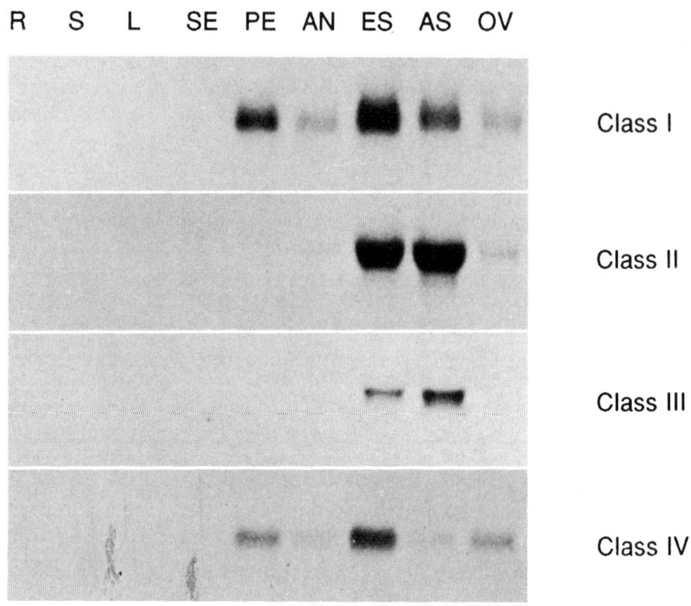

Legend: Northern blot analysis showing hybridization of Class I, II, III, and IV probes with RNA of roots (R), stems (S), leaves (L), sepals (SE), petals (PE), anthers (AN), styles/stigmas ESFD (ES), styles/stigmas ASFD (AS), ovaries (OV)

Sequence Analysis of the Clones

Further information about the different cDNA clones was obtained by determined their nucleotide sequence. Full length CDNA clones for class I and class III were obtained. The deduced amino acid sequences of both classes reveal the presence of putative signal peptides at the N-terminus of the respective proteins, suggesting that these proteins could be secretory proteins. The deduced amino acid sequences of classes I, II and III demonstrate that the corresponding genes encode proline-rich proteins. These three proteins also contain a high percentage of the amino acid serine and they all contain a few repetitions of the pentapeptide SPPPP, predominantly restriced to the N-terminal region. Proteins with similar motifs have been found to have the proline residues posttranslationally modified into hydroxyproline and to be glycosilated.

The hydroxyproline-rich glycoproteins (HRGP) have been shown to fall into three classes: hydroxyproline-rich lectins, arabinogalactans, and extensins (Tierney and Varner, 1987). The hydroxyproline-rich lectins are a group of hemaglutinating glycoproteins which appear to be limited to the Solanaceae. These proteins contain both a hydroxyproline/serine-rich and cysteine-rich domain (Showalter and Varner, 1987). The class I protein is also cysteine-rich and has repetitions of the dipeptides: tryptophan-proline, cysteine-proline and arginine-proline. Thus the class I protein has similarities to the lectins.

Arabinoglactan proteins (AGPs) are rich in hydroxyproline, serine and alanine (Showalter and Varner, 1987; Bacic et al., 1988). They are mainly localized extracellularly, although they are also associated with the plasma membrane and are major components of plant gums and exudates (Van Holst and Clarke, 1986). One suggestion is that arabinogalactan proteins may be involved in expression of identity of species, genus, cultivar, tissue or cell type (Van Holst and Clarke, 1986). They have been detected in pistil of many species (Gleeson and Clarke, 1980; Mau et al., 1982; Hoggart and Clarke, 1984; Van Holst and Clarke, 1986) including Nicotiana alata (Bacic et al., 1988). The synthesis of these pistils-AGPs is developmentally regulated in terms of concentration and the types of AGPs (Bacic et al., 1988) and they are located mainly in the extracellular matrix of the transmitting tissue (Sedgley et al., 1985). It has been suggested that they may play a role in fertilization (Hoggart and Clarke, 1984; Sedgley et al., 1985). In addition, by cross-reaction with a monoclonal antibody (MAC 207, Pennell et al., 1989), a family of glycoproteins related to the GAPs has been identified. The MAC 207 epitope is present in all cells of vegetative meristems, primordia and organs (stem, leaf and root) and throughout undeveloped flower buds. However, they are excluded from specific cells (male and female sporangia and gametophytes) of developing stamens and carpels (Pennell and Roberts, 1990). The authors propose that this change in AGP expression is specifically related to sexual reproduction. It is possible to postulate the existence of MAC 207 counterparts that would be specifically present in the different tissues involved with fertilization. The gene product of class II, which is rich in the amino acids proline, serine and alanine, may be a good candidate for such a pistil-specific protein.

The extensins are proteins rich hy hydroxyproline, serine, tyrosine and lysine and are the major protein component of plant cell walls (Tierney and Varner, 1987). It has been described that the pentapeptide SPPPP is a conserved repetitive sequence within the extensins, with some modifications in a chenopod extensin (Li et al., 1990). Considering this pentapeptide as a typical characteristic of the extensins, the proteins encoded by classes I, II and III genes can be designated as extensin-like proteins and are proposed to be novel types of extensins. These proteins are lysine-rich to different degrees, but have a very low content of tyrosine, unlike extensins. The class III protein is also rich in leucine and valine; of the three classes, it is the one with the most similarities to the extensins. We propose that the product of the class I genes are flower-specific extensin-like proteins and the products of class II and III genes are pistil-specific extensin-like proteins.

Future Perspectives

We have successfully isolated pistil-specific cDNA clones using a differential screening procedure. These clones correspond to different mRNAs that accumulate in the pistil with specific spatial and temporal profiles. To understand their role in the fertilization process, the localization of the corresponding proteins in the various tissues of the pistil will be analyzed. Another approach in understanding the function of the different cell types in the pistil, is to specifically destroy the cells in which these extensin-like mRNAs accumulate. These studies require the isolation of the extensin-like genes and the characteriziation of their cis-regulatory elements. The 5' region of the genes can be fused to the coding sequence of a ribonuclease and introduced in tobacco. These experiments could lead to female-sterile plants, in analogy with the male-sterile plants already induced by this strategy. (Mariani et al., I990) and could simulate the self-incompatible reactions (McClure et al., I989).

Acknowledgements

Dominique Rouana and Enno Krebbers are thanked for scientific discussion and critically reading the manuscipt. Vera Vermaereke, Karel Spruyt and Stefaan Van Gijsegem are thanked for preparing the illustration. M.H.S. Goldman is supported by the Brazilian Research Council (CNPPq), fellowship number 200746/88.0.

References

1. Bacic, A.; Gell, A.C.; Clarke, A.E. Arabinoglacian proteins from stigmas of Nicotiana alata. Phytochemistry 27:679-684; I988.
2. Bell, J.; Hicks, G. Transmitting tissue in the pistil of tobacco: light and electron microscopic observations. Planta 131:187-200; I976.
3. Cornish, E. C.; Anderson, M.A.; Clark, A.E. Molecular aspects of fertilization in flowering plants. Ann. Rev. Cell Biol. 4:209-228; I988.
4. Evans, P.T.; Holaway, B.L.; Malmberg, R.L. Biochemical differentiation in the tobacco flower probed with monoclonal antibodies. Planta; I988.
5. Gasser, C.S.; Smith, A.G.; Budelier, K.A.; Hinchee, M.A.; McCormick, S.; Horsch, R.B.; Shah, D.M.; Fraley, R.T. Isolation of differentially expressed genes from tomato flowers. In: Verma, D.P.S.; Goldberg, R.B., eds. Plant Gene Research. New York: Springer-Verlag; I988: p. 83-96. (Temporal and Spatial Regulation of Plant Genes; vol. 5).
6. Gasser, C.S.; Budelier, K.A.; Smith, A.G. Shah, D.M.; Fraley, R.T. Isolation of tissue-specific cDNAS from tomato pistils. The Plant Cell 1:15-24; 1989.
7. Gleeson, P.A.; Clarke, A.E. Arabinogalactans of sexual and somatic tissues of Gladiolus and Lilium. Phytochemistry 19:1777-1782; I980.
8. Goldberg, R.B. Plants: novel developmental processes. Science 240:1460-1467; I988.
9. Hoggart, R.M.; Clarke, A.E. Arabinogalactans are common components of angiosperm styles. Phytochemistry 23:1571-1573; I984.
10. Haring, V.; Gray, J.E.; McClure, B.A.; Anderson, M.A.; Clarke, A.E. Self-incompatibility; a self-recognition system in plants. Science 250:937-941; I990.
11. Li, X.; Kieliszewski, M.; Lamport, D.T.A. A chenopod extensin lacks repetitive tetrahydroxyproline blocks. Plant Physiol. 92:327-333; I990.
12. Malmberg, R.L.; McIndoo, K.; Hiatt, A.C.; Lowe, B.A. Genetics of polyamine synthesis in tobacco; developmental switches in the flower. Cold Spring Harbor Symp. Quant. Biol. 50:475-482; I985.

13. Mariani, C.; De Beuckeleer, M.; Truettner, J.; Leemans, J.; Goldberg, R.B. Induction of male sterility in plants by a chimaeric ribonuclease gene. Nature 347:737-741l 1990.
14. Mau, S.L.; Raff, J.; Clarke, A.E. Isolation and partial characterization of components of Prunus avium L. styles, including an antigenic glycoprotein associated with a self-incompatibility genotype. Planta 156:505-516; 1982.
15. McClure, B.A.; Haring, V.; Ebert, P.R.; Anderson, M.A.; Simpson, R.,J.; Sakiyama, F.; Clarke, A.E. Style self-incompatibility gene products of Nicotiana alata are ribonucleases. Nature 342:955-957; 1989.
16. Nasrallah, J.B.; Nasrallah, M.E. The molecular genetics of self-incompatibility in Brassica. Annu. Rev. Genet. 23:121-139; 1989.
17. Pennell, R.I.; Knox, J.P.; Scofield, G.N.; Selvendran, R.R.; Roberts, K.A. A family of abundant plasma membrane-associated glycoproteins related to the arabinogalactan proteins is unique to flowering plants. The Journal of Cell Biology 108:1967-1977; 1989.
18. Pennell, R.I.; Roberts, K. Sexual development in the pea is presaged by altered expression of arabinoglactan protein. Nature 344:547-549; 1990.
19. Sedgley, M.; Blesing, M.A.; Bonig, I.; Anderson, M.A.; Clarke, A.E. Arabinogalactan proteins are localized extracellularly in the transmitting tissue of Nicotiana alata Link and Otto, an ornamental tobacco. Micron and Microscopica Acta 16:247-254; 1985.
20. Shivann, K.R. Pollen-pistil interaction and control of fertilization. In: Johri, B. M., ed. Experimental Embriology of Vascular Plants; New York: Springer-Verlag; 1982; p. 131-'74.
21. Showalter, A.J.; Verner, J.E. Plant hydroxyproline-rich glycoproteins. In: Marcus, A., ed. The biochemistry of plants: a comprehensive treatise; New York: Academic Press; 1987; p. 485-520. (Molecular Biology; vol. 15).
22. Tierney, M.L.; Varner, J.E. The extensins. Plant Physiol. 84:1-2; 1987.
23. Tran Thanh Van, K. Direct flower neoformation from superficial tissues of small explants of Nicotiana tabacum. Planta 115:87-92; 1973.
24. Van Holst, G.; Clarke, A.E. Organ-specific arabinogalactan-proteins of Lycopersicon peruvianum (Mill) demonstrated by crossed electrophoresis. Plant Physiol. 80:796-789; 1986.

5

Inhibition of Flavonoid Biosynthesis in *Petunia* Anthers by Antisense RNA: A Novel Way to Engineer Nuclear Male Sterility

Ingrid M. van der Meer, Maike E. Stam, Arjen J. van Tunen, Joseph N.M. Mol,** and Antoine R. Stuitje*

* Present address: CPO, P.O. Box 16, 6700 AA Wageningen, The Netherlands
** Corresponding author

ABSTRACT

Inhibition of flower pigmentation in transgenic petunia plants was previously accomplished by expressing an antisense chalcone synthase (chs) gene under control of the cauliflower (CaMV) 35S promoter. This chimeric gene was not effective in inhibiting pigmentation in anthers due to an inappropriate cell type-specificity of the 35S promoter within this organ. By inserting the anther box, a sequence homology found in other genes expressed in anthers, the expression pattern driven by this promoter was modified. Using this modified 35S promoter to drive antisense chs gene expression inhibition of anther pigmentation was realized. Transformants with white anthers are male sterile due to a defect in pollen development, indicating that flavonoids may have an essential function in microsporogenesis. Hence, blocking the synthesis of natural compounds during anther development provides a novel way to introduce nuclear male sterility.

Flavonoids are phenylpropanoid-based secondary metabolites that have a key function in flower and fruit pigmentation, defense against phytopathogens1, protection against UV light 2, and in the induction of nodulation in legumes3. They also have been implicated in the regulation of auxin transport4 and resistance to insects5. We have studied the possible function of flavonoids in anther development and male gametogenesis by an antisense approach. It was shown previously that the expression of an antisense chalcone synthase (chs) cDNA driven by the constitutive viral CaMV35S promoter or the homologous chs promoter could lead to a dramatic decrease in corolla pigmentation in transformants6,7. However, in these experiments anther pigmentation was never affected. This can be explained by the fact that the CaMV35S and chs promoters control a different cell type-specific expression pattern within the anthers8, and that the activity of a transgenic chs promoter is only a fraction of that of the endogenous chs promoter7.

We attempted to change the cell type-specificity of the CaMV35S promoter by modifying it through insertion of the 'anther box', a highly conserved motif, present in the promoters of several flavonoid genes which are all active in immature anthers of petunia[9] (Fig. 1A).

The expression of the modified promoter was analysed in transgenic plants by using the b-glucuronidase (GUS) gene as a reporter. The presence of the anther box enhanced CaMV35S driven GUS activity in anthers relative to that in corolla approximately 3-fold (not shown). GUS expression was also monitored histochemically in cross-sections of anthers. The CaMV35S promoter drives a high GUS expression in virtually all cell types except for the sporogenic cells and the tapetum cells (Fig. 2B and C), which surround the loculi containing the sporogenic tissue (pollen) and function in nourishment and pigmentation of the latter[10,11]. Since flavonoid biosynthetic enzymes are localized in the tapetum tissue[12], this observation explains the inability of the CaMV35S promoter in our previous experiments to inhibit pigmentation in anther tissue of petunia[6,7]. The CaMV35S promoter with the anther box insertion drives GUS expression in tapetum cells on top of the basic CaMV35S-GUS expression pattern (Fig. 2D and E). Therefore, expression of the antisense chs cDNA driven by the modified CaMV35S promoter in transgenic plants should lead to reduction of anther pigmentation. Thirty five independent petunia (VR hybrid) transformants were obtained in total, of which 5 plants yielded flowers with a reduced anther pigmentation (Fig. 2F and G). Inhibition of anther pigmentation seems to be independent of the orientation of the inserted anther box, since white anthers could be obtained with constructs containing either orientation. RNase protection analysis of RNA from white and purple coloured anthers showed that inhibition of anther pigmentation is caused by a specific reduction (>90%) of the chs mRNA steady-state level (not shown).

Transformants with unpigmented anthers failed to produce mature seed pods after self-pollination and hence were male sterile. Most of the transgenic white pollen grains had a collapsed phenotype, and none of them were able to form pollen tubes on solidified germination medium (Fig. 2I) in contrast to transgenic purple coloured pollen grains (Fig. 2H). Examination of cross-sectioned anthers during development showed an arrest in development of pollen grains after the tetrade stage, at the early microspore stage in contrast to the fertile pollen grains (Fig. 2K and J). This stage in pollen development coincides with the onset of flavonoid gene expression[13]. Taken together, these data indicate a crucial function of flavonoids in pollen development at the early binucleate stage.

The white pollen phenotype proved to be a stable inherited dominant trait, since the progeny of a white pollen transformant crossed with Petunia hybrida (V30) showed a complete co-segregation of the antisense phenotype with the insert as analysed by Southern blot hybridization (not shown).

Two other findings support our observation that blocking of flavonoid biosynthesis in anthers effects male fertility. First, in studying co-suppression of chs genes[14], plants were recovered occasionally that had white anthers and sterile pollen (Taylor and Jorgensen, manuscript in preparation). Second, Coe et al.[15] described a transposon-induced white pollen mutant (whp) in maize which turned out to be male sterile. Taken together, this may indicate that flavonoids play a key role in male gametogenesis, although formally we cannot exclude that accumulation of precursors may inhibit pollen development. Additional experiments are required to assess how widespread this potential requirement for flavonoids in pollen development is throughout the plant kingdom.

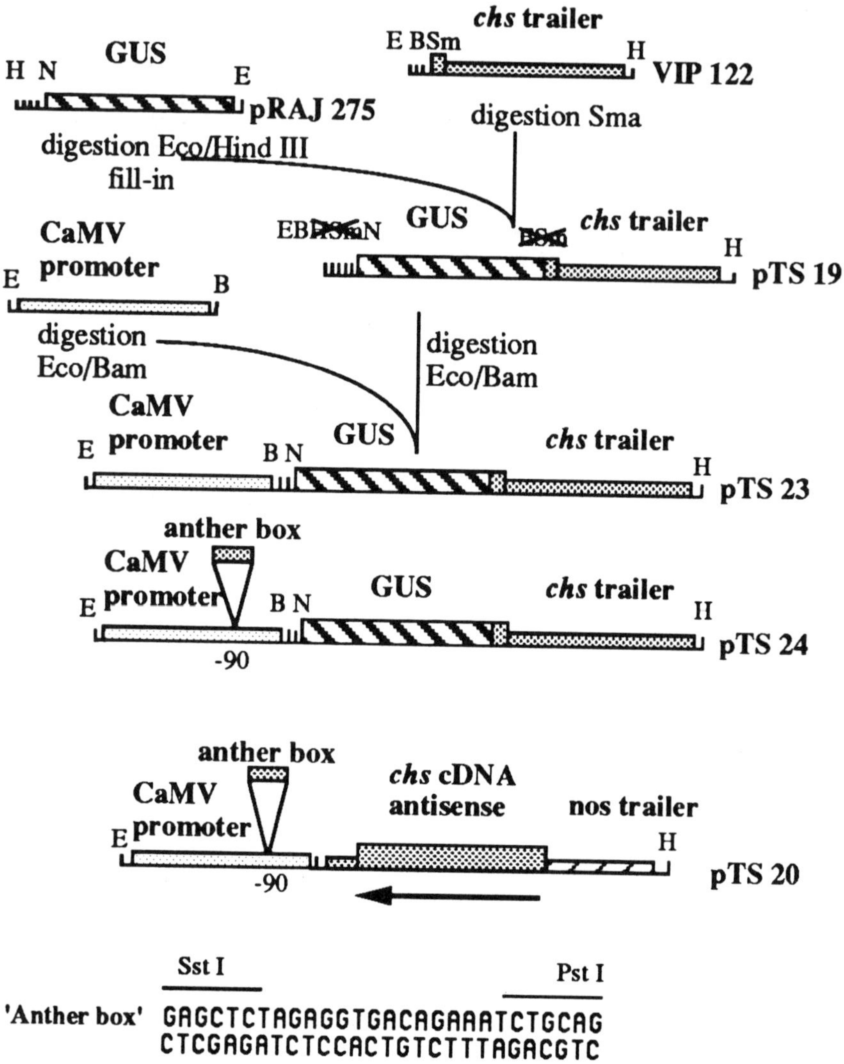
chs trailer
GUS
H N
E
E BSm
H
VIP 122
pRAJ 275
digestion Sma
digestion Eco/Hind III
fill-in
CaMV
promoter
E
B
GUS
chs trailer
H
pTS 19
digestion
Eco/Bam
digestion
Eco/Bam
CaMV
promoter
E
B N
GUS
chs trailer
H
pTS 23
anther box
CaMV
promoter
E
B N
GUS
chs trailer
H
pTS 24
-90
anther box
CaMV
promoter
E
chs cDNA
antisense
nos trailer
H
pTS 20
-90
Sst I
Pst I
'Anther box'
GAGCTCTAGAGGTGACAGAAATCTGCAG
CTCGAGATCTCCACTGTCTTTAGACGTC

Figure 1

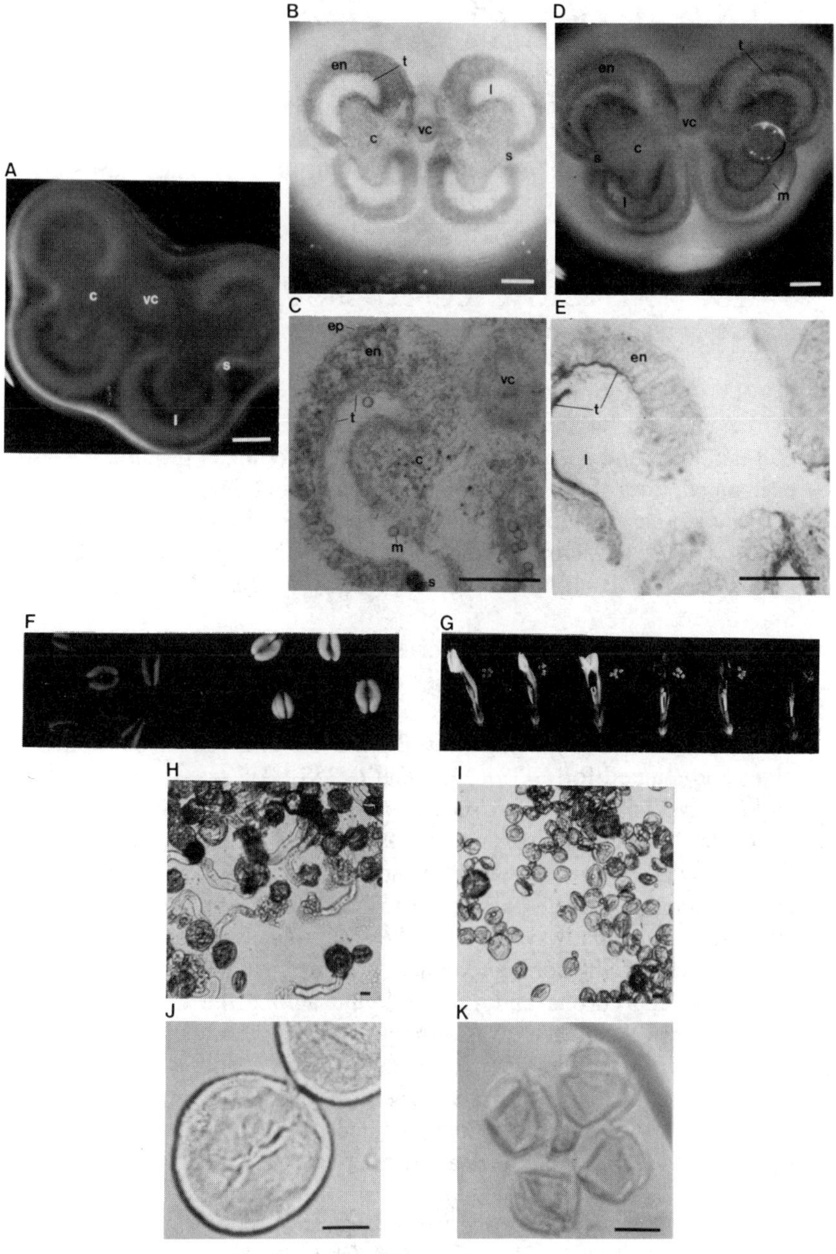

Figure 2

LEGENDS TO FIGURES

Fig. 1 Construction of the chimeric GUS and antisense chs genes. The GUS coding region of pRAJ 27517 was cloned as an EcoRI/HindIII fragment (filled in with klenow polymerase) into the Sma I site of VIP 122 that contains the chsA tail18, yielding pTS19. The CaMV35S promoter of VIP 1026 was inserted into pTS19 as an EcoRI/BamHI fragment resulting in pTS23.The oligonucleotides 5'-GAGCTCTAGAGGTGACAGAAATCTGCAG-3' and 5'-CTGCAGATTTCTGTCACCTCTAGAGCTC-3'wereannealedand 5' phosphorylated using T4 kinase. This synthetic anther box, derived from the chsA promoter18 was blunt-end ligated into pTS23 at the EcoRV site (position -90) yielding pTS24. The number and orientation of the inserted anther boxes was determined on a sequence gel by end-labeling and partial digestion. The modified promoter was exchanged for the CaMV35S promoter contained in VIP 102 (CaMV35S-chsA antisense-nos6). The chimeric constructs were inserted as EcoRI/HindIII fragments into the binary vector Bin 1919. Abbreviations: B=Bam HI; E=Eco RI; H=Hind III; N=Nco I; Sm=Sma I.

Fig. 2 Effect of modifying the CaMV promoter on GUS activity in anthers (A to E), and on antisense chs inhibition in anthers and pollen (F to K). Panels A,B, and D represent macroscopic images of transverse sectionedanthers derived from an untransformed control plant and stained for GUS (A), a CaMV35S-GUS transformant (B), and a CaMV35S/anther box-GUS transformant (D). Panels C and E show bright-field micrographs of stained anthers derived from a CaMV35S-GUS transformant (C) , and a CaMV35S/anther box-GUS transformant (E). Scale bars in A to E equal 300 mm. Panel F shows an image of normally coloured anthers (left side) next to transgenic white anthers (right side), and panel G represents the combinations of (un)pigmented corolla and anthers found among the CaMV/anther box-antisense chs cDNA transformants. In vitro germination of purple coloured fertile pollen, and non-coloured sterile pollen is shown in panels H and I respectively. Panels J and K show bright-field micrographs of aceto-carmine stained fertile (J), and sterile microspores (K) at the binucleate stage. Scale bars in H to K equal 10 mm.

Abbreviations: c = connectivum, en = endothecium, ep = epidermis, l = loculus, m = microspore, s = stomium, t = tapetum, vc = vascular cylinder.

Recently, nuclear male sterility has been engineered in higher plants by expressing fungal ribonucleases in tapetum cells, leading to their destruction16. The novelty of the antisense approach we used to introduce nuclear male sterility, lies within the fact that it has been obtained by blocking the synthesis of natural compounds.

Methods

Histochemical localization of GUS activity was performed essentially to Koes et al.8. To eliminate background GUS activity in anthers X-gluc staining solution was adjusted to pH 8.0. In vitro pollen germination was according to Bino et al.20. Pollen was incubated for 2 hr at 24°C in the dark and stained with 1% acetocarmine.

REFERENCES

1. Lamb, C.J., Lawton, M.A., Dron, M.áDixon, R.A. Cell 56, 215-224 (1989).
2. Schmelzer, E., Jahnen, W.áHahlbrock, K. Proc. natn. Acad. Sci. USA 85, 2989-299 (1988).
3. Long, S. Cell 56, 203-214 (1989).
4. Jacobs, M.áRubery, P.H. Science 241, 346-349 (1988). 5. Hedin,P.A.áWaage, S.K. In Progress in Clinical and Biological Research. Vol. 213 V. Cody, E. Middleton Jr., and J.B. Harborne, eds (New York: Alan R. Liss), pp. 87-100 (1986).
6. van der Krol, A.R. et al. Nature 333, 866-869 (1988).
7. van der Krol, A.R. et al. Molec. gen. Genet. 220, 204-212 (1990).
8. Koes, R.E., van Blokland, R., Quattrocchio, F., van Tunen, A.J.áMol, J.N.M. Plant Cell 2, 379-392 (1990).
9. van Tunen, A.J., Hartman, S.A., Mur, L.A.–ol, J.N.M. Plant molec. Biol. 12, 539-551 (1989).
10. Drews, G.N.áGoldberg, R.B. Trends Genet. 5, 256-261 (1989).
11. Wiermann, R. áVieth, K. Protoplasma 118, 230-233 (1983).
12. Beerhues, L., Forkmann, G., SchÜpker, H., Stotz G. áWiermann, R. J. Plant Physiol. 133, 743-746 (1989).
13. Koes, R.E., Spelt, C.E.áMol, J.N.M. Plant molec. Biol. 12, 213-225 (1989).
14. Jorgensen, R. Tibtech 8, 340-344 (1990).
15. Coe, E.H., Jr., McCormick, S.M.áModena, S.A. J. Hered. 72, 318-320 (1985).
16. Mariani, C., De Beuckeleer, M., Truettner, J., Leemans, J.áGoldberg, R.B. Nature 347, 737-741 (1990).
17. Jefferson, R.A., Kavanagh, T.A. áBevan, M.W. EMBO J. 6, 3901-3907 (1987).
18. van der Meer, I.M., Spelt, C.E., Mol, J.N.M.áStuitje, A.R. Plant molec. Biol. 15, 95-109 (1990).
19. Bevan M. Nucleic Acids Res. 12, 8711-8712 (1984).
20. Bino, R.J., Hille, J.áFranken, J. Plant cell Rep. 6,333-336 (1987).

6

Markers for the Y Chromosome of *Silene latifolia* and Their Use in Studying Pollen Tube Growth Rates

D.L. Mulcahy, S.B. Carroll, N.F. Weeden, and R. Kesseli

Developing Molecular Markers For the Y Chromosome

In Silene latifolia, females (XX) and males (XY) naturally differ from each other in many genes other than those located on the sex chromosomes. We masked differences which were not sex linked by pooling 18 females into one sample and 18 males into another (For the method, see Michelmore, et al. submitted). Differences between the two samples would presumably relate to the Y chromosome. DNA was extracted from the two samples and subjected to polymerase chain reaction amplification with 10 bp arbitrary primers (see Williams, et al. 1990). Any amplification from the bulked male DNA not found among those from the bulked female DNA was likely to include a Y chromosome sequence. Four putative Y chromosome markers were obtained and each was subsequently detected in all individual males tested. Y chromosome markers were not found in any of the individual females tested. These results suggest that we have reliable markers for the Y chromosome in our population of Silene latifolia. These markers, as is explained belwo, will facilitate future studies of pollen tube competition.

What Factors Influence Pollen Competition?

Studies of cultivated species have frequently indicated that pollen competition has an effect on the resulting sporophytic generation (Ottaviano and Mulcahy, 1989, for a review) but these effects are more difficult to detect in natural populations. (Snow, 1990). Quite possibly, sources of variance in natural populations obscure the effects of pollen competition. Young considers environmentally sources of variance on pollen competition in this volume (Chap.), and here we consider some genetic influences.

The Influence of Stylar Cytoplasm on Pollen Tube Growth

Hornby and Li (1975) applied mixtures of pollen in Lycopersicon esculentum, hoping to increase fruitset. They used two cultivars of tomato, cv. Immur Prior Beta (=IPB), exhibiting entire leaf margins (c/c) and cv. Bonnie Best (=BB), exhibiting the cut (C/C) leaves typical for tomato. The data shown in Table 1 were obtained.

Table 1. Influence of Stylar Genotype on Fertilizations Resulting From Application of Pollen Mixtures (*c* from IBP and *C* from BB) in *Lycopersicon esculentum*.

In each case, the female parent was an F_1 between cv. Immur Prior Beta and cv. Bonnie Best. The direction of the hybridization used to produce the female parent is indicated as the parent which donated cytoplasm to the F_1. (From Hornby and Li, 1975).

Female Parent	Offspring Observed		Offspring Ratio (≈3:1 expected)
	C-	cc	C- : cc
cC (IPB cyto.)	157	511	0.3 : 1.0 *1
Cc (BB cyto.)	484	101	4.8 : 1.0 *2

* deviation from expectations is significant at the 5% level.

1 an excess of cc progeny; 2 an excess of CC progeny.

If we assume that aliquots of the same pollen mixtures were applied to the two female parents, then the results become quite revealing. A pollen type may be significantly favored in the competition if it carries the same cytoplasm as that found in the female parent. As an alternative, or additional, explanation, pollen may be favored if it carries the nucleus corresponding to the cytoplasm type in the female parent. Thus the relationship between stylar cytoplasm and pollen cytoplasm (or nucleus) has a significant effect upon pollen competition.

The Combined Influence of Pollen Nuclei and Pollen Cytoplasm on Pollen Competition

Recent evidence demonstrates that nuclear genes which interact with the cytoplasm also can influence pollen competition. Van Damme (1991), studied the effect of a locus which restored male fertility in cytoplasmically male sterile Plantago coronopus. Since restoration of male fertility is sporophytic, in a plant which is heterozygous (Rf/rf) for the nuclear restorer of fertility, all pollen grains, even those not carrying the dominant restorer, can be fertile. Making pollinations with plants which were heterozygous for the restorer, Van Damme found that, with low pollen competition (30-40 grains per flower (each of which contains 5 ovules)) pollen not carrying the dominant restorer of male fertility sired 45% of the progeny, not significantly different from expectations. If pollinations were very heavy, however, the corresponding figure was 38%, significantly lower than 50%. Thus not only nuclear and cytoplasmic genetic components of the style and of the pollen, but also also pollen alleles which interact with the cytoplasm of the pollen, effect pollen competition.

Coadaptation Between Pollen And Style

Since pollen tube growth is influenced by several genetically variable factors, it seems likely that the style and pollen within a species (and probably within a population) would coevolve, becoming highly coordinated. In fact, Johnson and Mulcahy (1978) reported that the competitive ability of self pollen improved progressively from F_1 to F_{20} in Zea mays. Conversely, whenever either parent is a hybrid, or the pollination between different species, or even between widely separated populations, the coordination between pollen nuclei and cytoplasm or between components of the style and pollen could be severely disrupted.

X- and Y pollen As a Test System For the Genetic Determinants of Style-Pollen Interactions

Correns (1928) was the first to report that in Silene latifolia (= Lychnis alba and Melandrium album), pollen tubes carrying the X chromosome grow faster than do those carrying the Y chromosome. A recent reexamination of Correns' studies on competition between X- and Y pollen revealed some surprising facts (Carroll, 1990). Correns' major work on pollen competition and sex ratios was based on four female plants: one of which was Silene latifolia, another which was S. dioica, and two which were F_1 hybrids (one (S. latifolia x S. dioica) and one (S. dioica x S. latifolia)). All pollen was derived from a single male plant, an F_1 of (S. dioica ♀ x S. latifolia ♂). Thus, in each cross, the female parent was one of the four listed above and pollen competition was always between pollen carrying either the X chromosome of S. dioica or the Y chromosome of S. latifolia. With both the X and the Y pollen, the pollen cytoplasm was derived from S. dioica and the pollen autosomes were from the interspecific F_1 hybrid.

Given the nature of Correns' material, any coordination between nuclei and cytoplasms of style and pollen would presumably have been largely disrupted. In fact, Carroll (1990) reports that the Y pollen [derived from F_1 of (Silene dioica x S. latifolia)] performed significantly better in the pure S. latifolia styles (34.1% [390/1144] of the progeny from large pollen loads were male) than in the pure S. dioica styles (28.7% [152/529] of progeny were male) ($\chi^2 = 6.19$, $P < 0.05$, df = 1).

The materials used by Correns, while entirely appropriate for his purposes, should thus not be the basis for generalizations on the typical competition between X- and Y pollen tubes of Silene latifolia and S. dioica. Lawrence (1963a, 1963b) reported that, while Correns' observation of greater than expected proportions of female progeny after excess pollination was sometimes confirmed, this was by no means invariable. Furthermore, Carroll (1990) found that, (with very few exceptions), neither limited nor excess pollinations in a carefully studied population of S. latifolia, caused deviations from mendelian expectations. This is not to generalize from the two more recent studies but rather to indicate that the issue is not yet settled. More importantly, the differences between Correns' results and those of Lawrence (1963a, 1963b) and Carroll (1990) may reflect the genetic nature of the experimental material, hybrid and nonhybrid, respectively. Thus Correns' studies may point to a method for measuring the relative influences of style and pollen cytoplasmic and nuclear genotypes. Since Silene latifolia and S. dioica form a fertile interspecific hybrid, hybridizing them will allow many different permutations of style and pollen cytoplasmic and nuclear genotypes. Furthermore the presence of two discrete pollen genotypes (X and Y) greatly facilitates the analyses. Any comprehensive study of this topic will require that the gender of several thousand offspring be determined. This involves substantial labor but even more difficult is requirement that mortality before flowering must be kept to a minimum. However, now that molecular markers for the Y chromosome are available, several hundred plants can be analyzed in a day and early enough so that the effects of sex linked mortality can be minimized. This method, with Correns' pioneering insights, makes it possible to measure many of the genetic determinants of pollen tube growth rates.

References

Carroll, Steven B. 1990. An analysis of progeny sex ratios on dioecious Silene latifolia. PhD. dissertation; Univ. Mass., Amherst.

Correns, C. 1928. Bestimmung, Vererbung und Verteilung des Geschlechtes bei den hoheren Pflanzen. Hb. Vererb. 2:1-138.

Johnson C. M., and Mulcahy D. L. 1978. Male gametophyte in maize: II pollen vigor in inbred plants. Theor. Appl. Genet. 51:211-215.

Lawrence, C. W. 1963a. Genetic studies on wild populations of Melandrium. II. Flowering time and plant weight. Heredity 18:149-163.

Lawrence, C. W. 1963b. Genetic studies on wild populations of Melandrium. Heredity 18:149-163.

Michelmore, R. W., I. Paran, and R. V. Kesseli. (Submitted) Identification of markers linked to disease resistance genes by bulked segregant analysis: A rapid method to detect markers in specific genomic regions using segregating populations. Proc. Natl. Acad. Sci.

Snow, A. A. 1990. Effects of pollen load size and number of donors on sporophyte fitness in wild radish (Raphanus raphanistrum). Am. Nat. 136:742-758.

Van Damme. J. M.M. 1991. A restorer gene in gynodioecious Plantago coronopus subject to selection in the gametophytic and seedling stage. Heredity 66:19-28.

Williams, J. G. K., A. R. Kubelik, J. Livak, J. A. Rafalski, and S. V. Tingey. 1990. DNA polymorphisms amplified by arbitrary primers are useful as genetic markers. Nucleic Acids Research 18:6531-5.

7
Expression of a Rice Histone Gene during Pollen Developmental Transformations in *Hyoscyamus niger* (Henbane)

V. Raghavan, Changjun Jiang, and Rageeva Bimal

It is now well-established that pollen grains of angiosperms follow two diverse developmental pathways. One of these involves the familiar set of events leading to an asymmetric division of the pollen grain into a large vegatative cell and a small generative cell, germination of the pollen grain, division of the generative cell into two sperm and finally, double fertilization in the embryo sac, resulting in the fusion of one sperm with the egg nucleus and of the other sperm with the polar fusion nucleus. An alternate pathway of development displayed by pollen grains of a few angiosperms results int heir repeated division leading to the production of embryo-like structures (embryoids) and plantlets with the haploid set of chromosomes. This pehnomenon, known as pollen embryogenesis occurs when anthers excised at an appropriate stage of development--generally at the uninucleate stage--are cultured in a mineral salt medium with or without hormonal supplements. Thus, the pollen grain provides an interesting system to study the control of gene expression during its normal ontogeny and during induced embryogenesis (Raghavan, l986 for review).

In this study, we have monitored by in situ hybridization the subcellular distribution of a rice histone H3 gene during the development of the anther and pollen grains of Hyoscyamus niger (henbane) and during the transformation of the latter into embryoids. Preliminary experiments had indicated that 3H-labeled histone antisense RNA probe hybridized with certain layers of cells of the premeiotic anther wall of H. niger while sense RNA probe did not yield any hybridization signals. This led us to conduct more extensive experiments using 35S-labeled probes and dfferent stages of anther and pollen development.

Materials and Methods

Plants of H. niger were raised in the growth chamber as described (Raghavan, l978). For induction of embryogenic growth in pollen grains, flower buds (ca. 6 mm long) were sterilized in 12.5% Clorox for 5 min. followed by several washes in sterile distilled water. Anthers were excised under aseptic conditions and transferred to French square bottles containing 10 ml of a solidified medium supplemented with 2% sucrose (Bourgin and Nitsch, l967). Cultures were kept at 25 ± 2°C in a culture room provided with weak fluorescent light. For a protocol for anther culture in H. niger, see Raghavan (l990).

For histological studies, anthers at different stages of development excised from flower buds and open flowers and at different days after culture were fixed in 2% paraformaldehyde prepared in 0.05 M phosphate buffer (pH 7.0) at 4°C. After several washes in the buffer, samples were dehydrated in a graded series of ethanol and embedded in nonpolar glycol methacrylate (9 parts of butyl methacrylate, 1 part methylmethacrylate and 1% benzoyl peroxide) (Jamrich et al., 1984). Following polymerization of the monomer at 65°C overnight, the block was cut nominally at 1.0 lm thick on a Sorvall Porter-Blum microtome. Sections were mounted on drops of water on a slide, spread with chloroform vapor and dried on a slide warmer. Before annealing with the labeled probe, the slides were dipped in xylene to remove the plastic, passed through decreasing ethanol series to water, treated with 1.0 g/ml of proteinase K prepared in 100 mM Tris-HCl (pH 8.0), washed in water and dried.

A genomic DNA clone containing a 1.3 kb insert of the rice histone gene provided by Dr. Ray Wu (Cornell University) was restriced with BamH1 and was subcloned into pBS M13 vector. Linear templates of the antisense strand complementary to the histone mRNA and the sense strand to be used as a negative control were prepared by restriction of purified plasmid DNA by HindIII and EcoR1, respectively. Radiolabeled RNA probes were synthesized from the antisense and sense strands using T7 and T3 polymerases, respectively, as described earlier (Raghavan, 1989) except that 80 Ci uridine-5' [^{35}S] thiotriphosphate (^{35}S-UTP) was the label and 10 mM each of unlabelled ATP, CTP and GTP were used. The hybridization mixture contained Denhardt's solution, 10% dextran, 0.3 M NaCl, 20 mMTris-HCl (pH 8.0), 5 mM EDTA, 50% recrystallized formamide and 2.5 x 105 cpm/ll ^{35}S-labeled RNA transcripts. Other details of hybridization, washing and autoradiography are exactly as described (Raghavan, 1989). Autoradiographs we reexamined in a Zeiss Photamicroscope III with bright field and Nomarski optics.

Results and Discussion

A word of explanation is necessary for choosing to follow the expression of histone H3 gene. Histones are known to be relatively conserved through evolution and so, common sequences of histone genes may be found throughout the plant kingdom. Since anther and pollen development and pollen embryogenesis involve repeated cell divisions, there is reason to believe that histones will be needed for complexing with DNA.

The young anther of *H. niger* is a highly reduced structure consisting of an epidermis, a layer of cells constituting the endothecium, one or two middle layers and a layer of tapetum (Raghavan, 1981). Later, as the pollen mother cells undergo reduction division, the middle layers are disorganized, the tapetum disintegrates and the endothecium acquires radial thickenings. When sections of the anther primordiumare hybridized with ^{35}S-labeled antisense rice histone probe, a generally uniform distribution of the label is found throughout. Generally, a hypodermal cell in each lobe of the anther primordium designated as the archesporial initial divides periclinally to form a primary parietal cell and a primary sporogenous cell. The former gives rise to the wall layers and the tapetum, whereas the primary sporogenous cell generate a mass of cells constituting the secondary sprogenous cells. No differences were noted in the density of autoradiographic silver grains over these cells and the rest of the cells of the anther primordium. Apparently the initial events of microsporogenesis in *H. niger* do not appear to be associated with changes in histone mRNA abundance in the participating cells.

However, during later development of the anther, some striking differences are seen in the relative concentrations of silver grains in the wall layers and the tapetum. This results in a sequential accumulation of histone mRNA beginning with the epidermis on the outside and ending with the tapetum on the inside. When these changes in transcript abundance are related to the stages of microsporogeneis, a high silver grain density is observed in the epidermis at the stage of the secondary sporogenous cells. Later, concurrent with the separation of microsporocytes preparatory to meiosis, autoradiographs showed a higher number of silver grains over the endothecium than over the other cells of the anther wall. The tapetum began to accumulate the label at the stage of the secondary sporogenous cells, but a significant annealing of the label was found in this tissue during and immediately after meiosis. Thereafter the tapetal cells showed a decreased message abundance until they disintegrated at the stage of the binucleate pollen grain. These results show that histone transcripts hybridizable to a rice histone probe are present in the somatic cells of the anther and that expression of the gene in the anther follows a tissue-specific temporal pattern.

It was mentioned earlier that hybridization signals were distributed more or less uniformly in the cells of the anther primordium. However, compared to the primordial cells, there was a nearly four-fold increase in silver grain density in the secondary sporogenous cells up to the stage for formation of tetrads. This was followed by a decrease in the hybridization signals in the microspores and in the bicellular pollen grains. Coincident with starch accumulation, pollen grains gave strong hybridization signals and the increased density of silver grains was seen even in pollen grains collected from open flowers.

Previous studies from this laboratory have established that in cultured anthers of H. niger the generative cell of the bicellular pollen grain divides in the embryogenic pathway. Autoradiographs of sections of cultured anthers annealed with the antisense ^{35}S-labeled histone probe showed that compared to the normal uninucleate pollen grains, potentially embryogenic uninucleate pollen grains display an abundance of histone transcripts. As the genrative cell is cut off and as it begins to divide in the embryogenic pathway, there is a dense accumulation of silver grains in the newly formed cells, but the nondividing vegetative cell does not bind the probe. Histone mRNA continues to be present in the cells of the multicellular embroyoid. In a11 hybridization experiments described above, probe specificity was confirmed by the presence of a significantly higher number of silver grains generated with the antisense strand than with the sense strand.

In conclusion, molecular hybridization between histone mRNA of a dicot and a probe preapred from the genomic histone DNA clone isolated from a monocot confirms that the fidelity of this gene is highly conserved between the two major classes of angiosperms. Our results also show that histone mRNA is not necessarily localized in actively dividing cells as one would expect if they are used for complexing with DNA. Although there is an increase in histone mRNA abundance in pollen grains undergoing embryogenic divisions in cultured anthers, no claim is made that histone gene expression triggers embryogenic division; rather, histone gene may be one among a battery of genes that are activated during this developmental episode.

Acknowledgements

Part of the work described in this chapter was supported by grants from the National Science Foundation, U.S.A. to V.R. and from the ICSC-World Laboratory, Lausanne, Switzerland to C.J. and by a National Fellowship from the Government of India to R.B.

References

Bourgin, J.P; Nitsch, J.P. Obtention de Nicotiana halpoides a partir de'etamines cultvees in vitro. Ann. Physiol. Veg. (Paris) 9:377-382; l967.

Jamrich, M; Mahon, K.A; Gavis, E.R; Gall, J.G. Histone RNA in amphibian oocytes visualized by in situ hybridization to methacrylate-embedded tissue sections. EMBO J. 3: 1939-1943; 1984.

Raghavan, V. Origin and development of pollen embryoids and pollen calluses in cultured anther segments of Hyoscyamus niger (henbane). Amer. J. Bot. 65: 984-1002; 1978.

Raghavan, V. A transient accumulation of poly(A)-containing RNA in the tapetum of Hyoscyamus niger during microsporogenesis. Develop. Biol. 81: 342-348; 1981.

Raghavan, V. Embryogenesis in angiosperms. A developmental and experimental study. New York: Cambridge University Press; 1986.

Raghavan, V. mRNAs and a cloned histone gene are differentially expressed during anther and pollen development in rice (Oryza sativa L). J. Cell Sci. 92: 217-229; 1989.

Raghavan, V. Hyoscyamus spp. Anther culture studies. In: Bajaj, Y.P.S., ed. Biotechnology in Agriculture and Forestry. 12. Haploids in Crop Improvement 1, pp. 290-305. Berlin: Springer-Verlag; 1990.

8

Plastid Developmental Pathways in Some Angiosperm Reproductive Cells

Ettore Pacini, Philip E. Taylor, Mohan B. Singh, and R. Bruce Knox

Plants such as algae and mosses have only one plastid type i.e. the chloroplast, their number is very low per cell. From pteridophyta onwards other plastid types appear; in angiosperm at least five plastid types are recognized: proplastid, amyloplast, chloroplast, chromoplast, elaioplast (Schnepf, 1980). Some authors such as Whatley (1977) distinguish seven types. The number per cell in angiosperm is high and constant for each cell type (Butterfass, 1979). Plastid types are recognized by classical cytologists on the basis of their appearance under EM rather than their metabolism. Chloroplast is the only plastid type where we are reasonably sure that the morphology corresponds to the metabolic activity of this organelle engaged in the photosythetizing activity. The products of the photosynthesis, the starch can be present inside: a) chloroplasts; b) amyloplasts; c) proplastid. The period of storage lasts a few hours in case a), is undeterminate in case b); ranges from few hours to some months in case c). Sometimes it is difficult to distinguish proplastids from amyloplasts. We define plastids as amyloplasts when the starch fills the stroma and proplastids when only a few granules per plastid are present.

Plastid pathways

Angiosperm reproduction is characterized by extensive processes of differentiation. Plastids differentiate from their basic form and sometimes even dedifferentiate, more than once, according to the program of the cell (Heslop-Harrison, 1972). During reproduction we find cells which will continue their life, because they are the true reproductive cells, and others, ephemeral, whose task is to help the first ones in their development. Egg and sperm cells will continue their life after fecundation; calix, corolla, ovary walls, stigma, endosperm, cotyledons, tapetum and vegetative cell will die at the end of their task. Here we will compare the pattern of plastid differentiation of ephemeral and reproductive cells. The data, from personal observations and from the literature, are summarized in Tab. 1. Proplastids are always present at the onset of reproductive cells or tissue.

Calix and corolla

In the young flower bud calix and corolla have undifferentiated plastids, these may differentiate directly into chromoplasts or as chloroplast and later on, more frequently in the case of corolla, chromoplasts (Thomson & Whatley, 1980). There are only few cases where chromoplasts dedifferentiate into chloroplasts. Undoubtedly photosynthetically active chloroplasts are a source of energy during reproduction.

Tab. 1. Table depicting plastid pathways in some reproductive cells (explanation in the text).

----- sporadic pathway

\+ death of the cell or tissue

? eventual but not yet demonstrated pathway

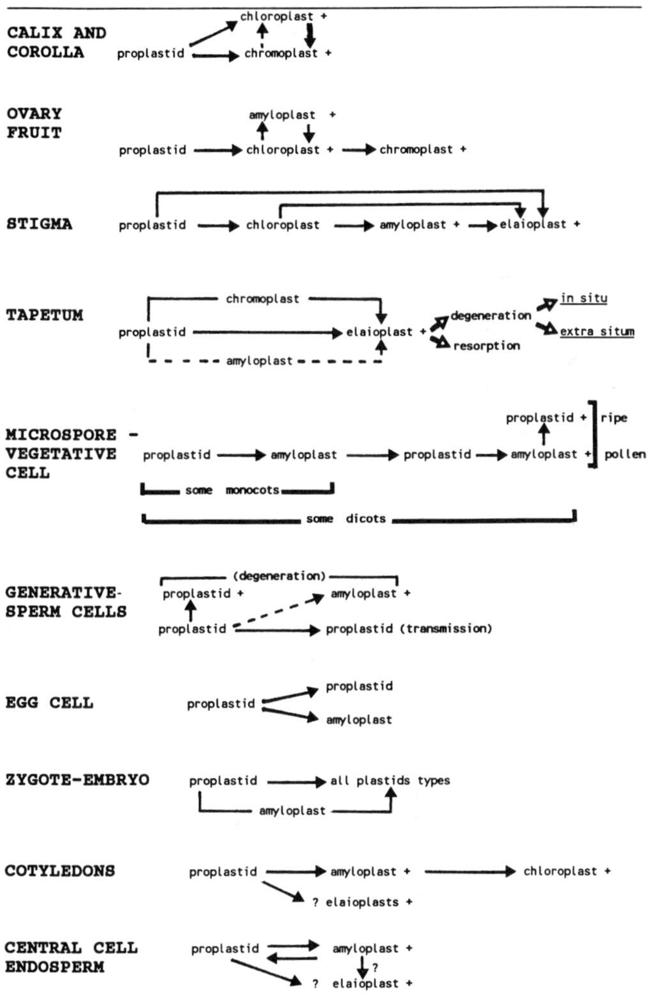

Ovary-Fruit

In the developing ovary and fruit proplastids may differentiate into amyloplasts or chloroplasts and these two types can transform into each other. In the fruit, expecially in the fleshy ones chloroplasts often differentiate into chromoplasts (Thompson & Whatley, 1980).

Stigma

In the stigma proplastids differentiate, directly or via chloroplasts, into amyloplasts. Proplastids, chloroplasts and amyloplasts may differentiate into elaioplasts. In the case of species having a wet stigma such as those belonging to the genuses *Acacia, Lotus* and *Trifolium*, plastids differentiate into chloroplasts and become elaioplasts after a transient chloroplast reproductive cells plastid 2 phase (Jobson et al., 1983). Plastids probably contribute to the formation of the exudate but this has not yet been observed directly. Nevertheless amyloplasts are present in the immature papillae cells of watermelon (wet stigma) with maximum accumulation one day prior to anthesis, the starch is hydrolized and elaioplasts, deduced by the plate but mentioned as "plastids" in the text, are present in the mature stigma (Sedgley, 1981).
Independently from the type of stigma the developmental pattern of stigmatic cell plastids is essentially the same.

Tapetum

Tapetal plastids are similar to those of microspore mother cells during the premeiotic phase, their patterns of differentiation starts to diverge at leptotene-pachitene (Pacini & Juniper, 1979; Pacini et al., 1986). Those of tapetum start to increase in number during tetrad-early microspore stage when they differentiate directly into elaioplasts (Pacini & Casadoro, 1981) or via chromoplasts (Pacini et al., 1991). In only a few instances tapetal plastids accumulate temporarily starch before differentiating into elaioplasts (Zavada, 1984). Elaioplasts are present in species with a typical anemophilous syndrome such as grasses, but their content is totally resorbed by developing pollen grains which start their engorgement phase just after tapetum degeneration (Pacini *et al.*, 1991). In enthomophilous species such as cherry (*Secretory tapetum*) and chicory (*Amoeboid tapetum*) elaioplasts, together with tapetal degenerating cytoplasm contribute to pollen kitt formation (Pacini *et al.*, 1986; Pacini & Keijzer, 1989). Pollenkitt will be deposited onto pollen exine just prior to anther opening. In these two cases plastids and tapetal cytoplasm degenerate *in situ*. In the case of some *cruciferae* such as Raphanus (Dickinson & Lewis, 1973), this degeneration occurs extra situm. At mid microspore stage the tapetal plasma membrane breaks and elaioplasts, together with other cytoplasmic components appear temporarily in the exine cavity before tryphine formation (Dickinson & Lewis, 1973).
Microspore-vegetative cell In the case of microspore-vegetative cell plastid behaviour seems to differ according to the taxonomy. In the case of grasses proplastids differentiate into amyloplasts only at early bicellular stage and starch is present in the ripe grains (Pacini *et al.*, 1991). In the case of dicots such as *Parietaria judaica, Prunus avium, Olea europaea, Lycopersicum peruvianum* proplastids differentiate into amyloplasts during

microspore stage, they undifferentiate into amyloplasts before the first haploid mitosis; they differentiate again into amyloplasts at bicellular stage (Pacini & Franchi, 1988). In some species these amyloplasts can persist in the ripe pollen, in others instead starch is hydrolized before pollen shedding. The products of starch hydrolisis are stored as insoluble polysaccharides in the cytoplasm and proplastids are few and scanty in the ripe pollen grains (Pacini & Franchi, 1988). The occurrence of starch and starchless pollen grains is a characteristic of a species and, in many instances, of the family (Baker & Baker, 1979); differences may occur in cultivated plants of those species having a long blooming period (Pacini & Franchi, 1988). It is not yet known as to the reason for this different storage of polysaccharides in the ripe pollen. The polysaccharidic reserves, independently from their cytological localization, are reasonably consumed during pollen tube growth.

Plastid inheritance

Plastids are undifferentiated at the first haploid mitosis in *Lycopersicum peruvianum* (Pacini & Juniper, 1984), *Prunus avium* (Pacini *et al.*, 1986), *Impatiens walleriana* and *I. glandulifera* (van Went, 1984); they are differentiated into amyloplasts in *Lilium longiflorum* (Nakamura & Miki-Hirosige, 1985). Independently from their differentiation stage they may or may not enter the generative cell. If they do not enter there is an "*a priori exclusion*". If they enter they can be eliminated at different stages of male gametophyte development "*a posteriori exclusion*". This exclusion can occur even after fertilization (Jensen, 1974). Hageman and Schroeder (1989) have named the mechanisms regulating male inheritance as exclusion, degeneration and transmission. Proplastids are always present in the progametic and gametic cells, there is only the exception of *Lolium perenne* (Pacini *et al.*, 1991) where generative cell proplastids differentiate into amyloplasts, simultaneously with those of the vegetative cell, but they are present only temporarily because they are destroyed by autophagic vacuoles. Plastid male inheritance was ultrasturcturally studied in only a few species and it was completely documented in *Plumbago zeylanica* (Russell, 1983, 1984, 1985).

If plastid inheritance is maternal the egg cell may contain proplastids or amyloplasts. In the case of *Spinacia oleracea* plastids are lacking from sperm cells (Theunis, 1990) and amyloplasts are present in the egg cell (Willms, 1981), the same occurs in *Triticum aestivum* (You & Jensen, 1985). Proplastids are present instead in the egg cell of *Agave parryi* (Tilton & Mogensen, 1979) and *Glycine max* (Folson & Petersen, 1984). If inheritance is biparental four possibilities exist, two with amyloplasts or proplastids simultaneously present in both gametes and two with assorted proplastids and amyloplasts present in the gametes. A case of plastids of different types present in the two gametes is that of *Oenothera erythrosepala (sin. O. lamarkiana)* where proplastids are present in the sperm cell and amyloplast in the female gamete (Meyer & Stubbe, 1974). In *Plumbago zeylanica* with biparental inheritance sperm cell and egg cell plastids are both undifferentiated at fecundation but their morphology is slightly different (Russell, 1985).

If the inheritance is uniparental there are no problems with respect to plastid differentiation; if it is biparental and they are different before gamete fusion, as in *Oenothera erythrosepala*, they must align before zygote division (Meyer & Stubbe, 1974).

Zygote-embryo

In *Oenothera erythrosepala* the number of plastids increase from 33-42 in the joung zigote to 68-82 in the mature zygote; their number decrease to 25-30 per cell in the globular embryo (Meyer & Stubbe, 1974). During the first phases of embryo growth proplastids may eventually differentiate into amyloplasts. In the ripe seed proplastids or amyloplasts can be present according to the embryo stage and the type of reserves. In only a few instances have chloroplasts been present in the ripe seed (Rascio et al., 1990). After germination when cotyledons emerge chloroplasts are commonly present.

Central cell-endosperm

Proplastids are present in the eight nucleated gametophyte before cellularization (Willemse & van Went, 1984) but amyloplasts are present, before fertilization, in the central cell of the following species: *Triticum aestivum* (You & Jensen, 1985) *Agave parryi* (Tilton & Mogensen, 1979), *Glicine max* (Folsom & Peterson, 1984), *Spinacia oleracea* (Wilms, 1981). In *Arachis hypogaea* (Periasamy & Sampoornan, 1984) amyloplasts of the central cell disappear at the third division of the nuclear endosperm. At least in this species it seems that the energy for the first phase of embryo growth derives from material accumulated in the gametophyte before fertilization. In the nuclear endosperm of *Eranthis hiemalis* plastids are undifferentiated during the nuclear phase (Pacini et al., 1972), after cellularization they differentiate into amyloplasts, and persist in the ripe seed (Pacini, unpublished data).

Plastids differentiation and dedifferentiation

From Tab. 1 emerges that plastid differentiation is not always unidirectional but sometimes, as in the case of microspore-vegetative cell, plastids differentiate and dedifferentiate more than once. The occurrence of amyloplasts means polysaccharides are successively consumed or stored. Plastids increase in number in several instances during reproduction, generally before a differentiation process into amyloplasts, ie. when it is necessary to create more regions ready to be filled by starch. Elimination of plastids is documented in the generative-sperm cells in some cases of maternal inheritance according to Hageman and Schroeder (1989). In many cases amyloplasts are the extreme differentiation of plastids. When starch within vegetative cell, cotyledons or endosperm is consumed the cells degenerate. Other final differentiation states are chloroplast, chromoplast and elaioplast. The temporaneous occurrence of chloroplasts in reproductive structures such as calix and ovary is a way to generate energy reserves in reproductive structures.

References

Baker, H.G.; Baker, I. Starch in angiosperm pollen grains and its evolutionary significance. Am. J. Bot. 66: 591-600; 1979.
Butterfass, T. Patterns of chloroplast reproduction:

A developmental approach to protoplasmic plant anatomy. Springer Verlag, Wien; 1979.
Dickinson, H.G.; Lewis, D. The formation of tryphine coating the pollen grains of Raphanus. Proc. R. Soc. Lond. B 184:149-165: 1973.
Folsom, M.W.; Peterson, C.M. Ultrastructural aspects of the mature embryo sac of soybean, Glycine max (L.) Merr.. Bot. Gaz. 145:1-10;1984.
Hageman, R.; Schroeder. M.B. The cytological basis of the plastid inheritance in angiosperms. Protoplasma 152:57-64;1989.
Heslop-Harrison, J. Sexuality of angiosperms. In: Plant Physiology (Steward F.C ed.), Academic Press, London, VIC, p. 133-189;1972.
Jensen, W.A. Reproduction in flowering plants. In: Dynamic Aspects of Plant Ultrastructure (Robards A.W. ed.) p. 481-503. Mc Graw Hill, London; 1974.
Jobson, S.; Knox, R.B.; Kenrick, J.; Dumas, C. Plastid development and ferritin content of stigmas of the legumes Acacia, Lotus and Trifolium. Protoplasma 116: 213-218;1983.
Meyer, B.; Stubbe, W. Das Zahlenverhaltnis von mutterlichen und vaterlichen Plastiden in den Zygoten von Oenothera erythrosepala Borbas (syn. Oe. lamarkiana) Ber. Deut. Bot. Ges. 89:29-39;1974.
Nakamura, S.; Miki-Hirosige H. Fine-structural study on the formation of the generative cell wall and intine-3 layer in a growing pollen grain of Lilium longiflorum. Amer. J. Bot. 72: 365-375; 1985.
Pacini,E.; Bellani, L.M.; Lozzi, R. Pollen, tapetum and anther development in two cultivars of sweet cherry (Prunus avium). Phytomorphology 36: 197-210;1986.
Pacini, E.; Casadoro, G. Tapetum plastids in Olea europaea L.. Protoplasma 106:289-296;1981.
Pacini, E.; Cresti, G.; Sarfatti, G. Incorporation of integumentary nuclei in Eranthis hiemalis endosperm and their disaggregation by endoplasmic reticulum. J. Submicr. Cytol. 4: 19-31;1972.
Pacini, E.; Keijzer, C.J. Ontogeny of intruding non - periplasmodial tapetum in the wild chicory, Cichorium intybus (Compositae). Pl. Syst. Evol. 167:149-164; 1989.
Pacini, E.; Franchi, G.G. Amylogenesis and amylolysis during pollen grain development. (Cresti, M.; Gori, P.; Pacini, E. eds.) Sexual reproduction in higher plants. Springer Verlag, Berlin; 1989: p. 181-186.
Pacini, E.;Taylor, P.E.; Singh M.B.; Knox RB. Development of plastids, including amyloplasts and starch granules, in pollen and tapetum of rye-grass Lolium perenne L. Ann. Bot. in press, 1991.
Pacini, E.; Juniper, B.E. The ultrastructure of pollen-grain development in the olive (Olea europaea). 2. Secretion by the tapetal cells. New Pthytol. 83: 165-174; 1979.
Pacini, E.; Juniper B.E. The ultrastructure of pollen grain development in Lycopersicum peruvianum. Caryologia 37: 21-50; 1984.
Periasamy, K.; Sampoornam, C. The morphology and anatomy of ovule and fruit development in Arachis hypogaea L. Ann. Bot. 53: 399-411; 1984.
Rascio, N.; Dalla Vecchia, F.; Casadoro, G. Behaviour of cotyledonal plastids during seed germination in some chloroembryophytes. Cytobios 64: 53-54; 1994.
Russell, S.D. Fertilization in Plumbago zeylanica: gametic fusion and fate of the male cytoplasm. Amer. J. Bot. 70: 416-434; 1983.
Russell, S.D. Ultrastructure of the sperm of Plumbago zeylanica. 2. Quantitative cytology and three-dimensional organization. Planta 162:385-391; 1984.
Russell, S.D. Preferential fertilization in Plumbago: Ultrastructural evidence for gamete-level recognition in an angiosperm. Proc. Nat. Acad. Sci. USA 82: 6129-6132; 1985.

Schnepf, E. Types of plastids: their development and intercoversions. In:Chloroplast. (Reinert,J. ed.) Springer Verlag, Berlin, Heidelberg, New York 1980, p. 1-28.
Sedgley, M. Ultrastructure and histochemistry of the watermelon stigma. J. Cell Sci. 48: 137-146; 1981.
Theunis, C. Ultrastructural analysis of Spinacia oleracea sperm cells isolated from mature pollen grains. Protoplasma 158:176-181; 1990.
Tilton, V.R.; Mogensen, H.L. Ultrastructural aspects of the ovule of Agave parryi before fertilization. Phytomorphology 29: 338-350; 1979.
Thomson, W.W.; Whatley J.M. Development of non green plastids. Ann. Rev. Plant Physiol. 31: 375-394;1980.
Went van, J.L. Unequal distribution of plastids during generative cell formation in Impatiens. Theor. Appl. Genet. 68:305-309; 1984.
Whatley, J.M. Variation in the basic pathway of chloroplast development. New Phytol 78: 407-420; 1977.
Willemse, M.T.M.; van Went, J.L. The female gametophyte. In: "Embryology of Angiosperms" (Johri B.M. ed.) Springer Verlag Berlin; 1984; p. 159-196.
Willms, H.J. Pollen tube penetration and fertilization in spinach. Acta Bot Neerl 30: 101-122; 1981.
You, R.; Jensen, W.A. Ultrastructural observations of the mature megagametophyte and fertilization in wheat (Triticum aestivum). Can. J. Bot. 63: 163-178; 1985.
Zavada, M.S. Pollen wall development of Austrobaileya maculata. Bot. Gaz. 145: 11-21; 1984.

9
In Vitro Polymerization of F-actin and Assembly of Microfilament Cytoskeleton of Pollen Actin

Lung-Fei Yen, Xiong Liu, and Guoqin Liu

Introduction

It is well known that actomyosin system plays an important role in the maintenance of cell shape and performing cell motility, such as mitosis, cytokinesis, cytoplasmic streaming, organelle movement and amoeboid movement in higher plants (Jackson, l982). Actin and myosin have been identified in higher plants by many authors with the use of biochemical and immunological methods (Condeelis, l974; Vahey et al., l982; Yenet al., l986; Tang et al., l989; Ma et al., l989; Villanueva et al., l990). But the quantity of actin is too low to be isolated for biochemical and physiological studies. Meagher and McLean (l990) pointed out that the concentrations of actin in any plant tissue are lower than they are even in animal cytoplasmic tissues. However, we found that pollen is an excellent source of actin in higher plants in which the quantity of actin is sufficient to do biochemical investigation.

The actin purified by our method is biologically active. The pollen actin is able to polymerize in vitro into actin filaments of which the electron micrograph shows typically helical appearance. In the partially purified form pollen actin can reassemble into microfilament cytoskeleton. Here we describe our recent investigations on the polymerization and reassembly of microfilament cytoskeleton of actin from corn pollen.

Materials and Methods

Pollen from corn (Zea mays L.) was used in this investigation. A corn plant is a rich source of pollen during its pollination stage. Corn pollen was collected from the corn field in the Experimental Station attached to Beijing Agricultural University during July and stored in a desiccator at -75 C. We applied the preparation of acetone powder used in purification of actin from skeletal muscle to the isolation of actin from cornpollen. The procedure is briefly described in the following. The detailed account will be published elsewhere. The corn pollen was ground with a buffer to a viscous material and extracted with cold acetone for four times to remove the fatty substances in pollen. After centrifugation the precipitate was dried. This is the acetone powder of corn pollen, just as the acetone powder of rabbit muscle. The acetone powder of pollen was fractionated by 50% saturated ammonium sulfate, DEAE-cellulose chromatography, polymerization and depolymerization and Sephacryl S-200 gel filtration respectively. The eluate after gel filtration was collected and stored at -20 C for further use.

The polymerization of pollen actin was carred out in the presence of 0.1M KC1 and 10 mM MgCl2 at 20 C. The UV-absorbance of actin was determined with a Shimadzu UV 240 spectrophotometer. The viscosity of the actin solution was determined with an Ostwald viscometer at 20 C. The ATPase of myosin was determined according to the method of Le Bel (l978) with modification.

The actin filament polymerized from pollen actin was examined under JEM 100CX electron microscope. The sample was negatively stained by 1% uranyl acetate. The light microscopy was carried out according to the method of Pena (l980). The sample was fixed by 2% gluraldehyde and stained by 0.2% Coomassieblue R-250.

Results

1. Purification and characterization of actin from corn pollen

The above method for purifying actin from corn pollen is much better than other methods to purify actin from plant tissues. Recently, Villanueva et al (l990) purified actin from soybean cultured cells and Jin et al (l991) purified actin from wheat endosperm both by DNase I affinity chromatography respectively. The yield of actin of our method is not only higher than the affinity chromatography method, but also its biological activity remained. Table 1 presents the yields of various steps of thepurification of actin form corn pollen.

Table 1. Purification and yield of actin from corn pollen

Step	Volume (ml)	Total Protein (mg)	Actin (mg)	Actin (1%)
Acetone powder of pollen	858	1097	211.8	19.3
5% saturated ammonium sulfate	63	172.3	51.4	29.6
DEAE-cellulose	46	96.4	24.5	53
Polymerization and depolymerization	4.5	11.9	9.0	75
Sephacryl S-200	10	7.1	6.5	91.5

*the pollen used in this experiment is 30 gm

The table shows that the content of actin in corn pollen is higher than other plant tissues. The actin approached 20% of total protein of pollen from corn. The yield of actin from corn pollen is 18-20 mg in average per 100 gm of pollen. The recovery of actin is about 3%. The yield of our method is comparable with that of Acanthamoeba (Gordon, Eisenber and Korn, l976) and human platelets (Rosenberg, Stracher and Lucas, l981). Figure 1 shows the purified actin after sephacryl S-200 gel filtration.

The molecular weight of actin from corn pollen is 43 KD as determined by SDA-polyacrylamide gel electrophoresis. The amono acid composition of pollen actin was determined and compared with those of muscle (Murphy 1971) and Acanthamoeba (gordon et al.,l976). The comparison indicates that the amino acid composition of pollen actin is very similar to those of muscle and Acanthamoeba actin (data not show). The circular dichroismspectrum of pollen actin was determined with spectropolarimeter and

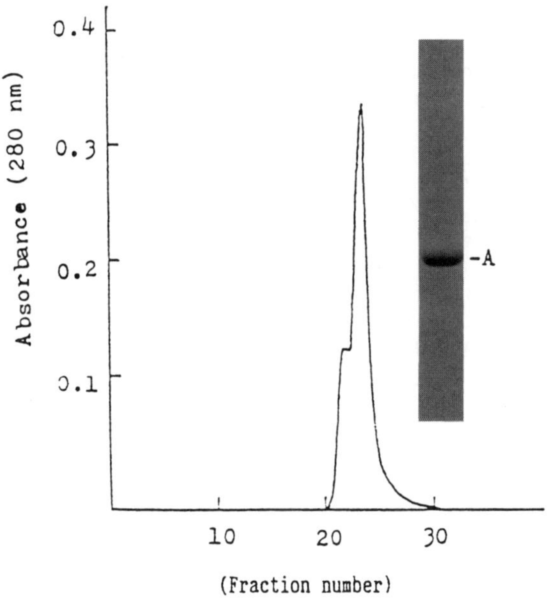

Figure 1. Sephacyl S-200 gel filtration of actin from corn pollen. A indicates actin.

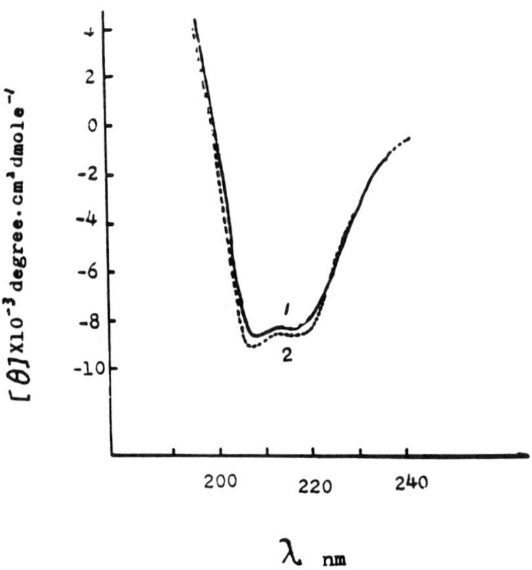

Figure 2. Circular dichroism of pollen actin compared with that of muscle actin. 1.. —— .pollen actin; 2. . . . . . .muscle actin.

compared with that of muscle actin. The C.D. spectrum of pollen actin coincides with that of muscle actin (Figure 2). The contents of -helices and -sheets between pollen actin and muscle actin are ery similar also. These results show that the actin from pollen has very similar biochemical properties with those of muscle and nonmuscle actins.

Polymerization of pollen actin in Vitro

The actin purified from corn pollen by our method is very active in its biological activities. Pollen actin is able to polymerize to actin biological activities. Pollen actin is able to polymerize to actin filament or F-actin. Figure 3 is the time course of the polymerization of pollen actin. In the presence of 0.1 M KC1 and 10 mM MgCl pollen actin was polymerized to actin filament. After 10 min of the addition of DC1 and MgC1 the polymerization is complete. During polymerization the UV-absorbance of the solution at 232 nm increased evidently (Figure3).

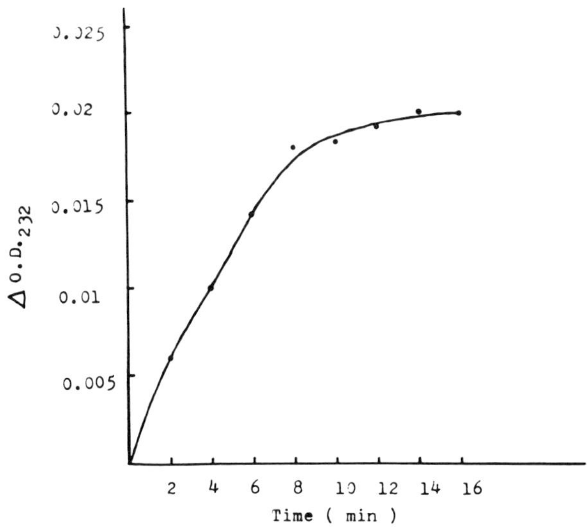

Figure 3. Time course of KC1 induced polymerization of actin from corn pollen.

During polymerization of pollen actin, the viscosity of the actin solution increased significantly (Figure 4). This is due to the globular protein of G-actin when polymerized to the fibrous actin filaments. Moreover, the specific viscosity of the pollen actin increased with the increase of the concentration of actin solution. The relationship between the specific viscosity and the concentration of actin is linear (Figure 5).

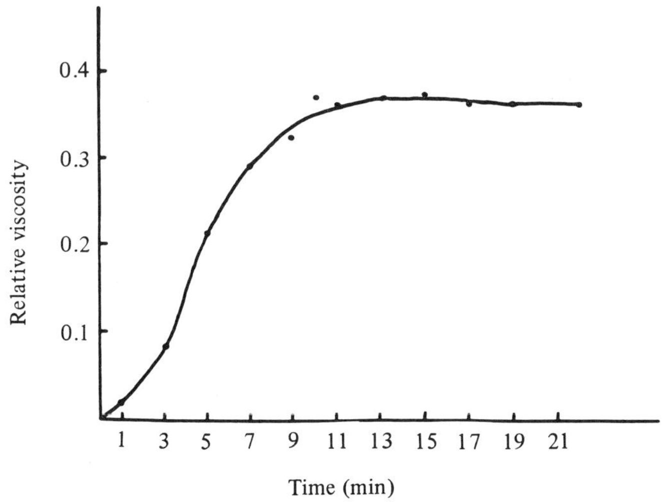

Figure 4. Viscosity change of pollen actin during the polymerization to F-actin. Pollen actin: 0.5 mg/ml.

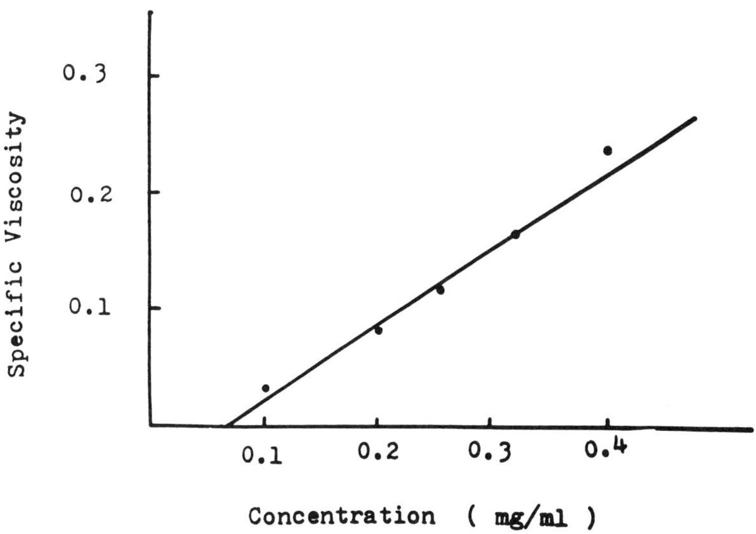

Figure 5. Polymerization of pollen actin. Plots of specific viscosity versus concentration of actin at 20 C.

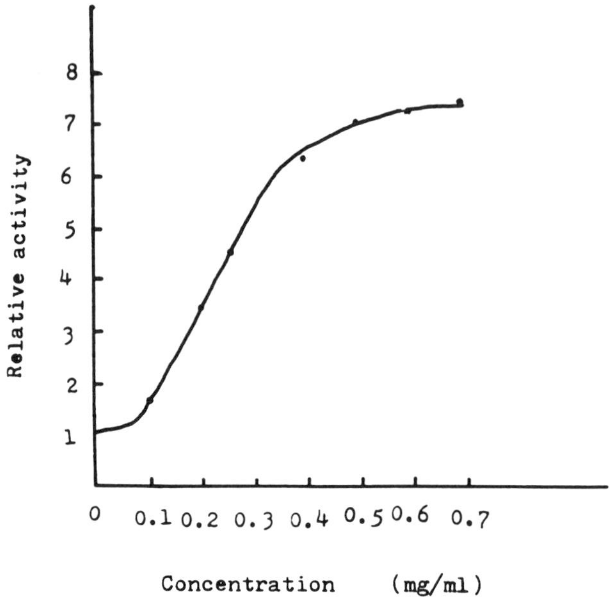

Figure 6. Activation of the activity of myosin ATPase by pollen F-actin.

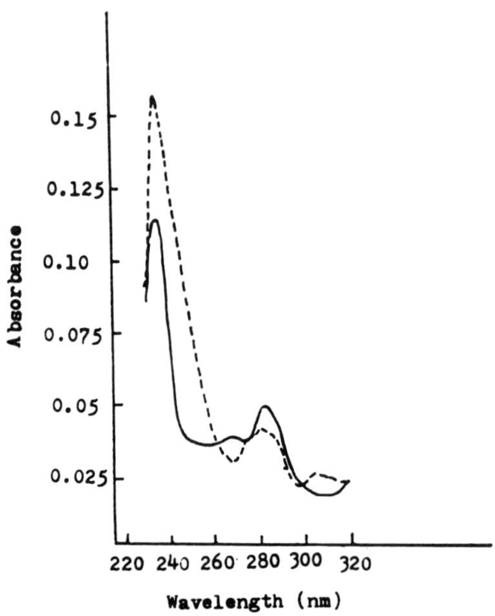

Figure 7 UV difference spectrum between F-actin and G-actin from corn pollen. Solid line _ pollen actin (0.3 mg/ml); Dotted line. . . .muscle actin (0.5 mg/ml).

The polymerized actin filament is able to activate the activity of mysin ATPase by several folds. We found that muscle myosin ATPase was activated by pollen F-actin by seven fold (Figure 6). This is identical with the result of Dictyostalium discoideum and other nonmuscle actin (Dyemura,Brown and Spudich, l978).

The UV difference spectroscopy between pollen F-actin and G-actin was determined with an UV dual wavelength spectrophotometer. The difference spectrum of pollen actin is shown in Figure 7. This is very similar to the difference spectrum of muscle actin determined by Higashi and Oosawa(l965). The difference spectrum of the denatured pollen actin by guanidinehydrochloride is similar to that of muscle actin also (figure not shown). This indicates that the polymerization of G-actin to F-actin is accompanied by a big conformational change, some of the tryptophane and tyrosine residues of actin molecule, and the conformation of the peptide chain is also changed.

The polymerized pollen actin filaments were examined under electron microscope. The pollen actin filament solution was added on the formavar film of the grid and negatively stained by 1% uranyl acetate. The sample was observed under JEM 100CX electron microscope. It shows that a lot of actin filaments appear on the micrograph (Figure 8). The diameter of the single filament is 7 n and its pitch of the helix is 36 nm. The actin filament consists of two right-handed helically wound strands composed of G-actin. This photograph is very similar to the electron micrograph observed in actin. This photograph is very similar to the electron micrograph observed in actin filament from BHK cultued cell by Koffer and Dickens (l989). This result indicates that pollen actin is able to reassemble to microfilament cytoskeleton in vitro. But if cytochalasin B were added to the actin filament preparation before staining, no filamentous network could be observed (figure not shown). This implies that the actin filaments are disrupted by cytochalasin B. The pollen actin filament is not only able to be observed under electron microscope, but also can be visualized under light microscope. According to the method of Pena (l980) the polymerized pollen actin was added on a glass slide, fixed with 2% gluraldehyde and stained with 2% Coomassie brilliant blue R250. Under light microscope, a blue filament network was observed (Figure 9). It is very similar to the microfilament cytoskeleton of fibroblast cells examined by Pena (1980).

Discussion

Pollen is a rich source of actin in higher plants. The high level of actin in pollen may be very significant in its role of biological functions. It is well known that the cytoplasmic streaming and organelle movement in the pollen tube is very vigorous during pollen tube growth. Cytoplasmic streaming and organelle movement depend on the actomyosin system as their motive forces and supporting networks. So actin plays a very important action in the physiological and biochemical behaviors of pollens. Because the growth of pollen tubes is very fast among plant cells, the polymerization and assembly of actin filament and microfilament cytoskeleton should be very active. Accompanying the growth of pollen tubes actin monomers should be polymerized continously.

Recently Meagher and McLean (l990) indicated that the typical methods for purifying actins from animals and protists such as cycle of polymerization and depolymerization and DNase I affinity chromatographyhave not yet proven quantitative in plant system. However, the application of preparing acetone powder to the purification of actin from corn pollen by us is successful. Not only the yield of actin is higher than other methods (eg. DNase I affinity chromatography), but also the biological activity of actin is also

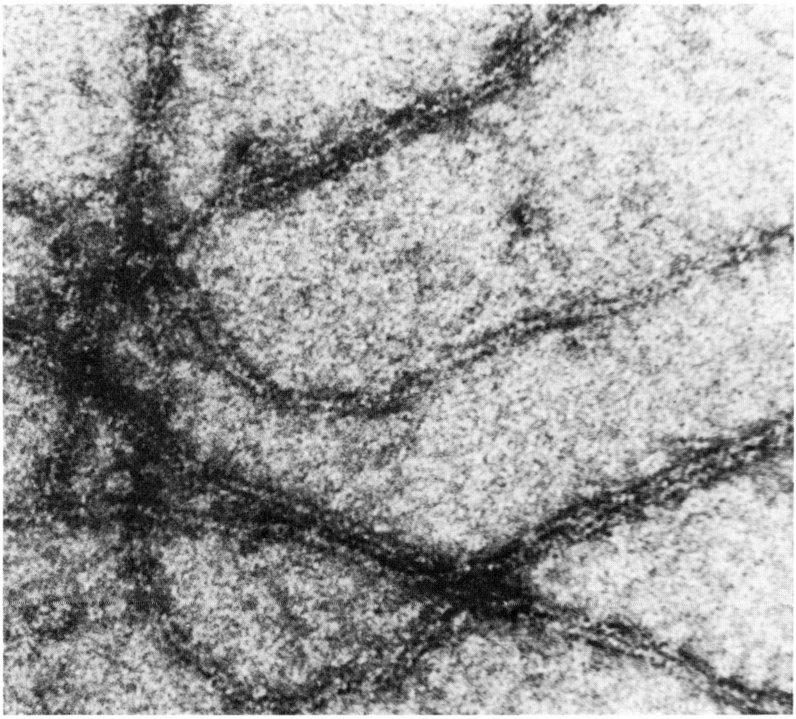

Figure 8. Electron micrograph of polymerized pollen actin. Pollen actin was assembled to microfilament in vitro. **Diameter of** actin filament; 7 nm;pitch of helices; 36 nm; JEM 100CX 350,000 X.

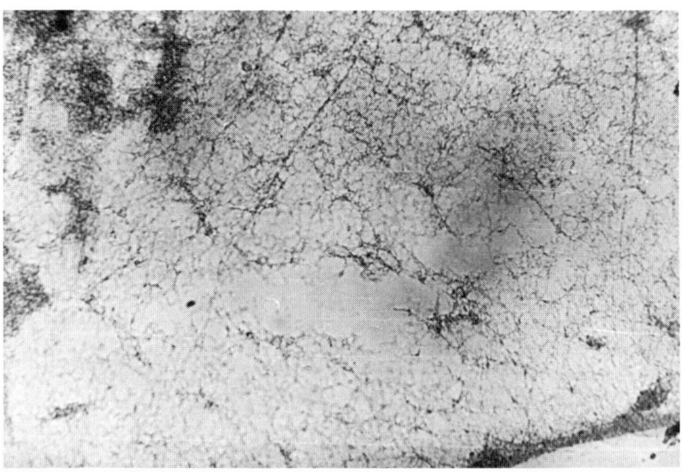

Figure 9. Light microscopy of the reassembed microfilament cytokeleton. 330X.

better than other methods. The pollen actin purified by our method reserves the biological activities to be polymerized to actin filament. The partially purified actin (fractions afterDEAE-cellulose chromatography) is able to reassemble into microfilament cytoskeleton (Figure 9). It is possible that some actin-binding protein (ABF), especially gelsolin, may be present in the partially purified actin of pollen. The investigation of gelsolin in pollen is undertaken in our laboratory. The reassembled microfilament cytoskeleton has the same biological and morphological properties as the microfilament in vivo, such as its spectroscopic, physiochemical and microscopic properties. The structure and property of actin in the animal and plant kingdom may be universal from the view point of evolution.

Acknowledgment

The authors wish to express their thanks to Prof. B.S. Yu of the Laboratory of Microscopy, and the Central Laboratories of Beijing Agricultural University for their help during this investigation. This work is supported by the National Natural Science Foundation of China.

References

1. Jackson, W.T. (l982) Actomyosin. In: Lloyd, C.W. ed. the Cytoskeleton inPlant Growth and Development. London: Academic Press. p. 3-29.

2. Condeelis, J.S. (l974) The identification of F-actin in the pollen tubeand protoplast of Amaryllis belladonna. Exp. Cell Res. 88:435-439.

3. Vahey, M.; Titus, M; Trautwein, R.; Scorillis, S. (l982) Tomato actinand myosin; Contractile proteins from a higher land plant. Cell Motility 2:131-147.

4. Yen, L. F.; Wang, X.Z.; Teng, X.Y.; Ma, Y.Z.; Liu, G.Q. (l986) Actin andmyosin in pollens and their role in the growth of pollen tubes, ChineseSci. Bulletin (Kexue Tongbao) 31: 267-272.

5. Tang, X.J.; Lancelle, S.A.; Hepler, P.K. (l989) Fluorescence microscopiclocalization of actin in pollen tubes; comparison of actin antibody andphalloidin staining. Cell Motility Cytoskeleton 12: 216-224.

6. Ma, Y.Z.; Yen, L.F. (l989) Actin and myosin in pea tendril. PlantPhysiol. 89:586-589.

7. Villanueva, M.A.; Ho, S.C.; Wang, J.L. (l990) Isolation andcharacterization of one isoform actin from cultured soybean cells. Arch.Biochem. Biophys. 277:35-41.

8. Jin, S.P.; Ling, J.; Zhang, W.C. (1991) Affinity chromotographicpurification of actin from immaturewheat endosperm and preliminary studieson its characteristics. Chinese Sci. Bulletin 36: 235-238.

9. Meagher, R.B.; McLean, M. (l990) Diversity of plant actins. Cell Motil.Cytoskeleton 16:164-166.

10. Le Bel, D.; Poirier, G.G.; Beaudoin,R.R. (l978) A convenient method ofATPase assay. Anal. Biochem. 85:86-89.

11. Pena, S.D.J. (l980) A new technique for the visualization of thecytoskeleton in cultured fibroblasts with Coomassie blue R250. Cell Biol.Intern. Rep. 4: 149-151.

12. Gordon, D.J.; Einsenberg, E.; Korn, E.D. (l976) Characterization ofcytoplasmic actiin isolated from Acanthamoeba castellanii by a new method.J. Biol. Chem. 251:4778-4786.

13. Rosenberg, S.; Stracher, A.; Lucas, R.C. (l981) Isolation andcharacterization of actin and actin-binding protein form human platelets.J.Cell Biol. 91:201-211.

14. Murphy, J.A. (l971) Circular dichroism of the adenine and6-mercaptopurine nucleotede complexes of actin. Biochemistry 10:3723-3728.

15. Uyemura, D.G.; Brown, S.S.; Spudich, J.A. (l978) Biochemical andstructural characterization of actin from Dictyostelium discoileum. J.Biol.Chem. 253:9088-9096.

16. Higashi, S.; Oosawa, F. (l965) Conformational changes associatedwith polymerization and nucleotide binding in actin molecules. J. Mol.Biol. 12:843-865.

17. Depue, R.H.; Rice, R.V. (l965) F-actin is a right handed helix. J. Mol.Biol. 12:302-303.

18. Koffer, A.; Dickens, J.J. (l987) Isolation and characterization ofactin from cultured BHK cells. J. Mus. Res. Cell Motil. 8: 397-406.

Self-incompatibility

10
The Role of the Pollen Grain Coating in Pollination and Self-incompatibility in *Brassica oleracea*

James Doughty, Andrew McCubbin, Fiona Hedderson, Carole Elleman, and Hugh Dickinson

Introduction

The presence of viscous coatings on the exine surface of many entomophilous pollen grains is well established and aspects of their synthesis (Heslop-Harrison, 1968; Dickinson and Lewis, 1973b; Dickinson, 1973) and enzymic properties (Knox and Heslop-Harrison, 1970; Vithange and Knox, 1976) have been reported. In some species, however, there is strong circumstantial evidence that these coatings are involved in systems of self-incompatibility (SI) (for review, see Dickinson, 1990). The SI mechanisms of flowering plants fall naturally into two groups; those in which pollen compatibility is gametophytically controlled by the haploid grain nucleus (e.g., Nicotiana sp.) and the other where it is sporophytically determined by the genes of the mother plant (e.g., Brassica sp.). There are a number of features of the latter type of SI which point to the incompatibility (S) allele product of the pollen being carried in the exine coating. First, the coat is derived from the tapetal cells (Dickinson and Lewis, 1973b) which are part of the sporophyte; second, this type of SI acts at the stigma surface, rather than deep in the style, suggesting that some superficial component might be involved. Third, a number of recent molecular studies also indicate a role for this layer in SI; for example, Guilluy et al. (1991) have demonstrated that PCR generated probes derived from female S-allele sequences hybridise with anther messenger RNA, while Nasrallah and Nasrallah (1989) have reported the presence of a low abundance transcript in the maturing anther with some homology with SLG sequences. Further, in recent and highly significant work, it has been shown that promotor sequences directing the expression of the female S-alleles in Brassica oleracea will not only induce the expression of reporter and other genes in the anther tapetum when transformed into other members of the Cruciferae (Toriyama et al., 1991), but will also cause expression in the mature pollen of Nicotiana (Thorsness et al., 1991), a species with gametophytic SI, where S-locus specific sequences would be expected to be expressed by the haploid genome of the pollen. Although a number of theories (e.g., Lewis, 1954) predict that the S-allele products expressed in both pollen and stigma should be similar, the male determinant of SI has remained highly elusive. Apart from the work cited above, attempts to identify sequences encoding the male determinant have either proved fruitless (C. Scutt, pers. comm.) or produced equivocal results. Certainly probes based on female sequences, such as those used by Guilluy et al. (1991), do identify certain species of anther mRNA, but they also hybridise with sequences expressed elsewhere in the plant. These confusing data may result from the expression of a Brassica equivalent of the SmPK 1 gene reported by Walker and Zhang (1990) in Zea mays. This gene shares strong sequence homology with the sequences encoding the S-locus specific glycoproteins (SLGs) of Brassica stigmas, but its expression appears not to be restricted to cells involved in reproduction.

In an alternative approach, the formation of the complex pollen coatings found in the Cruciferae has been examined in detail (Dickinson and Lewis, 1973a,b) and their behaviour investigated during pollination (Stead et al., 1979, 1980; Elleman and Dickinson, 1986; Dickinson and Elleman, 1985; Elleman and Dickinson, 1990). Most recently, methods have been developed for isolating the exine coating of Brassica whilst leaving the pollen viable and gametophytic proteins intact (Doughty et al., 1991) and we describe here the response of the stigmatic papillae to challenge by extracts from this coating. A number of polypeptides have been isolated from this layer and their interactions with stigmatic polypeptides - some of which are known to be secreted from the papilla into the coating - have been investigated.

Formation and structure of the pollen coating in Brassica

The formation of complex exine coatings from elements of the tapetum has been well documented (Dickinson and Lewis, 1973a,b; Dickinson, 1973; Murgia et al., 1991). In Brassica, the organisation of this layer - once applied to the pollen surface - presents a confusing aspect, apparently being composed of membranous elements, vesicles and crystaline inclusions (Dickinson and Lewis, 1973a). However, the use of anhydrous fixation methods (Dickinson and Elleman, 1985) has revealed the coating of Brassica to be largely homogenous at the time of pollen release. The coating matrix, which appears slightly mottled under the electron microscope, is invested by the so-called 'exinic outer layer' (EOL) (Gaude and Dumas, 1984) at all interfaces with the pollen wall. A distinct, membrane-like structure, termed the coating superficial layer (CSL) (Elleman and Dickinson, 1986) is also present on the outer surface of the coating. The function of these layers has yet to be fully established, but the EOL is envisaged to carry determinants which serve to anchor the first components of the coating as they are released from the tapetum (Gaude and Dumas, 1984). The function of the CSL is also unclear but, since it is the first component of the pollen to contact the stigmatic pellicle, it may be involved in activating the coating (Elleman and Dickinson, 1990). Interestingly, preliminary studies (S. Hiscock, pers. comm.) suggest that - like the pellicle - this layer possesses strong esterase activity.

When pollen coating is isolated from the exine surface and reconstituted in vitro it retains its homogenous, mottled appearance following 'dry' fixation (see Figure. 1). It does not, of course, feature either of the boundary layers described above.

Structural changes in the coating during pollination

Hydration of Brassica pollen, whether on the stigma or in aqueous fixations, is accompanied by a striking change in appearance of the pollen coating. The homogenous matrix is converted into an electron-opaque layer composed of membrane-like profiles and vesicles (Elleman and Dickinson, 1986). This conversion also involves physical changes, for regions of pollen coating converted in this way differ in their solubility in organic solvents (Elleman and Dickinson, 1990). While molecular evidence is not available, it is assumed that this "conversion" of the coating results from increased levels of free and bound water in the layer which then permit the reconstitution of bimolecular leaflets, presumably converted to micelles during dehydration of the grain in the anther. Identical changes also take place on hydration in vitro of isolated pollen coating (see Figure 2).

On pollination, the coat flows out from the exine to form an appresoria-like foot (Ockendon, 1972) over the papilla surface. The pollen tube then grows down into this foot and, through it, into the papilla wall (Elleman and Dickinson, 1990). In both Brassica oleraceae and Arabidopsis thaliana the presence of pollen induces the expansion of a specialised outer layer of the stigmatic wall (Elleman et al., 1991) and, when reconstituted coating is applied to the papilla surface - whether extracted mechanically or chemically (Doughty et al., 1991) - a similar reaction occurs (see Figure 3), indicating the coating to carry determinants capable of stimulating this response. Using dry fixation methods, profiles suggesting vesicles (see Figure 4) are often seen in this expanded wall layer; the direction of travel of these vesicles remains, however, unclear. Interestingly, when mechanically isolated coating, which often contains gametophytic proteins, is applied to the papillae the female cytoplasm will often synthesise callose (see Figure 5), suggesting that the papilla is capable of identifying and reacting to protein of gametophytic origin. Any relationship between these stigmatic responses to pollination and SI is difficult to determine for, although the expansion of the papilla cell wall occurs principally following compatible pollinations, self-pollinations - particularly those involving 'strong' alleles - cease development before wall reaction would have been expected to take place. Indeed, there is accumulating circumstantial evidence that, following slight hydration, the coating of self-grains often converts back to the homogenous matrix characteristic of the dehydrated grain (Elleman and Dickinson, unpublished results).

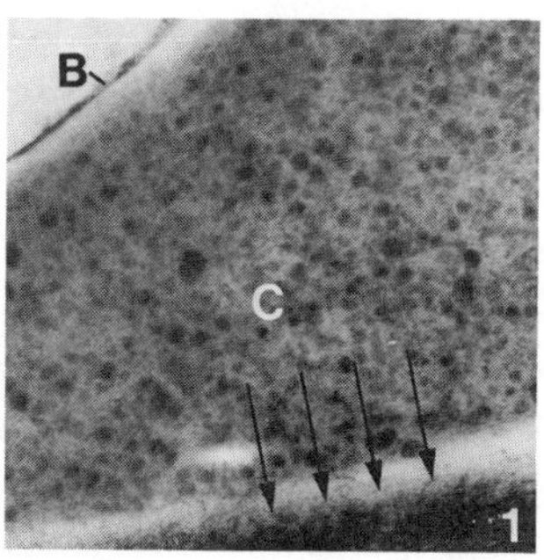

Figure 1. Isolated pollen coating (C), reconstituted *in vitro* and applied to the surface (arrows) of a stigmatic papilla. Hydration has not yet occurred, and the coat appears as a mottled matrix, devoid of any membranous structure. A boundary layer (B) is present which differs considerably from the CSL characteristic of the coating *in vivo*. Electron Micrograph (anhydrous fixation) x 65,000

Stigmatic reaction to pollination is common in plants with wet stigmas (Sedgley and Scolefield, 1980) and detailed studies of a number of species held to possess dry stigmas have also revealed such responses (Elleman and Dickinson, 1991). We remain ignorant both of the signals carried by the coating which stimulate this response, and of the nature of the molecules secreted by the stigmatic protoplast.

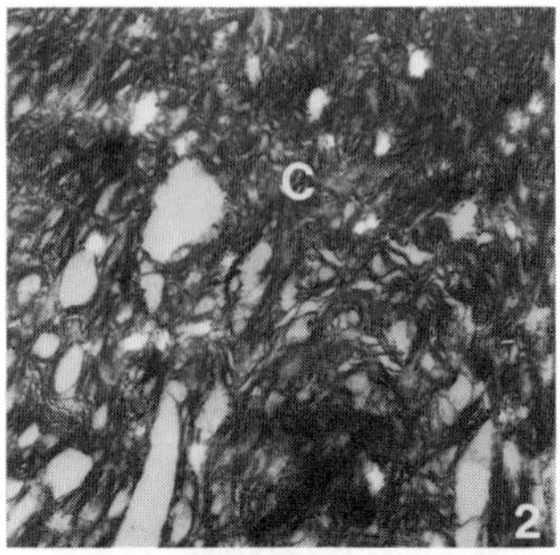

Figure 2. Material as shown in Figure 1, but following hydration. The reconstituted coating (C) has become reorganised into an electron-opaque layer containing numerous vesicles and membrane-like profiles. Electron Micrograph (anhydrous fixation) x 15,000

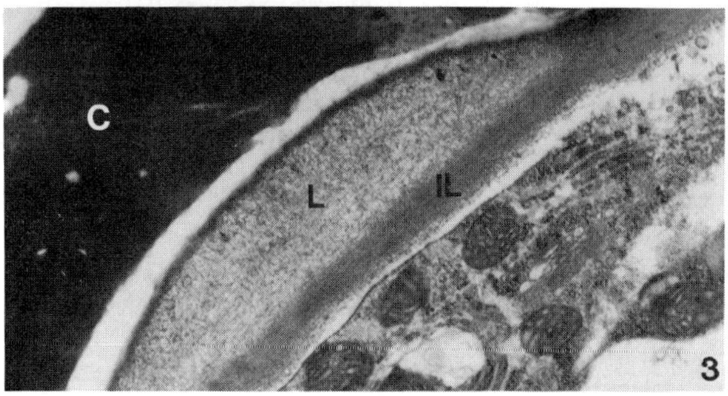

Figure 3. Expansion of the outer layer (L) of the stigmatic papilla wall following application of reconstituted pollen coating (C). The inner element of the wall (IL) does not respond to the presence of the coating. Electron Micrograph (modified Karnovsky fixation) x 20,000

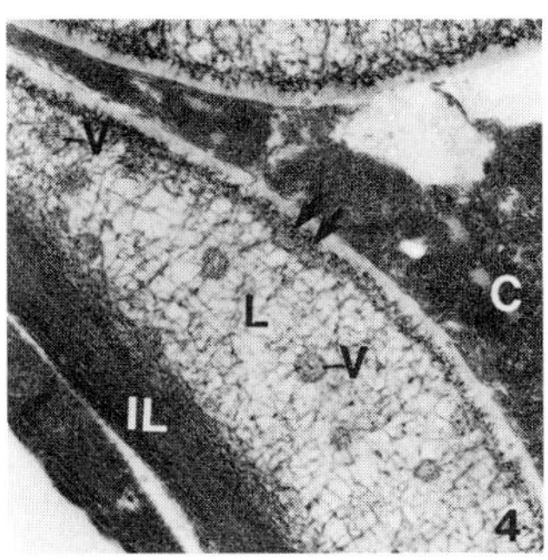

Figure 4. Vesicle-like profiles (V) in the expanded outer layer (L) of the papilla cell wall, following application of reconstituted pollen grain coating (C). The vesicles may be seen associated with the inner layer of the wall (IL) and beneath the cuticle (arrows). Electron Micrograph (anhydrous fixation) x 20,000

The nature and properties of the coating polypeptides

Pollen coatings in the Cruciferae have been known to contain lipid, carbohydrate and protein for many years (Dickinson and Lewis, 1973b) and since these layers are formed from the remains of the tapetum, it is not surprising that the activities of large numbers

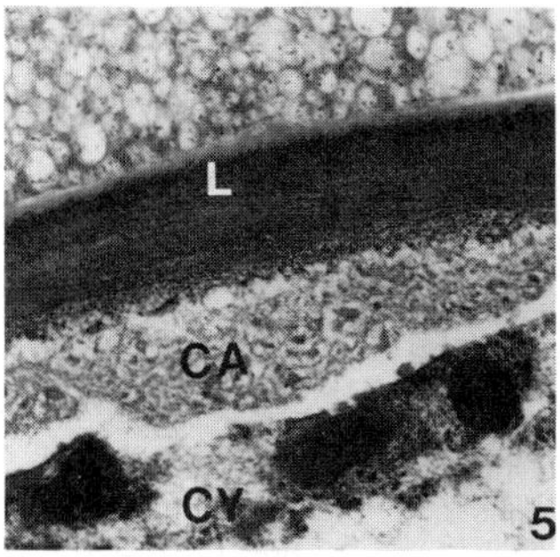

Figure 5. Synthesis of callose (CA) by the stigmatic papilla cytoplasm (CY) following the application of mechanically extracted pollen grain coating to the surface of the cell. This reaction occurs very rapidly, and precludes any expansion of the outer layer of the stigmatic cell wall (L). Electron Micrograph (anhydrous fixation) x 20,000

of enzymes have also been described (Knox and Heslop- Harrison, 1970; Vithange and Knox, 1976). More recently, the lipid composition of this layer in Brassica napus has been reported by Evans et al. (1990). However, to date all analyses of coating polypeptides have involved extraction of both sporophytic and gametophytic domains

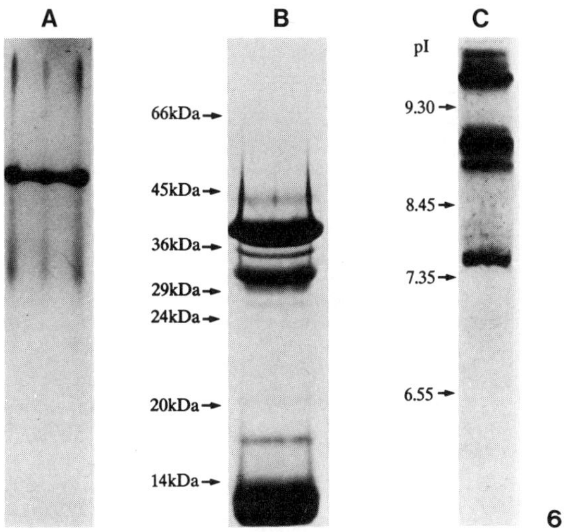

Figure 6. Analysis of pollen coat proteins (from plants homozygous for the S_{29} allele), extracted as described in Doughty *et al.* (1991), separated by acid native PAGE (lane A), SDS-PAGE (lane B) and isoelectric focusing between pH 3.5-10 (lane C). The positions of molecular weight markers (in kDa) are indicated for the SDS-PAGE separation, and isoelectric point markers (in pI units) for the isoelectric focusing.

of the pollen and it has thus proved impossible to identify positively which are associated with the pollen protoplast and which with the coat. New methods of coating isolation have been described by Doughty et al. (1991) and enable the study of the principal polypeptides of the coating. In Brassica these fall into two main families, one ranging between 10 and 15 kDa, and the other between 24 and 40 kDa (see Figure 6B). The pIs of these pollen coating polypeptides (PCPs) also fall into two distinct groups, one ranging between 8 and 9.5 and the other between 10 and 11 (see Figure 6C). Interestingly the major polypeptides migrate as a single band in acid native PAGE (see Figure 6A). More detailed studies, involving 2D electrophoresis suggests that each of the molecular weight families contains polypeptides from the two pI groupings. There is therefore the possibility that the two polypeptide families are related, with either the lower molecular weight family being degradation products of the larger proteins, or indeed the higher molecular weight group being composed of multimers of the 10-15 kDa polypeptides. In some extractions, polypeptides of higher molecular weights (c. 70 kDa) are also detectable.

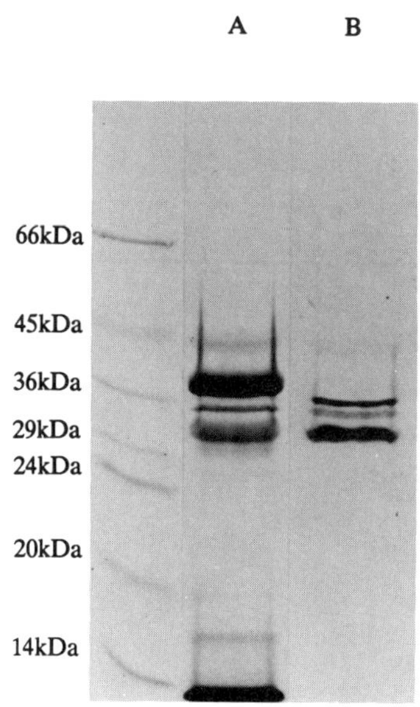

7

Figure 7. Coat proteins from pollen homozygous for the S_{29} allele following SDS-PAGE (lane A) and Concanavalin A affinoblotting (lane B). Molecular weight markers are shown in the left hand lane. Three proteins, ranging between 29 and 36 kDa are heavily glycosylated.

Many of the PCPs are heavily glycosylated and what appears to be a major band of 36 kDa in SDS-PAGE gels can be shown to consist of a glycosylated and non-glycosylated polypeptide (see Figure 7); whether they are related is unknown. The striking grouping of the PCPs and the fact that under acid gel conditions they run as a single band indicates a relationship between them. Preliminary peptide sequence analysis of the major 36 kDa PCP suggests that the N terminal region is composed of a number of proline-rich repeats which, while interesting, show little homology with sequences on current protein databases. All the lines of Brassica oleracea so far investigated, including several homozygous for different S-alleles contain very similar polypeptide profiles. Not surprisingly, a similar spectrum of proteins is also possessed by Brassica napus.

The method of extraction employed (Doughty et al., 1991) will tend to isolate PCPs with a high native affinity for lipids, and perhaps other classes of molecules. Some of the polypeptides show high levels of binding with other proteins; for example a glycoprotein of approximately 34 kDa binds strongly to gelatine. Since, on pollination, interactions are believed to occur between elements of the coating and molecules emanating from the stigmatic surface (Dickinson, 1990), the possibility that the PCPs could interact with stigmatic molecules was explored. Surprisingly, when mixtures are made of coating polypeptides and stigmatic extracts an interaction occurs which causes

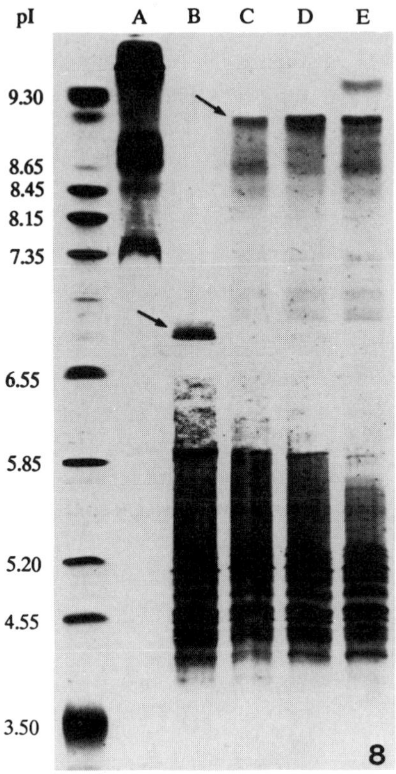

Figure 8. Interactions between extracts from stigmas homozygous for the S_{25} allele, analysed by isoelectric focusing from pH 3.5-10. The position of pI markers is shown in the left hand lane. Key to lanes: A) S_{25} pollen coat protein. B) S_{25} stigmatic extract, showing the position of the putative SLG (arrow). C) Interaction between S_{25} pollen coat and stigma extract, showing the disappearance of the putative SLG and the formation of a new band (arrow) not present in either pollen coat or stigma extract. D) Interaction between S_{29} pollen coating and S_{25} stigmatic extract. E) Interaction between S_{63} pollen coat protein and S_{25} stigmatic extract.

a reduction in level of one of the principal stigmatic glycoproteins - that normally identified from IEF gels as the SLG (see Figure 8). Close inspection of the IEF gel shows a new band, or range of bands, to be present in these gels but running at a higher pI (see Figure 8). When this band is cut from the gel and separated on SDS-PAGE its components can be identified as PCPs and the putative SLG (see Figure 9). The pollen polypeptides playing the major part in this interaction have yet to be fully characterised but experiments using a range of fractions from ammonium sulphate precipitation indicate only a small number to be involved.

The significance of these findings depends on whether the reactions between isolated PCPs and SLGs also occur in vivo. Most recently, methods have been devised by which pollen coat in situ can be challenged with stigmatic extracts. Surprisingly, the interaction between the putative SLG and the coating occurs even more strongly under these conditions, with the production of clearly defined interaction product (IP) bands identical with those seen using isolated PCPs (see Figure 10). Further, stigmatic extracts also seem to mobilise high levels of PCPs, many of which cannot be involved in the interaction with putative SLGs.

The fact that the only female protein involved in this interaction is the putative SLG is noteworthy, especially in view of the fact that Guilluy et al. (1990) have reported that monoclonal antibodies, raised to female SLGs, recognised polypeptides from whole pollen extracts of molecular weights equivalent to the two families of PCPs.

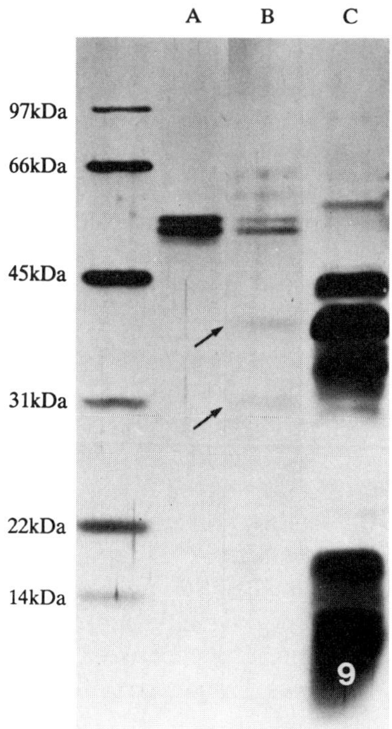

Figure 9. Analysis of the interaction product band formed following the mixture of S_{25} total stigmatic extract and pollen coat proteins from S_{63} (see Figure 8). The band was cut from the gel and then rerun on a gradient (7.5-15%) SDS-PAGE gel. Following electrophoresis, the band (lane B) separated to form a doublet of the same kDa as the putative SLG (as shown in lane A) and two smaller molecules (arrows) which were also represented in the pollen coating (lane C). Molecular markers were run in the left hand lane.

The nature of the interaction between the PCPs and female molecules is unknown. It is inhibited by high levels of the lectin concanavalin A and would appear to be sensitive to treatment with SDS and high temperatures. The interaction also occurs between PCPs and isolated putative SLG. To propose any linkage between these very general interactions and the highly specific events of SI it is clearly premature. However, interactions do differ between lines homozygous for different S alleles, and sometimes between 'self' and 'cross' interactions of the same line. Once the principal PCPs involved in this interaction have been identified, and the nature of the binding - if it is such - established, any linkage to SI may be explored more meaningfully. These events are not restricted to SI species, for as might be expected, the PCPs of Brassica napus also interact with the stigmatic glycoprotein to form a new 'band' at a higher pI. Since B. napus is an allotetrapoid formed from two SI species (B. campestris and B. oleracea), it remains possible that the female glycoprotein represents either 'SLGs' or S-locus related (SLR) glycoproteins (Lalonde et al., 1990) from one or both of the diploid progenitors.

The pollen coating of Brassica in pollenation and self-incompatibility

Following transfer of the pollen to the stigma surface by the insect vector, the grain coating participates in adhesion, conduction of water to the grain and the stimulation of a secretory response by the stigma. However, the molecular basis of these events, and indeed any role that the coating may play in SI, is only beginning to be understood.

Results presented here demonstrate the coating to contain only a small number of polypeptides, each present in relatively high abundance. Some of these PCPs are capable of binding to female molecules, principally that identified with the female

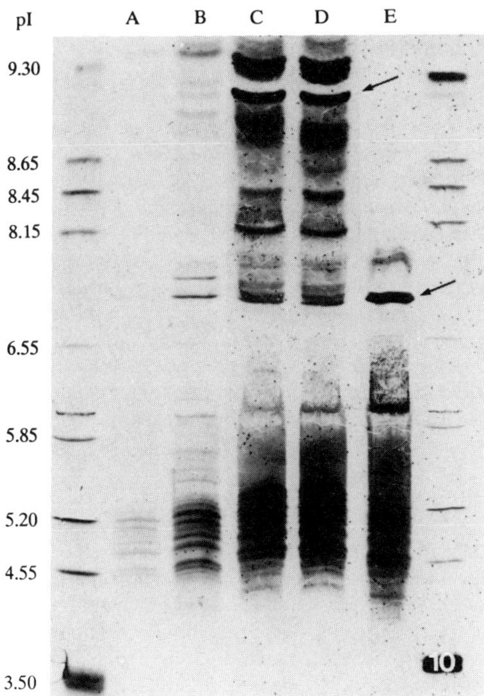

Figure 10. Interactions, run on an IEF gel, between extracts from stigmas homozygous for the S_{25} allele, and whole pollen grains carrying different S-alleles. Key to lanes: A) Proteins released from pollen in 0.2 M mannitol after 10 mins. B) Proteins released from pollen in 0.2 M mannitol after 30 mins. C) Interaction between S_{25} stigmatic extract and pollen grains carrying the same allele after 30 mins. D) Interaction between S_{25} stigmatic extract and whole pollen grains homozygous for S_{63} after 30 mins. E) Extract from stigma homozygous for S_{25} allele; the putative SLG (arrow) is clearly visible. The Interaction shown in lane C and D not only produces new families of molecules (i.e., that shown by arrow) but also seems to stimulate the release of pollen proteins. Isoelectric point markers are shown in the left and right hand lanes.

product of the S-locus. Little can be inferred from these observations save that this interaction - since it does occur in vivo - is likely to be related to pollination per se, rather than SI alone. The proposal that the determinants of SI have evolved from molecules with other functions has found wide acceptance, and is strengthened by more recent reports that SLRs, similar glycoproteins unlinked to the S-locus, are also expressed on the stigma (Trick and Flavell, 1989; Lalonde et al., 1990). A further layer of complexity has recently been overlaid on the SI system in Brassica by the discovery by Walker and Zhang (1990) that a domain of a putative serine/threonine kinase, encoded by the AmPK gene in Zea mays, has strong sequence homology with the SLGs. While similar kinases must presumably occur in Brassica it is far from clear what role, if any, they play in SI. However the fact remains that the majority of S-linked glycoprotein produced by the stigma is not associated with a kinase domain and is apparently transferred to the grain, suggesting that an interaction between this molecular and the male determinant is a primary step in the operation of SI. Simplistic models may therefore be constructed based on modification of the interaction by S-specific domains of the SGL and PCP(s). There may, of course, be parallel interactions between SLRs and their equivalent male targets. A situation could thus be envisaged whereby the interaction is inhibited if the SLG and a S-allele specific PCP are the products of the same S-locus, allowing the SLG to move, unaltered, into the grain and prevent further development.

How the SLG could inhibit pollen development remains unclear. Since the interaction with the coating-held male determinant would be S-allele specific, the receptor on the pollen protoplast could be gametophytically encoded. The homology between the SLG and kinases (Wallace and Zhang, 1990) is tantalising, and it is perhaps such a molecule in Brassica that detects the presence of SLG at the protoplast surface and transmits the information across the plasma membrane to the pollen cytoplasm. Protein synthesis inhibitors have been known to overcome SI for a number of years (Ferrari and Wallace, 1981; Sarker et al., 1988) but their site of action has not been established. It remains possible that the SLG transmits a signal, perhaps through a kinase, which results in an elevation of protein synthesis within the grain. Development of the pollen could be arrested either through the synthesis of a specific inhibitor, or by a more general disruption of cellular metabolism. Support for such a simplistic approach comes from a number of unexpected areas; for example interspecific pollinations within the Cruciferae display an interesting phenomena, in that crosses between self-compatible (SC) and SI species will often result in successful pollination when the female SC and the male SI, but not vice versa (S. Hiscock, unpublished results), indicating that the SI female secretes an 'inhibitor' which cannot be recognised and deactivated by the SC pollen. A mechanism of this type is, of course, not easily reconciled with currently accepted models the operation of SI - particularly the dimer hypothesis of Lewis (1954).

Acknowledgements

The authors thanks are due to the United Kingdom Agricultural and Food Research Council and to the European Commission for financial support. The assistance of Valerie Norris in preparing the manuscript is also gratefully acknowledged.

References

Dickinson, H.G. Cytobios 8: 25-40; 1973.
Dickinson, H.G.; Elleman, C.J. Micron and Microscopica Acta 16(4): 255-270; 1985.
Dickinson, H.G.; Lewis, D. Proc. R. Soc. Lond. (B), 183: 21-38; 1973a.
Dickinson, H.G.; Lewis, D. Proc. R. Soc. Lond. (B), 184: 149-165; 1973b.
Doughty, J.; Hedderson, F.; McCubbin, A.G.; Dickinson, H.G.
Elleman, C.J.; Dickinson, H.G. J. Cell Sci. 80: 141-157; 1986.
Elleman, C.J.; Dickinson, H.G. New Phytol. 114: 511-518; 1990.
Elleman, C.J.; Franklin-Tong, V.; Dickinson, H.G. (In the press); 1991.
Evans, D.E.; Sang, J.P.; Cominos, X.; Rothnie, N.E.; Knox, R.B. Plant Physiol. 93: 418-424; 1990.
Ferrari, T.E.; Wallace, D.A. Theor. Appl. Genet. 48: 243-249; 1976.
Guilluy, C.M.; Gaude, T.; Digonnet-Kerhoas, C.; Chaboud, A.; Heitzmann, P.; Dumas, C. In: Dale, B., ed. Mechanism of Fertilisation: Plants to Humans. NATO ASI Series, Series H, Vol. 45; 1990: p. 253-271.
Guilluy, C.M.; Trick, M.; Heitzmann, P.; Dumas, C. Theor. Appl. Genet. (In the press); 1991.
Heslop-Harrison, J. New Phytologist 67: 779-786; 1968.
Knox, R.B.; Heslop-Harrison, J. J. Cell Sci. 6: 1-27; 1970.
Lalonde, B.A.; Nasrallah, M.E., Dwyer, K.G.; Chen, C.H.; Barlow, B.; Nasrallah, J.B. Plant Cell 1: 249-258; 1989.
Lewis, D. Adv. Genet. 6: 235-285; 1954.
Murgia, M.; Charzynska, M.; Rougier, M.; Cresti, M. Sexual Plant Reprod. 4: 28-36; 1991.
Nasrallah, J.B.; Nasrallah, M.E. Annu. Rev. Genet. 23: 121-139; 1989.
Sarker, R.H.; Elleman, C.J.; Dickinson, H.G. Proc. Natl. Acad. Sci. USA 85: 4340-4344; 1988.
Sedgley, M.; Scholefield, P.B. Bot. Gaz. 141: 428-434; 1980.
Stead, A.D.; Roberts, I.N.; Dickinson, H.G. Planta 146: 211-216; 1979.
Stead, A.D.; Roberts, I.N.; Dickinson, H.G. J. Cell Sci. 42: 417-423; 1980.
Thorsness, M.K.; Kandasamy, M.K.; Nasrallah, M.E.; Nasrallah, J.B. Dev. Biol. 143: 173-184; 1991.
Toriyama, K.; Thorsness, M.K.; Nasrallah, J.B.; Nasrallah, M.E. Dev. Biol. 143: 427-431; 1991.
Trick, M.; Flavell, R.B. Mol. Gen. Genet. 218: 112-117; 1989.
Vithange, H.I.M.V.; Knox, R.B. J. Cell Sci. 21: 423-; 1976.
Walker, J.C.; Zhang, R. Nature (Lond.) 345: 743-746; 1990.

11

Molecular Genetics of Gametophytic Self-incompatibility in Solanaceae

Teh-hui Kao

Introduction

Self-incompatibility exhibited by members of the Solanaceae family is of gametophytic type and, in the simplest case, is controlled by a single multiallelic locus called the S-locus. (For a comprehensive treatise on the self-incompatibility system refer to the monograph by de Nettancourt, 1977.) Fertilization is blocked when the S-allele carried by the pollen matches one of the two alleles carried by the pistil. Incompatible pollen initially germinates and the tube penetrates the stigma, but its growth is subsequently inhibited before reaching half the length of the style. The tips of incompatible pollen tubes usually swell and burst open.

We are using solanaceous species as a model system to address the following questions about the gametophytic self-incompatibility system. 1. What is the molecular basis of S-allele mediated self-recognition? 2. What is the biochemical mechanism which inhibits the growth of self pollen tubes? 3. What is the molecular mechanism that generates allelic diversity at the S-locus? 4. What is the mechanism which has allowed the S-locus to maintain its extreme allelic polymorphism over the course of evolution? 5. What causes the breakdown of self-incompatibility in commercial cultivars? In this article, I will describe our efforts to address these questions, and summarize the major findings.

Primary Structural Features of S-Allele-Associated Proteins

Identification and characterization of S-allele products in the pistil and pollen are the first steps toward understanding the molecular basis of self-incompatibility. The strategy generally used to identify the S-allele products is to search for pistil- or pollen-specific proteins which co-segregate with their respective S-alleles in genetic crosses. S-allele-associated proteins (S-proteins) have been identified in the pistils of a number of solanaceous species: *Nicotiana alata* (Bredemeijer and Blass, 1981; Jahnen et al., 1989), *Lycopersicon peruvianum* (Mau et al., 1986), *Petunia hybrida* (Kamboj and Jackson, 1986; Broothaerts et al., 1989), and *Solanum tuberosum* (Kirch et al., 1989). We have identified 3 S-alleles each and their associated proteins from *Petunia inflata* (Ai et al., 1990) and *Solanum chacoense* (Xu et al., 1990a). The amino-terminal sequences of all 6 S-proteins have been determined and found to share sequence similarity with all the solanaceous S-proteins characterized by other groups.

To isolate cDNAs for various S-proteins, we first used differential hybridization to isolate cDNA clones for Sz-protein of N. alata. Sz-protein was previously identified by A. Kheyr-Pour (CNRS, Gif-sur-Yvette, France) and found to be the most abundant pistil-specific protein in flower buds one day before anthesis. We constructed a cDNA library using mRNA isolated from buds collected at that stage, and identified and sequenced the most abundant class of pistil-specific cDNA clones (Kheyr-Pour et al., 1990). The amino acid sequence deduced from this class of clones matches perfectly with the amino-terminal sequence of Sz-protein determined by Mau et al. (Mau et al., 1986), thus confirming its identity as Sz-cDNA clones. We have subsequently used Sz-cDNA as a hybridization probe to isolate cDNAs for 3 additional S-proteins of N. alata (Kheyr-Pour et al., 1990), 3 S-proteins of P. inflata (Ai et al., 1990), 2 S-proteins of Solanum chacoense (Xu et al., 1990), and more recently 1 S-protein of Lycopersicon peruvianum (Tsai et al., manuscript in preparation).

Table 1. Characteristics of Ribonuclease Activity of Three *P. inflata* S-Proteins

	S1	S2	S3
Specific Activity (A260 units/mg)	106	164	130
pH optimum	7.0	7.0	7.0
Mean temperature optimum	50^oC	50^oC	50^oC
Cleavage of poly (C)	+ + +	+ + +	+ + +
Cleavage of poly (U)	+ +	+	+
Cleavage of poly (A)	+	+	+ +
Inhibition by DEPC	Yes	Yes	Yes
Inhibition by EDTA or RNasin	No	No	No

+ + + indicates high activity; + + indicates moderate activity; + indicates low activity.

cysteines, all of which appear to be involved in intra-molecular disulfide bonds (Ai and Kao, unpublished results).

Amino acid sequences of 16 S-proteins from 5 solanaceous species have now been reported, including 9 by our lab, and 7 by other groups (Anderson et al., 1986, 1989; Clark et al., 1990; Haring et al., 1990). Various subsets of these 16 S-proteins have been compared to determine the primary structural features of the S-protein (Anderson et al., 1989; Ai et al., 1990; Clark et al., 1990; Haring et al., 1990; Kheyr-Pour et al., 1990; Xu et al., 1990b; Ioerger et al., 1991). The notable features are summarized below. 1. The sequences of S-proteins vary among themselves over a tremendously wide range, with sequence identity ranging from as low as 38% to as high as 93% (Ioerger et al., 1991; Tsai et al., manuscript in preparation). This high degree of sequence diversity is in sharp contrast to other multiallelic gene products where interallelic differences usually lie in a few amino acids. 2. Conserved, variable, and hypervariable residues are dispersed throughout the S-protein forming five conserved regions, designated C1 to C5, and two hypervariable regions, designated HVa and HVb (Ioerger et al., 1991). 3. The two hypervariable regions contain 11 of the 21 sites with the highest normed variability index (refer to Kheyr-Pour et al., 1990 for the definition of this index), and are also the most hydrophilic regions of the S-protein (Ioerger et al., 1991; Tsai et al., manuscript in preparation). These two characteristics suggest that they may be the primary determinant for S-allele specificity and involved in interaction with pollen. 4. Three of the five conserved regions, C1, C4, and C5, are hydrophobic, and may be involved in the formation of the core structure of the S-protein. 5. The two conserved regions C2 and C3 share extensive sequence similarity with the amino acids in the active site of two fungal ribonucleases, RNAse T2 and RNAse Rh (McClure et al., 1989; Haring et al., 1990; Ioerger et al., 1991). 6. There are eight conserved

Relevance of Ribonuclease Activity of the S-Protein to Its Biological Function

The finding of sequence similarity between solanaceous S-proteins and two fungal ribonucleases immediately suggested that S-proteins might be ribonucleases. Indeed, a number of N. alata S-proteins were shown by McClure et al. to have ribonuclease activity (McClure et al., 1989). We have subsequently shown that three P. inflata S-proteins, S1, S2, and S3, also have ribonuclease activity (Singh et al., 1991). Their enzymatic properties have been characterized and are summarized in Table 1. They have identical pH and temperature optima, comparable specific activity, and similar response to various inhibitors. They all show marked cleavage preference toward poly (C) over poly (A) or poly (U). However, they differ in their ability to cleave poly (U) and poly (A), perhaps reflecting the sequence diversity among them (20 to 25% diversity). The cleavage preference observed using artificial substrates most likely does not have any bearing on the biological specificity of the S-protein, because the three P. inflata S-proteins cleave all RNA substrates tested.

It is tempting to speculate that the ribonuclease activity of the S-protein is an integral part of the self-incompatibility reaction, that is, the allele-specific inhibition of pollen tubes is caused by specific degradation of RNA in self pollen tubes by the S-protein. Using an in vitro pollen germination system, Gray et al. have shown that purified S-proteins indeed enter pollen tubes. However, the uptake is not allele-specific because the S-proteins were found in the cytoplasm of both self and non-self pollen tubes (Gray et al., 1991). It is not clear then how the specific degradation of RNA in self pollen tubes, implicated in McClure et al.'s results (1990), is achieved.

One approach to addressing the question of whether the ribonuclease activity of the S-protein is essential to the self-incompatibility reaction is to change the two histidine residues in the active site of the S-protein to abolish its ribonuclease activity, and then to examine the effect on the biological function of the S-protein. Site-directed mutagenesis can be used to alter the two histidine codons, and the mutagenized S-gene can be introduced into plants via Agrobacterium-mediated transformation. The transgenic plants expressing the mutant S-protein can be tested for their ability to reject pollen bearing the same S-allele. Alternatively, the mutant S-protein can be produced in a heterologous system, and its ability to inhibit the growth of pollen carrying the same allele can be tested in an in vitro pollen germination system. We have undertaken the second approach, and have successfully expressed two alleles of P. inflata S-gene in a baculovirus expression system (Mu and Kao, unpublished results). Both glycosylated and unglycosylated forms of the S2- and S3-proteins have been produced, with the former exhibiting greater ribonuclease activity. Site-directed mutagenesis is now underway to produce mutant S-proteins for functional analysis.

The pistil of P. inflata appears to contain a large number of ribonucleases in addition to S-proteins. This is in contrast to leaf and petal tissues which contain only a few ribonucleases, as revealed by activity staining gels (Singh et al., 1991). We have purified a pistil ribonuclease, designated X2, and shown that, unlike the S-proteins, it has a cleavage preference toward poly (U), and also cleaves double-stranded RNA (Singh et al., 1991). Its amino-terminal sequence has been determined and found to share significant similarity with those of all the solanaceous S-proteins characterized so far (Meager and Kao, unpublished results). It thus appears that S-proteins might belong to a family of ribonucleases with common evolutionary origin.

Why the pistil is singularly rich in ribonucleases is a subject for speculation. It is possible that these ribonucleases play a role in the defense mechanism against infection by fungi or other pathogens. It would certainly be intriguing if it turns out that the self-incompatibility mechanism actually evolved by recruiting one of the "ancient" ribonucleases to use in defense against "invasion" by self pollen tubes.

Structure of the S-Gene and Molecular Organization of the S-Locus

Lewis' tripartite structure hypothesis predicts close linkage between the S-allele-specificity part, pollen-activity part, and style-activity part (Lewis, 1949, 1960). This hypothesis implies that the same S-allele-specificity part is expressed in pollen and pistil. So far, only the protein product of the pistil S-allele-specificity part (S-protein) has been identified. No pollen or pollen tube proteins have been found to be correlated with S-alleles, and attempts to find pollen or pollen tube mRNA species with sequence homology to the pistil S-gene have failed. It might appear that the pollen S-gene is separate from the pistil S-gene. However, it is difficult conceptually to envision how mutations generating a new allele specificity in both pollen and pistil could occur simultaneously and coordinately in two separate genes.

Now that cDNAs for several S-proteins have been cloned, we have set out to characterize the S-locus at the molecular level in order to identify the pollen S-gene, if it is different from the pistil S-gene, and to identify the putative genes for pollen- and style-activity parts. We first isolated genomic clones containing S1- and S3-alleles of the P. inflata S-gene by screening a genomic library using cDNAs for the three P. inflata S-proteins as probes. We have sequenced both clones and characterized the flanking regions of both alleles extensively (Coleman and Kao, manuscript submitted). Both genes contain a short intron which is located in the region encoding one of the two hypervariable regions of the S-protein. There are two surprising findings. First, there is no significant sequence homology in the immediate flanking regions of the two genes. Second, the flanking regions of the two genes are rich in repetitive sequences. So far, no coding information has been found in the immediate flanking regions of the pistil S-gene. Using chromosome walking as an approach to identify any linked genes might be made difficult by the presence of these repetitive sequences in the flanking regions.

Evolution of S-Allele Polymorphism

Comparison of amino acid sequences of 11 S-proteins from three genera of Solanaceae reveals that some alleles from the same species are less similar to each other than to alleles from other species (Ioerger et al., 1990). For example, the amino acid identity between Sz-allele of N. alata and S1- S2-, S3-, and S6- alleles of N. alata ranges from 44.5% to 48.7%; while the identity between Sz and three P. inflata S-alleles, S1, S2, and S3, ranges from 63.32% to 67.34% (Tsai et al., manuscript in preparation). An S-locus gene genealogy obtained from phylogenetic studies shows that S-allele polymorphism predates the divergence of the genera Nicotiana, Petunia, and Solanum (Ioerger et al., 1990). This analysis has recently been extended to include 17 S-protein sequences from 4 genera --- Lycopersicon plus the three genera mentioned above. These new results strongly reinforce the previous finding (Tsai et al., manuscript in preparation).

Our finding of the long existence of S-allele polymorphism (for example, Petunia and Solanum diverged 36 million years ago) is consistent with the general belief that self-incompatibility is an ancient trait responsible for the explosive success of angiosperms (Whitehouse, 1950). Since gametophytic self-incompatibility has been found in more than 60 families, we can trace its origin by expanding these molecular genetic studies to other families. It is highly likely that, if the evolution of gametophytic self-incompatibility is monophyletic, S-genes in other families can be identified on the basis of their sequence homology to the solanaceous S-genes.

Mechanisms for the Generation and Maintenance of S-Allele Polymorphism

The early appearance of the S-locus has given its alleles a long evolutionary time in which to diverge; this partially explains the extreme sequence diversity found in extant S-alleles. In addition, S-alleles are under selection pressure to diverge, because any new allele appearing in a population has reproductive advantage over extant S-alleles. This is because pollen bearing the new allele is more likely to land on stigmas carrying different alleles, and thus more likely to lead to a successful fertilization. The question, then, is how a new S-allele is generated from existing alleles.

To investigate whether recombination might be responsible for the generation of new alleles, we used a sensitive statistical test to analyze the nucleotide sequences of 12 S-alleles from 4 solanaceous species. We found an unusually low rate of intragenic recombination occurring at the S-locus (Clark and Kao, manuscript submitted). This finding, coupled with the unusual features of the flanking regions of the two P. inflata S-alleles, which have been described above, strongly suggests a recombination suppression mechanism operating at the S-locus to prevent homogenization of the S-alleles. The lack of recombination at the S-locus may also explain why a large number of S-alleles can be maintained in the population.

In view of this finding, that recombination is unlikely to be the mechanism for generating new alleles, we have proposed that accumulation of point mutations at sites which determine S-allele specificity of S-proteins results in the generation of new S-alleles. Since standard mutagenesis failed to generate new S-allele specificity, it seems that more than one amino acid substitution is required to alter S-allele specificity. It is interesting to note that the most homologous pair of S-proteins, S2-allele of P. inflata and S2-allele of P. hybrida, differ in only 13 amino acids. Whether all these 13 amino acids, or a subset of them, confer allelic identity remains to be investigated.

Breakdown of Self-Incompatibility in a Cultivar of P. hybrida

Self-incompatibility is widespread in wild solanaceous species, but most cultivated species are self-compatible, because self-incompatibility was purposely selected out in the early stages of their cultivation for the immediate benefit of producing inbred lines homozygous for desirable traits. The causes for the breakdown of self-incompatibility in cultivated species have not been determined at the molecular level. From his analysis of the self-incompatibility behavior of interspecific hybrids between self-incompatible L. peruvianum and self-compatible L. esculentum, Martin observed that introgression of S-alleles into self-compatible species is not sufficient to render the latter self-incompatible. He thus concluded that full manifestation of self-incompatibility requires the involvement of the S-locus as well as other unlinked modifier loci (Martin, 1961, 1968). The presence of modifier genes has also been implicated in other studies (Mather, 1943; Takahashi, 1973). It is generally believed that modifier genes interact epistatically with the S-locus to modulate the activity of the S-alleles in the self-incompatibility interaction. However, their identity has not been determined. It is apparent that, if we hope to be able to use gene transfer techniques to introduce the self-incompatibility trait to cultivated species to facilitate hybrid seed production, we must identify all modifier genes which are necessary for the full manifestation of self-incompatibility.

In order to study the nature of these modifier genes, and to eventually identify them, we have begun to examine the molecular basis of the breakdown of self-incompatibility in a cultivar of P. hybrida. We have first set out to determine whether it was mutation at the S-locus or another modifier locus (loci) which led to its self-compatibility phenotype. P. hybrida is very closely related to P. inflata, a wild self-incompatible species we have already studied extensively, and crosses between them are reciprocally compatible. We have identified two P. hybrida pistil proteins which segregated in F1 hybrids between P. inflata and P. hybrida. The amino-terminal sequences of these two proteins were determined and found to share significant similarity with those of S-proteins from self-incompatible solanaceous species (Ai et al., manuscript submitted).

With the identification of the two S-proteins in self-compatible P. hybrida, we can now determine the S-genotypes of all plants obtained in crosses involving P. hybrida, and correlate them with their self-incompatibility behavior. So far, our results suggest that there are at least two independent mutations in P. hybrida which caused the breakdown of self-incompatibility; one at the S-locus, the other at a modifier locus unlinked to the S-locus. These results are consistent with findings from earlier classical genetic studies that self-compatible mutations are mapped at the S-locus or other unlinked loci. The S-alleles carried by self-compatible species are usually referred to as Sf-alleles, or self-fertility alleles, because they fail to elicit a self-incompatibility response. Genetically, these alleles behave as null-specificity alleles, but it is not known whether they are null alleles or defective alleles. Our discovery of the expression of such a defective Sf-allele in the pistil of self-compatible P. hybrida shows that a null-specificity allele can be expressed. It will be of particular interest to compare the nucleotide sequence of the defective S-allele with those of functional S-alleles to determine the molecular basis of the mutation. It will also be interesting to find out whether the S-protein encoded by the defective S-allele has ribonuclease activity and can be taken up by pollen tubes.

Conclusions

Significant progress in the study of gametophytic self-incompatibility has been made since the mid-1980s when the first cDNA clone for an S-protein was identified in N. alata (Anderson et al., 1985). Through the efforts of several research groups, including ours, more than a dozen S-protein sequences from 6 solanaceous species have now been obtained. Comparison of the nucleotide and amino acid sequences of these S-alleles has already provided us important information on the structural features of the S-protein, the possible mechanism for generating new allelic specificity, and the evolutionary history of the gametophytic self-incompatibility system. Analysis of genomic sequences of two P. inflata alleles have revealed several structural features of the S-locus which suggest a recombination suppression mechanism operating at the S-locus. This mechanism may be primarily responsible for the maintenance of extreme S-allele polymorphism in a population.

Despite the impressive progress made so far in our understanding of several aspects of gametophytic self-incompatibility, there still remain many challenging questions. Foremost among them is the identity of the S-allele-specificity part expressed in the pollen and of the two activity parts implicated in the tripartite structure hypothesis. The molecular basis of the self-incompatibility interaction will not be fully understood until these components of the S-locus have been identified. Furthermore, it should also be emphasized that gametophytic self-incompatibility is a complex trait, as has already been convincingly demonstrated by earlier genetic studies, and reaffirmed by our recent study of hybrids between self-incompatible and self-compatible Petunia species. Full manifestation of self-incompatibility requires not only the presence of the functional S-locus but also modifier loci. Therefore, it is imperative that molecular genetic study be extended to the identification and characterization of all modifier genes involved in epistatic interactions with the S-locus.

Acknowledgement

I would like to thank members of my lab and Dr. Andrew G. Clark of the Biology Department for contributing to the results described here, and Craig Coleman and Dr. Anuradha Singh for stimulating discussion and comments on the manuscript. This work was supported by grants from the National Science Foundation (Grant Nos. DCB 86-16087 and DCB 89-04008), U.S. Department of Agriculture (Competitive Grant No. 90-37261-5560), and Pennsylvania Agricultural Experiment Station (Project No. 2997).

References

Ai, Y.; Singh, A.; Coleman, C. E.; . loerger, T. R.; Kheyr-Pour, A.; Kao, T.-h. Self-incompatibility in Petunia inflata: isolation and characterization of cDNAs encoding three S-allele-associated proteins. Sex. Plant Reprod. 3:130-138; 1990.

Anderson, M. A.; Cornish, E. C.; Mau, S.-L.; Williams, E. G.; Hoggart, R.; Atkinson, A.; Bonig, I.; Grego, B.; Simpson, R.; Roche, P. J.; Haley, J. D.; Penchow, J. D.; Niall, H. D.; Tregear, G. W.; Coghlan, J. P.; Crawford, R. J.; Clarke, A. E. Cloning of cDNA for a stylar glycoprotein associated with expression of self-incompatibility in Nicotiana alata. Nature 321: 38-44; 1986.

Anderson, M. A.; McFadden, G. I.; Bernatzky, R.; Atkinson, A.; Orpin, T.; Dedman, H.; Tregear, G.; Fernley, R.; Clarke, A. E. Sequence variability of three alleles of the self-incompatibility gene of Nicotiana alata. Plant Cell 1:483-491; 1989.Bredemeijer, G. M. M.; Blaas, J. S-specific proteins in styles of self-incompatible Nicotiana alata. Theor. Appl. Genet. 59:185-190; 1981.

Broothaerts, W. J.; Laere, A. v.; Witters, R.; Gisele, P.; Decock, B.; Damme, J. v.; Vendrig, J. C. Purification and N-terminal sequencing of style glycoproteins associated with self-incompatibility in Petunia hybrida. Plant Mol. Biol. 14:93-102; 1989.Clark, R. E.; Okuley, J. J.; Collins, P. D.; Sims, T. L. Sequence variability and developmental expression of S-alleles in self-incompatible and pseudo-self-compatible Petunia. Plant Cell 2:815-826; 1990.

Haring, V.; Gray, J. E.; McClure, B. A.; Anderson, M. A.; Clarke, A. E. Self-incompatibility as a model of cell-cell recognition in plants. Science 250:937-941; 1990.

1oerger, T. R.; Clark, A. G.; Kao, T.-h. Polymorphism at the self-incompatibility locus in Solanaceae predates speciation. Proc. Natl. Acad. Sci. USA 87:9732-9735; 1990.

1oerger, T. R.; Gohlke, J. R.; Xu, B.; Kao, T.-h. Primary structural features of the self-incompatibility protein in Solanaceae. Sex. Plant Reprod. 4: 81-87; 1991.

Kamboj, R. K.; Jackson, J. K. Self-incompatibility alleles control a low molecular weight, basic protein in pistils of Petunia hybrida. Theor. Appl. Genet. 71:815-819; 1986.

Kheyr-Pour, A.; Bintrim, S. B.; loerger, T.; Remy, R.; Hammond, S. A.; Kao, T.-h. Sequence diversity of pistil S-proteins associated with gametophytic self-incompatibility in Nicotiana alata. Sex. Plant Reprod. 3:88-97; 1990.

Kirch, H. H.; Uhrig, H.; Lottspeich, F.; Salamini, F.; Thompson, R. D. Characterization of proteins associated with self-incompatibility in Solanum tuberosum. Theor. Appl. Genet. 78:581-588; 1989.

Lewis, D. Structure of the incompatibility gene II. Induced mutation rate. Heredity 3:339-355; 1949.Lewis, D. Genetic control of specificity and activity of the S antigen in plants. Proc. Roy. Soc. London, Series B 151:468-477; 1960.

Martin, F. W. The inheritance of self-incompatibility in hybrids of Lycopersicon esculentum Mill. x L. chilense Dun. Genetics 46:1443-1454; 1961.

Martin, F. W. The behavior of Lycopersicon incompatibility alleles in an alien genetic milieu. Genetics 60:101-109; 1968.

Mather, K. Specific differences in Petunia. J. Genetics 45:215-235; 1943.

Mau, S.-L; Williams, E. G.; Atkinson, A.; Anderson, M. A.; Cornish, E. C.; Grego, B.; Simpson, R. J.; Kheyr-Pour, A.; Clarke, A. E. Style proteins of a wild tomato (Lycopersicon peruvianum) associated with expression of self-incompatibility. Planta 169:184-191; 1986.

McClure, B. A.; Gray, J. E.; Anderson, M. A.; Clarke, A. E. Self-incompatibility in Nicotiana alata involves degradation of pollen rRNA. Nature 347:757-760; 1990.

McClure, B. A.; Haring, V.; Ebert, P. R.; Anderson, M. A.; Simpson, R. J.; Sakiyama, F.; Clarke, A. E. Style self-incompatibility gene products of Nicotiana alata are ribonucleases. Nature 342:955-957; 1989.

Nettancourt, D. de. Incompatibility in angiosperms. Berlin, Germany: Springer-Verlag; 1977. Singh, A.; Ai, Y.; Kao, T.-h. Characterization of ribonuclease activity of three S-allele-associated proteins of Petunia inflata. Plant Physiol. 96:61-68; 1991.

Takahashi, H. Genetical and physiological analysis of pseudo-self-compatibility in Petunia hybrida. Japan. J. Genetics 48:27-33; 1973.

Whitehouse, H. L. K. Multiple-allelomorph incompatibility of pollen and style in the evolution of the angiosperms. Ann. Botan. New Series 14:198-216; 1950.

Xu B.; Grun P.; Kheyr-Pour A.; Kao, T.-h. Identification of pistil-specific proteins associated with three self-incompatibility alleles in Solanum chacoense. Sex. Plant Reprod. 3: 54-60; 1990a.

Xu, B.; Mu, J.; Nevins, D. L.; Grun, P.; Kao, T.-h. Cloning and sequencing of cDNAs encoding two self-incompatibility associated proteins in Solanum chacoense. Mol. Gen. Genet. 224:341-346; 1990b.

12
Molecular Genetics of Self-incompatibility in *Nicotiana alata*

Shaio-Lim Mau, Antony Bacic, Jane Murfett, Bruce McClure, Marilyn Anderson, and Adrienne Clarke

Introduction

Self-incompatibility is a genetically controlled mechanism which prevents inbreeding in plants (de Nettancourt, 1977). In many, but not all cases, it is controlled by a multi-allelic, single gene, the S-gene. The system operates to enhance outcrossing and to ensure that a plant is fertilized by a genetically distinct individual of the same species. There are two major types of self-incompatibility. The most widespread is gametophytic self-incompatibility which involves interaction of a product of the haploid genome of the male gametophyte (carried within the pollen grain) and a product of the diploid genome of the female tissue of the sporophyte, the pistil. In incompatible matings, as is the case when the S-allele carried by the haploid pollen matches either of the S-alleles present in the diploid style, pollen tube growth is arrested within the transmitting tract (Figure 1).

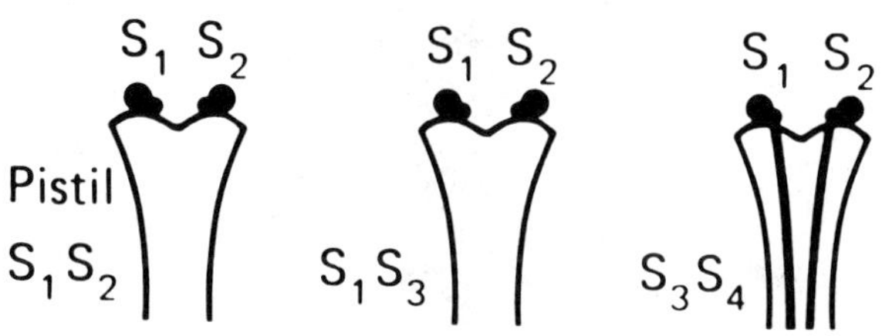

Figure 1. Behaviour of pollen in a single gene gametophytic system. The pollen parent genotype is S1S2. When an allele in the individual haploid pollen grain is matched with either allele in the diploid style tissues, growth of the pollen tube is arrested, usually in the style. For example, both S_1 and S_2 pollen are inhibited in a S_1S_2 style whereas the S_2-pollen will grow successfully through the S_1S_3 style. Where there is no match of alleles (eg. S_1S_2 pollen grains on a S_3S_4 pistil), the pollen tubes of both genotypes grow through the style to the embryo sac. [From Anderson et al., 1983, reproduced with permission from Alan R. Liss Inc., N. Y.].

The other system is sporophytic self-incompatibility in which the incompatibility reaction is mediated by factors carried by the pollen, but specified by the diploid tissues of the pollen parent. These pollen factors interact with a product of the female pistil, in some way which leads to the arrest of growth of the pollen tube. In the sporophytic systems the arrest occurs on the surface of the stigmatic papillae and is accompanied by deposits of callose, both in the tip of the emerging pollen tube and on the papilla surface at the point of contact with the pollen. In gametophytic systems, the arrest of tube growth occurs within the style. Styles of many plant species, for example plants of the family Solanaceae, have a central transmitting tissue consisting of longitudinal files of cells which are separated by mucilage at maturity. The pollen tubes grow extracellularly through the transmitting tissue and hence are in contact with the mucilage (Figure 2).

The molecular genetics of both the sporophytic and gametophytic systems have been studied. The approach in both cases was to clone cDNAs encoding the protein component of pistil glycoproteins which segregated with S-phenotype (S-glycoproteins).

Examples of style glycoproteins segregating with S-genotype are known for three families, the Solanaceae, Rosaceae and Paperavaceae (Cornish et al. 1988; Franklin-Tong

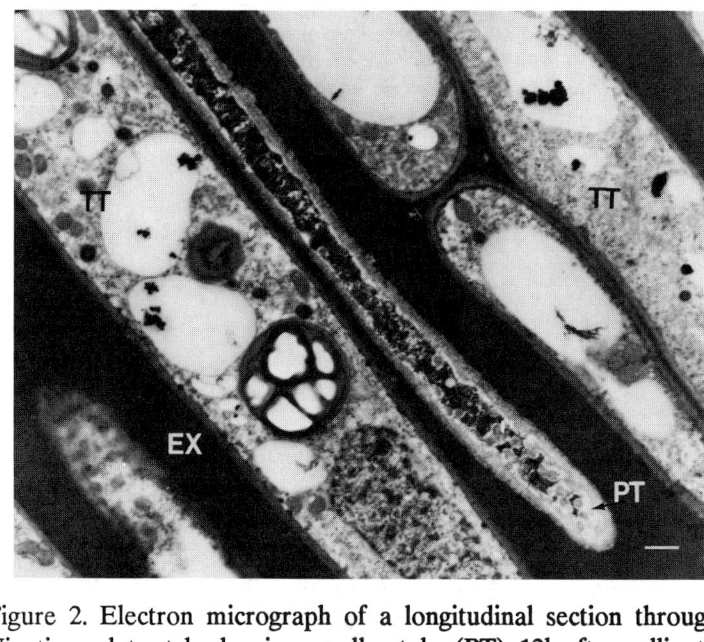

Figure 2. Electron micrograph of a longitudinal section through a Nicotiana alata style showing a pollen tube (PT), 12h after pollination, growing through the external matrix (EX) between transmitting tract cells (TT). Bar = 1μm. [From Gray et al., 1990, reproduced with permission from Plenum Publishing Co., New York].

et al. 1989) which have gametophytic self-incompatibility systems. Studies on the sporophytic self-incompatibility systems have been primarily on members of the Cruciferae, Brassica oleracea and B. campestris (Nasrallah, Nasrallah 1989).

In this chapter, we will concentrate on recent information regarding gametophytic self-incompatibility in the Solanaceae taking Nicotiana alata as the model system. This work and studies on self-incompatibility in other gametophytic systems has recently been reviewed by Haring et al. 1990, 1991; Ebert et al. 1989. Details of the molecular genetics of the sporophytic self-incompatibility systems can be found in references (Nasrallah áNasrallah 1989; Takayama et al. 1987; Haring et al. 1990).

Cloning of cDNAs encoding S-glycoproteins in the Solanaceae

The first S-glycoprotein isolated for N-terminal amino acid sequencing was the 32KD style glycoprotein that segregates with the S_2-allele of N. alata (Anderson et al. 1986). Using this sequence data, a cDNA encoding the protein component was obtained and then used to obtain cDNAs corresponding to the S_3 and S_6-alleles of N.alata.

The derived amino acid sequences from these cDNAs, showed a conserved hydrophobic region at the N-terminus of the mature protein (15 amino acids) and two conserved regions separated by regions with considerable variation, termed "hypervariable" regions. There were also 5' sequences encoding signal peptides,

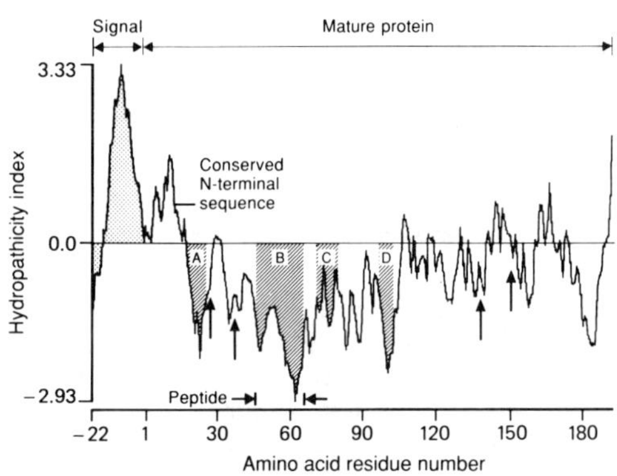

Figure 3. Hydropathy plot based on the *N. alata* S2-glycoprotein sequence and its alignment with cDNAs encoding other S-glycoproteins from *N. alata*, *P. inflata* and *S. tuberosum*. The hypervariable regions (A-D) are shown as ▒. Potential N-glycosylation sites are shown as↑. Region B contains the peptide sequence used to generate the S2-specific antibody. (Haring et al. 1990, reproduced with permission from the American Association of the Advancement of Science, Washington DC).

indicating that, as anticipated, the S-glycoproteins are secreted. From this sequence information, cDNAs corresponding to further S-alleles of N. alata (Anderson et al. 1989; Kheyr-Pour et al. 1990), S-alleles from Petunia inflata (Ai et al. 1990), Solanum chacoense (Xu et al. 1990) and Petunia hybrida (Reed Clark et al. 1990) were obtained.

Alignment of all these sequences (Haring et al. 1990) shows that 46 amino acids (22.4%) are conserved, including seven (of a maximum of ten) cysteine residues in each protein. There are two other conserved sequences each of which includes a histidine residue. There is also a region which represents a hyper-variable sequence in all sequences of S-glycoproteins from the Solanaceae. This data is summarized in Figure 3; details are presented in Haring et al. 1990.

Ribonuclease function of the S-glycoproteins

Sequence comparison of the S-glycoproteins of N.alata with those of the fungal RNases from Aspergillus oryzae (RNase T2) and Rhizopus niveus (RNase RH) showed limited, but important identities. Overall, 30 of the 122 amino acids conserved between three N. alata glycoproteins could be aligned with identical amino acids in the fungal RNases, and another 22 aligned with related amino acids. The most important feature of the alignment however, was the inclusion of two putative active site histidine residues and five cysteine residues in the sequences conserved between the fungal RNases and the S-glycoproteins. This finding lead to testing of the isolated S-glycoproteins for RNase activity and the finding that each was indeed an RNase (McClure et al. 1989). The specific RNase activities for the different S-glycoproteins ranged from 170-2200 (A_{260} units per min per mg) compared with that of fungal RNase T2 which was 3900 (A_{260} units per min per mg). The RNase activity of the S-glycoproteins accounted for 40-80% of the total RNase activity of the style extracts of different S-genotypes. Comparable style extracts from N. tabacum (self-compatible) had very low RNase activities (<1% that of N. alata extracts).

Role of S-RNases in arrest of pollen tube growth

The obvious question which follows from the finding that the S-glycoprotein in RNases, is that of whether the intrinsic RNase activity is directly involved in pollen tube growth. An alternative possibility is that the RNase activity is not related to the SI function of the S-glycoproteins, but that the enzyme was "recruited" during evolution for a role in self-incompatibility, quite unrelated to its catalytic activity. This question was approached by tracking pollen RNA in styles after both compatible and incompatible pollinations. This was achieved by growing plants in the presence of ^{32}P, and using the labelled pollen in test pollinations. Initially, pollen in both compatible and incompatible crosses germinates and the pollen tubes grow through the stigma and then intercellularly between the cells of the transmitting tract of the style. The difference between compatible and incompatible tubes becomes apparent in a zone of the style about 5-10mm below the stigma. Here the walls of incompatible tubes appear thickened and their growth is arrested; in contrast the growth of compatible tubes proceeds to the ovary. After test pollinations with ^{32}P-pollen, RNA was prepared from both stigma and style fractions of compatible and incompatible crosses. The first observation was that less ^{32}P-RNA was recovered from incompatible pollinations than from compatible pollinations. The second observation was that

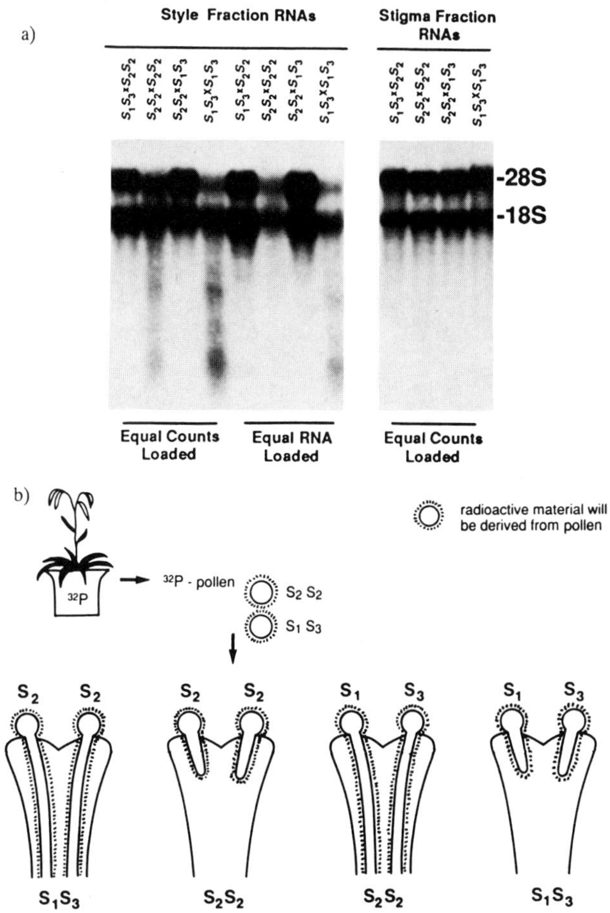

Figure 4. S-allele-specific degradation of pollen RNA within the style. a, Experimental design. N. alata plants were grown with ^{32}P-orthophosphate to obtain ^{32}P-pollen which was applied to pistils of unlabelled flowers. Reciprocal crosses were performed involving the S-genotypes shown. b, Analysis of RNA prepared from crosses shown in (a). RNA was fractionated in a 2% agarose-formaldehyde gel, blotted on to a nylon membrane and autoradiographed at -70°C with an intensifying screen for

7 days. The stigma and style fractions were treated separately. The gel lanes containing the style fraction RNAs were loaded both by equal ^{32}P counts and by equal amounts of RNA as shown. ^{32}P-RNA from the style fractions derived from the compatible crosses (S_1S_3 x S_2S_2; S_2S_2 x S_1S_3) consists largely of intact 28S and 18S rRNA. In contrast, ^{32}P-RNA from the style fractions derived from the self-incompatible crosses (S_2S_2 x S_2S_2; S_1S_3 x S_1S_3) shows degradations of the 28S and 18S rRNA as well as accumulation of lower-Mr species. This pattern is seen both in lanes loaded for equal ^{32}P counts or equal amounts of RNA. ^{32}P-RNA from the stigma fractions shows intact 28S and 18S RNA from all crosses. (From McClure et al., 1990, reproduced with permission from Macmillan Magazines Ltd., London).

ribosomal RNAs were apparently intact in extracts from compatible crosses but were degraded in incompatible crosses (Figure 4). The low Mr products were recovered in the poly (A)- fraction and were not present in the poly (A)+ nor the LiCl supernatant. Messenger RNA would be found in the poly (A)+ fraction and transfer RNA would be found in the LiCl supernatant, indicating that the low Mr products in the poly (A)- fraction represented degradation products of rRNA. This finding suggests that expression of self-incompatibility may be mediated by degradation of pollen rRNA in incompatible pollen tubes during their growth within the style (McClure et al. 1990).

Degradation of rRNA in incompatible pollen tubes would be a very effective process to cause arrest of pollen tube growth, because it seems that there is no transcription of ribosomal genes in mature pollen. This was demonstrated first for Tradescantia pollen (Mascarenhas 1975) and subsequently for N. alata pollen (McClure, unpublished observations). This lack of capacity to synthesize rRNA leaves pollen extremely vulnerable as degradation of rRNA by S-RNases would lead directly to depletion of the capacity for protein synthesis and hence to arrest of tube growth.

This explanation is consistent with all the experimental data but does not exclude the possibility that the degradation of rRNA is a secondary effect following some other inhibitory mechanism.

If the explanation is correct, then several other issues arise. What is the nature of the allelic specificity? The S-RNases do not have any obvious specificity for polymeric substrates as they effectively degrade RNA from many sources. For example, RNA from pollen genotypes S_6S_6 was equally well digested in vitro with purified style S-RNases from genotypes S_2S_2 and S_6S_6 as well as RNase A. In each case the pattern of products was similar (McClure et al. 1990). There is some base preference at the dinucleotide level (McClure, unpublished observations) but this could not account for the specificity of the multiple S-alleles. The lack of base specificity also agrues against the possibility that the S-RNases function in a way analogous to the DNA restriction endonucleases. The observation that there were no differences in the range or relative concentrations of the proteins synthesized in the presence or absence of S-RNase in vitro also indicates that it is unlikely that specific mRNAs are preferentially degraded by S-RNases (Gray et al. 1991).

One suggestion to account for the S-allele specificity of interaction between the pollen tube and the style, is exhibited in the mechanism by which the S-RNases move across the wall and the membrane of the pollen tube, into the cytoplasmic compartment. In this case, the expression of the pollen part of the S-gene would be as a "receptor" for the style S-RNase. This "receptor" may or may not be a receptor in the classical sense as developed in studies of receptors on the surfaces of animal cells. In the case of the pollen tube, the cell wall overlays the membrane and provides a potential barrier which may or may not be selective, to access of extracellular material to the plasma membrane.

Uptake of S-RNases by pollen tubes

As a first approach to the question of whether there is any allele-specific uptake of S-RNases by pollen, isolated S-RNases were incubated with in vitro grown pollen and their uptake followed. The S-RNases were taken up by the pollen and were present in both the walls and the cytoplasm of the pollen tube after 3h incubation, but were apparently concentrated in a specific layer of the pollen grain intine close to the plasma membrane (Gray et al. 1991). There was some indication that less S_2-RNase may be taken into S_3-pollen than into S_2-pollen; that is, that there may be some allelic specificity in uptake. However, this observation requires further work to support it; furthermore, the work is restricted to in vitro grown pollen, and may not truly reflect the in vivo situation. Nonetheless, the finding that S-RNases can be taken into pollen tubes and that they are not degraded, either within the wall or the cytoplasm, is consistent with an active role in pollen tube growth arrest. The S-RNases also inhibited translation of pollen RNA as total protein biosynthesis with in vitro grown pollen tubes, as measured by ^{35}S-methionine incorporation, was depressed by contact with S-RNases (Gray et al. 1991). These data are all consistent with, but certainly do not prove that the specificity of the interaction between the products of the pollen and pistil S-alleles, involves specific uptake of the S-RNase from the extracellular matrix of the style transmitting tract into the pollen tube.

Nature of the pollen products of the S-gene

The experiments described suggest that the pollen S-gene product may act as a "receptor" protein. Understanding the nature of both the pollen part of the S-gene and its product is the next major target of research in this field. We have tested a number of possible approaches and hypotheses relating to the problem.

1. Is the pollen S-gene homologous with the style S-gene? This is suggested by Lewis' hypothesis (1949, 1960) that the s-gene is tripartite, with a pollen part, a style part and a specificity part. Northern analysis involving pollen RNA and style S-cDNAs showed no hybridization. At reduced stringency there was some hybridization, but it was non-specific. These experiments indicate that if a homologous transcript exists, it must be of very low abundance.
2. If the style S-cDNA hypervariable region encodes (even in part) allelic specificity, then that region might be expected in the pollen S-gene. This was tested using the antibody raised to a synthetic peptide corresponding to the hypervariable region in a Western analysis of pollen proteins. No binding of the antibody to pollen proteins was detected, indicating that if the pollen product containing this region does exist, it must be in very low abundance.
3. Does the pollen S-gene share some conserved sequences with the style S-gene? This was tested using the polymerase chain reaction with primers corresponding to conserved regions. Although products were obtained, there was no allele specific

product detected, suggesting that this route would probably not lead to the allele specific pollen product.

4. Is the S-gene different from that of style, but closely linked? This was tested by sequencing 5.2Kb of the S6-genomic DNA sequence. Although an unidentified open reading frame (URF) was identified which was distant from that corresponding to the S-glycoprotein, no RNA homologous to this (URF) was detected. Thus this URF is not the pollen product of the S-gene. This does not rule out the possibility of a pollen product of the S-gene being different and more distant from the style S-gene product. If the products of the S-locus in pollen and style are indeed distinct, they must be closely linked as the two functions have not been separated by conventional breeding.

These experiments indicate that there is no simple, straightforward entry point to establishing the nature of the pollen S-gene.

There are a number of other unanswered questions relating to the growth of the pollen tube. What is the nature of the wall components of pollen tubes? How is the biosynthesis of the wall components of the tube coordinated? What is the normal process of pollen tube nutrition? What is the process of uptake of materials by the pollen tube? What is the relationship between the plasma membrane and the pollen tube wall?

In a broader context, the relationship of the S-RNases to other RNases with special (non-catalytic) biological actions is of interest (D'Alessio et al.1991). These RNases such as angiogenin and bovine semen RNase are secreted RNases for which the relationship between RNase activity and physiological action is unclear. D'Alessio et al. (1991) suggest that "evolution found the RNase structural scaffold; including the RNase active site, particularly convenient for the engineering of molecules which assist or interfere with, key biological events such as organogenesis, cancer and immunomodulation." It seems that the S-RNases probably do act as cytotoxins, but the detailed mechanism for their biological action is still to be established.

References

Ai, y., Singh, A., Coleman, C.E., loerger, T.R., Kheyr-Pour, A. and Kao, T.-H. Self-incompatibility in Petunia inflata: Isolation and characterization of cDNAs encoding three S-allele-associated proteins. Sex. Plant Reprod. 3:130-138; 1990.

Anderson, M.A., Hoggart, R.D. and Clarke, A.E. The possible role of lectins in mediating plant cell-cell interactions. In: Chemical taxonomy, molecular biology and function of plant lectins. New York, Alan R. Liss Inc.; 1983: p143-161.

Anderson, M.A., Cornish, E.C., Mau, S.-L., Williams, E.G., Hoggart, R., Atkinson, A., Bünig, I., Grego, B., Simpson, R., Roche, P.J., Haley, J.D., Penschow, J.D., Niall, H.D., Tregear, G.W., Coghlan, J.P., Crawford, R.J. and Clarke, A.E. Cloning of cDNA for a stylar glycoprotein associated with expression of self-incompatibility in Nicotiana alata. Nature 321:38-44; 1986.

Anderson, M.A., McFadden, G.I., Bernatzky, R., Atkinson, A., Orpin, T., Dedman, H., Tregear, G., Fernley, R. and Clarke, A.E. Sequence variability of three alleles of the self-incompatibility gene of Nicotiana alata. The Plant Cell 1:483-491; 1989.

Cornish, e.c., Anderson, m.a. and Clarke, a.e. Molecular aspects of fertilization in flowering plants. Annual Review Cell Biology, 4:209-228; 1988.

D'Alessio, G., Di Donato, A. Parente A. and Piccoli, R. Seminal RNase; a unique member of the ribonucleases superfamily. TIBS 16: 104-106; 1991.

de Nettancourt, D. Incompatibility in Angiosperms. In: Frenkel, R., Gall, G.A.E. and Linskens, H.F., eds. Monographs in theoretical and applied genetics. Berlin, Springer-Verlag; 1977.

p.Ebert, P.R., Anderson, M.A., Bernatzky, R., Altschuler, M. and Clarke A.E. Genetic polymorphism of self-incompatibility in flowering plants. Cell 56:255-262; 1989.

Franklin-Tong, v.e., Ruuth, e., marmey, p., lawrence, m.j. and franklin, r.c.h. New Phytol. 112: 307;1989.

Gray, j.e., Mcclure, b.a., Haring, V., anderson, m.a. and Clarke, a.e. Self-incompatibility as a model for cell-cell recognition in flowering plants. In: Herrmann, R.G., ed. Plant Molecular Biology. New York: Plenum Publishing Co.; 1990.

Gray, j.e., Mcclure, b.a., bÜnig, i., anderson, m.a. and Clarke, a.e. Action of the style product of the self-incompatibility gene of Nicotiana alata (S-RNase) on in vitro grown pollen tubes. The Plant Cell 3:271-283; 1991.

Haring, V., McClure, B.A. and Clarke, A.E. Molecular aspects of self-incompatibility in the Solanaceae. In: Dennis, E.S. and Llewllyn, D., eds. Advances in Plant Gene Research. Vol 7. Vienna, Springer-Verlag; 1991.

Haring,V., Gray, j.e., McClure, b.a.,Anderson, M.A. and Clarke, a.e. Self-incompatibility: a self-recognition system in plants. Science 250:937-941; 1990.

Kheyr-Pour, A., Bintrim, S.B., loerger, T.R., Remy, R. Hammond, S.A. and Kao, T.-H. Sequence diversity of pistil S-proteins associated with gametophytic self-incompatibility in Nicotiana alata. Sex. Plant Reprod. 3:88-97; 1990.

Lewis, D. Incompatibility in flowering plants. Biol. Rev. 24:472-496; 1949.

Lewis, D. Genetic control of specificity and activity of the S-antigen in plants. Proc. Roy. Soc. Lond. [Biol.] 151:468-477; 1960.

Mascarenhas, j.p. The biochemistry of angiosperm pollen development. Bot. Rev., 41: 259-314; 1975.

McClure, B.A., Haring, V., Ebert, P.R., Anderson, M.A., Simpson, R.J., Sakiyama, F. and Clarke, A.E. Style self-incompatibility gene products of Nicotiana alata are ribonucleases. Nature 342:955-957; 1989.

McClure, B.A., Gray, J.E., Anderson, M.A. and Clarke, A.E. Self-incompatibility in Nicotiana alata involves degradation of pollen rRNA. Nature 347:757-760;1990.

Nasrallah, j.b. and Nasrallah, m.e. The molecular genetics of self-incompatibility in Brassica. Ann. Rev. Genet. 23: 121-139; 1989.

Reed Clark, K., Okuley, J.J., Collins, P.D. and Sims, T.L. (1990). Sequence variability and developmental expression of S-alleles in self-incompatible and pseudo-self-compatible Petunia. The Plant Cell 2:815-826;1990.

Takayama,S. Isogai, A., Tsukamoto, C., Ueda, Y., Hinata, K., Okazaki. and Suzuki, A. Sequences of S-glycoproteins, products of the Brassica campestris self-incompatibility locus. Nature 326:102-105; 1987.

Xu, B., Mu, J., Nevins, D.L., Grun, P. and Kao, T.-H. Cloning and sequencing of cDNAs encoding two self-incompatibility associated proteins in Solanum chacoense. Molec. gen. Genet. 224:341-346; 1990.

13
Recognition, Signals, and Pollen Responses in the Incompatibility Reaction in *Papaver rhoeas*

Veronica E. Franklin-Tong, Glenn J. Thorlby, Michael J. Lawrence, and F. Christopher H. Franklin

Introduction

The process of pollination and fertilization in flowering plants involves a series of interactive events between male and female cells. One of the earliest stages in the process of fertilization is the recognition, and acceptance or rejection, of pollen grains alighting on the stigma of the recipient plant. Self-incompatibility (SI) involves these processes. We are particularly interested in the mechanism whereby incompatible pollen is inhibited, and the events that occur during a SI reaction.

Pollination

Figure 1 illustrates the floral morphology of *P. rhoeas*. The stigmatic disc sits directly on top of the gynoecium, since there is no style. The stigma is dry and consists of a series of stigmatic rays, which are lined with uninucleate papillae (Heslop-Harrison and Shivanna, 1977). The structure of the pollen grain has recently been described by Cresti et al. (1990). Pollen landing on the stigmatic papillae produces pollen tubes which penetrate the papillae (Elleman et al., in press) and grow along the stigmatic rays, aggregating in long "ropes", until they reach the central transmitting tract in the gynoecium (see figure 1).

Incompatible pollen is rapidly inhibited on the stigma surface. In most

Figure 1. Gross morphology of the pistil, and pollination in *P. rhoeas*.

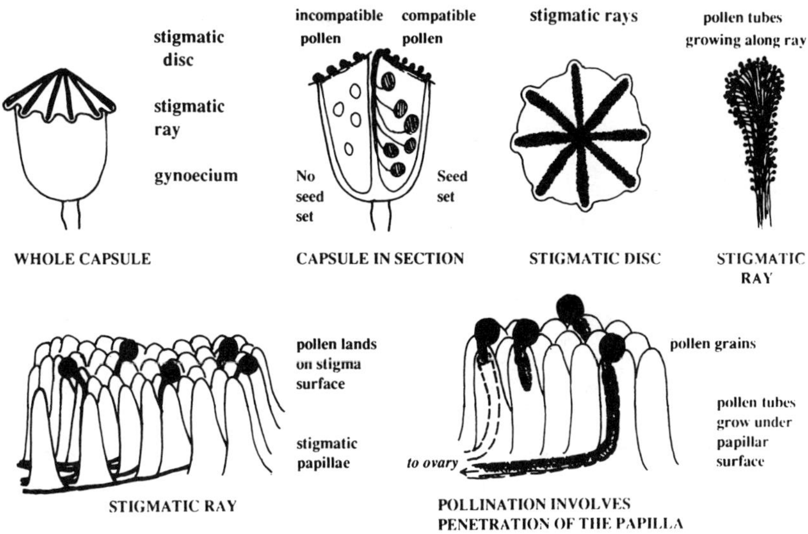

cases it occurs during, or just after germination. There is typically deposition of callose inside the pollen grain; thus inhibition may take place in many cases before tube emergence. Pollen tubes that do emerge are typically very short, often distorted, and there is a heavy deposition of callose at the pollen tube tip.

The Genetics of SI in *P. rhoeas*

SI in *P. rhoeas* is homomorphic and is controlled by a single, multi-allelic gene, (S-), which is gametophytically controlled. The genetics of self-incompatibility (SI) in *P. rhoeas* was worked out by Lawrence and his co-workers (Lawrence, Afzal & Kenrick, 1978). Self-fertilization is prevented when pollen carrying an S-allele which is identical to that carried by the stigma on which it lands (incompatible) is discriminated from pollen carrying an S-allele which is not carried by the stigma (compatible). Since control is gametophytic, three types of pollination may be observed: incompatible, fully-compatible and half-compatible. Thus, genetically, the SI system in *P. rhoeas* is identical to that in *Nicotiana alata* and, indeed, this type of gametophytic SI has often been termed the "Nicotiana" system.

While most species with a gametophytic system of SI have their pollen inhibited in the style, *P. rhoeas* is unusual (but not unique), since inhibition of its pollen is stigmatic. In addition, *P. rhoeas* has a dry stigma, whereas most species with this gametophytic control have been found to have wet stigmas (Heslop-Harrison and Shivanna, 1977). These differences make the study of the cellular and molecular biology of this species worthwhile.

An *In Vitro* System for *P. rhoeas*

We have developed an *in vitro* system for the S-gene product encoded by the stigma in poppies (Franklin-Tong et al., 1988). This system has enabled us to reproduce the SI reactions *in vitro* using stigmatic extracts. Using the appropriate extract:pollen S-genotype combinations, the extracts elicit, *in vitro*, the three classes of the self-incompatibility reaction that are normally found in *in vivo* pollination in *P. rhoeas* (see table 1). These active molecules appear to be S-allele-specific, tissue-specific and developmentally controlled. Thus, like other isolated S-associated components, those from *P. rhoeas* fulfil the criterion of their association with the expression of self-incompatibility. There is, therefore, good evidence that these S-specific inhibitory components are products of the S-gene and actually function in the SI reaction.

Characterization of the Stigmatic Components

The *in vitro* system has allowed us to characterize and partially purify the components from three S-genotypes of *P. rhoeas*, since it enables us to monitor S-specific activity during fractionation (Franklin-Tong et al., 1989). Concanavalin A-(conA) sepharose affinity chromatography revealed that the active molecules from *P. rhoeas*, like other S-associated components are glycoproteins. Gel filtration, preparative isoelectric focussing and 2D electrophoresis were used to identify the allelic components of the S-protein. The putative S-gene products are glycoproteins; they have apparently the same molecular weight of 22kD and have variable pIs, which are attributable to different S-alleles (Franklin-Tong et al., 1989).

The Stigmatic S-gene Product is Not a Ribonuclease (RNase)

It has recently been shown that the S-linked glycoproteins from Solanaceaeous species *N. alata, Lycopersicon peruvianum, Petunia hybrida* and *Solanum tuberosum* are ribonucleases (McClure et al., 1989; Clark et al., 1990). Since *P. rhoeas* has a genetically identical SI system to *N. alata*, we decided to investigate whether our stigmatic component is a RNase. We found several differences between *N. alata* and *P.rhoeas* with respect to ribonuclease activity (Franklin-Tong et al., 1991). These differences are summarized in Table 2.

Table 1. *In vitro* system for *P. rhoeas*: effect of stigma extracts on two pollen genotypes (modified from Franklin-Tong et al, 1988).

eluate	pollen genotype S_1S_2	S_1S_3	eluate means
control (c)	88.63	80.38	84.50
fully-compatible (++)	77.63	75.38	76.50
half-compatible (+)	50.75	40.63	45.69
incompatible (-)	15.22	14.38	15.25

Analysis of variance of the above data

item	df	MS	significance
extracts	3	80 224.45	***
c *vs* ++	1	2560.00	*
++ *vs* -	1	37 362.66	***
++ *vs* +	1	37 976.41	***
+ *vs* (++ & -)	1	0.83	N.S.

We found that stigmas from *P. rhoeas* have levels of RNase activity which are several hundred-fold lower than those found in styles of *N. alata;*even lower than those found in styles of self-compatible *N. tabacum.* Immature stigmas exhibit ribonuclease activity at a level indistinguishable from that of the mature self-incompatible stigmas; thus, ribonuclease activity does not appear to be correlated with acquisition of self-incompatibility. We also found that S-specific pollen inhibitory activity is not correlated with ribonuclease activity, since some fractions which had almost identical ribonuclease specific activities, had very different pollen inhibitory activities (see table 2). As a result of these studies we believe that SI in *P. rhoeas* does not involve RNase activity (Franklin-Tong et al., 1991). We, therefore need an alternative model for the mechanism of SI in this species.

Cellular Processes in Pollen involved in SI

Experiments involving incorporation of metabolic inhibitors into the *in vitro* system has revealed several items of interest. Firstly, although germination of pollen of *P. rhoeas* is independent of protein synthesis, pollen tube growth is not. Secondly, neither glycosylation or transcription is required for germination or pollen tube growth. The independence of pollen tube growth from

Table 2. Comparisons of ribonuclease and pollen inhibitory activities.

Species Fraction	RNase activity $min^{-1}\ mg^{-1}$	Pollen tube length % control length compatible	incompatible
N. alata mature style	281.00	-	-
P. rhoeas mature stigma	1.16	-	-
P. rhoeas mature stigma	1.14	-	-
P.rhoeas immature stigma	1.05	-	-
P. rhoeas conA-UB	0.34	76%	77%
P. rhoeas conA-B	0.35	92%	38%

Pollen tubes were measured after 3 h growth (n=12). Modified from Franklin-Tong et al, 1991.

transcription suggests that no new RNA is synthesized during germination or pollen tube growth. Thirdly, both glycosylation and transcription in the pollen are required for an incompatible response which results in complete pollen tube inhibition (Franklin-Tong et al., 1990). For example, in an incompatible reaction, the mean pollen tube length was 27% of the control, while the same reaction in the presence of the transcription inhibitor actinomycin D, gave pollen tubes which were 59% of the control length (see Franklin-Tong et al., 1990). This requirement for transcription appears to be specific to an incompatible response. Thus, we have shown that the SI response in *P. rhoeas* requires *de novo* RNA transcription in the pollen.

Pollen Proteins Associated with an Incompatible Response

Since it appears that pollen gene expression is induced during the SI response, we have investigated the possibility of identifying proteins associated with the response. RNA from ungerminated, germinated, compatible and incompatible pollen was translated and the products analysed on 2D gels. As predicted from the previous experiments, there were no detectable differences between the samples from ungerminated and germinated pollen or from pollen in which a compatible response had been elicited. This confirms that no new RNA is synthesized during germination and pollen tube growth in *P.rhoeas* (Franklin-Tong et al., 1990). In contrast, incompatible pollen contains novel translation products which are absent in the other samples. These new RNA transcripts are specific to an incompatible reaction and are presumably produced in pollen as a consequence of the SI response (Franklin-Tong et al., 1990). This observation, taken together with the ability of actinomycin D to alleviate the incompatibility response, is a clear indication that the *de novo* transcription of pollen genes play an important role in the recognition and/or inhibition of incompatible pollen in this species. There also appears to be "down-regulation" of some transcripts in an incompatible response. These transcripts are seen at quite high levels in dry, germinated and compatible pollen, but their levels appear to be reduced in incompatible pollen.

Identification of Pollen Response Genes

Since new RNA transcripts are specifically produced in incompatible pollen as a consequence of the SI response, we have searched for, and identified a number of pollen "response" genes. A cDNA library, generated from pollen challenged with an incompatible stigma extract was differentially screened to identify pollen "response" genes. Some of these are up-regulated (see figure 2) and some are "down-regulated" as a result of the incompatible response. The differential expression of these genes has been examined in order to reveal the level of transcription of these pollen genes in an incompatible and compatible reaction. The "down regulated" genes are expressed at high levels in mature dry, ungerminated and compatible pollen, but at low levels in incompatible pollen. The up-regulated genes are expressed at high levels in incompatible pollen and at low levels, or not at all in compatible pollen.This provides convincing evidence that these transcripts are truly up- and "down"-regulated during an incompatible response. Expression of these genes has also been examined using slot blots of PCR generated cDNA from incompatible, compatible and dry pollen, stigma and leaf tissue and probed with the clones. Expression of these genes appears to be pollen-specific, since no signals were detected with leaf or stigma. The up- and "down"-regulated clones gave similar patterns to those described above.

Thus, the genes which relate to the transcripts which we previously identified on 2D gels have been cloned. The pattern of gene expression has shown that they are clearly associated with the SI response. We assume that the up-regulated genes are transcriptionally regulated, since the experiments with actinomycin D have shown that transcription is non-essential to normal growth, but required for an incompatible response. Similarly, the 2D gels showed that new transcripts were produced specifically in an incompatible response. It appears likely that the down-regulated genes are post-transcriptionally regulated, since the transcripts are present in pollen at quite high levels and their level drops to a low level in incompatible pollen. This would argue for the specific degradation of these RNA species. We are currently characterizing these genes further. We have preliminary evidence that these genes are specifically expressed in incompatible *in vivo* pollinations, providing further evidence that what we are looking at is involved in the SI

Figure 2. Secondary screen of up-regulated pollen response genes.

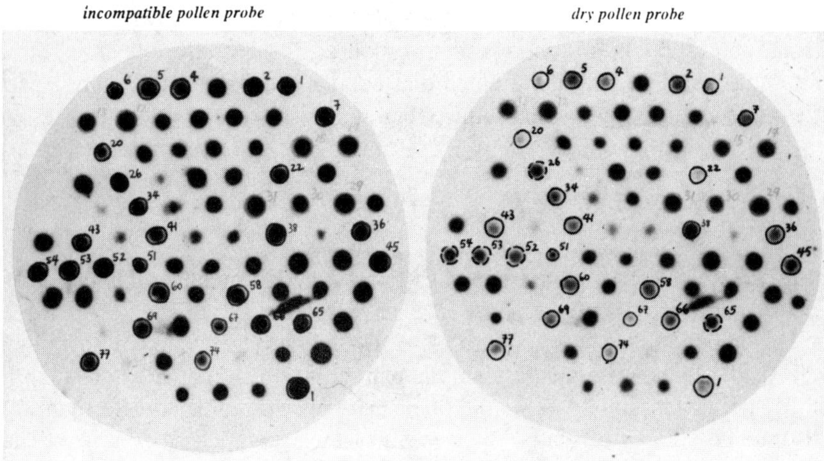

response. Information as to the nature of these pollen genes should substantially improve our knowledge as to how SI operates in this species.

The events occurring in the SI response, as far as we know them, suggest that gene expression in incompatible pollen is switched on as the result of an external stimulus from the stigma molecule. If this is the case, it suggests a role for signal transduction in SI.

Signalling in Self-Incompatibility

The switching on or off of specific genes within the pollen provides important evidence that a signalling system is involved in the SI response. It is possible that the SI response may involve a classic signal transduction system, since there is some evidence that suggests that signal transduction may play a role in cell-cell signalling in plant cells. One major mechanism for signal transduction in animal cells, which appears likely to be used by plant cells, is the "phosphatidylinositol response" (Michell, 1975). This pathway is of particular interest to us, since it leads to activation of protein kinase C, which causes gene expression *via* protein phosphorylation. Since we get gene transcription, presumably as the result of a signal, we have begun to explore the possibility of a role for signal transduction in the SI reaction of *P. rhoeas*.

Preliminary results have shown a rapid and transient phosphorylation of several pollen proteins when pollen is challenged, with distinct differences between a challenge with an incompatible and a compatible stigma extract (see figure 3). Some proteins are newly phosphorylated within 1 minute of the SI

Figure 3. Changes in phospho-protein profiles during a SI reaction.

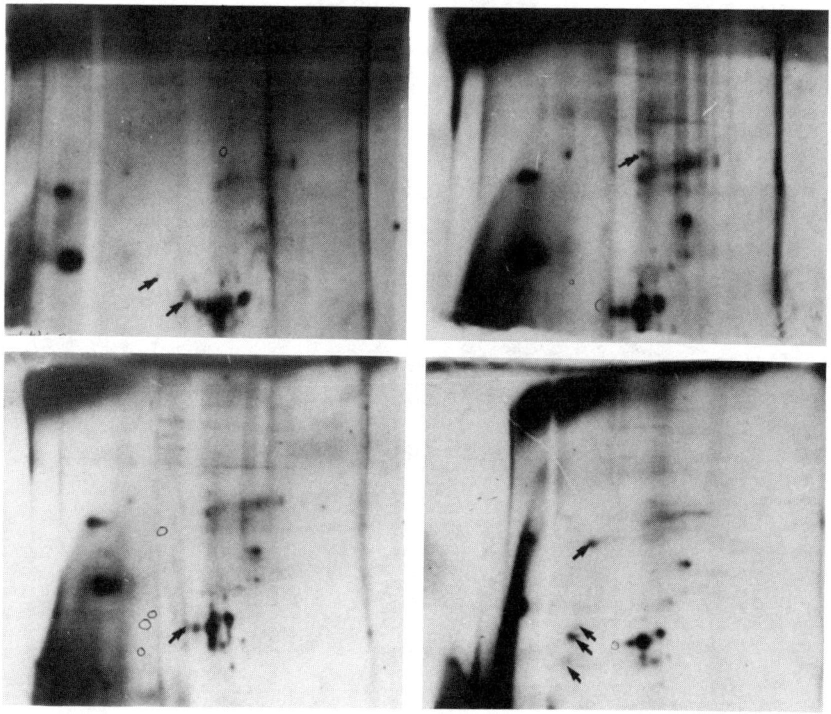

Autoradiograph of 2D gels of pollen phosphoproteins. a. incompatible reaction, 1 min.; b. compatible reaction, 1 min.; c. incompatible reaction, 6 min.; d. compatible reaction, 6 min.

interaction, and there is evidence that some of these beginning to be dephosphorylated within 6 minutes. There are also some phosphophoproteins which appear in the compatible pollen by 6 minutes after challenge, which are not present in incompatible pollen. This suggests that there is a role for protein kinases and phosphatases in the SI reaction. These observations are consistent with the idea that signal transduction may be involved in the SI response. Thus, there is some evidence which suggests the involvement of signal transduction molecules in pollen germination, growth and the SI response.

A Model for the Mechanism of SI in *P. rhoeas*

Our evidence of activation of genes specifically as a consequence of an incompatible reaction provides the basis for a model for the mechanism of the SI response. This is outlined in Figure 4. Presumably the stigmatic glycoproteins act as signal molecules; these interact with receptors (which

Figure 4. A model for SI in *Papaver rhoeas*.

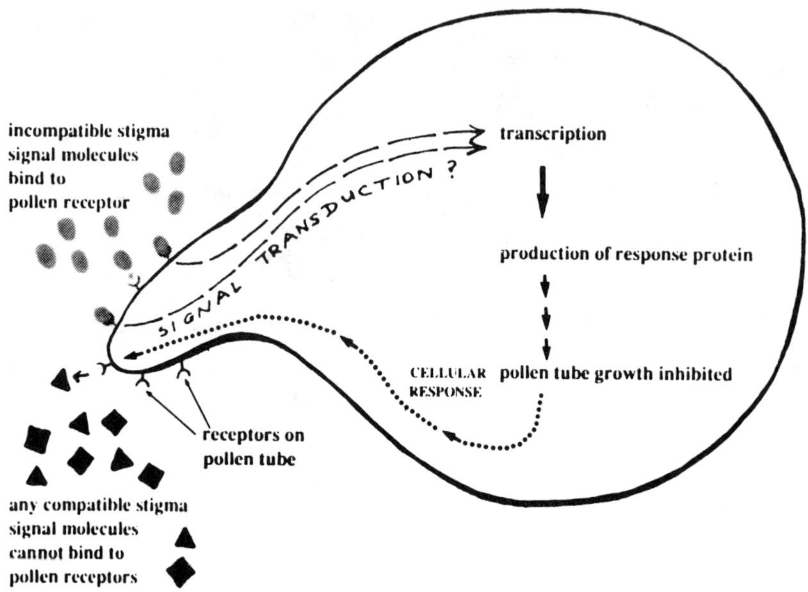

have S-specificity) on the pollen tube. Thus, we postulate that only in an incompatible pollination does the stigmatic signal molecule bind to the pollen receptor; in a compatible interaction the receptor does not have the correct specificity. When a signal molecule binds the receptor, triggering a cascade of signal transduction mechanisms inside the pollen. The exact pathways are, as yet, unknown, but evidence suggests a role for protein kinases and phosphatases. We know some events that take place inside the pollen, presumably as a result of this message. Gene expression in incompatible pollen is altered and these genes have been identified and isolated. Transcription products would lead to the cellular response, involving inhibition of pollen germination or pollen tube growth.

Acknowledgements

Much of this work was funded by the A.F.R.C. Plant Molecular Biology Programme. This work would not have been possible without our hard-working gardening staff and technical assistance from Antony Jones. We would also like to thank Janet Lord and Jon Ride for useful discussion and guidance.

REFERENCES

Clark, K.R.; van Laere, A.; Witters, R.; Preaux G.; Decock, B.; van Damme, J.; Vendrig, J.C. Purification and N-terminal sequencing of style glycoproteins associated with self-incompatibility in *Petunia hybrida*. Plant Mol. Biol. 14: 93-102; 1989.

Cresti, M.; Milanesi, C.; Salvatici, P.; van Aelst, A.C. Ultrastructural observations of *Papaver rhoeas* mature pollen grains. Bot. Acta 103: 349-354; 1990.

Elleman, C.E;, Franklin-Tong, V.E.; Dickinson, H.G. Pollination in species with dry stigmas. The nature of the stigma surface and pollen tubes. Submitted: New Phytologist.

Franklin-Tong, V.E.; Lawrence, M.J.; Franklin, F.C.H. An *in vitro* bioassay for the stigmatic product of the self-incompatibility gene in *Papaver rhoeas* L. New Phytol., 110: 109-118; 1988.

Franklin-Tong, V.E.; Ruuth, E.; Marmey, P.; Lawrence, M.J.; Franklin, F.C.H. Characterization of a stigmatic component from *Papaver rhoeas* L. which exhibits the specific activity of a self-incompatibility (S-) gene product. New Phytologist, 112: 307-315 1989.

Franklin-Tong, V.E., Lawrence, M.J. & Franklin, F.C.H. Self-incompatibility in *Papaver rhoeas* L: inhibition of incompatible pollen tube growth is dependent on pollen gene expression. New Phytologist 116: 319-324; 1990.

Franklin-Tong, V.E.; Atwal, K.K.; Howell, E.C.; Lawrence, M.J.; Franklin, F.C.H. SI in *Papaver rhoeas* L.: there is no evidence for the involvement of stigmatic ribonuclease activity. Plant, Cell, and Environment 14; 1991.

Heslop-Harrison; Y.; Shivanna, K.R. The receptive surface of the Angiosperm stigma. Annals of Botany 41: 1233-1258; 1977.

Lawrence, M.J.; Afzal, M.; Kenrick, J. The genetical control of self-incompatibility in *Papaver rhoeas* L. Heredity 40: 239-253; 1978.

McClure, B.A.; Haring, V.; Ebert, P.R.; Anderson, M.A.; Simpson, R.J.; Sakiyama, F.; Clarke, A.E. Style self-incompatibility gene products of *Nicotiana alata* are ribonucleases. Nature 342: 955-957; 1989.

Michell, R.H. Inositol phospholipids and cell surface receptor function. Biochim. Biophys. Acta 415: 81-147; 1975.

14
Self-incompatibility Products of the Male Partner in *Brassica oleracea*

A.A. Friry, C.M. Guilluy, P. Heizmann, T. Gaude, and C. Dumas

Introduction

In Brassica, self-incompatibility is controlled by the S-locus whose the products expressed in the female stigmatic tissue have been identified and characterized (Nasrallah and Nasrallah, 1984; Takayama et al., 1986). The gene encoding these S molecules (the SLG gene for S locus glycoproteins) has been shown to belong to a multigenic family consisting of about 12 homologous copies among which at least two other members are expressed, namely the S-locus related genes (SLR1 and SLR2, see Lalonde et al., 1989; Boyes et al., 1991). By contrast to the numerous data now gathering on the female products of the S locus, almost nothing is known of the expression of the S locus in the male tissues. In this paper, after a brief presentation of some models classically used to explain self-incompatibility (SI) in Cruciferae, a sporophytic system as defined in Nettancourt (1977), attempts to identify the S products of the male partner are reported. Finally, a working hypothesis is formulated about the physiological response of self-incompatible pollen-stigma rejection.

Emerging Models on Self-Incompatibility in Brassica

The D. Lewis'model (1963): a classical genetical model.

In a recent Meeting of the Genetical Society (Norwich, UK, 1988), Dan Lewis after several talks on S.I. summarized ideas about his genetical model by these words: "dream or reality?". The tripartite structure proposed for the S gene, including the "same and unique allele specific (=coding) segment" expressed in the male and female tissues, under the control of pollen or stigma specificity segments (=tissue specific promoters), still remains the most realistic model for self-incompatibility. This model has one great advantage over others: it is amenable to experimental verification.

The J.Heslop-Harrison's model (1975): a cellular interacting system

This model is based on physiological and cytological observations leading to a sort of dialogue between pollen recognition factors and a read-out system formed by the so-called pellicle (Mattsson et al., 1974) during the initial step of the pollination. This model takes into account the tapetal origin of the pollen coat , i.e., the possible location of recognition factors, and the rejection reaction or callosic response (Heslop-Harrison, 1975), as well as the role of exine-held fractions associated with SI (Heslop-Harrison et al., 1974)

This elegant model underlines the necessity to identify precisely the interacting surfaces between the two partners: pollen wall and stigma pellicle in both compatible and incompatible situations. (see reviews Gaude and Dumas, 1987, for example).

Pollen-stigma chemical adhesion or specific activation

During the past ten years several trials have been carried out in order to find some evidence to explain the pollen-stigma adhesion and its putative direct role in SI phenomena (see review Dumas et al., 1984).

Another explanation emerged from the lectin mechanism to illustrate the idea of a specific attachment. Some of us tested this hypothesis (Gaude et al., 1983) and if some agglutinins have been demonstrated by a pollen-rosetting technique using red blood cells, we failed to inhibit them by a range of mono-and disaccharides suggesting that the agglutinin may be membrane-bound, possibly in the plasma membrane or in the exinic outer layer (Gaude and Dumas, 1984). This idea has been developed by different authors but significant evidence is still lacking.

Pollen hydration control

Several groups tried to link SI expression and the control of pollen hydration. Based on some ideas coming from original experiments and model from Heslop-Harrison in grasses (1979), we have firstly proposed a functional hypothesis during the International Botanical Congress at Sydney in 1980 (see Dumas and Gaude, 1983).This model was based on several observations realized on pollen grain and pollen wall, especially the characterization of a membrane-like structure in the outest part of the exine .We termed such a structure: exinic outer layer (E.O.L.according to Gaude and Dumas, 1984).Such a membrane-like structure has also been reported in Brassica with a different terminology : membrane-like coating superficial layer (C.S.L. according to Elleman and Dickinson, 1986).

Probably the most convincing interpretation arises from elegant microtechniques developed on Brassica pollen visualised directly with a video system. In this study, Sarker et al. (1988) demonstrate that pollen hydration kinetics seems to be strictly dependent upon the S genotype of the pistil. Moreover, they showed that the hydration step is controlled by proteinic molecules rather than sugar residues. Yet, glycosylation is obviously necessary for SI.phenomenon With their evidence, Sarker et al., (1988) proposed a functional model for the control of pollen hydration which may confirm our earlier hypothesis. But, further investigations remain necessary to elucidate this critical step of control of pollen hydration.

Attempts to identify the S products of the male partner

The dimer hypothesis of Lewis (1963) predicts that pollen should carry molecules near-identical to those found in the stigma and early investigations were thus directed to the identification of pollen glycoproteins which resembled the S locus specific glycoproteins of the stigma. By using isoelectric focusing, a technique that proved particularly suited to the identification of the stigmatic S proteins, Nishio and Hinata (1978) separated 30 protein bands of different pIs from homogenates of Brassica pollen. However, no unequivocal differences could be detected between pollen carrying different S-alleles. As the self-incompatibility response results from an identity of S alleles in pollen and stigma (see review De Nettancourt, 1977), some structural homologies between S products of the same allele (at the protein and mRNA levels) should be found. Adopting this hypothesis and a sophisticated immunological approach, we have used monoclonal antibodies raised to the S6 SLG of Brassica oleracea stigmas to probe Western blots containing pollen proteins (Gaude et al., 1988; Guilluy et al., 1990). Intriguingly, the anti-S6 SLG antibody bound to the same family of protein bands, whatever the S allele possessed by the pollen. The pollen cross reacting material was present in the extracts at extremely low levels, and the polypeptides involved were of molecular masses of 36.5, 34 and 11-12 kDa. Interestingly, no reaction was observed at the 54-67kDa molecular mass range that corresponds to the mass of stigmatic SLGs. This data suggests that some commonality exists between pollen proteins and stigma SLGs and that the S specific molecules of the pollen are not identical to those of the stigma. These pollen polypeptides constitute putative candidates for the products of the S locus expressed in the male tissues.

At the genetic level, cDNA probes encoding stigma S-products failed to detect homologous sequences in the anther tissue (Scutt et al., 1990). Considering the apparently low level of male components of self-incompatibility, as suggested by our immunological study (Gaude et al., 1988), we have taken advantage of the very high sensibility of the PCR technique to investigate the presence in the anther of mRNAs homologous to SLG products.Fragments of the SLG cDNA present in the mature stigma of Brassica oleracea were amplified with specific primers bordering the 5' region of the SLG-6 sequence (figure 1). This PCR product presents a size of 472 bp which corresponds to that expected from the published SLG-6 sequence. It has been shown by Southern analysis that this amplified fragment shares some sequence homology with the BS29-2 clone isolated by Trick and Flavell (1989) which contains a SLG cDNA. No hybridization was observed with the BS29-1 clone representing a SLR1 probe.

Used as a probe on RNA dot-blots, this fragment reveals a strong hybridization signal in mature stigma and a weak one in young anthers. Northern analysis with the same probe indicates the presence of the abundant SLG RNA of about 2 kb in the mature stigma, but no band is observed in the male tissue (figure 2).

DNA amplification performed on anther cDNA at the microspore stage with the previous 5'-SLG-specific primers, gave rise to a fragment whose size is 472 bp, as expected. Since differential hybridization of this PCR product with both BS29-1 and BS29-2 clones only showed a signal with the SLG probe, the amplified fragment derived from anther cDNA are closely related to the SLG but not to the SLR1 sequence. Used as a probe in northern hybridization, this amplified region detects the stigmaticaly expressed SLG and a faint band of about 4 kb is observed in the anther RNA (figure 2). No hybridization is revealed in other tissues, either leaves or anther at the trinucleated stage only.

These results show that a 4 kb SLG-homologous RNA is expressed in the anther at the beginning of their development; the transcription of the corresponding gene is drasticaly lowered at a later developmental stage. This sequence represents the best candidate for being the transcript of the gene encoding the male S-product. As the size of the male

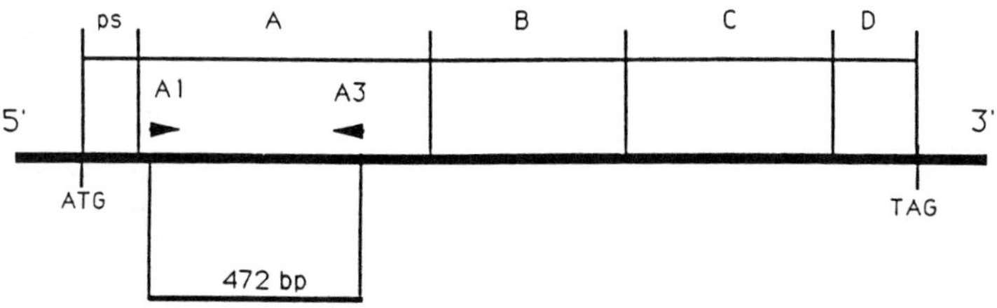

Figure 1 : PCR primer positions along the SLG-6 sequence (Nasrallah et al., 1987). A, B, C and D are the variable and conserved regions described for the SLG gene. A1 and A3 are the primers used to amplify a fragment of 472 base pairs.

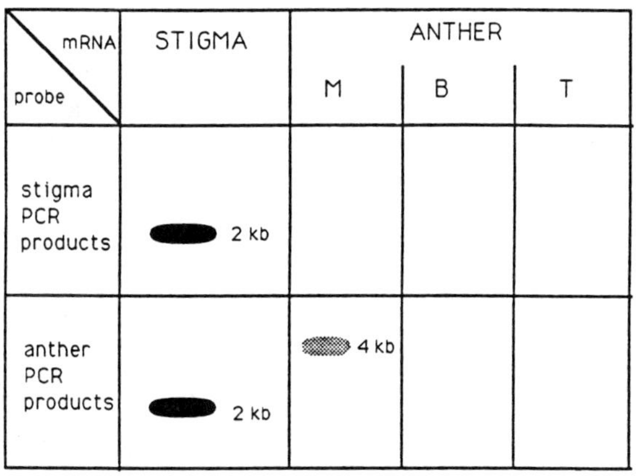

Figure 2 : Schematic representation of hybridization of mRNAs with labelled PCR products from stigma and anther cDNAs.
M : mRNAs from anther at the microspore stage.
B : mRNAs from anther at the bicellulate pollen stage.
T : mRNAs from anther at the tricellulate pollen stage.

and female RNA is quite different (4 kb vs 2 kb), the male S-gene must be homologous but not identical to the female one. This is in agreement with the fact that no cross-hybridization is detected in the anther with the stigma amplified probes, as well as with our previous immunological data (Gaude et al., 1988).

The absence of the 4 kb RNA in anther at a late developmental stage is consistent with the model of sporophytic expression of the male S-gene in the tapetal cells and migration of the corresponding products in the exinic layer of the pollen-coat during the tapetum degeneresence (Heslop-Harrison, 1968). This has been confirmed by the analysis of the expression of a promoter SLG13:GUS fusion in transgenic Arabidopsis thaliana (Toriyama et al., 1991) : GUS activity is detected in the tapetal cells of the anther at the microspore stage and gradually decreases to disappear in buds containing binucleate pollen.

The results obtained in transgenic N. alata with the same fusion (Thorsness et al., 1991) indicates, as proposed by Heslop-Harrison (1968), that the male S-gene is expressed post-meiotically in the haploid pollen grains.

The low quantity of the male S-gene transcripts could indicate a role of signal molecule implied in the pollen-pistil recognition event.

SLG promoter is functional in both male and female tissues. This result indicates that the expression of the gene (s) encoding the male and female S-products share sequence homology or are identical. In that later case, it is possible that through different post-transcriptional processing events, the SLG gene encodes different proteins in the two interacting cell types.

Discussion

Several approaches using different methodologies and yielding complementary results converge to depict the expression of the SLG gene in the anthers of Brassica.

1) The SLG gene is active in both male and female sexual organs, in agreement with the tripartite structure of the S-locus gene predicted by Lewis (1963). The potential products of the coding segment in the anthers have been detected with our PCR amplifications; these data demonstrate that sequences highly homologous to the stigma SLG message are expressed in low amounts in anther tissues.

2) The transformation of Arabidopsis and Nicotiana (Toriyama et al., 1991; Thorsness et al., 1991) with reporter genes controlled by the SLG promoter provides quite different results; it does not concern the S-coding segment but only its promoter region controlling the coding SLG sequence. It is probable that dissection of the large 3.65 kbp fragment upstream of the coding SLG sequence, would allow to further delimit promoter subdomains, each specific of the tapetum at the microspore stage of development or of the stigma surface in mature buds of crucifer flowers (and of pollen grains and style in solanaceae plants). The male S-products would not be identical with those from the stigma, since the two mRNAs differ by their sizes; complete sequencing of this male SLG mRNA will be necessary to ascertain its nature and potential function in pollen-stigma recognition.

A working hypothesis can be formulated about the physiological response of self-incompatible pollen-stigma rejection, supported by the observation of protein phosphorylation events during pollen rejection (C. Franklin, cited in Dickinson, 1990). The activation of a pollen protein-kinase could be involved through dimerization (or tetramerization according to the original hypothesis of Lewis) of its extracellular domain with a stigma SLG glycoprotein. This kinase activation would result in a cascade of metabolic events yielding the formation of callose in the pollen tube and the arrest of its growth.

The present hypothesis is consistent with the demonstration that in Brassica the self-incompatible pollen arrest can be reversed for hours after the beginning of pollination by inhibitors of protein synthesis or glycosylation [Sarker et al., 1988]; This is clearly different from the model proposing the involvement of pollen RNA degradation in the case of the gametophytic self-incompatibility system of Nicotiana alata [McClure et al., 1990], which is probably an irreversible way of pollen rejection.

Acknowledgements

This work was supported by grants from the INRA, from the MRT and from the EEC. Aline Friry was recipient of a CIFRE grant from Orsan.

References

Becker R.S. and Knight K.L,. Somatic diversification of immunoglobulin heavy chain VDJ genes: evidence for somatic gene conversion in rabbits. Cell, 63(1990), 987-997.

Boyes D.C., Chen C-H, Tantikanjana T., Esch J.J and Nasrallah J., Isolation of a second S-locus-related cDNA from Brassica oleracea: Genetic relationships between the S locus and two related loci. Genetics 127(1991) 221-228.

Clarke A.E. and Knox R.B., Plants and immunity. Dev.Immunol., 31(1980), 571-589.1980-

Dickinson H.G., Self-incompatibility in Flowering Plants. BioEssays, 12(1990) 1-7.

Dumas C. and Gaude T., Stigma-pollen recognition and pollen hydration Phytomorphology, 31(1983), 191-201.

Dumas C., Knox R.B. and.Gaude T., Pollen-pistil recognition: New concepts from electron microscopy and cytochemistry. Internat.rev.cytol., 90 (1984), 239-272.

Edelman G.M., CAMs and Igs: Cell adhesion and the evolutionary origins of immunity, Immunological Reviews, 100(1987), 11-45.

Elleman C.J. and Dickinson H.G., Pollen-stigma interactions in Brassica. IV. Structural organization in the pollen grains during hydration. J.Cell Sci., 80(1986),141-157.

Gaude T. and Dumas C. Molecular and cellulart events of self-incompatibility. In"Pollen: Cytology and Development"(K.LGiles and Prakash J.eds), Internat.Rev.Cytol. 107(1987), 33-366, Academic Press, Orlando.

Gaude T. and Dumas C., Molecular aspects of the Self-Incompatibility systems in Brassica and Nicotiana. Botanica Acta. 103(1990), 323-327.

Gaude T.,Fumex B. and Dumas C., Are lectin-like compounds involved in stigma-pollen adhesion and /or recognition in Populus and Brassica? In: "Pollen: Biology and Perspectives for Plant Breeding" (D.L.Mulcahy+.Ottaviano eds), pp.265-272.Elsevier, New York,1983

Gaude T., Nasrallah M.E. and Dumas C., Use of monoclonal antibodies to determine putative S-products of pollen in Brassica oleracea. Heredity 61(1988), 317-318.

Guilluy C-M, Gaude T., Digonnet-Kerhoas C., Chaboud A., Heizmann P. and Dumas C., New data and concepts in Angiosperm fertilization. In "Mechanism of Fertilization" (B. Dale ed.), pp.254-270. Springer-Verlag, Berlin, 1990.

Guilluy C-M., Trick M., Heizmann P.and Dumas C., PCR detection of transcripts homologous to the Self-Incompatibility Gene in anthers of Brassica. Theor. Appl. Genet., (1991) in press.

Heslop-Harrison J., Pollen wall development. Science, 161(1968) 230-237.

Heslop-Harrison J., Incompatibility and the pollen-stigma interaction. Ann.Rev.Plant Physiol.26(1975), 403-425.

Heslop-Harrison J., An interpretation of the hydrodynamics of pollen. Amer.J.Bot.66(1979), 737-743.

Heslop-Harrison J., Knox R.B. and Heslop-Harrison Y. Pollen-wall proteins: Exine-held fractions associated with theincompatibility response in Cruciferae. Theor.Appl.Genet.44(1974), 133-137.

Hodgkin T., Dickinson H.G. and Lyon G.D., Self-incompatibility in Brassica oleracea: a recognition system with characteristics in common with plant-pathogen interactions? In:"Eukaryote Cell Recognition"(G.P.Chapman, C.C.Ainsworth and C.J.Chatham eds),Cambridge Univ.Press, Cambridge," pp.257-274, 1988.

Lalonde B.A., Nasrallah M.E., Dwyer K.G., Chen C-H, Barlow B. and Nasrallah J.B., A highly conserved Brassica gene with homology to the S-locus- specific- glycoprotien structural gene. Plant Cell 1(1989) 249-258.

Lewis D., A protein dimer hypothesis on incompatibility. Proceedings IInd.Int. Congress of Genetics, La Haye (1963). Genetics today, 3(1965) 657-663.

Nettancourt D. De, Incompatibility in Angiosperms. Springer-Verlag, Berlin, pp.1-230, 1977.

Nishio T. and Hinata K., Stigma proteins in self-incompatible Brassica campestris L. and self-compatible relatives, with special reference to S-allele specificity. Japan. J. Genet. 53(1978) 197-205.

Mattsson O., Knox R.B., Heslop-Harrison J. and Heslop-Harrison Y. Protein pellicle of stigmatic papillae as a probable recognition site in incompatibility reactions. Nature, 247(1974), 298-300.

McClure B.A., Gray J.E., Anderson M.A. and Clarke A.E., Self-incompatibility in Nicotiana alata involves degradation of pollen RNA. Nature, 347(1990) 757-760.

Nasrallah J.B. and Nasrallah M.E., Electrophoretic heterogeneity exhibited by the S-allele specific glycoproteins of Brassica. Experientia 40(1984) 279-281.

Nasrallah J.B., Kao T.H., Chen C.H., Goldberg M.L. and Nasrallah M.E.. Amino-acid sequence of glycoproteins encoded by three alleles of the S-locus of Brassica oleracea. Nature 326(1987) 617-619.

Sarker R.H., Elleman C.J. and Dickinson H.G., Control of pollen hydration in Brassica requires continued protein synthesis, and glycosylation is necessary for intraspecific incompatibility. Proc. Natl. Acad. Sci. USA, 85(1988) 4340-4344.

Scutt C.P., Gates P.J., Gatehouse J.A., Boulter D. and Croy R.R.D., A cDNA encoding an S-locus specific glycoprotein from Brassica oleracea plants containing the S5 self-incompatibility allele. Mol. Gen. Genet., 220(1990) 409-413.

Takayama S., Isogai A., Tsukamoto C., Ueda Y., Hinata K., Okazaki K. and Suzuki A., Isolation and some characterization of S-locus-specific glycoproteins associated with self-incompatibility in Brassica campestris. Agric. Biol. Chem. 50(1986) 1365-1367.

Thorsness M.K.,Kandasamy K.K., Nasrallah M.E. and Nasrallah J.B., A Brassica S locus gene promoter targets toxic gene expression and cell death to the pistil and pollen in transgenic Nicotiana. Devel. Biol. 143(1991) 173-184.

Toriyama K., Thorsness M.K., Nasrallah M.E. and Nasrallah J.B., A Brassica S locus gene promoter directs sporophytic expression in the anther tapetum of transgenic Arabidopsis. Develop. Biol., 143(1991) 427-431.

Trick M. and Flavell R.B., A homozygous S genotype of Brassica oleracea expresses two S-like genes. Mol. Gen. Genet., 218(1989) 112-117.

15

Gametophytic Incompatibility in *Petunia Hybrida*: Molecular Aspects of Male Gametophyte and Female Sporophyte Interactions

Lidia Kovaleva

In Petunia hybrida, the incompatibility reaction is mediated by both the haploid genome of the male gametophyte and the diploid genome of the female sporophyte tissues of pistil. Pollen carried to the stigma surface of the pistil produced the tubes that penetrate the stigma and grow through the style to the ovary.In plants with a style like in Petunia the pollen tubes grow between the cells of the central transmitting tract. These cells are arranged in longitudinal files, which are separated by mucilage. The arrest of incompatible pollen tubes occurs within the transmitting tissue of the style (Linakens, 1975). The rejection of incompatible pollen tubes (8 - 16 hours after self-pollination) is accompanied by higher sucrose and fructose but not glucose levels. Hence compatible pollen tubes absorb more carbohydrates from the style tissues than incompatible tubes. Glucose produces callose deposited in the tubes, particularly in the incompatible pollen tubes.

The self-incompatibility mechanism in Petunia develops during flower development and begins to function just before bud opening. After pollination of pistils occurring four, three or two days before bud opening, the seeds set in all cases. The injection of inhibitors into buds resulted in overcoming self-incompatibility and setting of seeds (Kovaleva et al, 1978). The results obtained suggest that specific synthesis of RNA and protein takes place during the formation of self-incompatibility.

We showed that both male and female organs of Petunia have a rich spectrum of antigens, the pollen and stigma surface in particular. S-antigen (S-glycoprotein) was detected in the pistils (only in the style tissues) and we postulated its involvement in the incompatibility mechanism (Kovaleva, 1983). It is shown that the synthesis of S-antigen takes place the day before bud opening and retains its activity during first hours after pollination. Hence location and timing of S-antigen appearance are related to the incompatibility reaction, however its function in the pollen-pistil interactions is yet unclear.

Now the situation is also unclear with another factor of the incompatibility mechanism-isozyme of ribonuclease, revealed also in Petunia pistils (Kovaleva et al, 1975). RNAase activity in the incompatible pistils the day before bud opening was two-fold higher than in compatible pistils. S-RNAase isozyme retains its activity during pollination. The injection of oligomycin resulted in overcoming self-incompatibility and absence of S-RNAase isozyme. These data implicate S-RNAase isozyme in the mechanism of gametophytic self-incompatibility in Petunia. Recently McClure et al.(1989) presented evidence that complementary DNAs for the style S-glycoprotein have homology with two final RNAases. These authors put forward a model in which the gametophytic self-incompatibility system in Nicotiana alata through a cytotoxic mechanism, directed against pollen RNA. In this model, pollen bearing the same S-allele as the style is selectively susceptible to S-RNAases. But the corresponding S-allele product in the pollen have not yet been identified. For this reason some other hypotheses remain actual including ligand-receptor model (Nettancourt, 1977).

The special recognition reaction in the self-incompatibility system seems to be due to the interaction of glycoprotein or proteins of the pollen surface and complementary macromolecules on the cell surface of the stigma or style. Lectins (carbohydrate -binding proteins) and their receptors have been the likely candidates for these recognition macromolecules. The physiological role of lectins in plant cells is yet unclear. The presence of lectins in various tissues of the same plant and the differences in properties of lectins permit to suggest that they may perform various functions in different tissues. The properties of these unique protein macromolecules have led to a number of hypotheses concerning their possible involvement in defence against pathogens, in packaging or mobilization of the storage materials, in cell wall elongation and in cell-cell recognition.

There is experimental evidence that several glycoprotein in Gladiolus pistils bind to the lectin Con A(Knox et al, 1976). We demonstrated that the stigma surface of Petunia has an affinity for the Con A, detected by binding of FITC-ConA to the stigma surface (fluorescence microscopy). It may be concluded that in the case of overcoming Petunia self-incompatibility by treatment with Con A lectin bounds carbohydrate determinants of specific macromolecules (perhaps glycoprotein) taking part in the recognition mechanism of pollen-pistil interactions. To study the location of lectin in reproductive organs of Petunia we used the immune-chemical analysis with Con A -antiserum and poly specific anti-pollen and anti-pistil antisera. The lectin is revealed in different Petunia reproductive organs: in both female and male tissues of flower before pollination. Moreover Con A- activity was detected in pistils during growth of the pollen tubes independently on the pattern of pollination (compatible or incompatible). The assumed functions of this lectin can not be regarded as convincing. In Petunia reproductive elements Con A seems to participate in stored carbohydrate utilization in the course of flower growth and pollination.

Our preliminary investigations have demonstrated that pollen and pistils of Petunia contain both soluble and membrane-bound lectins. The agglutinating activity of pistil lectins during incompatible pollination differs as compared to compatible pollination. A further approach to the elucidation of glycoprotein-lectin interactions in pollen-pistil system is to establish details of the lectin specified and their receptors.

Finally, I would like to attract the attention to the signal transduction. Evidence for the existence of signalling between the pollen and pistil of Petunia were provided: 1) by the stimulation of protein synthesis within ovules following pollination (Deurenberg, 1976), 2) by electrical potential in self- and cross-pollinated pistils (Linskens and Spanjers, 1973), 3) by changes in Ca++ concentration during compatible and incompatible pollination (Kovaleva, unpublished results). The chemical composition of the signal molecules is unknown. Some indications point to the possible role of cell wall oligosaccharide in eliciting pollen-pistil interactions (Lewis, 1980). A similar suggestion may be put forward concerning oligosaccharide which influence the reproductive morphogenesis.

References

Deurenberg J.J.M(1976) Activation of protein synthesis in ovaries from Petunia hybrida after compatible and incompatible pollination. Acta Bot Neerl, 25: 221-226.

Knox R.B, Clarke A.E., Harrison S., Smith P., Marchalonis J.J. (1976) Cell recognition in plants: determinants of the stigma surface and their pollen interactions. Proc. Natl. Acad. Sci. USA, 73: 2788-2792.

Kovaleva L.B. (1983) Incompatibility protein factor in Petunia hybrida. Dokl. AN SSSR, 272:1017-1020.

Kovaleva L.V., Milyaeva E.L., Chailakhyan M.Kh. (1978) Overcoming self-incompatibility by inhibitors of nucleic acid and protein metabolism. Phytomorphology 28:445-449.

Kovaleva L.V., Musatova N.A.(1975) RNAase activity in Petunia pistils. Dokl. SSSR 223:762-765.

Lewis D.H.(1980) Are there inter-relations between the metabolic role of boron, synthesis of phenolic phytoalexins and the germination of pollen? New Phytol 84:261-270.

Linskens H.F., Spanjers A.W.(1973) Changes of the electrical potential in the transmitting tissue of Petunia styles after cross- and self-pollination. Incompatibility News, 3:81-85.

McClure B.A., Haring V., Ebert P.R., Anderson M.A., Simpson R.J.,Sakiyama F., Clarke A.E. (1989) Style self-incompatibility gene products of Nicotiana alata are ribonucleases. Nature, 342: 955-957.

Nettancourt D de (1977) Incompatibility in angiosperms. Springer. Berlin.

16
Self-incompatibility in *Antirrhinum*

Andrew McCubbin, Rosemary Carpenter, Enrico Coen, and Hugh Dickinson

Introduction

Self-incompatibility (SI) systems have been reported in almost half of the families of all flowering plants (de Nettancourt, 1977). Most commonly, SI is regulated by a single multiallelic locus, with the compatibility of the pollen with respect to the stigma being controlled by the haploid nucleus (East, 1940). The genus Antirrhinum possesses such a gametophytically-regulated monofactorial SI system and, when pollen and pistil share SI (S) alleles, tube growth is arrested in the style. The significance of SI systems to plant breeders, combined with the fact that they represent some of the best defined systems of intercellular interaction in plants has resulted in their being intensively studied (Anderson et al., 1986; Nasrallah et al., 1985). For example, molecular and biochemical research has identified the female S-allele products of members of the Solanaceae as being highly charged glycoproteins with molecular weights of between 20-35 kDa (for review, see Ebert et al., 1989). Many putative S-alleles have been cloned and sequenced, and it has been recently discovered that all of the S-glycoproteins in the Solanaceae have ribonuclease activity (McClure et al., 1990), leading to the proposal that this type of SI may involve a step in which pollen tube RNA is selectively degraded. Although progress has been made in our understanding of the S-allele products of the pistil in gametophytically-controlled SI (GSI), and also in the sporophytically-controlled systems where pollen compatibility is regulated by two alleles (Bateman, 1952; Nasrallah et al., 1970), little is known of the organisation of the S-locus and of S-allele expression in the pollen.

There are two features which render the genus Antirrhinum well suited for the study of GSI. First, it is a member of the Scrophulariaceae, a family distinct from, but closely related to the Solanaceae. Secondly, and most importantly, species of Antirrhinum possess well-defined families of active transposable elements (Tns) (for review, see Coen and Carpenter, 1986). Tn mutagenesis may thus be used to generate mutants in all domains of the S-locus, including that responsible for gene expression in the pollen. All the strategies so far employed in the study of SI have relied on the cloning cDNA from stigmatic message (Nasrallah et al., 1985; Anderson et al., 1986); thus the only route to the identification of male-expressed genes has been through the study of promoter sequences (Toryama et al., 1991; Thornsness et al., 1991). The work reported here represents the first phase of a major project intended to generate a range of self-compatible (SC) Tn mutants, in which the different domains of the S-locus are affected in isolation. Such plants would not only provide information on S-allele expression, but also on the general structure of the S-locus itself. Since the lines of Antirrhinum majua carrying well-characterised Tns are SC, the component of the research reported here has been focused on (i) generating families of Antirrhinum hybrids exhibiting SI, (ii) examining the inheritance of this trait and identifying putative S-allele products in the pistil, and (iii) carrying out a large Tn mutagenesis experiment using the plants generated in (i).

S-allele segregation

A number of populations of hybrids between SC A. majus and SI A. hispanicum were generated to study S-allele segregation and to provide material for the identification of putative S-allele products. Initially, the work centred on the alleles S_3S_4 (arbitrary nomenclature, since S-alleles have not been characterised in A. hispanicum) which were introduced from a single A. hispanicum plant. This individual was crossed with A. majus (S_FS_F) to generate an F_1. F_1 plants were then intercrossed and a selection of F_2 populations grown. As suspected, SI progeny were only detected in a proportion of these F_2 populations and it was consequently presumed that where SI progeny were generated, the F_1 parents possessed different S-alleles; conversely, if no SI progeny were detected, the F_1 genotypes were assumed to carry identical alleles. However, only 10% of the F_2 were found to be SI (as opposed to the 25% expected); furthermore, the majority of these were backcross compatible with the original A. hispanicum parent, although they should have carried identical alleles. This phenomenon was investigated further through a comprehensive range of pollinations between the S_3S_4 A. hispanicum, the two F_1 genotypes generated from it, the SI F_2 progeny, and the progeny generated from these crosses. From this study it emerged that the S_4 allele was highly unstable, apparently giving rise to an allele with a new specificity which was consistent within - but not between - populations of almost every generation. Some 10% of the progeny, however, retained the S_4 specificity. This finding is of considerable interest and the mechanism responsible may well have resulted in the rather confused picture of SI inheritance in Antirrhinum found in the literature (Brieger, 1935). The results obtained may have been generated by any one of a number of mechanisms, ranging from strong background effects to the activation of the natural mechanism of S-allele generation by factors derived from the actively transposing lines of A. majus used to generate the hybrids.

These rather confusing results indicated that S-allele segregation would have to be investigated by another route. This was achieved by crossing F_1 genotypes with SI hybrids carrying different alleles. In progeny generated in this way all SI genotypes and no SC genotypes should carry the allele possessed by the F_1 plant. Thus populations of plants could be generated to study the segregation of putative S-gene products with their respective alleles.

Putative stylar S-allele products

The breeding programme described above generated populations of plants containing S-alleles in the combinations of S_3S_F and S_4S_F (S_3 and S_4 being derived from A. hispanicum, and the S_F allele from A. majus). When stylar extracts from these plants were analysed using isoelectric focusing (IEF) the only consistent difference between the two protein spectra was a polypeptide with an isoelectric point (pI) of 10 which appears to be unique to the S_3S_F genotypes. To determine the point in flower development when this polypeptide first appears, and to discover where in the pistil it is located, proteins were isolated from pistils at different stages of development, and from different portions

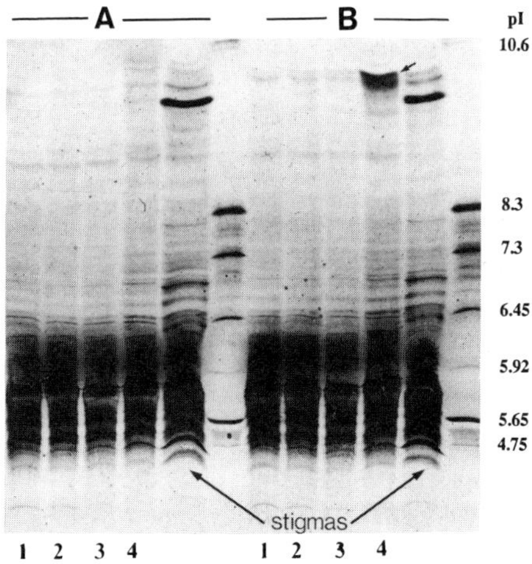

Figure 1. Isoelectric focusing of aqueous stylar extracts of genotype S_3S_F (lanes A) and S_4S_F (lanes B) polypeptides isolated from different regions of the pistil, extending from the base to the top of the style (1-4) and also from stigmas, as indicated.

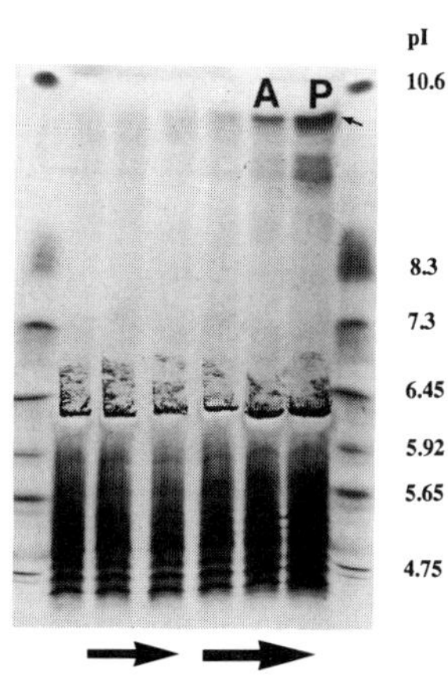

Figure 2. IEF of aqueous extracts of stylar polypeptides during development. Buds were sampled at one day intervals (in direction of large arrows) up to anthesis (A), and post-anthesis (P). A predominant polypeptide band becomes visible in these extracts following anthesis (small arrow).

of mature pistils. IEF of extracts from developing pistils indicate this protein to increase dramatically in abundance at anthesis (Fig. 1). Similar analysis of different parts of the pistil suggest that it is located principally at the top part of the style, just beneath the stigma (Fig. 2). This polypeptide therefore shares many characteristics with known S-allele products from members of the Solanaceae (Mau et al., 1986).

Analysis of the same extracts using SDS-Page showed similar spatial and developmental profiles in a protein of 27 kDa for the S_3S_F allele combination, and a polypeptide of 24 kDa for the S_4S_F alleles. Interestingly, all the F_1 genotypes generated in this programme (arbitrarily classified as S_{1-6}) show a characteristic polypeptide in the 18-27 kDa range. All these polypeptides have strong affinity for the lectin

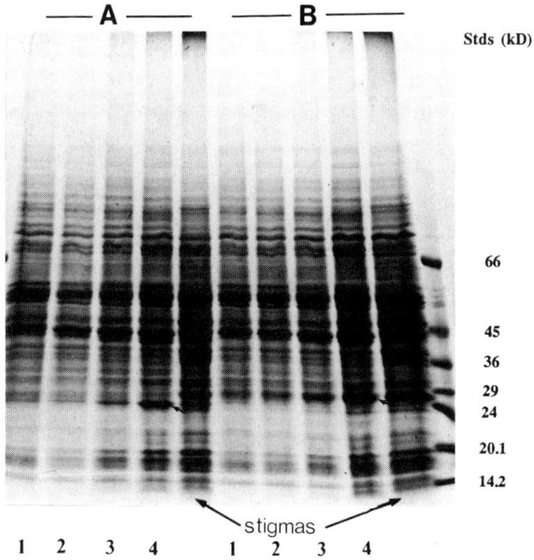

Figure 3. Polypeptides from aqueous extracts of styles of S_3S_F (lanes A) and S_4 and S_F (lanes B) genotypes separated on SDS-PAGE (5-15%). The individual lanes are as shown for the IEF analysis in Figure 1. Putative S-allele product polypeptides are indicated by small arrows in lane 4 of each genotype.

Concanavalin A (as determined by Concanavalin A-peroxidase localisation) and hence are most probably glycoproteins. Further analysis of the putative S_3 S-locus glycoprotein using 2D electrophoresis has demonstrated the bands characteristic of both IEF and SDS-PAGE separations to represent the same polypeptide. It is noteworthy that none of the other putative S-glycoproteins are detectable after IEF, which is a characteristic shared with many of the putative S-locus glycoproteins in the Solanaceae (Cornish et al., 1987). Current studies are focused on determining whether these polypeptides possess ribonuclease activity.

The generation of plants for transposon mutagenesis

An extensive breeding programme was undertaken with the aim of generating suitable lines for Tn tagging. In addition to the production of SI lines, containing Tns from A. majus, it was considered important to establish whether the Tns were capable of active transposition in these plants. For this reason a number of marker alleles, Niv^{rec-98}, Pal^{rec-2} (alleles affecting anthocyanin biosynthesis) and Olive (a mutation affecting chlorophyll biosynthesis) were introduced into the A. hispanicum x A. majus hybrid lines. Plants segregating both for SI and the recessive marker alleles were thus selected in the F_2 generation. Not surprisingly, considerable variation was discovered in the rates of transposition within this F_2 progeny, and only genotypes with high transposition frequency were chosen for the subsequent experiments.

The Tn mutagenesis experiment itself involved growing a range of S-genotypes at 15°C from prefloral initiation to optimise transposition. These plants were intercrossed.

Subsequently, the seed was collected and 6500 progeny were grown at the John Innes Centre for Plant Science Research over the summer of 1990. Once mature, the plants were self-pollinated and screened for SC. 77 individuals showed some sign of breakdown in SI, and 16 of these have been chosen for further studies. Analysis of the parentage of these 16 plants suggests that the effects observed could have resulted from as few as 7 transposition events. Genetical and molecular analysis of these plants is currently under way.

Acknowledgements

This work was funded by the IK Agricultural and Food Research Council, and by the Gatsby Foundation. The authors wish to thank Valerie Norris for assistance in preparation of the manuscript.

References

Anderson, M.A., Hoggart, R.M. and Clarke, A.E. (1983). The possible role of lectins in mediating cell-cell interactions. In: Goldstein, R.J. and Etzler, M. (eds.) Chemical Taxonomy. Molecular Biology and Function of Plant Lectins. Alan R. Liss, pp. 143-161.

Anderson, M.A., Cornish, E.C., Mau, S-L., Williams, E.G., Hoggart, R., Atkinson, A., Bonig, I., Grego B., Simpson, R., Roche, P.J., Haley, J.D., Niall, H.D., Treager, G.W., Coghlan, J.P., Crawford, R.J. and Clarke, A.E. (1986). Cloning of cDNA for a stylar glycoprotein associated with expression of self-incompatibility in Nicotiana alata. Nature **321**, 38-44.

Bateman, A.J. (1952) Self-incompatibility in angiosperms I. Theory. Heredity **6**, 285-310.

Brieger, F.G. (1955). The inheritance of self-sterility and peloric flower shape in Antirrhinum. Genetica **17**, 385-408.

Coen, E. and Carpenter, R. (1986). Transposable elements in Antirrhimum majus generators of genetic diversity. TIG Nov. 292-296.

Cornish, E.C., Pettitt, J.M., Bonig, I. and Clarke, A.E. (1987). Developmentally-controlled, tissue-specific expression of a gene associated with self-incompatibility in Nicotiana alata. Nature **326**, 99-102.

East, E.M. (1940). The distribution of self-sterility in the flowering plants. Proc. Am. Phil. Soc. **82**, 449-518.

Ebert, P.R., Anderson, M.A., Bernatsky, R., Altschuler, M. and Clarke, A.E. (1989). Genetic polymorphism of SI in flowering plants. Cell **56**, 255-262.

Mau, S-L., Williams, E.G., Atkinson, M.A., Cornish, E.C., Grego, B., Simpson, R.J., Kyeur-Pour, A. and Clarke, A.E. (1986). Style proteins of a wild tomato (Lycopersicon peruvianum) associated with expression of self-incompatibility. Planta **169**, 184-191.

McClure, B.A., Gray, J.E., Anderson, M.A. and Clarke, A.E. (1990). SI in Nicotiana alata involves degradation of pollen rRNA. Nature **347**, 757-760.

Nasrallah, J.B., Kao, T.H., Goldbert, M.L. and Nasrallah, M.E. (1985). A cDNA clone encoding an S-locus specific glycoprotein from Brassica oleracea. Nature **318**, 263-267.

Nasrallah, M.E., Barber, J.T. and Wallace, D.M. (1970). Self-incompatibility proteins in plants detection, genetics and possible mode of action. Heredity **25**, 23-27.

Nasrallah, M.E. and Nasrallah, J.B. (1986). Molecular biology of self-incompatibility in plants. Trends Genet. **2**, 239-244.

de Nettancourt, D. (1977). Incompatibility in angiosperms. Springer-Verlag, Berlin, Heidelberg and New York.

Thorsness, M.K., Kandasamy, M.K., Nasrallah, M.E. and Nasrallah, J.B. (1991). The Brassica S-locus gene promoter targets toxic gene expression and cell death to the pistil and pollen of transgenic Nicotiana. Dev. Biol. **143**, 173-184.

Toriyama, K., Thorsness, M.K., Nasrallah, J.B. and Nasrallah, M.E. (1991). A Brassica S-locus promoter directs sporophytic expression in the anther and tapetum of transgenic Arabidopsis. Dev. Biol. **143**, 427-431.

17

Cross-incompatibility in Maize – A Case of a 3 Locus Interaction by Non-allelic Male and Female Genes

Peter A. Peterson and Abdul Rashid

Introduction

Sexual incompatibility represents one class of mutants affecting the male gametophyte. This rejection of viable pollen by a fertile pistil takes many forms. It could reject its own pollen or foreign pollen with specific genotypes. This is a very common phenomena in flowering plants and is known in about 1/2 of the total angiosperms species (Darlinton and Mather, 1949; Brewbaker, 1959). The case for cross-incompatibility involving interspecific and intergeneric crosses has been adequately reported in a number of species (Sarker, Elleman and Dickinson, 1988; Darnell and Lyrcna, 1989). Some cases of cross-incompatibility are unilateral in nature, for example, (Boyle and Stinart, 1986). In these cases the crosses are compatible in one direction but the reciprocal were incompatible.

Such unidirectional incompatibility in maize was first reported by Demerec (1929). Further efforts in the study of the *ga1* were investigated by both Schwartz (1950) and Nelson (1952) to add to the cases of unilateral cross-incompatibility.

A similar case of unidirectional cross-incompatibility was observed in our maize cytogenetics nursery (Sukhapinda and Peterson, 1983). In this case, there is a specific female that when crossed by a specific male, the resulting ears show reduced seed setting (RSS). These data supporting this cross-incompatibility will be presented.

Results

Genetic Analysis

RSS female cytoplasm

This unilateral incompatibility has been identified as a reduced seed setting (RSS). To test the possibility that the incompatibility was due to cytoplasm, the RSS female lines were used in crosses as a male parent. The choice of the RSS male line as the female parent in the plant crossing scheme was based on the fact that this cross is compatible (Sukhapinda, 1981) and if cytoplasmic, would not be transmitted to the progeny. Further the RSS male lines cytoplasms are different from the RSS female lines cytoplasm. In the cross, the *F2* plants were grown and tested for the presence of cross-incompatibility action by pollinating with the pollen from the RSS male lines. The expectation was that if this incompatibility was caused by cytoplasmic factors, then the RSS factor(s) will not be present in the *F2* plants and their crosses with the RSS male lines will not show the RSS ears. The results showed that of the 26 plants tested, 6 showed the RSS phenotype. The cross-incompatibility expectation is that with cytoplasmic control, none should show it. Thus, this result demonstrated that the factors causing this unidirectional cross-incompatibility were transmitted to the progeny through the RSS female line that was used as a male parent. Thus the RSS control is not cytoplasmic. This case of unilateral cross-incompatibility has recently been dissected and determined (Rashid and Peterson, 1991a,b).

Genetic control: In the RSS female

By eliminating cytoplasmic control as an option, the next question is to determine the number of genes responsible for this incompatibility reaction in the female parent.

In all these strategies, the segregating genotype is tested by a known female or male incompatibility genotype. The RSS female line was crossed as a male to a RSS male line for transformation of this female line. It is known that the RSS male line is not incompatible when crossed by any other RSS male line. Thus in the *F2* and in the backcross, the identification of NSS vs. RSS could be determined. Of the 26 plants observed in the cross, 6 were RSS. This is close to the expectation of a single pair of segregating alleles. In the backcross of 54 plants, 28 were RSS which agrees with 1/2 expected from the heterozygote. Thus, from this series of crosses it is quite evident that the unidirectional cross-incompatibility reaction of RSS in the female line is controlled by a single recessive gene and it is evident that in its presence in the homozygous recessive form, will make that line cross-incompatible as a female parent when used against the RSS male line. This female gene is now designated *cif* (cross incompatible female).

Genetic control: In the RSS male

A similar strategy was used for the determination of the genetic control in the RSS male. The same plants (RSS male as female) were crossed but in this case by the RSS female line. The pollen of the *F2* and *Bc1* plants was then tested on the RSS female plants. The resulting segregation of NSS and RSS show the incompatibility reaction recessive to the compatibility reaction. This was previously shown by Sukhapinda (1981). It is evident that in our tests of the 215 plants in the *F2* population, 16 showed RSS in tests on the female RSS line. In the backcross population, there were 53 plants and 17 showed the RSS phenotype in this test. These results according to these segregating ratios indicate that two genes are controlling the incompatibility reaction in the male side. The expectation for two genes segregating among the 215 would be 13.44 for the *F2* and the expectation in the backcross would be 13.25. These results agree with the expectation. The results of these crosses indicate that two genes are then associated with the RSS male parent and are designated as *cim1* (cross incompatible male) and *cim2*. The wildtype form is designated $+^{cim1}$ $+^{cm2}$.

It is obvious that the imcompatible cross is not complete. There are some seeds produced and these could be considered "escapes". In order to determine the genetic nature of the escaped seed arising in the cross of the RSS female x RSS male, an *F2* population of each of the escaped seeds was obtained. Each of the *F2* plants originating from these escaped seed was crossed onto a RSS female and by a RSS male. The determination of this cross was to distinguish whether the alleles of RSS male incompatible reaction have mutated or is this due to contamination. Thus in the cross of the RSS female by the pollen of the *F2* plants, if the same genes are present as in the original cross, there will be a segregation of RSS and NSS in the progeny. On the other hand, namely, using the *F2* as a female by the RSS male, it could be determined if the RSS female incompatibility reaction has mutated and if it is not mutated, then there will a segregation of RSS and NSS. In the six escaped seed tested in this *F2* population, one of the six could be determined to be the result of accidental selfing. The other five seeds tested show the presence of both RSS male and RSS female genes in their *F2* progenies. The results from tests of these latter five seeds indicates that neither the RSS male nor the RSS female genes had mutated.

Time and place of the compatibility reaction

It was previously determined by Bdliya (1984) that the incompatible pollination does not adversely effect the pistil physiology and that normal seed setting can be obtained with the subsequent pollination with compatible pollen. This would indicate that the incompatibility reaction occurs before the fertilization of the ovules. Thus, the incompatibility might take place at some time before fertilization.

Using scanning electron microscopy it was clear that the incompatibility pollen grains do germinate on the incompatible silks. Following germination, the pollen tubes do grow down to the silk and penetrate the silk through the silk hairs.

However, inside the silk, the growth of the pollen tube of the RSS male on RSS female shows a decided difference from that of the normal pollen tube growth. At six hours, the pollen tubes have already passed through the silk hairs, both in the compatible and in the incompatible crosses. At 12 hours, the pollen tubes have reached into the transmitting tracks of the silk body and are growing towards the ovary. However, the difference appears here in that in the incompatible crosses, the pollen tubes had swollen tips. At 18 hours after pollination, the compatible crosses were still growing in the transmitting tracks of the silk tissues. In the incompatible cross, the pollen tube growth appears to be arrested and the tips of the swollen tubes were considerably swollen.

Gametophytic vs Sporophytic

The next question to be determined is whether the control is gametophytic or sporophytic. The determination of this question is identified with a cross of the RSS male (♀) by a RSS female. These genotypes are then crossed to an RSS female. The resulting genotypes are crossed onto an RSS male. The genotypes present if the control is gametophytic are

$$\frac{+^{cim1}}{+^{cim1}} \qquad \frac{+^{cim2}}{+^{cim2}}$$

and if sporophytic, are

$$\frac{+^{cim1}}{+^{cim1}/cim1} \qquad \frac{+^{cim2}}{+^{cim2}/cm2}$$

To distinguish the different genotypes, the cross of these progeny is made to the RSS female. The results of this cross should yield all NSS if gametophytic and, if it is sporophytic, some progenies, approximately 1/4, will show the RSS. This strategy is being followed to determine the options between gametophytic and sporophytic.

Discussion

It has been presented that this unilateral cross-incompatibility in maize is controlled by three pairs of alleles. There is one in the female side, that is, *cif*, and two on the male side, *cim1* and *cim2*. In this case, the incompatibility reaction is controlled not by the same but different genes. This is in contrast to others (Lewis and Crowe, 1958; Keulmans, 1984; Boyle and Stinart, 1986) where the same locus controls self-incompatibility.

Unlike the previously described unidirectional incompatibility in maize (Nelson, 1952; Ashman, 1975; Kermicle and Allen, 1990), the RSS system is more closely related to the general incompatibility system found in the Gramineae in which, generally, more than one locus controls the incompatibility reaction. These genes are non-allelic and segregate independently of each other.

There are numerous studies supporting the thesis that the incompatibility reaction need not be controlled by one locus but by two or more loci. The Clark *et al.*, (1990), Xu *et al.*, (1990), and the Moore and Nasrallah (1990) studies indicate that pollen and pistil control of incompatibility may likely not be ascribed to the same loci.

The model that best supports this three-locus-RSS unidirectional cross-incompatibility involves female encoded glycoproteins which have ribonuclease activity (McClure *et al.*, 1990) that degrade rRNAs in the pollen or pollen tubes. The pollen genes would encode transmembrane receptor proteins, located in the pollen or pollen tube walls which would regulate the reaction by having bonding specificities for the glycoproteins. By regulating the intake of these glycoproteins into the pollen or pollen tubes, the incompatibility reaction becomes allele specific. Thus, in this RSS unidirectional cross-incompatibility case in our study, the *cif* locus produces a glycoprotein in the silk and the *cim1* and *cim2* male loci are expressed in the pollen. The active receptor in the model proposed would be found from a dimer of the two male proteins. Thus, the *cif* encoded glycoprotein cannot bind to the receptor of the hybrid protein or that of the $+^{cim1}$ and $+^{cim2}$ wildtype. This latter genetic composition would result in NSS.

1. Ashman, R. B. Modification of cross-sterility in maize. J. Hered. **66**:5-9; 1975.
2. Bdliya, P. M. Transmission genetics of the incompatibility factors in nonreciprocal, partial cross-incompatible maize. M.S. Thesis. Iowa State Univ., Ames, Iowa; 1984.
3. Boyle, T. H.; Stimart, D. P. Self-incompatibility and interspecific incompatibility: relationships in intra- and interspecific crosses of *Zinnia elegans* Jacq. and *Z. Angustifolia* HBK (Compositate). Theor. Appl. Genet. **73(2)**:305-315; 1986.
4. Brewbaker, J. L. Biology of the angiosperm pollen grain. Ind. J. Genet. Plant Breed. **19**:121-133; 1959.
5. Clark, K. T.; Okuley, J. J.; Collins, P. D.; Sims, T. L. Sequence variability and developmental expression of *S*-alleles in self-incompatible and pseudo-self-compatible *Petunia*. Plant Cell **2**:815-826; 1990.
6. Darlington, C. D.; Mather, K. The elements of genetics. George Allen and Unwin Ltd., London. 1949.
7. Darnell, R. L.; Lyrene, P. M. Cross-incompatibility of two related rabbiteye blueberry cultivars. HortScience **24(6)**:1017-1018; 1989.
8. Demerec, M. Cross-sterility in maize. Z.I.A.V. **50**:281-291; 1929.
9. Kermicle, J. L.; Allen, J. O. Cross-incompatibility between maize and teosinte. Maydica **35(4)**:399-408; 1990.
10. Keulemans, J. Results of five years of pollination experiments on sweet cherries. Acta. Hortic. **149**:103-108; 1984.
11. Lewis, D.; Crowe, L. K. Unilateral inter-specific incompatibility in flowering plants. Hered. **12**:233-256; 1958.
12. McClure, B. A.; Gray, J. E.; Anderson, M. A.; Clarke, A. E. Self-incompatibility in *Nicotian alata* involves degradation of pollen rRNA. Nature **347**:757-760; 1990.
13. Moore, H. M.; Nasrallah, J. B. A brassica self-incompatibility gene is expressed in the stylar transmitting tissue of transgenic tobacco. Plant Cell **2**:29-38; 1990.
14. Nelson, O. E. Non-reciprocal cross-sterility in maize. Genet. **37**:101-124; 1952.
15. Rashid, A.; Peterson, P. A. The RSS system of unidirectional cross-incompatibility in maize: 1. Genetics. J. Heredity (manuscript submitted); 1991a.
16. Rashid, A.; Peterson, P.A. The RSS system of unidirectional cross-incompatibility in maize. 2. Cytology. Genome (manuscript submitted); 1991b.
17. Sarker, R. H.; Elleman, C. J.; Dickinson, H. G. Control of pollen hydration in Brassica requires continued protein synthesis, and glycoylation is necessary for intraspecific incompatibility. Proc. Natl. Acad. Sci. (USA) **85**:4340-4344; 1988.

18. Schwartz, D. The analysis of a case of cross-sterility in maize. Proc. Nat. Acad. Sci. **36**:719-724; 1950.
19. Sukhapinda, K. Nonreciprocal partial cross-incompatibility in maize. Ph.D. Thesis. Iowa State University, Ames, Iowa; 1981.
20. Sukhapinda, K.; Peterson, P. A. Nonreciprocal partial cross-incompatibility in maize. Can. J. Genet. Cytol. **25**(3):270-277; 1983.
21. Xu, B.; Grun, P.; Kheyr-Pour, A.; Kao, T. H. Identification of pistil specific proteins associated with three self-incompatibility alleles in *Solanum chacoens*. Sex Plant Reprod. **3**:54-60; 1990.

18 Effects of Style Age on the Performance of Self and Cross Pollen in *Campanula rapunculoides*

A.G. Stephenson, J.A. Winsor, T.E. Richardson, A. Singh, and T.-H. Kao

Self-Incompatibility (S-I) involves a complex set of interactions between the microgametophyte and the pistil resulting in the recognition and inhibition of self pollen. Because S-I is regulated by the action of one or more S-genes, plant scientists have tended to think of self-incompatibility as a qualitative trait: i.e., a species is either self-compatible or it possesses an S-gene and is self-incompatible. In reality, however, S-I is often a quantitative trait. Moreover, the degree of S-I is known, at least in some species, to vary among specific S-alleles, with the composition of the pollen load (pure self-pollen vs. mixtures of self and cross pollen), with external environmental conditions such as temperature, and with internal stylar conditions such as the age of the flower (e.g. Jenkins, 1931; Ascher and Peloquin, 1966; Dane and Melton, 1973; Fearon et al., 1983; Bowman, 1987; and Knox et al., 1987).

Preliminary studies of a natural population of Campanula rapunculoides (Campanulaceae) by our group have indicated that some individuals become more self-fertile as their flowers age. Consequently, we began to investigate changes in self and cross pollen tube growth, seed set, and style protein composition in young and old pistils of C. rapunculoides. In regard to the protein composition of the styles, we were particularly interested in determining if C. rapunculoides produces S-locus-specific glycoproteins similar to those implicated in the inhibition of self-pollen in the Solanaceae (see Haring et al., 1990; Kao, this book), and, if so, in determining if the concentration of the S-specific glycoprotein changes with floral age.

Methods and Results

Campanula rapunculoides is a naturalized perennial herb that is locally abundant along roadsides and wasteplaces throughout much of the northeastern United States. It perennates by means of rhizomatous rootstocks that produce rosettes. At the time of flowering each rosette sends up a shoot that produces a racemose inflorescence of 10-40 protandrous flowers that mature acropetally. The blue bell-shaped flowers are bumblebee pollinated. Peak flowering occurs in July and in early autumn the mature fruits dehisce and scatter the seeds. The fruits from cross pollinated flowers contain approximately 45 seeds (for details see Richardson and Stephenson, 1989; Richardson et al, 1990).

To examine the effects of floral age on cross and self pollen tube growth, 11 plants were selected from a natural population of C. rapunculoides growing in Centre County, Pa. We covered the entire inflorescence on these plants with cheesecloth bags to exclude pollinators and we randomly assigned 3 flowers on each inflorescence to each of 4 pollination treatments: A) outcrossed pollination at the onset of stigma receptively (day 1), B) outcrossed pollination 4 days after the onset of stigma receptively (day 4), C) self pollination on day 1, D) self-pollination on day 4. We then collected one pistil from each pollination treatment on each plant at 6, 12 and 24 hours after the pollination and examined pollen tube growth using fluorescent microscopy (Martin, 1959). Pollen tube growth was scored from 1-5, where 1 = no tubes penetrating the style and 5 = many tubes having grown completely through the style.

An analysis of variance revealed that pollen tube length increases for both pollination treatments from 6 to 24 hours after pollination (Table 1). More importantly, the analysis revealed that cross pollen tubes grow significantly faster through 1 day old styles than do self pollen tubes but there are no significant differences between the growth of cross and self pollen tubes in 4 day old styles (Table 1). Together these results indicate that the styles of older flowers lose their ability to retard the growth of self pollen tubes.

In order to examine the effects of floral age on seed set following self pollination, the rosettes (clones) of 14 plants (genotypes) from our natural population were transplanted into pots (one rosette per pot), given a cold treatment (4°C for 90 days) and then placed into a greenhouse. The flowers from each rosette were then assigned to one of four treatments: A) stigma and style harvested and frozen on day 1, B) stigma and style harvested and frozen on day 4, C) self-pollinated on day 1, or D) self-pollinated on day 4. The stigmas and styles were used for the molecular analyses discussed below. The mature fruits resulting from the self-pollinations on each genotype were harvested; the fruits from the 1 and 4 day old pistils from each genotype were stored in separate envelopes; and, later, the seeds were counted. Because some of the fruits in each envelope dehisced their seeds, mean seed number per fruit was determined by dividing the total number of seeds produced by a pollination treatment on each plant by the total fruits produced by the pollination treatment.

Table 1. Analysis of variance for self and cross pollen tube growth (Pollination treatment) at 6, 12 and 24 hours after pollination (time).

SOURCE OF VARIATION	df	TYPE III SS	F	P
1 DAY OLD PISTILS				
Time	2	31.25	18.3	0.0001
Pollination Treatment	1	5.38	6.3	0.016
Plant	10	9.71	1.1	0.357
4 DAY OLD PISTILS				
Time	2	8.95	3.68	0.040
Pollination Treatment	1	0.01	0.01	0.929
Plant*	7	30.22	3.55	0.009

*Complete data sets were not obtained for 3 plants

A chi-square test reveals that the probability of producing a mature fruit is not independent of the age of the flower at the time of self-pollination (Table 2). The 4 day old pistils are significantly more likely (59% vs. 48%) to produce a mature fruit than are 1 day old pistils. In addition, the mature fruits from the 4 day old pistils produce significantly more seeds per fruit than do the 1 day old pistils. When both fruit production and seed number per fruit are considered together, each self-pollination on a 4 day old pistil produces almost 2.5X the number of seeds as a self-pollination on a 1 day old pistil. Moreover, self-pollinations of 1 day old pistils in a greenhouse probably overestimate seed production from self-pollinations of 1 day old pistils under field condition. Under field conditions the removal of pollen by bees during the male phase of the protandrous flowers hastens the onset of the female phase (Richardson and Stephenson, 1989). In the greenhouse, pollen is not removed from the flowers and the male phase persists for 2-3 days longer than in the field. In short, a 1 day old pistil in the field is younger than a 1 day old pistil in the greenhouse.

In order to determine if C. rapunculoides produces S-locus specific glycoproteins such as those implicated in the inhibition of self pollen in the Solanaceae, we gathered three types of evidence. First, we collected the stigmas and styles from 1 and 4 day old pistils, leaves and flower buds from the greenhouse plants and ground them separately. We then subjected the extract to SDS-polyacrylamide electrophoresis as described by Ai et al. (1990). We found an abundant protein (molecular weight approximately 32kD) in both 1 and 4 day old pistils, but not in extracts from leaf and bud samples. The molecular weight, the relative abundance and the temporal and spartial distribution of this protein is similar to those reported for S-locus specific glycoproteins in the Solanaceae (see Haring et al., 1990; Kao, this book).

Table 2. Fruit and Seed Production from Self-Pollinations on One and Four Day Old Pistils of *C. rapunculoides*.

Age of Pistil	Number of Pollinations	Number of Mature Fruits(%)[a]	Number of Plant	Mean Seed Per Fruit Per Plant[b]
One Day Old Pistils	356	171 (48%)	13*	25.9±3.7
Four Day Old Pistils	334	196 (59%)	14	49.5±9.4

*One plant did not produce and mature fruits

[a]Mature vs. Failed Pollinations: chi-square = 7.84; df = 1; $p<0.01$

[b]$t=2.34$; $df=25$; $p<0.025$

Second, in order to determine if any of the stylar proteins from C. rapunculoides cross-reacts with antibodies to Petunia S-locus specific glycoproteins, we separated the proteins from 1 and 4 day old pistils on SDS polyacrylamide gel (see above), transferred the proteins to a nitrocellulose membrane and produced a conjugated antibody that we stained with 3,3'-diaminobenzidine (DAB). We found that the conjugated antibody corresponding to the 32kD protein reacted strongly with the stain.

Third, recent studies have revealed that the S-locus specific glycoproteins of Nicotiana alata (Solanaceae) have RNase activity and that the RNase activity has been implicated in the inhibition of self-pollen (McClure et al., 1989; 1990). Consequently, we tested pistil extracts of C. rapunculoides for RNase activity (see Blank et al., 1982). A positive response to the RNase activity specific stain, toluidine blue, appeared at 32kD.

Together, the three types of molecular evidence gathered to date indicate that the molecular properties of self-incompatibility in C. rapunculoides are similar to that reported for the Solanaceae. Additional studies are underway to further establish and to explore these similarities. These molecular data, however, do not tell us why the incompatibility breaks down in older pistils. Consequently, we wanted to examine the pattern of protein synthesis within stigmas and styles of C. rapunculoides. We collected fresh 1 and 4 day old pistils and leaf material from several different plants growing in the greenhouse, and placed the materials in wells containing ^{35}S labeled methionine and cysteine. The wells were kept humid and at room temperature for 18 hours. We ran an SDS-polyacrylamide gel as described above, dried the gel, and placed it under X-ray film for 20 hours at -80°C, and then developed the film. The 32kD band was prominent on the autoradiograph of pistil material but absent in leaf material. The intensity of the band was greater in the 1 day old pistils than in the 4 day old pistils indicating greater levels of synthesis in younger flowers.

Discussion

Our results, though admittedly preliminary, seem to demonstrate a reduction with increasing floral age in the ability of some C. rapunculoides plants to prevent self-fertilization. This appears to result from a loss in the ability of these plants to inhibit self-pollen tube growth through styles of older flowers. Preliminary investigations into the molecular basis of this breakdown raises the possibility that the synthesis of a potential S-locus-specific protein takes place primarily in young flowers and that the protein may become less effective in older flowers. The necessity of pooling material from several plants for each analysis prevents us from addressing the molecular variation among plants. Current work using clonal material will enable us to investigate variation among individual C. rapunculoides plants at the molecular level.

Our findings also have important ecological and evolutionary implications. In natural populations of C. rapunculoides, the breakdown of S.I. with floral age may enable plants to promote outcrossing early in the duration of a flower while permitting self-fertilization later if cross-pollination has not occurred. A flexible breeding system may be particularly important for plants that inhabit open (ephemeral) habitats. When populations are small and/or when pollinator activity is limited, C. rapunculoides would essentially be a selfer. However, when populations are large and when pollinators are active, it would be essentially an outcrosser. Although we need to know much more about variation in the degreé of S.I. in natural populations, it is becoming increasingly obvious that S.I. is not a qualitative trait nor do S.I. plants possess a fixed, outcrossing mating system.

Acknowledgments

This research has benefitted from support by NSF Grants BSR-83-15612 A01 and BSR-88-18184 to AGS and JAW and DCB-86-16087 to THK.

References

Ai, Y; Singh, A.; Coleman, C.E.; Ioerger, T.R.; Kheyr-Pour, A.; Kao, T.-H. Self-incompatibility in Petunia inflata: isolation and characterization of cDNAs encoding three S-allele-associated proteins. Sex. Plant Reprod. 3:130-138; 1990.

Ascher, P.D.; Peloquin, S.J. Effect of floral aging on the growth of compatibile and incompatibile pollen tubes in Lilium longiflorum. Amer. J. Botany 53:99-102; 1966.

Blank, A; Sugiyama, R.H.; Dekker, C.A. Activity staining of nucleolytic enzymes after sodium Dodecyl Sulfate--Polyacrylamide Gel Electrophoresis: Use of aqueous isopropanol to remove detergent from gels. Analytical Biochemistry 120:267-275; 1982.

Bowman, R.N. Cryptic self-incompatibility and the breeding system of Clarkia unguculata (Onagraceae). Am. J. Bot. 74:471-476; 1987.

Dane, F.; Melton, B. Effect of temperature on self- and cross-compatibility and in vitro pollen growth characteristics in alfalfa. Crop Sci. 13:587-591; 1973.

Fearon, C.H.; Hayward, M.D.; Lawrence, M.J. Self-incompatibility in ryegrass. VI. Self seed-set and incompatibility genotype in Lolium perenne L. Heredity 50:169-177; 1983.

Haring, V.; Gray, J.E.; McClure, B.A., Anderson, M.A.; Clarke, A.E. Self-incompatibility as a model of cell-cell recognition in plants. Science 250:937-941; 1990.

Jenkins, T.J. Self-fertility in perennial ryegrass (Lolium perenne L.) Bull. Welsh Plant Breed Stn Ser H 12:100-119; 1931.

Kao, T.-H. Molecular genetics of gametophytic self-incompatibility in the Solanaceae. In: Mulcahy, D.; Bergamini-Mulcahy, G.; Ottaviano, E., eds. Angiosperm Pollen and Ovules; Basic and applied aspects. New York: Springer-Verlag;

Knox, R.B.; Gaget, M.; Dumas, C. Mentor pollen techniques. Int. Rev. Cytol. 107:315-332; 1987.

Martin, F.W. Staining and observing pollen tubes in the style by means of fluorescence. Stain Tech. 34:125-128; 1959.

McClure, B.A.; Haring, V.; Ebert, P.R.; Anderson, M.A.; Simpson, R.J.; Sakiyama, F.; Clarke, A.E. Style self-incompatibility gene products of Nicotiana alata are ribonucleases. Nature 342, 955-957; 1989.

McClure, B.A.; Gray, J.E.; Anderson, M.A., Clarke, A.E. Self-incompatibility in Nicotiana alata involves degradation of pollen rRNA. Nature 347, 757-760; 1990.

Richardson, T.E.; Hrincevich, A.; Kao, T.-H; Stephenson, A.G. Preliminary studies into age-dependent breakdown of self-incompatibility in Campanula Rapunculoides: Seed set, pollen tube growth, and molecular data. Pl. Cell Incompat. Newsl. 22:41-47; 1990.

Richardson, T.E.; Stephenson, A.G. Pollen removal and pollen deposition affect the duration of the staminate and pistillate phases in Campanula rapunculoides. Amer. J. Bot. 76:532-538; 1989.

Morphology and Structure

19
Gene Expression during *Brassica napus* Pollen Development

Diego Albani, Illimar Altosaar, Paul G. Arnison, and Steven F. Fabijanski

The development of pollen is a complex process that requires the coordinate participation of various tissue types and specific gene expression patterns. Both gametophytic (pollen) and sporophytic (anther, tapetum) tissues participate in this process. Of the many nutrients and macromolecules that are involved in the ultimate production of pollen, some are made within the pollen grain while others are contributed by the tapetal tissue. Recently more attention has been paid to this process and molecular and cytological studies have been undertaken in different plant species. Our study focuses on pollen development in Brassica napus, an important oilseed crop. Here we describe some of the molecular events that occur during the formation of pollen grains in this plant species.

Structural studies of the development of the male gametophyte of B. napus have been conducted by Polowick and Sawhney (1986) and Grant et al. (1986). The aim of our study is a characterization of the changes in gene expression in developing microspores through analyses of protein patterns and pollen-expression of both constitutive and pollen-specific genes. The characterization of the molecular events involved in the formation of pollen is a study that is not only scientifically interesting, but is of commercial interest as well. Plant breeding has used male sterility as a means for the efficient production of hybrid varieties. The cross of genetically dissimilar parents can often lead to heterosis or hybrid vigour. This phenomenon has led to the production of new varieties with increased uniformity, strength and overall productivity. Male sterility can derive from changes in a variety of physiological processes. The male sterile trait however is invariably associated with a breakdown of the pollen developmental process. Defining the individual molecular events that take place during normal pollen development can contribute to our understanding of how the whole process is orchestrated within the anther. A better knowledge of this process will suggest new and more efficient strategies for the control of male fertility in agronomically important plants.

Flower bud size can be used as an efficient index of the developmental stage of pollen in B. napus. Proteins and RNA were extracted from microspores of each stage and the RNA samples were in vitro translated using a rabbit reticulocyte system. In vivo labelling of pollen proteins was also performed. Unlabelled pollen proteins, in vivo labelled proteins and proteins made from the in vitro translation of pollen RNA were analyzed by one dimensional SDS electrophoresis. Northern blot analyses were performed on the total RNA isolated from pollen of the different stages.

Analysis of Pollen Gene Expression at the Protein Level

In this study we divided pollen development into six stages which are based on flower bud sizes ranging from approximately 1.5 mm in length for stages 1 to approximately 6.5 mm in length for stage 6. These six stages cover almost the entire period of pollen development starting from tetrad formation (stage 1) to late bi-cellular pollen grains (stage 6). The mostly mature tri-cellular pollen present in open flowers was not considered in this study. As shown in Figure 1A, coomassie blue staining of the SDS gels of pollen protein extracts reveals a relatively uniform polypeptide profile throughout the developmental stages analyzed. The majority of the bands are almost equally represented in all the stages and only a few variations are evident. A wide band corresponding to a polypeptide of approximately 58 KD (arrow c) is strongly specific to stage 1 (tetrad formation), another protein of about 49 KD (arrow b) is specific to the early stages 2 (uni-nucleate microspores) while a band of about 35 KD (arrow a) appears only in the later stages (bi-cellular pollen grains, post mitosis) showing a maximum intensity at stage 6. In addition, some minor bands representing polypeptides of high molecular weight (>150 KD) are seen maximally at stage 4 and 5. However, apart from these few variations, it is clear that the majority of the abundant proteins in Brassica napus pollen are uniformly present through the entire developmental period and therefore are not developmentally regulated with regards to stage specific accumulation.

Microspores are not reliably labelled by incubation with amino acid precursors. For this reason the synthesis of proteins during the development of pollen was analyzed by in vivo labelling of the whole inflorescence. In addition, the use of whole organs reduces the possibility of changing physiological conditions away from normal development, thus avoiding artifacts due to cellular stresses. However, this technique does not allow for high levels of incorporation of labelled precursors, hence a relatively large sample was loaded on the gel. The autoradiogram shown in Figure 1B illustrates the pattern of pollen proteins synthesized throughout all the six developmental stages. This autoradiogram is typical of the results obtained in three separate experiments.

Unlike the patterns of the unlabelled pollen proteins, the profiles of the proteins in vivo labelled at different stages appear to be more heterogeneous and clearly more bands are stage specific. For instance, polypeptides of approximately 95 KD (arrow f) and 49 KD (arrow c) are mostly synthesized at stage 3 and 4 at which time pollen undergoes the first mitotic division. A protein of about 64 KD (arrow e) is characteristic of the later stages and shows maximum synthesis at stage 6 while a polypeptide of approximately 58 KD (arrow d) is present only at stage 1 (tetrad formation) and apparently coincides with the protein observed in the unlabelled protein pattern (Figure 1A). Other examples can be seen. A protein of approximately 38 KD (arrow b) is predominantly synthesized in the early stages (uni-nucleate microspores) while a polypeptide of about 30 KD (arrow a) appears only in the late stages. However, with perhaps the exception of the 58 KD protein, the developmental changes in the patterns of the in vivo synthesized proteins do not correlate with the few changes observed in the total extracted protein. This implies that the in vivo synthesized protein profiles represent lower abundance proteins that do not accumulate to high levels in microspores and are only transiently expressed and rapidly turned over.

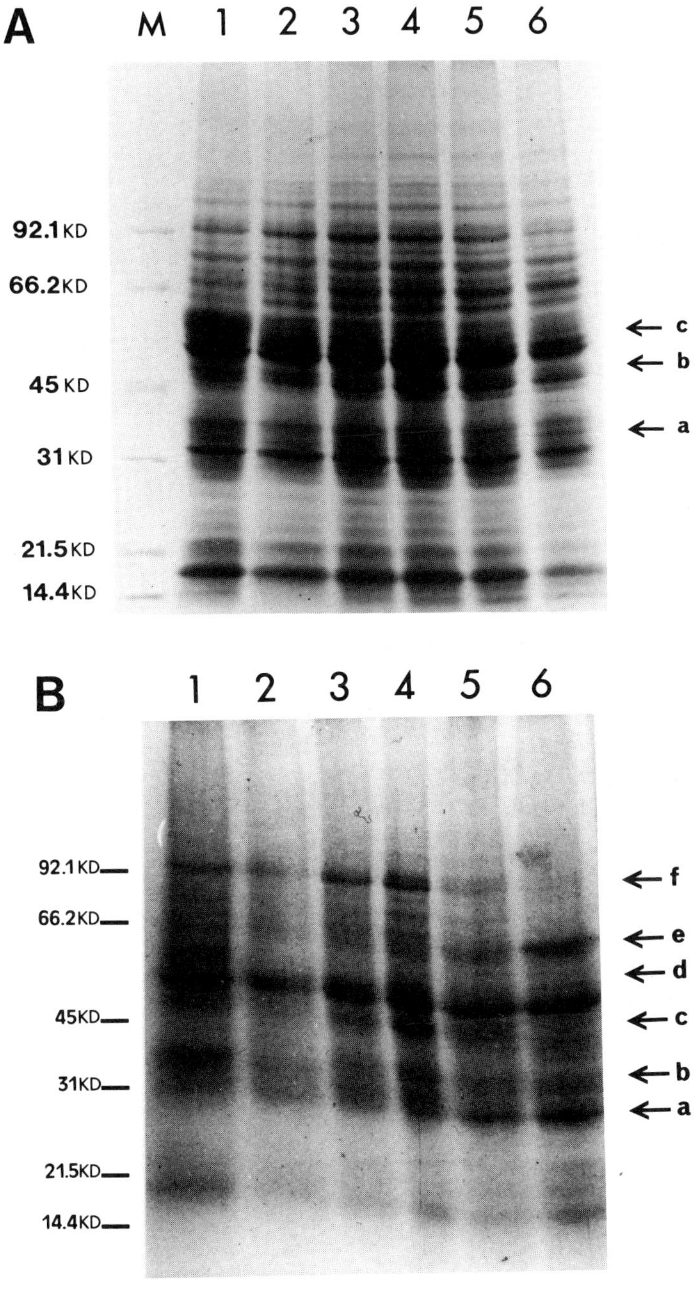
A
M 1 2 3 4 5 6
92.1 KD
66.2 KD
45 KD
31 KD
21.5 KD
14.4 KD
c
b
a
B
1 2 3 4 5 6
92.1 KD
66.2 KD
45KD
31KD
21.5KD
14.4KD
f
e
d
c
b
a

Figure 1.

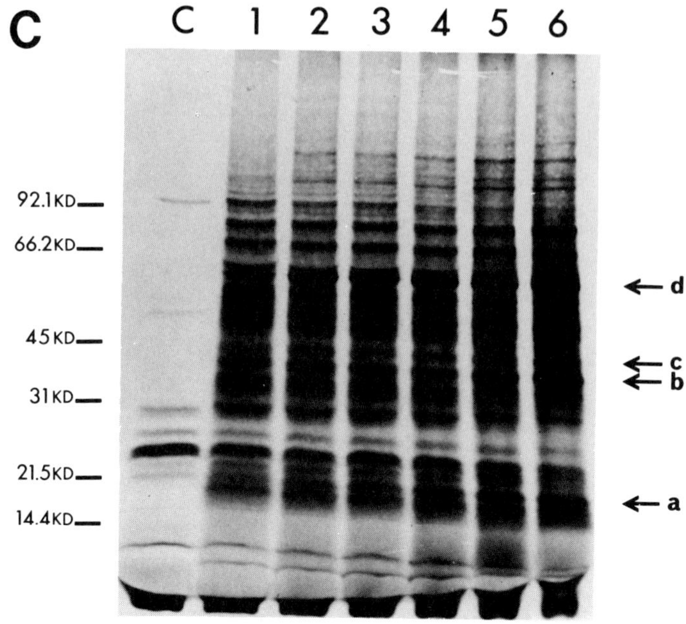

Figure 1. Profiles of total proteins, *in vitro* synthesised proteins and *in vitro* translation products during *B. napus* pollen development. Figure 1A. SDS-gradient gel electrophoresis of the total proteins extracted from pollen at six developmental stages. Each lane contains approximately 90 μg of proteins. The arrows a, b and c indicate three developmentally regulated changes in band patterns (see text). Lane M contains molecular weight markers. Figure 1B. Autoradiogram of the SDS-gradient gel analysis of the proteins synthesized *in vivo* in pollen at six developmental stages. Each lane contains approximately 7,000 cpm of labelled proteins. Exposure was for three weeks at -70°C. Arrows d to f indicate developmentally regulated proteins (see text). Molecular weight markers are indicated. Figure 1C. Autoradiogram of the SDS-gradient gel analysis of the *in vitro*

system. Exposure was for two days at -70°C. Arrows a, b, c and d show increasing RNA products (see text). Molecular weight markers are indicated.

Taken together, these results show that the fraction of abundant proteins which is contained in pollen does not vary significantly during the different stages of microspore development and probably largely represents pollen structural and/or house-keeping proteins. The proteins which are more clearly developmentally regulated are those present in comparatively smaller amounts which more likely have specific enzymatic or regulatory functions. These proteins were more evident in labelling experiments where proteins being actively synthesized are more visible than proteins that accumulate.

Analysis of Pollen Messenger RNA Patterns by in vitro Translation

In order to study developmental events more closely, the total cellular RNA was extracted from B. napus pollen of the same six stages as in the protein study and was analyzed by in vitro translation. Translational activity was lower for RNA isolated from pollen at early stages and increased with time. RNA from pollen at stage 6 was translated 4 to 6 times more efficiently (based on cpm inc./μg total RNA) than RNA from pollen at stage 1. As shown in Figure 1C, the pattern of the in vitro synthesized products does not show clear correlation with either unlabelled or in vivo labelled protein

Some bands corresponding to products of approximately 60, 45, 43 and 23 KD (arrows d, c, b and a) together with some minor bands in the high MW region (>100 KD), appear to increase in intensity throughout development. The translation product of about 60 KD increases very strongly and becomes the major protein from stage 2 onwards. These changes in the profiles of RNA translation products suggest that there is a significant portion of mRNA that accumulates during pollen development. The lack of correlation between the translation products and labelled pollen protein profiles suggests that these accumulating RNAs are largely stored and likely code for products that are required during germination and pollen tube growth.

Analysis of the expression in pollen of a few constitutive and pollen-specific genes

The level of pollen-specific gene expression is estimated to be only in the order of 10-20% of the total gene expression in pollen (Mascarenhas, 1989). It is clear that this limited gene expression can have profound effects on the development of a cell. A few examples of pollen-specific genes have been recently described (Hanson et al., 1989; Twell et al., 1989; Brown and Crouch, 1990; Wing et al., 1990). All of them correspond to genes which are expressed late during pollen development and whose transcripts accumulate in mature pollen. We have isolated pollen-specific genes from B. napus by a combination of both genomic and cDNA cloning (Albani et al., 1990). We used this approach to specifically isolate genes expressed in early pollen development that do not accumulate to significant levels in mature pollen.

We have studied 3 pollen-specific genes in detail. None of these genes is expressed in other tissues, all show significant stage specific regulation and all are members of small multigene families. One gene family, called Bp4 (Albani et al., 1990), is composed of 10 to 15 members that code for low molecular weight proteins whose amino acid sequence and composition are indicative of a possible structural function. We believe therefore that these proteins are most likely associated with the pollen wall. Another gene family we have studied, called Bp19, (Albani et al., 1991) codes for proteins that share extensive homology with pectin esterases, enzymes involved in pectin degradation.

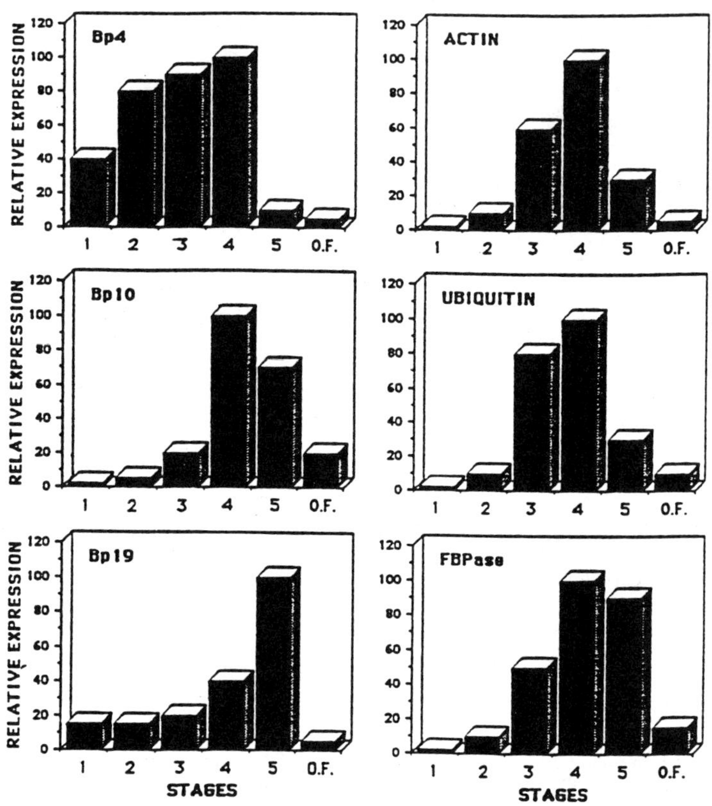

Figure 2. Patterns of expression during *B. napus* pollen development of the three pollen-specific genes Bp4, Bp10, Bp19 and the three constitutive genes actin, ubiquitin and fructose-1,6-bisphosphatase (FBPase). The relative expression is referred to as a percentage of the maximum expression level.

The last pollen-specific gene family we have studied is referred to as the Bp10 family and consists of a number of closely related members which show sequence similarity to ascorbate oxidase. We have found that pollen-specific genes homologous to Bp10 exist in other plant species. Further studies on this gene and its appearance in other plant species are presently under way. Additionally we have studied genes that are not only expressed in pollen, but in other tissues as well. These included actin, ubiquitin and a gene coding for fructose-1,6-bisphosphatase, a key regulatory enzyme in the gluconeogenic pathway leading to sucrose accumulation.

We examined the expression of these constitutive and pollen-specific genes during microsporogenesis by northern blotting. For this study we divided pollen development into six stages which are slightly different from those used for the previous protein analyses. The first five stages correspond to microspores isolated from flower buds of sizes increasing from ca. 1.5mm in length for stage 1 to ca. 6.5mm in length for stage 5 while stage 6 (O.F.) corresponds to pollen isolated from open flowers and consists of mostly mature trinucleate pollen. These results are summarized in Figure 2 and demonstrate that there is a complex regulation of gene activity during pollen development. Ubiquitin and actin show similar patterns of expression suggesting that they are all active over a particularly extended period of time with a maximum activity apparently coincident with the first microspore mitosis and early binucleate pollen stages. Fructose-1,6-bisphosphatase shows maximum expression at slightly later stages. The genes of the Bp 4 family are extensively expressed throughout microsporogenesis and are apparently turned off as pollen matures. The Bp10 genes are instead expressed in a more stage specific manner, maximally in the stages corresponding to early binucleate microspores. The Bp19 gene family is activated slightly later than the Bp4 family and accumulates to near maturity; however, in the very latest stage of pollen development (open flowers) the expression declines considerably. These three pollen-specific gene families show clearly different expression patterns all of which are however consistent with the patterns of genes being important for the formation of the pollen grain, rather than genes whose RNA products are destined for storage and utilization during pollen germination. It is tempting to speculate as to the role of these genes in pollen development, given the homology which two of the products show to proteins of known function. An investigation of their function could successfully rely on the use of antisense RNA technology. Antisense inhibition of these genes will lead to pollen development abnormalities, hence helping in defining their role. Certainly this approach has the capacity to discover key genes in the developmental process.

The differentially regulated expression in microspores of both constitutive and pollen-specific genes argues that pollen development follows a complex program whose implementation takes place over the normal cellular expression of genes required for growth and maintenance. The precise temporal expression of these genes is central to the proper completion of this developmental process. We are interested in the control of this expression and have analyzed the sequence of the putative regulatory regions of these pollen-specific genes. Interestingly, some of these genes have promoter sequences in common while others do not, indicating that a "pollen-box" principle or a common protein binding region is not definable. There is obviously a more complex and subtle form of regulation that takes place, since both pollen-specific and housekeeping genes show stage specific regulation. One of the emerging areas of study is the determination of the DNA sequences that are not only involved in pollen-specific gene expression, but expression of the housekeeping genes in pollen as well. The regulatory signals for both pollen and constitutive expression must be intertwined in the promoter region of these genes. The hierarchy of regulation is not a simple issue. For multigene families such as actin or ubiquitin, it will be of interest to know if pollen specific isoforms exist. For single genes, the portions of the promoters recognized in pollen will be a primary area of study. All of these different aspects of the control of gene expression in pollen are presently under investigation in many laboratories. The study of pollen gene expression is emerging as an area of excitement in molecular biology and will certainly continue for some time.

References

1. Albani, D.; Robert, L.S.; Donaldson, P.A.; Altosaar, I.; Arnison, P.G.; Fabijanski, S.F. Characterization of a pollen-specific gene family from Brassica napus which is activated during early microspore development. Plant Mol. Biol. 15:605-622; 1990.

2. Albani, D.; Altosaar, I.; Arnison, P.G.; Fabijanski S.F. A gene showing sequence similarity to pectin esterase is specifically expressed in developing pollen of Brassica napus. Sequences in its 5' flanking region are conserved in other pollen-specific promoters. Plant Mol. Biol. 16: 501-513; 1991.

3. Brown, S.M.; Crouch, M.L. Characterization of a gene family abundantly expressed in Oenothera organensis pollen that shows sequence similarity to polygalacturonase. Plant Cell 2:263-274; 1990.

4. Grant, I.; Beversdorf, W.D.; Peterson, R.L. A comparative light and electron microscopic study of microspore and tapetal development in male fertile and cytoplasmic male sterile oilseed rape (Brassica napus). Can. J. Bot. 64: 1055-1068; 1986.

5. Hanson, D.D.; Hamilton, D.A.; Travis, J.L.; Bashe, D.M.; Mascarenhas, J.P. Characterization of a pollen-specific cDNA clone from Zea mays and its expression. Plant Cell 1: 173-179; 1989.

6. Mascarenhas, J.P. The male gametophyte of flowering plants. Plant Cell 1: 657-664, 1989.

7. Polowick, P.L.; Sawhney, V.K. A scanning electron microscopic study on the initiation and development of floral organs of Brassica napus (cv. Westar). Amer. J. Bot. 73: 254-263; 1986.

8. Twell, D.; Wing, R.; Yamaguchi J.; McCormick, S. Isolation and expression of an anther-specific gene from tomato. Mol. Gen. Genet 217: 240-245; 1989.

9. Wing, R.A.; Yamaguchi, J.; Larabell, S.K.; Ursin, V.M.; McCormick, S. Molecular and genetic characterization of two pollen-expressed genes that have sequence similarity to pectate lyases of the plant pathogen Erwinia. Plant Mol. Biol. 14: 17-28; 1990.

20

Pollen Mitochondrial Proteins in Fertile and Male Sterile *Nicotiana sylvestris*

Rosine De Paepe, Varonique Vitart, Philippe Chetrit, Arlette Forchioni, Rena Remy, and Fernand Vedel

Introduction

Flower development is a complex process, and little is known concerning molecular events related to gamete formation (Mascarenhas, 1990). Male sterility, i.e. the inability of plants to produce viable pollen, may be controlled by the nuclear genome alone, or by interactions between mitochondrial and nuclear genes (Hanson áConde, 1985). Mitochondria are semi-autonomous organelles, the genome of which encodes less than 10% of the several hundred mt polypeptides. The remaining proteins encoded by nuclear genes are then imported (Tzagoloff áMyers, 1986). Mt systems in plants can be distinguished from those of other organisms by larger heterogeneous genomes, with complex gene expression (QuÄtier áVedel, 1977; Palmer áShield, 1984; Levings áBrown, 1989).

To date, only two phenotypes, the non-chromosomal stripe trait (NCS) in maize, and cytoplasmic male sterility (cms), found in a large number of species, have been related to the plant mt genome (Levings áBrown, 1989). Specific mt DNA sequences and polypeptides have been characterized in cms maize and Petunia. In both cases, these sequences are chimeric genes resulting from recombination and their transcription is developmentally regulated (Forde áLeaver, 1980; Dewey et al., 1986; Young áHanson, 1987; Dewey et al., 1987; Nivison áHanson, 1989). Up to now, mt transcription and in organello protein synthesis have only been studied in somatic cells, and nothing is known about the expression of sterility in pollen itself. Mt polypeptide maps from pollen are lacking and they are rare for other plant tissues (RÄmy et al., 1987; Rios et al., 1991). Here, we describe the isolation of pollen mitochondria from a fertile Nicotiana sylvestris botanical line. Two dimensional electrophoresis of mt proteins revealed the presence of additional polypeptides in pollen compared to leaves. We also found differences between fertile N. sylvestris and several male sterile mutants, either nuclear or cytoplasmic, obtained by protoplast culture.

Material and Methods

Plant material and pollen isolation. N. sylvestris original fertile line plants and four male sterile mutants (two nuclear and two cytoplasmic) obtained by protoplast culture (Li et al., 1988; De Paepe et al., 1990), were grown in a greenhouse, under 16 h light. 200 to 500 flowers, according to genotype, were harvested one or two days before anthesis. Anthers were collected in a buffer containing 0.4 M sucrose, 0.01 M KH_2PO_4 (pH 7.6),

0.001 M EGTA, 0.5% (w/v) bovine serum albumin, 2 ml/1 2-mercaptoethanol, 0.2 g/l PMSF and disrupted in a Potter grinder. Pollen were separated from somatic cells by filtration through a 100 mm net and then recovered on a 25 mm net (De Paepe et al., 1990).

Preparation of pollen mitochondria.
Pollen were crushed with mortar and pestle as described previously (De Paepe et al., 1990). Pollen debris, nuclei and protoplasts were pelleted by centrifugation at 4000 g for 5 min, mitochondria were then pelleted by centrifugation of the supernatant at 12,000 g for 15 min . The mt fraction was further purified by centrifugation on a three layer Percoll gradient (Boutry et al., 1984).

Electron microscopy.
Pollen mt pellets were fixed for 2 h in 2% glutaraldehyde in 0.05 M sucrose, 0.05 M Na phosphate (pH7), and post-fixed in 1.5% O_sO_4 in the same buffer for 90 min at 0ſC. The suspension was preembedded in fibrinogen, dehydrated through a graded series of alcohol and propylene oxide, and then embedded in epon (Vedel áQuÄtier, 1974).

In organello protein synthesis.
Mitochondria were isolated aseptically from pollen of 250 flowers or from 1 g of young green leaves and purified as indicated above. Plant material was previously sterilized for 5 min with 2% (w/v) calcium hypochlorite. In organello protein synthesis was performed according to Boutry et al. (Boutry et al., 1984), in an incubation medium containing 12 mCi of [35S] methionine (1000 Ci/mM, Amersham) and 400 mg of erythromycin, a potent inhibitor of bacterial and plastidial translation (Newton áWalbot, 1985). Mitochondria were then pelleted and lysed with SDS- Laemmli buffer and the samples run on a 12% SDS-polyacrylamide minigel (Biorad) at 170 V for 1 h. After Coomassie blue staining, the gels were dried and exposed to Amersham b max film.

Two-dimensional SDS-PAGE.
Mitochondria were purified as described above. Mt proteins from 5 g of leaves or from pollen of 100 fertile flowers were precipitated in 80% (v/v) acetone, solubilized, loaded on isoelectric focusing gel rods with urea, CHAPS and carrier ampholytes pH 5-8 (RÄmy et al., 1987), and run at 500 V for 15 h. Gel rods were submitted to second dimension analysis on 12% SDS-polyacrylamide slab gels at 130 V for 5 h. Gels were stained with Coomassie brilliant blue. Comparison between fertile and male sterile plants was performed by loading pollen mt proteins corresponding to 30 and 300 flowers respectively, and by staining the second-dimension gel with the silver nitrate procedure described in Damerval et al. (1987). For each genotype, at least five independent sets of protein extracts were analysed.

Results

Isolation of pollen mitochondria
As expected from previous data (Boutry et al., 1984), the mt fractions were recovered at the 25%/45% Percoll gradient interface. Electron microscopy examination revealed that these fractions contained intact mitochondria, with outer double membrane and internal cristae (Fig.1). In organello protein synthesis showed that isolated pollen mitochondria

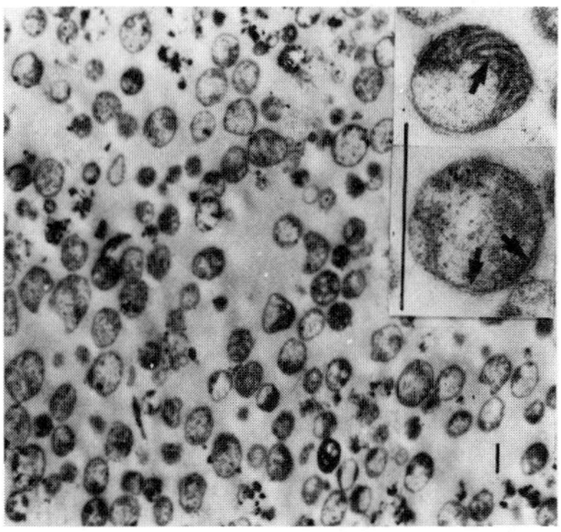

Figure 1

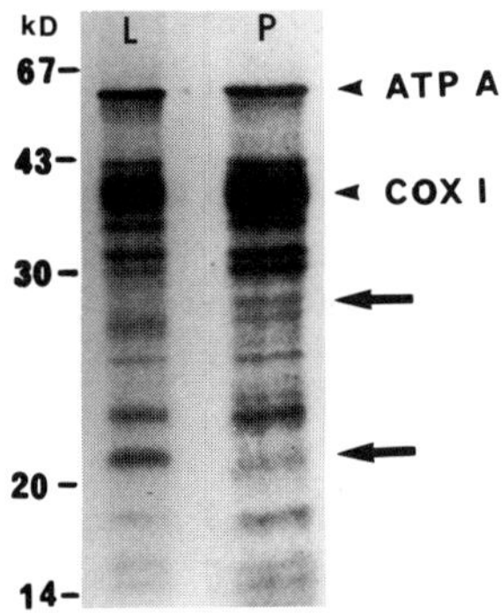

Figure 2

Figure 1 Electron micrograph of isolated mitochondria from N. sylvestris pollen. Bar : 1 mm.

Figure 2 In organello protein synthesis of leaf (L) and pollen (P) mitochondria. Arrows show differences between leaf and pollen profiles.

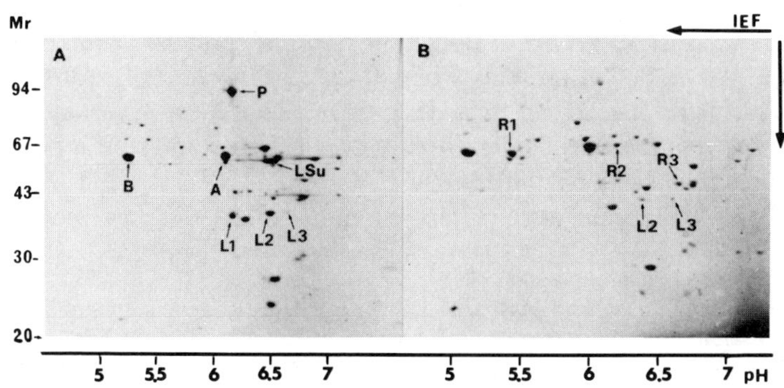

Figure 3 2 D SDS-PAGE of leaf and pollen mt proteins from fertile N. sylvestris gels were stained with Coomassie blue.A. Leaf. Mt proteins corresponding to 5 g of leaf tissue were loaded on the gel. P : glycine decarboxylase P subunit (nuclear) ; A : ATPase A (mitochondrial) ; B : ATPase B (nuclear) ; LSu : large subunit of the RubiscoB. Mature pollen. Mt proteins obtained from the pollen of 100 fertile flowers were loaded on the gel. Arrows show reproductive differences between leaf and pollen (12 extractions each).

were able to produce polypeptides. A few differences were observed between leaf and pollen syntheses (Fig.2).

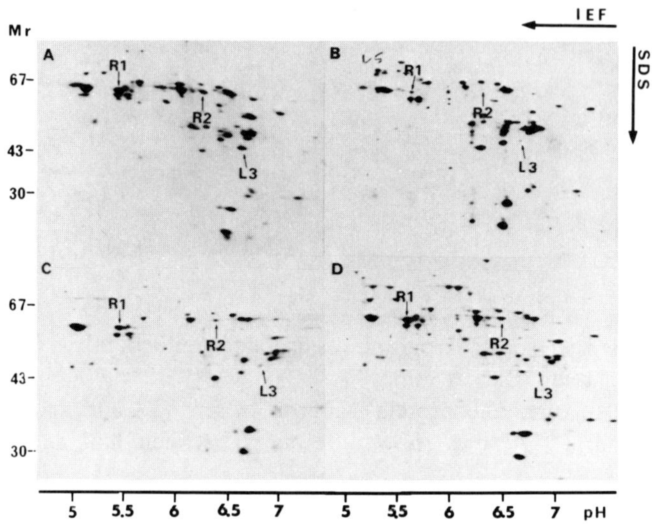

Figure 4 2 D SDS-PAGE of pollen mt proteins from fertile and male sterile plants. Pollen proteins from 30 fertile flowers, and 300 sterile flowers, were loaded on gels, that were stained by silver nitrate. Arrows show consistent differences between fertile and sterile pollen (at least 5 replications).
A : fertile pollen, B : mst4 pollen, C : mst3 pollen, D : cms1 pollen.

Mitochondrial proteins from fertile pollen

Mt proteins from pollen and leaves of fertile N. sylvestris were compared by 2-D SDS-PAGE analysis and Coomassie blue staining. In leaf mitochondria, (Fig. 3A) a few polypeptides (ATP A, ATP B and glycine decarboxylase P subunit) were identified from previous studies in other plants (RÄmy et al., 1987, Rios et al., 1991, Hack áLeaver, 1983). Pollen and leaf patterns could be distinguished by several differences. First, several L spots such as L2 (36 kD, pI 6.4) were present at lower levels on pollen gels, or were undetectable, such as L1 (35 kD, pI 6.1). By contrast, the L3 spot (36 kD, pI 6.7) was more apparent and several polypeptides undetectable in leaf mitochondria, were observed. These additional polypeptides were called R (reproductive). Major R spots were R1, (53 kD, pI 5.4), R2 (55 kD, pI 6.2) and R3 (40 kD, pI 6.7). Mt protein patterns from anther somatic tissues contained the R1 and R3 spots, but lacked the R2 spot (not shown).

Mitochondrial proteins from sterile pollen

2 D SDS-PAGE profiles of pollen mt proteins from the fertile parental line and four male sterile mutants - cms1, cms2 (cytoplasmic mutants), mst3 and mst4 (nuclear mutants) - obtained by protoplast culture were compared (Fig 4). Protein yields from sterile pollen being lower than from fertile pollen, a silver staining procedure was used, which is much more sensitive than Coomassie blue. All male sterile pollen show differences compared to fertile pollen. In mst4 profiles several differences were observed, for instance, R1 spot was reduced, and R2 and L3 were hardly detectable. In mst3 and both cms, R1, R2 and R3 spots were present, while L3 was reduced in mst3 and undetectable in cms.

Discussion

This is the first description of pollen mitochondrial proteins. Yield from a hundred flowers of N. sylvestris is enough to perform either in organello mt protein synthesis or Coomassie stained 2 D SDS-PAGE. We found that the abundance of some leaf mt polypeptides was altered in pollen mitochondria, and, on the other hand, that some mt polypeptides were specific to the reproductive pathway. The R2 polypeptide is present in pollen only, while R1 and R3 occur in both anther somatic cells and pollen. We checked that the presence of R spots could not be explained by contamination of mitochondria with plastids or peroxysomes. The differences observed between leaf and pollen could be due to either mitochondrial or nuclear gene expression. Different patterns of in organello mt protein synthesis were previously reported, either between different organs from the same plant, or between fertile and cms plants (Forde áLeaver, 1980; De Paepe et al., 1990; Newton áWalbot, 1985). In N. sylvestris, differences between pollen and leaf in organello mt protein synthesis indicated that mt expression is somewhat different in leaf and pollen. However no relationship can be established with the changes evidenced by 2 D SDS-PAGE. In contrast to in organello protein synthesis, 2 D SDS-PAGE electrophoresis will allow us to prepare enough R polypeptides for microsequencing, antibody production and further gene identification.

Acknowledgements

We thank Dr. J. Brangeon for help in electron microscopy, and Dr A. de Courcel for critical reading of the manuscript.

References

Boutry, M.; Faber, A. M.; Charbonnier, M.; Briquet, M. Micro-analysis of plant mitochondrial protein synthesis products. Plant Mol Biol 3 : 445-452; 1984

Damerval, C.; Le Guilloux, M.; Blaisonneau, J.; De Vienne, D. A simplification of Heukeshoven and Dernick's silver staining of proteins. Electrophoresis 8 : 158-159; 1987.

De Paepe, R.; ChÄtrit, P.; Vitart, V.; Ambard-Bretteville, F.; Prat, D.; Vedel, F. Several nuclear genes control both male sterility and mitochondrial protein synthesis in Nicotiana sylvestris protoclones. Mol Gen Genet 222 : 206-210; 1988.

De Paepe, R.; Koulou, A.; Pham, J. L.; Brown, S. C. Nuclear DNA content and separation of N. sylvestris vegetative and generative nuclei at various stages of male gametogenesis. Plant Sci 70 : 255-265; 1988.

Dewey, R. E.; Leavings, III. C. S.; Timothy, D. H. Novel recombinations in the maize mitochondrial genome produce a unique transcriptional unit in the Texas male sterile cytoplasm. Cell 44 : 439-449; 1986.

Dewey, R. E.; Timothy, D. H.; Levings, III. C. S. A mitochondrial protein associated with cytoplasmic male sterility in the T cytoplasm of maize. Proc Natl Acad Sci USA 84 : 5374-5378; 1987

Forde, B. G.; Leaver, C. J. Nuclear and cytoplasmic genes controlling synthesis of variant mitochondrial polypeptides in male sterile maize. Proc Natl Acad Sci USA 77 : 418-422; 1980.

Hack, E.; Leaver, C. J. The a-subunit of the maize F1-ATPase is synthesised in the mitochondria. EMBO J 2 : 1783-1789; 1983.

Hanson, M. R.; Conde, M. F. Functioning and variation of cytoplasmic genomes : lessons from cytoplasmic-nuclear interactions affecting male fertility in plants. Int Rev Cytol 94 : 213-267; 1985

Levings, III. C. S.; Brown, G. G. Molecular biology of plant mitochondria. Cell 56 : 171-179; 1989.

Li, X. Q.; ChÄtrit, P.; Mathieu, C.; Vedel, F.; De Paepe, R.; RÄmy, R.; Ambard-Bretteville, F. Regeneration of cytoplasmic male sterile protoclones of Nicotiana sylvestris with mitochondrial variations (1988). Curr Genet 13 : 261-266; 1988.

Mascarenhas, J. P. Gene activity during pollen development. Ann Rev Plant Physiol Plant Mol Biol 41 : 317-338; 1990.

Newton, K. J.; Walbot, V. Maize mitochondria synthesize organ-specific polypeptides. Proc Natl Acad Sci USA 82 : 6879-6883; 1985.

Nivison, H. T.; Hanson, M. R. Identification of a mitochondrial protein associated with cytoplasmic male sterility in Petunia. The Plant Cell 1 : 1121-1130; 1989.

Palmer, J. D.; Shield, C. R. Tripartite structure of the Brassica campestris mitochondrial genome. Nature 307 : 437-440; 1984.

QuÄtier, F.; Vedel, F. Heterogeneous population of mitochondrial DNA molecules in higher plants. Nature 268 : 365-368; 1977

RÄmy, R.; Ambard-Bretteville, F.; Colas des Francs, C. Analysis by two-dimensional gel electrophoresis of the composition of pea mitochondria isolated from different tissues. electrophoresis 8 : 528-532; 1987.

Rios, R.; De Buyser, J.; Henry, Y.; Ambard-Bretteville, F.; RÄmy, R. Two-dimensional electrophoretic comparison of mitochondrial polypèptides from different wheat (Triticum aestivum L) tissues. Plant Sci (in press); 1991

Tzagoloff, A.; Myers, A. M. (1986) Genetics of mitochondrial biogenesis. Ann Rev Biochem 55 : 249-285

Vedel, F.; QuÄtier, F. Physicochemical characterization of mitochondrial DNA from potato tubers. Biochim Biophys Acta 340 : 374-387; 1974.

Young, E. G.; Hanson, M. R. A fused mitochondrial gene associated with cytoplasmic male sterility is developmentally regulated. Cell 50 : 41-49; 1987.

21
Pollen Specific Gene Expression during Development and Germination in Tobacco

Koen Weterings, Wim Reijnen, Jan Schrauwen, Marinus van Herpen, Peter de Groot, Margryt Oldenhof, and George Wullems

INTRODUCTION

The pollen grain represents the male part of the sexual reproduction cycle of the higher plant. In itself it combines a transport unit and a producer of sperm cells. Thus the pollen grain or male gametophyte is the effector of the active dispersal of the plant's genotype. Because of it's crucial function energetic biochemical, cytological and physiological research has been performed on this highly differentiated haploid organ. Molecular biological processes lying at the basis of pollen development have been studied only in recent years (Giles and Prakash, 1987; Mascarenhas, 1990).

The developing gametophyte expresses sets of genes in a stage specific regulated manner. The expressed genes can be divided into three different groups based on the timing of appearance of their transcripts (Schrauwen et al., 1990). One set is expressed throughout development. mRNAs of the second group are preferentially present around mitosis. Two cDNA clones originating from this group have been isolated (Albani et al., 1991). Transcripts of the third group of genes start accumulating after the first haploid mitosis. The majority of pollen specific cDNA clones which have been isolated belong to this group (Mascarenhas, 1990).

Immediately after the pollen has been captured by the stigma a cascade of processes initiates the growth of the pollen tube. The tube will guide the sperm cells towards the ovule or female gametophyte. Thus the double fertilization is effected (Giles and Prakash, 1987). During the germination and tube growth the presynthesized mRNA of the pollen grain is assumed to be translated (Mascarenhas and Mermelstein, 1981). Synthesis of mRNA in both vegetative and generative nuclei, however, also is initiated (Haskell and Rogers, 1985).

In order to understand the processes underlying pollen development and germination at a molecular level, pollen specific gene expression is studied. For this purpose we isolated a pollen specific cDNA clone (NTPc303) from Nicotiana tabacum. NTPc303 represents a mRNA which is abundant in mature pollen (Weterings et al., 1991).In this chapter the characteristics of this cDNA clone are presented.

The specificity of expression of NTP303 -for both tissue and developmental stage- has been the subject of a thorough examination. It involved Northern blot analysis and high resolution in situ localization. To learn more about the possible function of NTPc303 we also have analyzed the nucleotide sequence. The deduced amino acid sequence has been used to scan databases for homologous sequences.

RESULTS

Screening of the cDNA library

A cDNA library was constructed in λgt10 against mRNA from pollen at anthesis. A part of this library was screened using single stranded cDNA made against mRNA from leaf or mRNA from pollen. Based on their differential hybridization properties several clones were initially isolated.

The clones could be divided into two cross-hybridizing groups. Clone 303 (λ303) -a representative of one of the two groups- was subcloned in pBluescript (Stratagene; pNTPc303) for further analysis. NTPc303 is 1960 base pairs in length.

Localization of expression of NTP303

In order to establish it's tissue specificity NTPc303 was hybridized to a Northern blot containing RNA from vegetative and RNA from generative tissue. No hybridization could be detected to RNA from leaf, stem or root nor to RNA from petals, sepals, filaments or pistils. NTPc303 only hybridized to RNA from pollen, confirming it's tissue-specificity. The position of the band that appeared in the lane containing RNA from mature pollen corresponds to a RNA molecule of approximately 2100 nucleotides.

An indication of the expression level of NTP303 was obtained by rescreening a representative portion of the cDNA library. Approximately 0.25% of the clones present in the library cross hybridized to NTPc303 implying that NTP303 transcripts are abundant in pollen (Mascarenhas, 1990)

In situ hybridization was used to assess more accurately the location of the NTP303 transcripts inside the pollen grain. This might provide further information concerning the possible function of NTP303. The small size of the pollen grain (30μm) harbouring the sickle-shaped generative cell (5*20μm) requires an optimal detection of the hybridization signals. The confocal laser scanning microscope (CLSM) provides this requisite. The CLSM images the reflection of the silver grains (hybridization signal) ,in the light sensitive emulsion, permeated through the pollen section, in the in-focus field only. By this method a 10μm specimen can be optically divided in the z-direction in sections at 1μm distance from each other (Houtsmuller et al., 1990). Applying this technique to sections of mature pollen grains revealed that NTP303 transcripts are restricted to the cytoplasm of the vegetative cell. No hybridization of labeled pNTPc303 transcripts could be detected within the generative cell or vegetative nucleus.

Conserved expression of NTP303

If NTP303 is part of a gene expression program important during the development or germination of the pollen grain it probably will be evolutionary conserved and consequently be present in pollen of a variety of plant species. To test this we have hybridized NTPc303 to a Northern blot containing RNA from pollen of (in order of decreasing relationship) *Solanum spec.*, *Digitalis purpurea* (both dicots) and *Lilium longiflorum* (a monocot). Homologous transcripts appear to be present in the pollen grains of all tested species. The taxonomic relationship to the tobacco pollen is reflected by the intensity of the hybridization signal.

Developmental expression of NTP303

Northern blot analysis revealed that the first NTP303 transcripts are detectable after the haploid mitosis of the microspore. RNA was isolated from pollen at the uni-nucleate, early-, mid- and late bi-nucleate stage. As shown in no hybridization of uni-nucleate pollen RNA to labeled NTPc303 could be detected. At the early bi-nucleate stage, however, probing with NTPc303 did generate a hybridization signal. Hereafter the signal increased indicating that NTP303 transcripts are piling up during the last phases of pollen development. Interestingly a similar, stage specific accumulation pattern could be observed in a Northern blot assay containing RNA from *in vitro* ripening pollen.

Expression of NTP303 during germination and pollen tube growth

During germination of the pollen grain NTP303 continues to be transcribed. The hybridization signal obtained with RNA from pollen imbibed for one hour is 1.5 times as intense as compared to the signal from RNA from mature pollen, when probed to NTPc303. The steady state level of NTP303 persistantly rises during pollen tube growth until two hours after imbibition (relative expression=273% as compared to mature pollen). NTP303 transcripts still can be detected twenty hours after the pollen have been allowed to germinate.

To investigate the presumptive transcription of NTP303 during germination a pulse labeling assay was performed. [5,6-^{3}H]uridin was added to the germination medium as a result of which *de novo* transcribed RNA will be labeled. NTP303 RNA is specifically isolated by liquid hybridization to antisense pNTPc303 transcripts followed by RNase-T_1 treatment. Using gel electrophoresis and fluorography the newly formed NTP303 RNA is visualized. As control reactions either sense pNTPc303 transcripts or t-RNA was used in the liquid hybridization and applied to the gel. Also total RNA treated with RNase-T_1 was applied to the gel to check for protected secundairy RNA structures. On the resulting fluorograph only one band appears. It is localized in the lane containing the RNA hybrids obtained from the pulse labeling assay. This band must be the resultant of *de novo* transcription of NTP303 as all control reactions were negative.

Nucleotide sequence of NTPc303

The NTPc303 sequence has been determined by sequencing both strands. The largest open reading frame is 1662 nucleotides in length. It encodes a putative polypeptide of 554 amino acid residues. The peptide has a calculated molecular weight of 62 kD and a predicted isoelectric point of 9.5.

The hydropathy plot of the presumed protein encoded by NTPc303 indicates that -except for the N-terminal region- the protein is relatively hydrophilic. The N-terminal region carries a signal peptide of 21 amino acid residues conforming to the rules formulated by von Heyne (von Heyne, 1986)

Sequence comparison

A comparison with the NBRF protein database and with the EMBL database sequences revealed that the deduced amino acid sequence of NTPc303 shows significant similarity to two protein sequences. A sequence identity of 29% (75% considering conservative substitutions) was found with the precursor of ascorbate oxidase from Cucumis sativus (Ohkawa et al., 1989). Furthermore a homology of 25% (75% including conservative substitutions) was found with the precursor of the enzyme laccase from Aspergillus nidulans (Aramayo et al., 1990).

Both enzymes belong to the group of multi-copper oxidases capable of catalyzing the reduction of molecular oxygen to water. For this purpose the ascorbate oxidase sequence contains clusters of copper binding histidine residues (Messerschmidt et al., 1989). Sequence homology between ascorbate oxidase and laccase on the one hand and the amino acide sequence of NTPc303, on the other is poor in these regions. Therefore it is unlikely for the gene product of NTP303 to fullfill a function similar to one of both enzymes.

DISCUSSION

Northern blot analyses have demonstrated that NTPc303 is a pollen specific cDNA clone. Supplementary, preliminary results from in situ localization experiments have affirmed this at the cellular level. Furthermore, considering the facts that the transcripts are abundantly present and that homologous transcripts have been found in pollen of various other species, it can be deduced that NTP303 transcripts probably play an important role in pollen.

The expression pattern of NTP303 shows some similarities to the developmental expression patterns of other pollen specific clones (Mascarenhas, 1990). The rise of the hybridization signal of NTP303 late during pollen development probably is due to the overall accumulation of mRNA in maturing pollen grains (Tupy, 1982).

NTP303 is regulated concerted with a group of other genes in pollen of tobacco. Their transcripts appear after the first haploid mitosis and accumulate during the maturation of the microgametophyte (Schrauwen et al., 1990). This supposes that the regulatory mechanism for these genes is the same. Concluding from the fact that transcripts homologous to NTPc303 are present in pollen from other species it is very probable that this regulatory mechanism is evolutionary conserved.

Gene expression of the NTP303 gene initiates as the nursing tapetal cells already have decayed (Koltunow et al., 1990). This suggests that the expression of NTP303 is regulated by the haploid gametophyte itself and does not involve gene products of the surrounding diploid sporophytic tissue. The observation that a similar accumulation pattern of NTP303 transcripts could be observed in in vitro ripening pollen reinforces this hypothesis.

In situ hybridization has exactly pinpointed the NTP303-RNA to the cytoplasm of the vegetative cell. So there is reason to believe that NTP303 will perform some vegetative kind of function in the pollen grain. The fact that transcription of the NTP303 gene can be observed during germination makes it highly probable that this function has something to do with pollen tube growth. More clues about the possible function of NTP303 can be found by analysis of it's nucleotide sequence. This has revealed that the presumed protein coded by NTP303 is hydrophilic and contains a leader peptide which suggests that the mature protein will be localized extracellularly. No homology has been found with any known protein directly involved in degradation of style material so perhaps the NTP303 gene product may just perform an opposite function. Taken together these results indicate that NTP303 most probably will be active in cell wall synthesis and thus aid in the growth of the pollen tube.

Because of it's conceivably important function and it's high transcription level during maturation and germination the NTP303 gene will be used in further studies on promoter activity and regulation. To this end we have used NTPc303 to trace a genomic clone from a genomic library. The gene plus it's 5' promoter region is currently being sequenced.

REFERENCES

Albani, D.; Altosaar, I.; Arnison, P.G.; Fabijanski, S.F. A gene showing sequence similarity to pectin esterase is specifically expressed in developing pollen of Brassica napus. Sequences in its 5' flanking region are conserved in other pollen-specific promoters. Plant Mol. Biol. 16:501-513; 1991.

Aramayo, R.; Timberlake, W.E. Sequence and molecular structure of the Aspergillus nidulans yA (laccase l) gene. Nucl. Acids Res. 18:3415-3415; 1990.

Giles, K.L.; Prakash, J., editors. International Review of Cytology; vol. 107. Pollen: Cytology and development. Academic press Inc.; 1987.

Haskell, D.W.; Rogers, O.M. RNA synthesis by vegetative and sperm nuclei of trinucleate pollen. Cytologia 50:805-809; 1985.

von Heyne, G. A new method for predicting signal sequence cleavage sites. Nucl. Acids Res. 14:4683-4690; 1986.

Houtsmuller, A.B.; Oud, J.L.; van der Voort, H.T.M.; Baarslag, M.W.; Krol, J.J.; Mosterd, B.; Mans, A.; Brakenhoff, G.J.; Nanninga, N. Image processing techniques for 3-D chromosome analysis. J. Microscopy 158:235-248; 1990.

Koltunow, A.M.; Truettner, J.; Cox, K.H.; Wallroth, M.; Goldberg, R.B. Different temporal and spatial gene expression patterns occur during anther development. The Plant Cell 2:1201--1224; 1990.

Mascarenhas, J.P. Gene activity during pollen development. Ann. Rev. Plant Physiol. Mol. Biol. 41:317-381; 1990.

Mascarenhas, J.P.; Mermelstein, J. Messenger RNAs: their utilization and degradation during pollen germination and tube growth. Act. Soc. Bot. Pol. 50:13-20; 1981.

Messerschmidt, A.; Rossi, A.; Ladenstein, R.; Huber, R. X-ray crystal structure of the blue oxidase Ascorbate oxidase from Zucchini. Analysis of the polypeptide fold and a model of the copper sites and ligands. J. Mol. Biol 206:513-529; 1989.

Ohkawa, J.; Okada, N.; Shinmyo, A.; Takano, M. Primary structure of cucumber (Cucumis sativus) ascorbate oxidase deduced from cDNA sequence: Homology with blue copper proteins and tissue-specific expression. Proc. Natl. Acad. Sci. USA 86:1239-1243; 1989.

Schrauwen, J.A.M.; de Groot, P.F.M.; van Herpen, M.M.A.; van der Lee, Reynen, W.H.; Weterings, K.A.P.; Wullems, G.J. Stage-related expression of mRNAs during pollen development in lily and tobacco. Planta 182:298-304; 1990.

Tupy, J. Alterations in polyadenylated RNA during pollen maturation and germination. Biol. Plant. 24:331-340; 1982.

Weterings, K.; Schrauwen J.; Reijnen, W.; Wullems, G. Analysis of pollen specific gene expression in tobacco. J. Cell Biochem. Suppl. 15A:141; 1991.

22
Concurrent Development of the Ovule Structures in Some Flowering Plants

T.B. Batygina, I.I. Shamrov, G.E. Titova, and V.E. Vasilyeva

The ovary, ovule and embryo sac are a complex integrated system. It is characterized by a continuously changing state of its integrity in the process of development. The processes occurring in the ovule are a start for its further transformation into a seed.

In continuation of the system approach elucidation (Batygina & Vasilyeva, 1981, 1983; Batygina, 1984a,b) we have suggested a strategy for a successful study of this dynamic system: 1- investigation of the ovule and seed structures in dynamics and in their interrelation with the surrounding tissues of the ovary; comparison of the stages and rates of their development with processes of sporo-, gametophyto- and embryogenesis; 2 - comparison of the kinetics of morphogenetic and physiological-biochemical processes in the ovule and ovary; 3 - comparison of the regularities of ovule and ovary differentiation in situ, in vivo and in vitro; 4 - choice of model plant species, contrasting in their embryological features, ecotype, anthecology and ecology of germination in relation to plant position in the phylogenetic system of flowering plants.

The present communication is an attempt to illustrate some of these statements and elucidate morphogenetic and morphophysiological correlations and critical periods in the ovule and seed development.

Under investigation were the representatives of the various orders of flowering plants: at the base of the phylogenetic tree - Nelumbonales, Nymphaeales, Ceratophyllales and at a higher level of the system - Gentianales and Poales. Species studied belong to different ecotypes: hydatophytes - Ceratophyllum demersum L., Nuphar lutea (L.) Sm., hydrophyte - Nelumbo nucifera Gaertn., mesophytes - Triticum aestivum L., Gentiana cruciata L. The species also contrast in their embryological features, such as: type of ovule, number of integuments, types of development of embryo sac, embryo, endosperm, type of germination, etc.

To carry out a comparative embryological analysis of the data available six stages were suggested in the ovule development in flowering plants - from the ovule at the stage of megasporocyte to mature seed (Nikiticheva, 1990, 1991). We think it pertinent to consider here two other stages, namely the ovule primordium at the stage of archesporial cell and germinating seed.

System approach to ovule and seed investigation showed that the structures form, function and interchange at a certain stage of sporo-, gametophyto- and embryogenesis (Fig. 1). Moreover, the data obtained support the current concept that ovule evolution from a more primitive crassinucellate to a more advanced tenuinucellate type leads to decrease in the number of structures providing for the embryo normal development. For example, crassinucellate ovule in Ceratophyllum, Nelumbo, Nuphar is characterized by long-term basic structures: by two many-layered integuments, the well-developed nucellus differentiating into a number of structural elements (nucellar cap, parietal tissue, etc.) and hypostase. However the parietal tissue in Nelumbo and Nuphar is destroyed during the embryo sac development (Batygina et al., 1980, 1982, 1983; Shamrov, 1981, 1990; Winter, 1987; Titova, 1988; Winter, Shamrov, 1981).

In the tenuinucellate ovule of Gentiana the nucellus presenting only by the epiderma degenerates as early as the stage of the 4-nuclear embryo sac. The parietal tissue and nucellar cap are absent (Shamrov, 1988). Early degeneration of the nucellus is compensated by the development of the multilayered integument, the inner epiderma of which differentiates into a integumentary tapetum (Shamrov, 1990).

The ovule in Triticum has features of both crassinucellate and tenuinucellate types. It possesses a vigorous nucellus that disintegrates as early as the stage of globular embryo, and ephemeral hypostase (Batygina, 1974). Parietal tissue lacks in the ovule. The ovule in Triticum should, presumably, be related to an intermediate type of ovule, as has been reported by Norstog (1974) for Hordeum vulgare. Taken together, the family Poaceae is characterized by heteromorphy of nucellus, a fact showing that ovule evolution occurs already within itself.

It is of interest to note here that in the species examined, except Gentiana cruciata, the embryo in mature seed contains a plumule of diverse degree of development. As known the embryo of monocotyledons is characterized by a well-developed plumule. This feature combined with other characteristics bears witness to a relation between nymphaeous plants (Nymphaeales, Nelumbonales, Ceratophyllales) and monocotyledons, fact shedding further light on the problem of monocotyly.

To characterize the concurrent development of ovule structures, we performed a histochemical analysis of their polysaccharides, total proteins and tannins. In crassinucellate ovules of Nuphar, Nelumbo and Ceratophyllum starch deposition is correlated basically with the fertilization. But in tenuinucellate ovule of Gentiana it starts earlier - from the stage of primordium with archesporial cell, reaching maximum at the moment of fertilization.

It should be noted that in crassinucellate ovules the site of starch appearance and the time of its absorption are varied. In Ceratophyllum, for example, starch accumulation in all structures of the ovary and ovule (integument, nucellus) attaining its maximum by the onset of zygote division. Subsequently starch quantity falls dramatically almost up to its complete disappearance at the stage of globular embryo. It is noteworthy that from the earliest stages of embryogenesis starch is present in the embryo itself, then reaches its maximum by the embryoderma differentiation in globular embryo.

In Nelumbo starch deposition occurs in the cells of external integument following fertilization, it reaches its maximum during late organogenesis in the embryo (differentiation of 2-3 leaf primordia in the plumule). Then the starch absorption is observed. In contrast to Ceratophyllum, starch deposition in Nelumbo embryo occurs at the stage of formed cotyledon lobes and the first leaf blade of the plumule.

Histochemical reaction for tannins showed their presence in crassinucellate ovule of Ceratophyllum, Nelumbo and Nuphar exclusively. Again as in the case of starch, the examined species showed different sites and periods of tannin occurrence. In Ceratophyllum tannins occur in hypostase at the octant stage and persist as long as mature seed. In Nelumbo tannin deposition begins in chalazal, funicular and, partly, outer integumental cells during embryo sac maturation and persists again until the stage of mature seed.

Described morphogenetic correlations and the data of the dynamics of starch, tannins and total proteins have led us to distinguish some critical periods in the ovule and seed development. It seems to exist common and taxon-specific critical periods (stages).

The common critical periods for crassinucellate ovules are: before fertilization - megasporocyte before meiosis, mature embryo sac; after fertilization - zygote, globular embryo with embryoderma, early organogenesis, late organogenesis, mature seed.

Apart from the above two common critical stages, we can distinguish another stage, specific for Nelumbo and Nuphar, namely, the stage of 2-, 4-nuclear embryo sac, which is characterized by a maximum development of nucellar cap and the beginning of parietal tissue degeneration.

Distinction of critical periods in seed development is supported by data on the dynamics of growth and soluble carbohydrates in the lotus and hornwort embryo (Batygina, Vasilyeva, 1981; Shamrov, 1981; Vasilyeva et al., 1987).

The critical periods in the development of tenuinucellate ovule are those of: before fertilization - ovule primordium with archesporial cell, prophase I meiosis, functioning megaspore, mature embryo sac; after fertilization: zygote, early globular embryo, beginning of organogenesis in embryo and mature seed. This division into periods is supported by data on the dynamics starch, dextrins and proteins (Shamrov, 1990; Shamrov, in press).

The difference between the species is that each critical point corresponds to a different combination and state of structures and a different pattern of deposition. This leads probably to a diverse manifestation of embryo autonomy, when the embryo is capable of autotrophic development (Batygina, Vasilyeva, 1981; Vasilyeva, in press).

Research into the genesis of ovule structures with regard to contrasting species showed that embryonal processes occurring in the system ovary - ovule - embryo sac are correlated in space and time. Disturbances in time and space coordination due to absence of fertilization, for example in Nelumbo, were accompanied by the decrease in tannin content in the ovule chalazal part (Vasilyeva et al., 1988). Interestingly, this also leads to a considerable reduction in the wet mass and content of saccharose up to its complete disappearance in the ovule, ovary and receptacle. The ovules, in which fertilization did not occur seem to cease functioning as attractive centres.

Analysis of the feature of ovule morphogenesis in the examined species leads us to conclude that specificity of the development of individual structures (obturator, operculum, nucellar cap, etc.) in the early stages is presumably due to ecology of pollination. Distinctions in the development of ovule structures after fertilization, particularly in the structure of mature seed (degree of embryo differentiation, localization of storage substances, structure of seed coat, etc.) appear to be largely due to dissemination patterns, ecology of seed and fruit germination.

Cited Literature

Batygina, T.B. 1974. The embryology of wheat: (In Russ.), (Leningrad, USSR:Kolos).

Batygina, T.B. 1984a. About necessity of improving of modern classifications of ovules types in Angiosperms. Abstr. 1-st All-Union Confer. on plant anatomy. Leningrad, USSR:Nauka, 18.

Batygina, T.B. 1984b. The ovary and ovule is a complex integrated system (concurrent development of structures and some aspects of the substances transport in reproductive structures). Proc. All-Union Symp. "Morpho-physiological aspects of female structures development in ovulated plants." Telavi, USSR, 6.

Batygina, T.B., Kolesova (Titova) G.T., Vasilyeva, V.E. 1983. Embryology of Nymphaeales and Nelumbonales. III. Embryogenesis of Nelumbo nucifera. Bot. J, v. 68, N 3, 311-325 (In Russ.).

Batygina, T.B., Kravtsova, T.I., Shamrov, I.I. 1980. The comparative embryology of some representatives of the orders Nymphaeales and Nelumbonales. Bot. J., v. 65, N 4, 1071-1087 (In Russ.).

Batygina, T.B., Shamrov, I.I., Kolesova (Totova), G.E. 1982. Embryologyt of the Hymphaeales and Nelumbonales. II. The development of the female embryonic structures. Bot. J., v. 67, N 9, 1179-1195 (In Russ.).

Batygina, T.B., Vasilyeva, V.E. 1981. Experimental study of embryo differentiation in angiosperms. Acta Soc. Bot. Polon., v. 50, N 1, 257-263.

Batygina, T.B., Vasilyeva, V.E. 1983. The expediency of system approach to the problem of embryonic differentiation in the angiosperms. Ontogenesis, v. 14, N 3, 304-311 (In Russ.).

Nikiticheva, Z.I. 1990. Dynamics of seed structures development in Angiosperms. Abstr. XI Intern. Symp. "Embryology and seed reproduction." Leningrad, USSR, 112.

Nikiticheva, Z.I. 1991. Dynamics of seed structures development in Angiosperms. Proc. XI Intern. Symp. "Embryology and seed reproduction." Leningrad, USSR.

Norstog, K. 1974. Nucellus during early embryogeny in barley: fine structure. Bot. Gaz., v. 135, 97-103.

Shamrov, I.I. 1981. The embryology of the family Ceratophyllaceae in connection with its systematic position. Diss. Leningrad, USSR (In Russ.).

Shamrov, I.I. 1988. Ovule development and structural characteristics of embryo sac in some members of Gentianaceae family. Bot. J., v. 73, N 2, 213-222 (In Russ.).

Shamrov, I.I. 1990. The ovule Gentiana cruciata (Gentianaceae): structural-functional aspects of development. Bot. J., v. 75, N 10, 1363-1379 (In Russ.).

Titova, G.E. 1988. The embryology of some aquatic plants (Nelumbo nucifera Gaertn., Cabomba caroliniana A. Gray, Trapa astrachanica (Fler.) Winter.). Diss. Leningrad, USSR (In Russ.).

Vasilyeva, V.E., Batygina, T.B., Titova, G.E. 1987. Morpho-physiological correlations in the development of the reproductive structures of Nelumbo nucifera Gaertn. Phytomorphology, v. 37, N 4, 349-357.

Vasilyeva, V.E., Titova, G.E., Ermakov, I.P., Morozova, N.M., Chochya, K.A. 1988. Time and space concordance of the growth, differentiation and metabolism in the developing seed of Nelumbo nucifera Gaertn. Proc. 10th Symp. "Sexual Reproduction in Higher Plants." Eds. M. Cresti, P. Gori, E. Pacini. Springer-Verlag, Berlin e.a., 487.36.

Winter, A.N. 1987. The embryology of the genus Hydrostemma (Barclaya) in connection with its systematic position. Diss. Leningrad, USSR (In Russ.).

Winter, A.N., Shamrov, I.I. 1991. Ovule and embryo sac development in Nuphar lutea (Nymphaeaceae). Bot. J., v. 76, N 3, 74-86 (In Russ.).

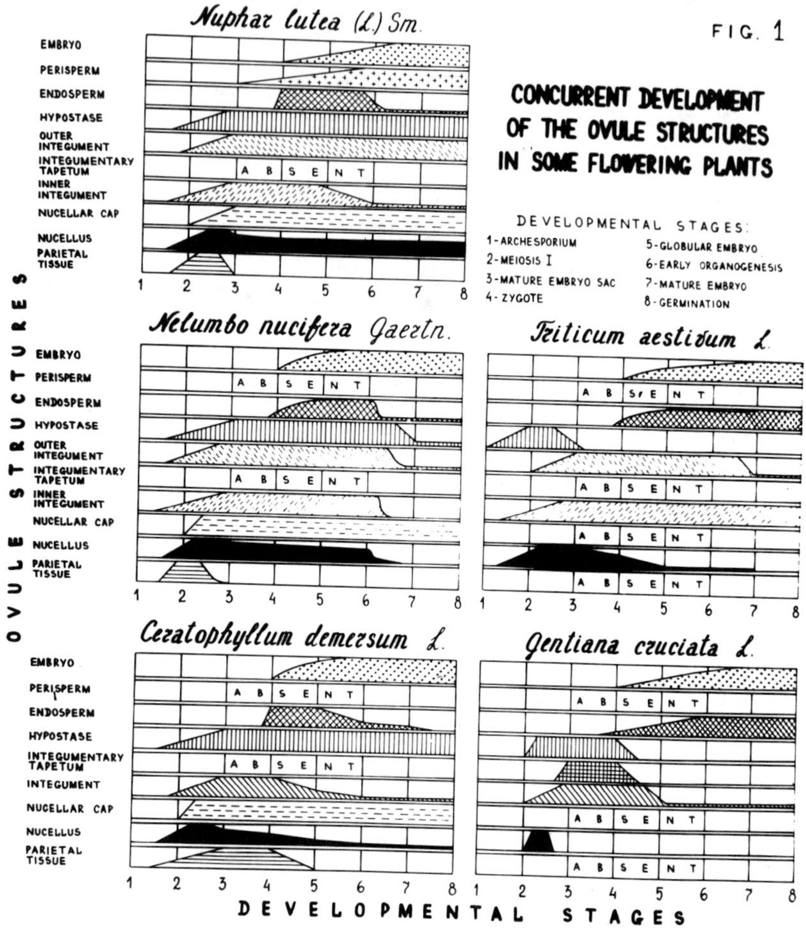

FIG. 1
CONCURRENT DEVELOPMENT OF THE OVULE STRUCTURES IN SOME FLOWERING PLANTS
DEVELOPMENTAL STAGES:
1 - ARCHESPORIUM
2 - MEIOSIS I
3 - MATURE EMBRYO SAC
4 - ZYGOTE
5 - GLOBULAR EMBRYO
6 - EARLY ORGANOGENESIS
7 - MATURE EMBRYO
8 - GERMINATION
Nuphar lutea (L.) Sm.
Nelumbo nucifera Gaertn.
Triticum aestivum L.
Ceratophyllum demersum L.
Gentiana cruciata L.
OVULE STRUCTURES
EMBRYO
PERISPERM
ENDOSPERM
HYPOSTASE
OUTER INTEGUMENT
INTEGUMENTARY TAPETUM
INNER INTEGUMENT
INTEGUMENT
NUCELLAR CAP
NUCELLUS
PARIETAL TISSUE
ABSENT
DEVELOPMENTAL STAGES

23
Recent Advances in Clearing Techniques for Study of Ovule and Female Gametophyte Development

J.M. Herr, Jr.

Introduction

In the past twenty years, several clearing techniques have been developed to replace or supplement traditional section methods for the study of ovule and female gametophyte development in angiosperms. A common feature of these new approaches is the use of specific chemical treatments to make the tissue of the ovules translucent with minimal extraction of protoplasmic content. At the level of transmission light microscopy, the ovular tissues are structurally unaltered by the process. Although the operative mechanism is not fully understood, apparently cellular organelles which ordinarily differ in their refractive properties become more closely uniform in this respect. The effect is a very low contrast among cellular components of cleared tissue examined with bright-field optics. Phase contrast and Nomarski interference optics, however, intensify the small refractive differences among organelles and so reveal features of cellular structure.

Procedures: Origins and Guidelines

The technique introduced by Herr (1971) is based on a mixture of lactic acid, chloral hydrate, phenol, clove oil, and xylene (2:2:2:2:1, by weight). Because of the proportions of the ingredients, this mixture became known as four-and-a-half (4½) clearing fluid. Some fixed ovules treated with this clearing fluid become sufficiently translucent to permit an examination of their structural and developmental features with phase contrast or Nomarski interferences optics. Several modifications of the original fluid were introduced, and the most widely successful of these (Herr, 1974) provided for the addition of benzyl benzoate (1 part by weight) to the original mixture (BB-4½). Both fluids take effect through altering the index of refraction of cellular components with minimal extraction of cell content. Other clearing fluids that operate on this principle were soon to appear. Crane (1978) introduced the use of methyl salicylate as a clearing fluid suitable for the study of apomixis in ovules of Zephyranthaceae, and Young et al. (1979) made use of this method in detecting aposporous apomixis in grass ovules. Stelly et al. (1984) used methyl salicylate to clear ovules selectively stained with Mayer's hemalum. This method introduced sufficient contrast in the cleared tissues for examination with bright-field optics. Methyl salicylate has proved especially effective as a clearing fluid for members of Solanaceae. The most recently developed clearing fluid, a mixture of benzyl benzoate and dibutyl phthalate (2:1, v/v) was introduced by Crane and Carman (1987) for the study of apomixis in the Australian wheatgrass, *Elymus*.

A central guiding principle that emerges from these innovations is that no clearing technique has universal application. New methods undoubtedly will appear in the future, and present methods will be further modified.

Inadequacy of a selected clearing technique may not be related to the type clearing fluid employed, but rather to other aspects of the procedure. Often, for example, ovules appear brown as a result of natural tannins or from oxidized products of fixation. The coloration, which tends to obscure the cleared image, often can be removed by treating the ovules for 10 to 20 hours in Stockwell's solution (Johansen, 1940; Herr, 1985; Palser et al., 1989).

Perhaps the most frequent cause of failure with any of the techniques mentioned is pressure on cleared ovules imposed by the cover glass. Pressure compacts the tissues and destroys the cleared image. The effect is especially hazardous for ovules made fragile in 4½ type fluids. Although methyl salicylate tends to harden ovules, the image is destroyed when they are flattened by pressure of a cover glass for several hours (Young et al. 1979). Microscopists of the mid-19th century recognized the potential ill effects of cover glass pressure on specimens cleared and mounted in solutions of calcium chloride. In the technique employed by Schleiden (1849) and Schacht (1853), paper strips of appropriate thickness were glued to a slide, and the specimen was mounted between them in a drop of calcium chloride solution. The cover glass placed over the paper strips was raised slightly above the specimen. Early in the development of the 4½ clearing technique, the need for support mounts to eliminate pressure on the ovules was recognized. Special slides, referred to as Raj slides and similar to those used by Schleiden and Schacht, were prepared using permanently affixed cover glasses instead of paper strips as the support mounts (Herr, 1971; Smith, 1973). Palser et al. (1989) did not find support mounts necessary probably because of the large number of ovules under each cover glass (150-250). In cases where ovules are quite small, the sheer buoyancy of the fluid floats the cover glass above the specimen so eliminating a need for support mounts. Still, users of clearing techniques for ovule studies should be aware of the adverse effects of cover glass pressure on otherwise completely adequate preparations.

Clearing Fluid Modifications

The addition of benzyl benzoate to the original 4½ clearing fluid noted earlier greatly extended the range of usefulness for the technique. However, the suitability of one fluid over the other in any specific situation cannot be predicted. Ovules from closely related taxa often exhibit dissimilar responses to the two fluids. Both fluids turn dark brown and become ineffective in two to four weeks in the light at room temperature. Refrigerated in dark bottles, these fluids maintain their effectiveness for over two years. The principal cause for the adverse changes as the fluids age, viz., the photooxidation of clove oil, prompted a search for a replacement for this unstable ingredient. Attention soon focused on dibutyl phthalate as a possible substitute, since Crane and Carman (1987) have shown it to be effective in combination with benzyl benzoate. Dibutyl phthalate in several features resembles eugenol, the principal constituent of clove oil. Both substances are benzene derivatives of similar density, index of refraction, and solubility in organic solvents. They are, however, distinctive in molecular structure and in some physical properties. Unlike clove oil or eugenol, dibutyl phthalate is highly resistant to photooxidation which enhances its value as a clearing agent. An exact weight substitution of dibutyl phthalate for clove oil produces a clearing fluid equal or superior to the standard mixtures for some applications. It does not readily darken with age at room temperature, and it can be produced at a fraction of the cost of the clove oil-based fluids. The stability of the dibutyl phthalate fluid under storage conditions is apparent when it is compared to clove oil-based fluids (Fig. 1 A). Tube one (left to right) contains fresh dibutyl phthalate-based fluid (BB-DP-4½), tube two, the same type fluid refrigerated for over a year, and tube three, fluid kept at room temperature for over a

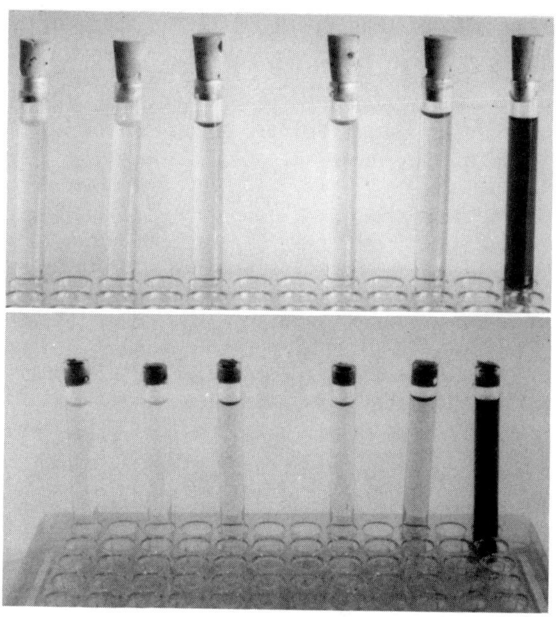

a) b) Figure 1 **A,B**. Six tubes of clearing fluid (each figure); tubes 1-3, dibutyl phthalate-based fluid; tubes 4-6, clove oil-based fluid. **A** Tube 1, fresh; tube 2, refrigerated for one year; tube 3, room temperature for one year; tube 4, fresh; tube 5, refrigerated for two years; tube 6, room temperature for four weeks. **B** Six tubes of **A** after two years of refrigeration.

year. Tube four contains fresh clove oil-based fluid (BB-4½). The same fluid in the fifth tube was refrigerated for over two years, while that in the sixth tube was maintained at room temperature for only four weeks. Fluids in tubes two and five are from batches which were periodically brought to room temperature for use. The stability of BB-DP-4½ over BB-4½ is apparent (Tube 3 cf Tubes 5 and 6). However, tube five clearly indicates that BB-4½ can be maintained in a useful, little oxidized state by refrigeration. The six tubes of figure 1 A were refrigerated continuously for an additional two years and then photographed a second time (Fig. 1 B). The value of refrigera-tion in preserving both types of fluid is evidenced by the absence of appreciable change during the interval between figures 1 A and 1 B.

The relative effectiveness of both fluids was tested on ovules during megasporogenesis and female gametophyte development in *Ludwigia hexapetala*, *L. uruguayensis*, *Ilex opaca*, and *Cassia occidentalis* and on the internal cell layers of the massive integument and the cellular endosperm in immature seeds of *Ilex opaca*. In this diverse material, the two fluids proved nearly identical in effectiveness as is exemplified by an ovule of *Ilex opaca* at the 4-nucleate stage of female gametophyte development (Fig. 2 A cf Fig. 2 B). The only consistently noticeable difference in the clearing images provided by these fluids is their effect on cytoplasmic inclusions and is especially obvious in cleared megasporocytes of *Ludwigia hexapetala* throughout the course of meiosis I (Fig. 2 C cf Fig. 2 D). From prophase I onward, the megasporocyte contains clusters of cytoplasmic inclusions at the chalazal and micropylar poles of the cell. Starch and probably protein, the components of the clusters, exhibit greater phase density in megasporocytes cleared with BB-DP-4½ (Fig. 2 C) than in those cleared in BB-4½ (Fig. 2 D). That dibutyl phthalate may serve as an effective substitute for clove oil in 4½ type fluids appears established by the observations reported here.

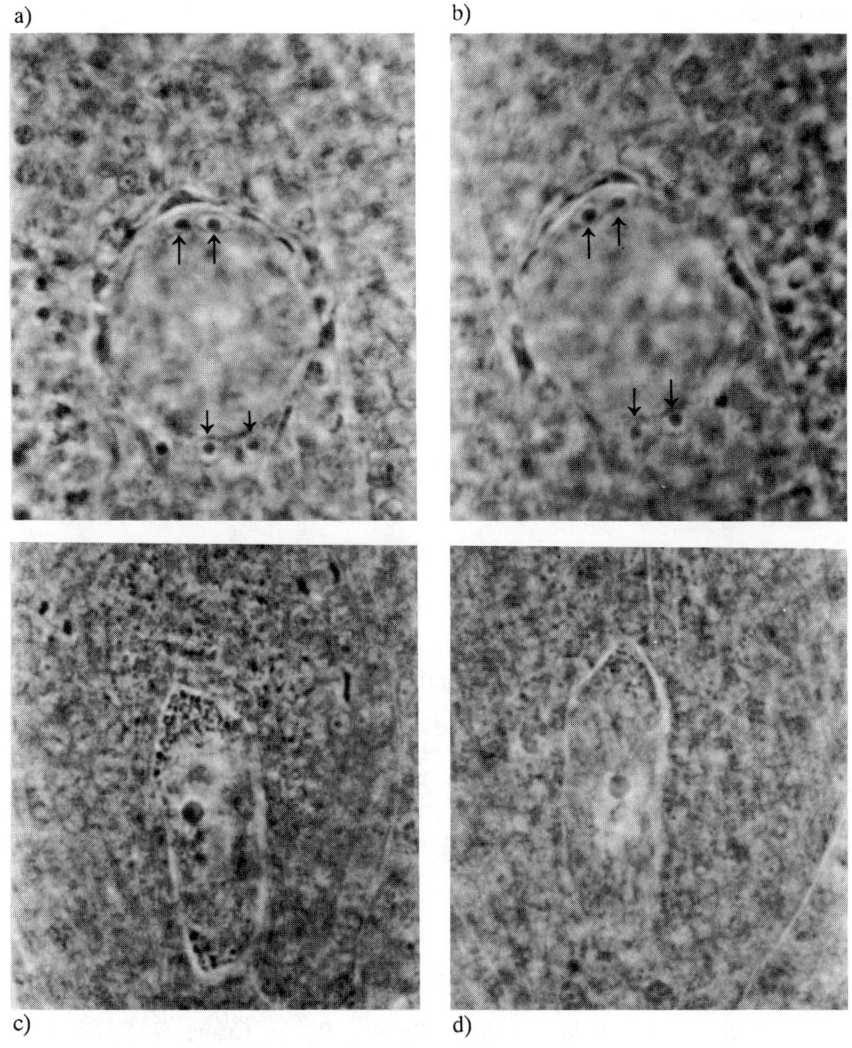

Figure 2 **A,B.** Four-nucleate female gametophyte of Ilex opaca (arrows indicate nuclei); **C,D** megasporocyte of Ludwigia hexapetala. All X619. **A,C** Ovule cleared in dibutyl phthalate-based clearing fluid. **B,D** Ovule cleared in clove oil-based clearing fluid. In the megasporocyte, inclusions in micropylar and chalazal clusters show greater phase density in BB-DP-4½ (**C**) than in BB-4½ (**C**).

References

1. Crane, C. F.; Carman, J. G. Mechanisms of apomixis in *Elymus rectisetus* from eastern Australia and New Zealand. Am. J. Bot. 72:477-496; 1987.
2. Herr, J. M. Jr. A new clearing-squash technique for the study of ovule development in angiosperms. Am. J. Bot. 58:785-790; 1971.
3. Herr, J. M. Jr. A clearing-squash technique for the study of ovule and megagametophyte development in angiosperms. In: Radford, A. E.; Dickison, W. C.; Massey, J. R.; Bell, C. R., eds. Vascular plant systematics. New York: Harper & Row Publishers;1974.
4. Herr J. M. Jr. The removal of phlobaphenes for improved clearing of sections and whole structures. Am. J. Bot. 72:817; 1985.
5. Johansen D. A. Plant microtechnique. New York,NY: McGraw-Hill; 1940.
6. Palser B. F.; Rouse J. L.; Williams E. G. Coordinated timetables for megagametophyte development and pollen tube growth in *Rhododendron nuttalli* from anthesis to early postfertilization. Am. J. Bot. 76:1167-1202; 1989.
7. Schacht H. The microscope in its special application to vegetable anatomy and physiology. London: Samuel Highley; 1853.
8. Schleiden M. J. Principles of scientific botany (translated by Lankester E). London: Longman, Frown, Green, and Longmans; 1849.
9. Smith B. B. The use of a new clearing technique for the study of early ovule development, megasporogenesis, and megagametogenesis in five species of *Cornus*. Am. J. Bot. 60:322-338; 1973.
10. Stelly D. M.; Peloquin S. J.; Palmer R. G.; Crane C. F. Mayer's hemalum-methyl salicylate: a stain-clearing technique for observations within whole ovules. Stain Technol. 59:155-161; 1984.
11. Young B. A.; Sherwood R. T.; Bashaw E. C. Cleared-pistil and thick-sectioning techniques for detecting aposporous apomixis in grasses. Can. J. Bot. 57:1668-1672; 1979.

24
Comparisons of Early Ovule Development through Megagametogenesis in Rapid-cycling *Brassica rapa* L. and *B. campestris* L. Using the Herr Clearing Techniques

Bruce B. Smith

The study of early ovule development and debates between Hofmeister and Schleiden on the nature of pollen tube development occured in the mid-nineteenth century. Giovanni Battista Amici's accidental discovery of pollen-tube development in Portulaca oleracea put him squarely opposed to Schleiden's interpretations and furthered investigations by others who entered this exciting new research of that period (Maheshwari, 1950). Although thousands of articles have been printed about pollen and seed development only a small fraction of flowering plant species have been investigated (Davis,1963).

Attention has been directed to the minute size of the megagametophyte in relation to the sporophytic generation and the taxonomic importance that could be given to these variations (Herr,1967;Smith,1973). The development of the techniques employed in these studies was an attempt to make the megagametophyte more accessible to studies for the purpose of recording these subtle variations (Herr,1971;Smith,1975,1989a,1989b). Dr. J.M.Herr Jr., University of South Carolina, Columbia, South Carolina 29208 has a record of this activity.

Since most curricula omit these studies, I recognized the potential for helping students learn more about this subject using rapid-cycling Brassicas(RCBr). Having worked with B. campestris L., I felt certain that the Herr clearing procedures could easily be adapted to rapid-cycling B.rapa L.(Smith,1981,1989a,1989b).

Development of Rapid-Cycling Brassica

Rapid-cycling B.rapa L. was developed from Brassica with a normal 6-12 month life cycle. The average life cycle of B.rapa L.(RCBr) is approximately 35 days with flower development occurring on the 13th day. RCBr and five other species of Brassica are available from the Crucifer Genetics Cooperative (Williams & Hill,1986).

The Wisconsin Fast Plants Manual™ contains over two dozen laboratory topics which provides students an opportunity to work with living plants (Carolina Biological Supply, Burlington, N.C. USA).

Why Study RCBr?

The purposes of this study were: (1) to determine the applicability of the Herr Clearing fluid for studying early ovule development through megagametogenesis for this newly developed RCBr, and (2) having studied other mustards including the wild mustard B.campestris L. which is synonymous to B.rapa I wanted to compare the ontogenetic events of the new variety to my previously published study of B.campestris L..

Materials and Methods

Using the Wisconsin Fast Plant growing system (Carolina Biological Catalog,Burlington, NC,USA), rapid-cycling B.rapa L. seeds were planted and whole inflorescence were collected on days 13-18 after planting. Inflorescence were fixed in FPA_{50}(formalin, propionic acid, ethyl alcohol (ETOH) 50%) for a minimum of 24 h and stored in 70% ETOH. Whole inflorescence were dehydrated to 100% ETOH and placed directly in one or two versions of the Herr clearing fluid. Clearing fluid variety number one was composed of lactic acid, phenol, diisobytyl phthalate (clove oil substitute), propionic acid, and xylene, 2:2:2:2:1 by weight. This particular version of the Herr clearing fluid has never been tried and has the advantage of not using the ingredient chloral hydrate which is a controlled substance requiring federal USA permits to use. The second variety did contain the chloral hydrate as usual. The entire procedure from planting to complete analysis of ovule development through megagametogenesis can be completed in less than three weeks.

Early Ovule Development

Ovules begin to grow from the placenta on the 13th day after the seeds are planted. A bitegmic condition is noticeable soon after the ovule grows from the placenta. The ovule is crassinucellate with more than one hypodermal archesporial cell developing into a primary parietal cell and a megaspore mother cell (Fig.1). Nucellar tissue contains more than one megaspore mother cell (MMC) commencing the meiotic process (Fig.2).

Megasporogenesis

Multiple MMC's create multiple dyads. As the second meiotic division progresses, the central dyad survives while the other dyads are dissolved. The chalazal spore enlarges to become the embryo-sac and the remaining components of the linear tetrad are crushed. Even though the chalazal spore is the functional spore, the micropylar spore of the linear tetrad is often the last to be crushed (Fig.3).

Megagametogenesis

Cellular volumn enlarges after the first mitotic division (Fig.4). Two pairs of nuclei align at the extreme ends of the embryo-sac and often at right angles to the nucellar axis (Figs.5,6). Remaining spores divide to form 2-nucleate stages. The ovule is campylotropous in shape but the nucellus is amphitropous. This causes the embryo-sac

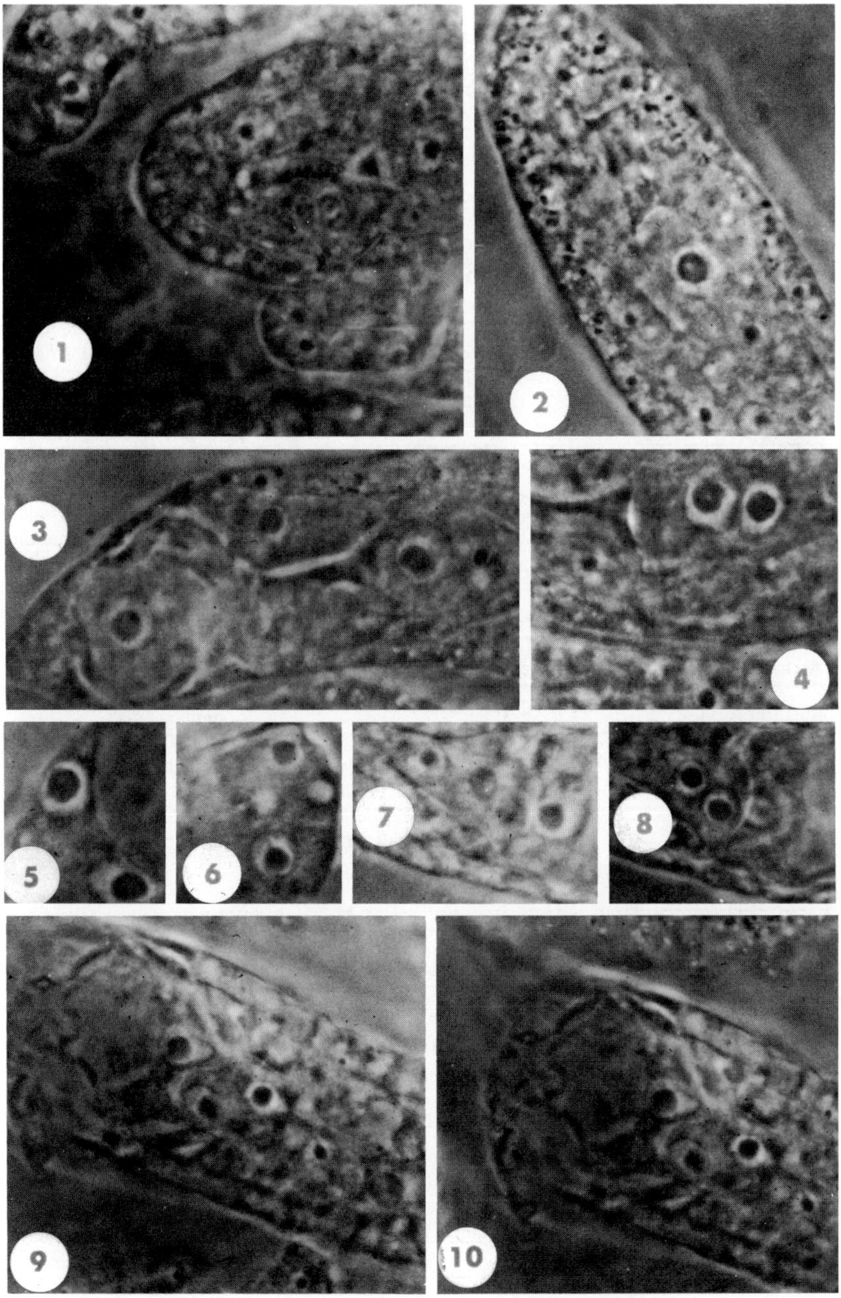

Figures 1-10. Early ovule development through megagametogenesis of rapid cycling *B.rapa L.*. All figures X 1385. Figures 5,6 represent the same 4-nucelate stage. Figures 7,8 are of the micropylar end of the same mature 8-nucleate stage. Figures 9,10 are the same mature 8-nucleate stage isolating the chalazal nuclei.

to be bent at its center. Using oil immersion, the nuclei of 4-and 8-nucleate stages are not in the same focal plane. As a result, more than two photographs are needed to show both the micropylar and chalazal activity. The mature embryo-sac contains 8 nuclei and would therefore be referred to as Polygonum Type. In the micropylar end there are three closely arranged nuclei of about the same size with a fourth and larger nucleus toward the center. A large central vacuole separates four other nuclei in the extreme opposite or chalazal end of the embryo-sac. These four nuclei are loosely arranged with one of the four also arranged more closely to the center of the embryo-sac (Figs.7,8,9,10).

Table 1. Comparisons between *B.rapa* L. and *B.campestris* L. for early ovule development through megagametogenesis

	B.rapa L.(RCBr)[a]	B.campestris L.[b]
Seed to flower	12-13 days after planting	annual,fruits June-Oct.,NE USA
Plant size	3-4 dm tall	3-9 dm tall
Inflorescence	raceme	similar
Ovule shape	epitropous	similar
Ovule integuments	bitegmic	similar
Archsporial cell(s)	several enlarge asynchronously	similar
Megasporocyte	several	similar
Megasporogenesis	several tetrads form, 1 survives, chalazal spore functional	similar
2-nucleate stage	chalazal spore divides,nuclei aligned with the micropylar-chalazal axis	similar
4-nucleate	nuclei align at right angles to micropylar-chalazal axis	similar
8-nucleate	3 antipodals in the narrowed end, two synergids flank a larger egg nucleus, 2 polar nuclei near egg	similar

a Smith, present study.

b Smith,Bruce and Johnathan Lightner.1989. Early ovule development, megasporogenesis, and megagametogenesis in Brassica campestris L.(Cruciferae).J.Pa.Acad.Sci.63(3):161-165

Discussion

Rapid-cycling B.rapa L. lends itself well to clearing. The gametophyte is often omitted in phylogentic considerations. Even though phylogenetic interpretation was not the intent of this study, ontogenetic differences between a wild species and a newly developed variety using a method that provides information quickly certainly allows for this if enough taxa are studied. The similarities and differences noted qualitatively between rapid-cycling B.rapa L. and the previously studied B.campestris L.(synonym B.rapa L.) can be seen in Table I.

With the exception of plant size, the newly developed rapid-cycling B.rapa L. has similar features to that of wild B.campestris L.(synonym B.rapa L.). To look more carefully at the differences, I would recommend further studies of quantitative variations that might exist between and among these newly developed varieties of Brassica.

References

1. Davis,G.L. Systematic embryology of the angiosperms. New York:John Wiley & Sons,Inc.;1966.
2. Herr,J.M.Jr. On the nature of variation.Phytomorphology.17(4):200-207;1967.
3. Herr,J.M.Jr. A new clearing squash technique for the study of ovule development in angiosperms.Amer.J.Bot. 58(8):785-790;1971.
4. Maheshwari,P. An introduction to the embryology of angiosperms. New York:McGraw-Hill Book Company;1950.
5. Smith,B.B. The use of a new clearing technique for the study of early ovule development, megasporogenesis, and megagametogenesis in five species of Cornus L. Amer.J.Bot. 60(4):322-338;1973.
6. Smith, B.B. A quantitative analysis of the megagametophyte of five species of Cornus L. Amer.J.Bot. 62(4):387-399;1975.
7 Smith,B.B. Early ovule development, megasporogenesis, and megagametogenesis of Capsella bursa-pastoris (L.) Medicus using Herr fluid and clearing agent. Proc.Pa.Acad.Sci. 50:130-132;1981.
8. Smith,B.B. and S.Gilbert. Early ovule development, megasporogenesis, and megagametogenesis in Draba verna L. (Cruciferae). Proc.Pa.Acad.Sci. 63(2):88-92;1989.
9. Smith,B.B. and J.Lightner. Early ovule development, megasporogenesis,and megagametogenesis in Brassica campestris L.(Cruciferae). Proc.Pa.Acad.Sci. 63(3):161-165;1989.
10. Williams,P.H.and C.B.Hill. Rapid-cycling populations of Brassica. Science.232:1385-1389;1986.

25
Maturation of the Pollen of Cotton

William A. Jensen and Cherie L.R. Wetzel

Pollen has a highly significant mission: It must develop a tube and deposit two sperm within the embryo sac. In the case of cotton, Gossypium hirstum L., the pollen tube is four to five centimeters long and the growth of the tube takes 14 to 15 hours. During this growth there is no clear evidence that the tube absorbs nutrients from the style. Consequently the mature pollen is both large and filled with reserve food. The mature cotton pollen has a remarkable architecture in which the endoplasmic reticulum (ER) has greatly enlarged cisternae that contain reserves of proteins. (Fisher, et al. 1968; Jensen et al. 1968). The enlarged ER forms pockets that are lined with lipid bodies and filled with dictyosome vesicles. The surface of the ER is lined with densely packed ribosomes. Thus, the cotton pollen grain has a remarkable supply of higher energy reserves and abundant raw material for rapid growth. But, how do these reserves accumulate and when during the maturation of the grain are they formed?

These questions are the bases for the present work on the maturation of the pollen grain of cotton. To answer these questions we undertook a study of the last nine days of pollen development concentrating on the condition of cytoplasm of the vegetative cell. Standard electron microscope procedures were used combined with cytochemical methods and observation with a high voltage electron microscope. We studied the deposition of starch, non-starch polysaccharides, lipids, proteins, and ribonucleic acid (RNA).

One of the first reserves to accumulate is starch. At first, the plastids that will accumulate starch appear little different from the mitochondria, but then starch accumulation begins and both the amount and the number of plastids increase. By three days before anthesis the density of amyloplasts is very high. Later, as the ER cisternae begin to dilate, amyloplasts are separated from each other. By anthesis the actual number of plastids appears reduced, but the individual grains are larger. Total starch reserves are high in mature pollen.

The same time starch is accumulating in the plastids, lipid bodies are forming in the cytoplasm. These lipid bodies lie adjacent to the ER cisternae and initially may be completely surrounded by the cisternae. A proliferation of the ER occurs at the same time lipid bodies become apparent and it is clear that a relationship exists between the lipid bodies and the ER throughout their development.

Related to both starch and lipid accumulation in the maturing pollen cytoplasm is the increase in dictyosome vesicles containing polysaccharide. It appears likely that the dictyosomes of cotton pollen first produce vesicles that are used in the formation of the intine of the pollen wall. Later, the dictyosome vesicles in the cytoplasm increase in number and are retained in the cytoplasm. The switch from secretion to accumulation is marked by a change in the numbers of cisternae per dictyosome. In the first phase, where the vesicles are taking part in the formation of the intine, the number of cisternae per dictyosome is 5 to 7, while in the later stages, where the vesicles accumulate in the cytoplasm, the number of cisternae per dictyosome increases to 7 or 8 and over. In the final maturation of the pollen, the dictyosome-derived vesicles become concentrated in pockets formed by the ER.

During the period of starch and lipid accumulation, large numbers of polyribosomes are observed both attached to the outer surfaces of the ER cisternae and free in the cytoplasm. The morphological evidence suggests a high rate of protein synthesis is occurring and that much of this protein is accumulating in the cisternae of the ER. This accumulation results in the rapid expansion of the ER and this in turn causes formation of the ER pockets. What is of interest is that the ER pockets form only in the last hours before the pollen is released. Stereographic observations show that the ER does, in fact, form true pockets lined with ribosomes and containing lipid bodies and dictyosome vesicles. Moreover, the ER is filled with protein, thus, making the ER pockets remarkable storage units containing all the substances necessary for pollen tube growth.

Why should such pockets form? The answer to this question is not readily apparent. The ER pockets form only hours before the grain germinates and dissipate soon after gemination begins. Only further studies on cotton pollen germination will answer the role the ER pockets play in the growth of the pollen tube. But, cotton pollen, as all other pollen grains are growth machines primed to start when they land on the right stigma. What happens to the ER pockets and the other reserves on germination must await further research.

Literature Cited

Fisher, D.B., W.A. Jensen, and M.E. Ashton. 1968. Histochemical studies on pollen: storage pockets in the endoplasmic reticulum (ER). Histochemie 13: 169-182

Jensen, W.A., D.B. Fisher, and M.E. Ashton. 1968. Cotton embryogenesis: the pollen cytoplasm. Planta 81: 206-228.

26

Pollen Tube Growth in Culture. I. Control of Morphology and Generative Nucleus Division in Cultured Pollen Tubes of *Nicotiana*

M. Read, A. Bacic, and A.E. Clarke

Pollen tubes are undoubtedly interesting cells. They are unusual in growing by tip-extension, with an ordered zonation of organelles, and are one of the few types of differentiated higher-plant cell that can be grown successfully in single-cell culture. Cultured pollen tubes thus form excellent material for studying the fundamental mechanism of tip-growth (Vasil 1987; Steer áSteer, 1989), the metabolic pathways of individual cells, and cell wall deposition and maturation. The pollen tube is unusual in containing the (1«3)-º-linked glucan, callose, in its walls and septa (plugs) without being wounded, and we are thus using pollen-tube membranes to study the control of callose biosynthesis. Lastly, the pollen tube is the male gametophyte, and in many species bears the receptor for the pistil self-incompatibility (SI) product (Haring et al., 1990); pollen tubes can therefore be used in bioassays for SI (Jackson áLinskens, 1990).

Pollen tubes are, classically, easy cells to grow in culture if the pollen grains are bicellular, but the tubes produced are generally much shorter than those observed in the style, and have many morphological abnormalities. Cultured pollen tubes have also been compared to incompatible pollen tubes (tubes from an SI pollination) because of their typically slow and disorganised growth, lack of callose plug deposition, and generally undivided generative cell (Rosen, 1971; de Nettancourt et al., 1974; Kahn áDeMason, 1988). The standard set for healthy pollen-tube growth in culture is of course more rigorous than that for other cells, because pollen tubes leave behind a permanent record of their growth, and if tip-growth is disrupted even briefly this will remain visible as an obvious kink or swelling in the tube. Confidence in the physiological health of cultured pollen tubes is particularly important both in the study of callose biosynthesis (as wounding or disruption of the plasma-membrane is generally sufficient in other cells to induce callose deposition), and in the development of a bioassay for SI (where tubes of different growth-rates and morphologies need to be compared, and reduced specificity of inhibition can be a problem: Jahnen et al., 1989). The characteristics of pollen-tube growth in the style thus need to be replicated in vitro before culture conditions can be judged acceptable.

The growth patterns of the two types of pollen (bicellular and tricellular) are qualitatively different (Brewbaker, Majumder, 1961; Brewbaker 1967), and it has been suggested that bicellular pollen (including Nicotiana) passes though two phases of growth in the pistil (Rosen 1971; Herrero, Dickinson, 1980; Mulcahy, Mulcahy, 1982, 1983). With bicellular pollen, phase 1 (stigmatic) is slower, autotrophic, no callose plugs are deposited, and the generative cell remains undivided, while phase 2 (stylar) is faster, heterotrophic, regular callose plugs are produced, and the generative cell divides to produce two sperm cells. Culture media are normally not sufficiently complex to allow the transition to the second phase in vitro, and in addition it has been postulated that pollen-tubes exhibiting gametophytic SI (GSI) only become sensitive to SI signals at the end of the first phase of their growth and so also do not progress to extensive second-phase growth (Brewbaker, Majumder, 1961; Rosen, 1971; Vasil, 1977; Herrero, Dickinson, 1980, 1981); hence the apparent similarity between cultured and SI pollen-tubes. Plants exhibiting GSI are found widely in the Solanaceae, Liliaceae, Leguminoseae and Rosaceae, among other families, and GSI is generally associated with bicellular pollen and stylar inhibition of pollen-tubes.

Hoekstra and Bruinsma (1979) and Capkova-Balatkova et al. (1980) used inhibitors to distinguish the growth patterns of the two types of pollen; bicellular pollen required active protein translation for continued growth and thus was susceptible to inhibition by cycloheximide. Growth of bicellular pollen is therefore potentially also susceptible to inhibition by the RNAse activity of SI proteins (McClure et al., 1989).

Incompatible pollen tubes from genera of the Solanaceae are shorter than compatible (uninhibited) tubes, are more kinked, have swollen tips and disrupted callose deposition, and eventually burst (de Nettancourt et al., 1974; Cresti et al., 1979; Herrero, Dickinson 1981). Generative cell division appears to occur in SI tubes in some cases: in Petunia and Campanula (Brewbaker, Majumder, 1961) and Citrus (Kahn, DeMason, 1988) stylar arrest occurs before division, while in Lotus and Theobroma arrest is ovarian and generative cell division occurs first (Brewbaker, Majumder, 1961). In Nicotiana alata (Bonig, personal communication) at least some of the generative nuclei in incompatible pollen tubes have divided by 48 hr. Incompatible tubes are packed with organelles and reserve materials and do not appear to be starving to death; rather, there appears to be disruption of their tip organisation.

It is generally presumed that pollen tubes gain some of their nutrition from the stylar matrix through which they grow, although pollen tubes of species with short styles may contain sufficient stored material in the grain to permit the full extent of their pollen-tube growth, and other species, such as Lilium, have pollen that can grow well on sugar-free medium (Dickinson 1968). Exogenous sugars are readily incorporated into pollen-tube walls (Roggen, Stanley, 1971; Read et al., 1989), and sucrose is normally included in pollen-tube growth media. The general requirement of boron for pollen-tube growth has also long been recognised. Brewbaker found that Ca^{2+} ions were a growth factor needed for pollen-tube growth at low pollen concentrations, and in addition added K^{+} and Mg^{2+} ions to the standard medium (Brewbaker, Kwack, 1963). Tupy et al. (1983) showed that amino-acids allowed growth to continue after 18 hr, as the tubes become more dependent on an exogenous source of nitrogen, and Tupy, Rihova (1984) then found that the addition of a non-toxic buffer (MES) prevented acidification of the medium and allowed still longer growth. However, growth was measured by mass, rather than by length or some index of health such as tube tip morphology, generative nucleus division or callose plug deposition.

Our research requires pollen tubes cultured in vitro well into the second phase of their growth, both for bulk wall and membrane preparations and also in case sensitivity to SI proteins is not expressed in young tubes. We have thus made further improvements to the pollen-tube growth medium (Read, Bacic, Clarke, manuscript in preparation). Although other workers may of course not require such an extent of pollen-tube growth, we believe that a complete medium, allowing growth at rates approaching those found in the style, and with healthy morphologies, is likely to produce physiologically more relevant conclusions at both short and long time-points.

Our initial amendments involved reduction in the sucrose concentration, and its partial replacement by PEG-4000, to give better incorporation of radioactivity from labelled sugars into the cell walls (Read et al, 1989; Jahnen et al, 1989). Further improvements to the medium now include incorporation of optimised levels of NH_4Cl or amino-acids; addition of Fe^{2+} ions or, better, buffered levels of Cu^{2+} ions; and inclusion of specially purified and tested PEG-4000 or -6000. Our medium is optimised for the growth of N. tabacum, because this species produces healthier tubes in culture than does N. alata, the tubes of N. tabacum being thinner, longer and straighter, and containing ladders of callose plugs. Cultured pollen tubes of N. tabacum can therefore now be grown at approximately 200 um/hr. The tube tip is thin and straight-sided, producing straight growth of a thin-walled tube (Figure 1); the cytoplasm is even and iridescent, with a long 'cap-block' (Iwanami, 1959); callose plug deposition starts at about 8-10 hr, and by 30 hrs the tubes contain ladders of up to 25 regularly spaced plugs; and generative cell division occurs synchronously at 12-14 hr, and by 18 hr 90% of the tubes contain a pair of compact sperm nuclei (Figure 2). It is difficult to measure tube length accurately at times after 30 hr (Figure 3), but the tubes continue to grow for over 48 hr and until clearly over 1 cm in length.

We are therefore able to observe, between 8 & 14 hr in culture under our conditions, some of the phenomena claimed to be associated with the development of mature, 'phase-2' pollen tubes, which is not normally achieved with pollen tubes grown in vitro (Mulcahy, Mulcahy, 1982). These include callose plug deposition, nuclear division, and heterotrophy (the acquisition of dependence on exogenous reduced nitrogen and metal ions, and an increased uptake of sucrose). However, uncoupling of some of these 'phase-2' phenomena is also observed: the tubes do not show a significant acceleration in growth rate at this point, as occurs in the style (Herrero, Dickinson, 1980; Mulcahy, Mulcahy, 1982), and withholding Cu^{2+} inhibits growth and callose plug deposition but does not inhibit generative cell division. These observations are thus evidence against there being a single signal for entry into the second phase of pollen-tube growth.

The self-compatible species Nicotiana tabacum and N. sylvestris grow well in our optimised medium, but the SI species N. alata does not, and produces thicker and less straight tubes, with thicker walls and very few callose plugs. We have thus not yet been able to test with confidence the effect of SI proteins on the growth and morphology of healthy older N. alata pollen tubes. We are able to conclude, however, that the SI

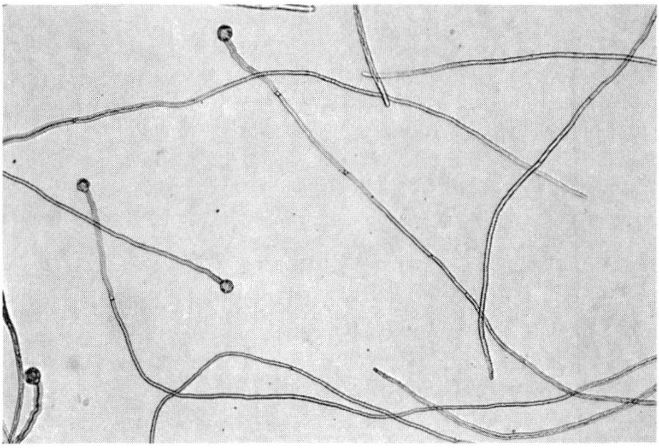

Figure 1: Pollen tubes of N. tabacum after 12 hr growth; 74x.

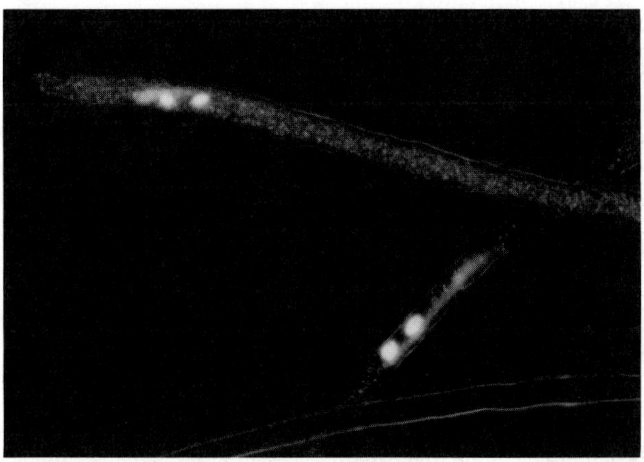

Figure 2: Pollen tubes of N. tabacum after 18 hr growth, stained with DAPI for DNA and viewed in combined fluorescence and dark-field optics. Note the single, elongate, diffuse vegetative nucleus and two rounded sperm nuclei in each pollen tube; 230xmorphology observed *in vivo* cannot be replicated in vitro by removal of sucrose or amino-acids, and therefore that SI arrest of pollen-tube

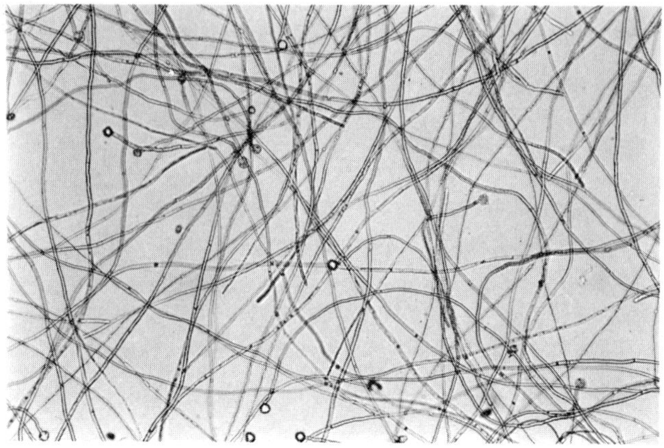

Figure 3: Pollen tubes of N. tabacum after 24 hr growth. Initial pollen grain concentration, 0.1 mg/ml. 33x.

growth is probably not mediated by an inability to metabolise external nutrients.

With these improvements it is now possible to grow reliably in culture pollen tubes of Nicotiana tabacum that extend at 20-30% of the in vivo rate, have healthy, straight-sided tips, deposit a regular ladder of callose plugs, and undergo synchronous generative cell division; they appear to have entered the 'second phase' of pollen-tube growth. These pollen tubes are suitable for the preparation of cell membranes and walls. It does appear, though, that the medium needs to be optimised independently for each species of Nicotiana.

References

Brewbaker, J.L.; Majumder, S.K. Cultural studies of the pollen population effect and the self-incompatible inhibition. Amer. J. Bot. 48: 457-464; 1961

Brewbaker, J.L.; Kwack, B.H. The essential role of calcium ion in pollen germination and pollen tube growth. Amer. J. Bot. 50: 859-865; 1963

Brewbaker, J.L. The distribution and phylogenetic significance of binucleate and trinucleate pollen grains in the angiosperms. Amer. J. Bot. 54: 1069-1083; 1967

Capkova-Balatkova, V.; Hrabetova, E.; Tupy, J. Effects of cycloheximide on pollen of Nicotiana tabacum in culture. Biochem. Physiol. Pflanzen. 175: 412-420; 1980

Cresti, M.; Ciampolini, E.; Pacini, G.; Sarfatti, G.; van Went, J.L.; Willemse, M.T.M. Ultrastructural differences between compatible and incompatible pollen tubes in the stylar transmitting tissue of Petunia hybrida. J. Submicr. Cytol. 11: 209-219; 1979

Dickinson, D.B. Rapid starch synthesis associated with increased respiration in germinating lily pollen. Plant Physiol. 43: 1-8; 1968

Haring, V.; Gray, J.E.; McClure, B.A.; Anderson, M.A.; Clarke, A.E. Self-incompatibility: a self-recognition system in plants. Science 250: 937-941 (1990)

Herrero, M.; Dickinson, H.G. Pollen tube growth following compatible and incompatible intraspecific pollinations in Petunia hybrida. Planta 148: 217-221; 1980

Herrero, M.; Dickinson, H.G. Pollen tube development in Petunia hybrida following compatible and incompatible intraspecific matings. J. Cell Sci. 47: 365-383; 1981

Hoekstra, F.A.; Bruinsma, J. Protein synthesis of binucleate and trinucleate pollen and its relationship to tube emergence and growth. Planta 146: 559-566; 1979

Iwanami, Y. Physiological studies of pollen. J. Yokohama Municipal Univ. Series C-34 (Biol.13) 116: 1-134; 1959

Jackson, J.F. áLinskens, H.F. Bioassays for incompatibility. Sex. Plant Reprod. 3: 207-212; 1990

Jahnen, W.; Lush, W.M.; Clarke, A.E. Inhibition of in vitro pollen tube growth by isolated S-glycoproteins of Nicotiana alata. Plant Cell 1: 501-510; 1989

Kahn, T.L.; DeMason, D.A. Citrus pollen tube development in cross-compatible gynoecia, self-incompatible gynoecia, and in vitro. Can. J. Bot. 66: 2527-2532; 1988

McClure, B.A.; Haring, V.; Ebert, P.R.; Anderson, M.A.; Simpson, R.J.; Sakiyama, F.; Clarke A.E. Style self-incompatibility gene products of Nicotiana alata are ribonucleases. Nature 342: 955-957; 1989

Mulcahy, G.B.; Mulcahy, D.L. The two phases of growth of Petunia hybrida (Hort. Vilm-Andz.) pollen tubes through compatible styles. J. Palynol. 18: 61-64; 1982

Mulcahy, G.B.; Mulcahy, D.L. A comparison of pollen tube growth in bi- and tri-nucleate pollen. Mulcahy, D.L.; Ottaviano, E., eds. Pollen: Biology and its Implications for Plant Breeding; Elsevier Biomedical; 1983

de Nettancourt, D.; Devreux, M.; Laneri, U.; Cresti, M.; Pacini, E.; Sarfatti, G. Genetical and ultrastructural aspects of self and cross incompatibility in interspecific hybrids between self-compatible Lycopersicon esculentum and self-incompatible L. peruvianum. Theor. Appl. Genet. 44: 278-288; 1974

Read, S.M.; Rasmussen, R.; Bacic, A.; Clarke, A.E. Metabolic analysis of cell wall biosynthesis in pollen tubes of Nicotiana alata. Fry, S.C.; Brett, C.T.; Reid, J.S.G., eds. Fifth Cell Wall Meeting, Edinburgh, U.K.; Scottish Cell Wall Group, Edinburgh; 1989; Abst 147

Roggen, H.P.; Stanley, R.G. Autoradiographic studies of pear pollen tube walls. Physiol. Plant. 24: 80-84; 1971

Rosen, W.G. Pollen-pistil interaction in Lilium. Heslop-Harrison, J., ed. Pollen: Development and Physiology; Butterworth, 1971: pp. 239-260

Steer, M.W.; Steer, J.M. Pollen tube tip growth. New Phytol. 111: 323-358; 1989

Tupy, J.; Hrabetova, E.; Capkova, V. Amino acids and bivalent cations in the growth of tobacco pollen in mass culture. Plant Sci. Lett. 30: 91-98; 1983

Tupy, J; Rihova, L. Changes and growth effect of pH in pollen tube culture. J. Plant Physiol. 115: 1-10; 1984

Vasil, I.K. Physiology and culture of pollen. Int. Rev. Cytol. 107: 127-174; 1987

27

Pollen Tube Growth and Ovule Penetration in *Gasteria verrucosa* (Mill.), H. Duval

Michiel T.M. Willemse and Martinie A.W. Franssen-Verheijen

Pollen tubes of Gasteria verrucosa germinate in the stigmatic exudate and penetrate the solid cell walls in the tip of the style. In the direction of the ovarythis solid part gradually becomes liquid and the pollen tubes grow in this fluid with an increasing velocity to reach the locule of the ovary. Here papillar cells excrete an exudate. Because of a different composition of this exudate, if compared with the stylar exudate, the pollen tube growth in the ovary is delayed (Willemse & Franssen-Verheijen,1988).

In the loculus of the ovary the placental fluid partly covers the anatropic ovules including the micropyles. The micropyle may contain an exudate with a different composition than the placental fluid. In this chapter the use of the proteins in the placental fluid by pollen tubes in vivo and in vitro as well as the function of the micropylar exudate in the penetration of the pollen tubes are studied.

The use of placental fluid proteins by pollen tubes

The placental fluid is composed of carbohydrates and proteins and can easily be collected by centrifugation. The fluid of about 24 ovaries is diluted with water and concentrated with an ultrafree 10.000 NM NL filter to about 10 μl. From this sample with native proteins 2 μl was used for electrophoresis. Using the Pharmacia System IsoElectro Focussing, pH 4-6.5 was the first dimension and an 8-25% gradient PAGE gel was used for the second dimension. Silver staining was used to detect the proteins and the concanavaline A/Horse radish peroxydase reaction to state the presence of glycoproteins.

When comparing the two-dimensional patterns of placental fluid without and with pollen tubes in vivo, proteins as well as glycoproteins appear to be affected by enzymes from the pollen tubes (Willemse, 1992). Using in vitro germinating pollen with the addition of about 0.1% placental fluid, the placental fluid is more consumed. As shown in figure 1, the most remarkable difference between the placental fluid with and without in vitro germinating pollen is the breakdown of the rows of glycoproteins and the appearance of a new protein in this region.
Addition of different concentrations of placental fluid to a gently moving germination medium, componed of 5.5% sucrose, 2.5% glucose, 2.5% mannitol solved in a Brewbaker and Quack medium with 300 ppm boric acid at room temperature influences the pollen tube growth. Table 1 shows this influence on the pollen tube length after 40 minutes of germination.

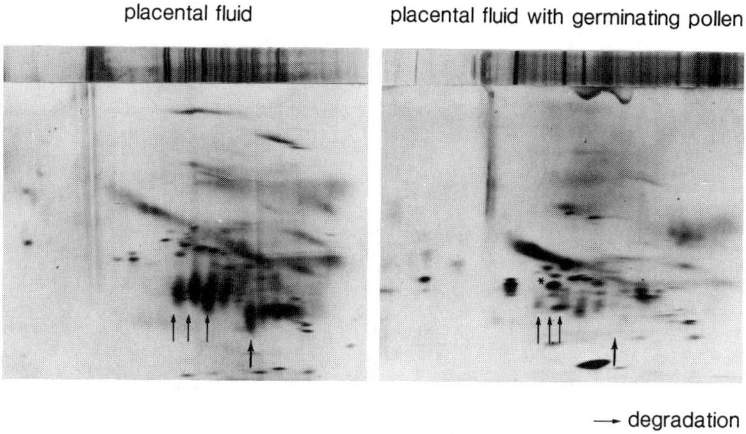

Figure 1. Degradation of placental fluid by in vitro germinating pollen. The arrows point to the disappearing glycoproteins. * = new protein.

Table 1. Percentages of pollen without tube, short pollen tubes (till the length of 3x the pollen diameter), medium pollen tubes (till 20x the pollen diameter) and long pollen tubes (>20x the pollen diameter) respectively after germination in germination medium or medium enriched with 0.5 respectively 0.8 % placental fluid. The number of counted pollen per experiment is between 170-190.

	Control medium:	Placental Fluid 0.5%:	Control medium:	Placental Fluid 0.8%:
No pollen tube:	27	18	16	8
Short pollen tube:	14	5	22	4
Medium pollen tube:	51	65	60	48
Long pollen tube:	8	12	2	40

More long pollen tubes are formed after addition of the placental fluid. This indicates a higher speed of pollen tube growth, irrespective of the added concentration of placental fluid. This increase in speed in-vitro is in contrast with the retarded pollen tube growth in-vivo.

The placental fluid is used by the pollen tubes and influences the velocity of pollen tube growth in vivo by speed reduction and in vitro by acceleration. In both cases glycoproteins including the probably released carbohydrates, and proteins are used, but their function in relation to the pollen tube growth is still unclear.

The micropylar exudate

During the receptive period of the stigma a micropyle may contain an exudate. As is shown in figure 2, some of the ovules possess abundant exudate while others show only very few or nothing.

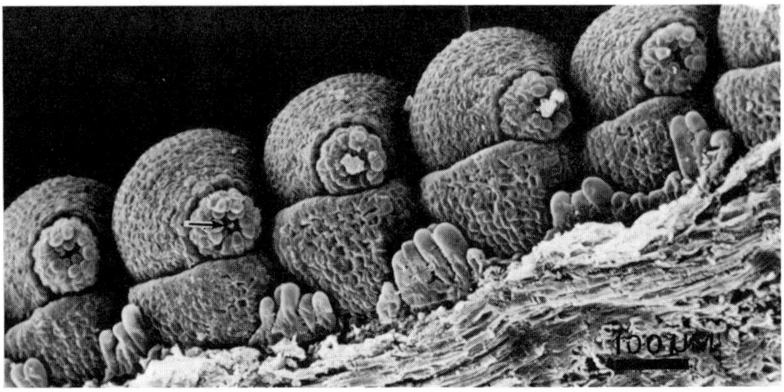

Figure 2. A row with ovules with and without micropylar exudate, the arrow points to a small quantity of exudate.

The micropyle is funnel-shaped and the exudate appears as a small droplet or hose. The exudate seems to be viscous because of its presence after some fixation and washing steps. The micropylar exudate can be produced by the inner integument, the nucellar tissue or the egg apparatus. Tilton (1980) shows the existence of micropylar exudate in Ornithogalum produced by the nucellar cap cells and inner integument which also occurs in Paspalum (Chao, 1977).

The micropylar exudate of Gasteria reacts on different histochemical stains and in some cases a comparable reaction can be observed in the filiform apparatus of the synergids. Table 2 summarizes the histochemical tests.

These tests show that the chemical compositions of the micropylar exudate and the filiform apparatus are partly comparable. However, some of the stains have to be repeated during different developmental stages of the filiform apparatus.

Tilton (1980) reports the production of a mucilaginous substance found in the micropyle in Senecio which is of synergid origin after Mottier (1893). So the possibility that the micropylar exudate partly originates from the filiform apparatus may not be excluded. Franssen-Verheijen and Willemse (1990) showed a change in the ultrastructure of the filiform apparatus during the period of stigma receptivity in Gasteria.

Table 2. Reaction of some histochemical stains on the micropylar exudate and filiform apparatus of one ovule.

Stain:	Micropylar exudate:	Filiform apparatus:
General:		
Saffranine	+ +	+ +
Carbohydrates:		
PAS	+	+
Calcofluor white	+	?
Primuline	+	?
Aniline blue	-	+
Proteins:		
Amido black	+	+
Coomassie blue	+	?

The micropylar exudate and pollen tube penetration

During pollen tube penetration of the micropyle the exudate remains present as is shown in figure 3. It shows that both the intensity of the stain and the position of the droplet change. A use of the exudate by the pollen tube cannot be excluded.

Although not all ovules may contain micropylar exudate (see figure 3) and sometimes pollen tubes do not penetrate micropyles with exudate or penetrate micropyles without light microscopical visible exudate, a relation between the presence of the micropylar exudate and the pollen tube penetration may exist. Considering the appearance of the micropylar exudate during the flower development, there is a preference to form the exudate at first in the basal part of the ovary. Figure 4 shows the percentage of ovules with exudate during the flower development.

From this pattern of appearance of the micropylar exudate it can be seen that the upper part of the ovary contains the ovules with the most freshly formed droplets at the time of stigma receptivity, which is marked by presence of the stigmatic exudate.

Counting the rate of pollen tube penetration in the micropyles after crossing experiments with abundant pollen a preference appear to exist to penetrate the ovules in the upper part of the ovule. Here also multiple penetration often occurs but the most basal ovules in the ovary are seldom penetrated (Naaborgh & Willemse 1992). So the preference of pollen tube or multiple pollen tube penetration in the upper part of the ovary may be an indication that this micropylar exudate is in a more attractive condition than in the lower part of the ovary.

The mucilaginous micropylar exudate seems to have a function in the penetration of the micropyles. If this exudate originates from the filiform apparatus and contains an attraction substance, in Gasteria the pollen tube attraction is regulated by the synergids.

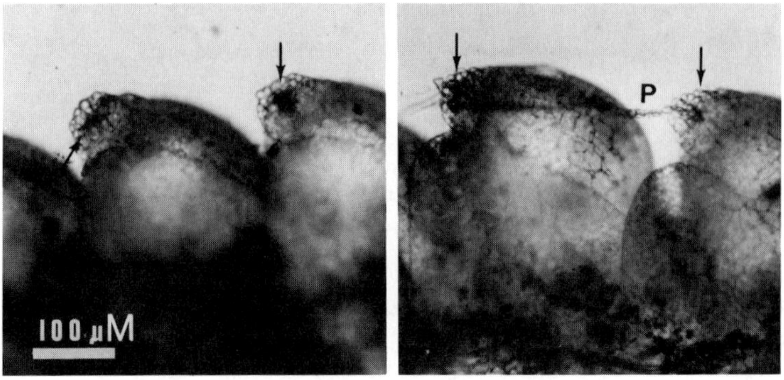

Figure 3. Fresh ovules stained with a 0.1% Coomassie blue solution in water. The arrows point to the micropylar exudate present before and after pollen tube penetration. P = Pollen tube.

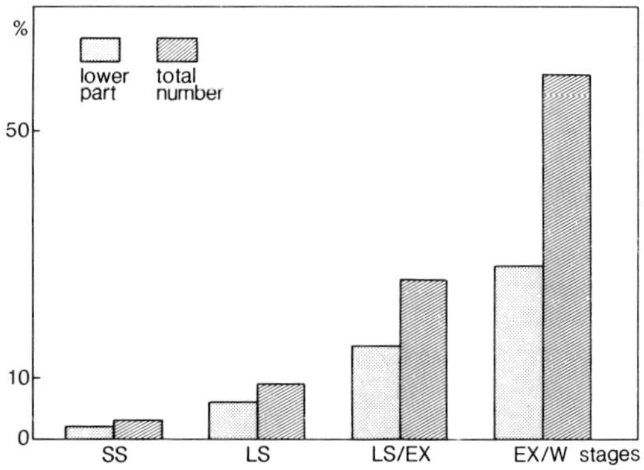

Figure 4. Percentages of ovules with micropylar exudate during the flower development presented in the number of ovules in the basal part of the ovary versus the total number. The developmental stages of the flower are SS = short style, SL = long style, SL/EX = long style - stigma with exudate, EX/W = stigma with exudate and wilted perianth.

The total number of counted ovules per stage is between 110 and 140.

Acknowledgement

The authors thank Dr.Ir. C.J. Keijzer for critical reading the manuscript and Mrs G.G. van de Hoef-Espelo for typing of the manuscript and Mr. A.B. Haasdijk and P. Snippenburg for the illustrations and Mr. S. Massalt for the photographs.

Literature

Chao, C.Y. A periodic acid - Schiff's substance related to the directional growth of pollen tube into embryo sac in Paspalum ovules. Am. J. Bot. 58: 649-654; 1971.

Franssen-Verheijen, M.A.W. & Willemse, M.T.M. The ovule of Gasteria verrucosa at receptivity of the stigma: an ultrastructural study. Acta Bot. Neerl. 39 (1): 53-63; 1990.

Mottier, D.M. On the embryo sac and embryo of Senecio aureus L. Bot. Gaz. (Chicago) 18: 245-253; 1893.

Naaborgh, A.T. & Willemse, M.T.M. The ovular incompatibility system in Gasteria verrucosa (Mill.) H. Duval. Offered to Euphytica.

Tilton, V.R. The nucellar epidermis and micropyle of Ornithogalum caudatum (Liliaceae) with review of these structures in other taxa. Can. J. Bot. 58: 1872-1884; 1980.

Willemse, M.T.M. & Franssen-Verheijen, M.A.W. Pollen tube growth and its pathway in Gasteria verrucosa (Mill.) H.Duval. Phytomorphology 38:127-132; 1988.

Willemse, M.T.M. Protein patterns in the pollen tube pathway of Gasteria verrucosa (Mill.) H.Duval. Will be submitted to the Journal of Plant Sexual Reproduction; 1992.

Physiology and Biochemistry

28
Membrane Behaviour and Stress Tolerance in Pollen

Folkert A. Hoekstra, John H. Crowe, Lois M. Crowe, and Danielle G.J.L. van Bilsen

Temperature and water stress are the most important of stresses imposed on pollen during maturation in the anther and after shedding. During dehydration, osmotic and salt stress become important, and enormous mechanical forces are exerted during the shrinking of the cell. Tolerance to desiccation, which is common among pollen species, requires the proper tactics of the cells towards both dehydration and rehydration. For long term survival in the dry condition a special biochemical composition may be required, which curtails free radical damage. Dry pollen usually ages slowly, particularly in the cold. In contrast, both high temperatures and elevated water contents considerably accelerate ageing. Because reduced vigour due to the above mentioned stresses usually coincides with increased solute leakage during imbibition, we focus on the conformational and compositional status of the membrane phospholipids.

Phase Changes in Phospholipids (PLs) with Drying

In 1967, Chapman et al. demonstrated for the first time that the gel-to-liquid crystalline transition temperature (T_m) of dipalmitoylphosphatidylcholine(DPPC) increases during dehydration (Fig. 1). This is due to the decrease of the lateral spacing of the polar head groups of the PLs, which gives increased opportunities for Van der Waals' interactions between their hydrocarbon chains (reviewed in Crowe and Crowe, 1988a). This was also found for the more fluid, unsaturated PLs such as dilinoleoyl-PC (Lynch and Steponkus, 1989; Fig. 1), which are characteristic for plants. The introduction of a Fourier Transform IR-spectroscopic method for *in situ* measurement of phase behaviour of membrane lipids (Crowe et al., 1989a; Crowe et al., 1989b) enabled us to produce the pollen curves of Fig. 1. Apparently T_m values of membrane PLs in pollen also increase with dehydration. This means that during dehydration at room temperature, the originally fluid membranes in pollen may become rigid. Subsequent rehydration leads to melting. When liquid crystalline and gel phase PLs coexist during the

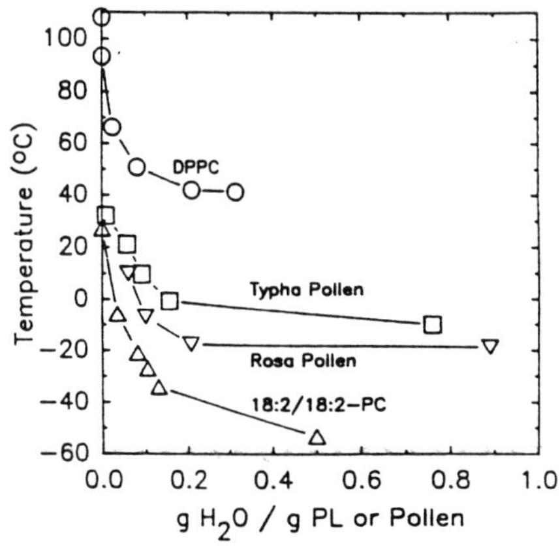

Fig. 1. Phase diagrams of DPPC (= 16:0/16:0-phosphatidylcholine), 18:2/18:2--PC and pollen phospholipids *in situ*. Above the curves, lipids are in liquid crystalline phase, below, in gel phase.

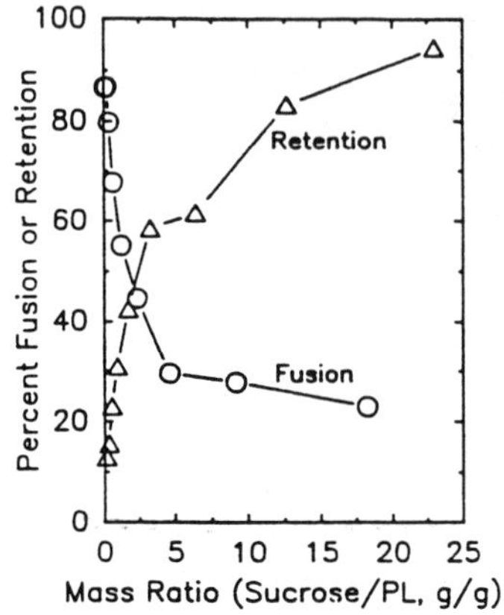

Fig. 2. Effects of sucrose on fusion and solute retention of liposomes prepared from *Typha* pollen phospholipids, upon a cycle of dehydration and rehydration.

phase transition, liposomes or membranes become leaky (Hammoudah et al. 1981). In addition, specific PLs such as phosphatidylethanolamines (PEs) may assume a non-bilayer configuration upon dehydration, the inverted hexagonal phase (H_{II} phase; Luzzati and Husson, 1962). This prompted Simon (1974) to hypothesize that H_{II} phase is the cause of the leakage observed during imbibition of dry seed. Although no evidence exists that H_{II} phase occurs in desiccation tolerant, dry seeds and pollen (McKersie and Stinson, 1980; Priestley and De Kruijff, 1982), we discuss this matter in more detail in a separate section.

In order to circumvent temporary loss of membrane integrity during drying and rehydration, it would be preferable if phase changes do not occur at all, or at least are postponed to very low water contents. How cells may accomplish this is discussed in the next section.

Role of Sugars and Fatty Acid Unsaturation in Phase Behaviour

Dry, desiccation tolerant organisms contain large amounts (10-25% of DW) of certain disaccharides, trehalose in the case of nematodes, *Artemia* cysts, bacteria and yeasts (Crowe et al., 1984), and sucrose in the case of pollen and seeds (Hoekstra and Van Roekel, 1988; Hoekstra et al. 1989a; Koster and Leopold, 1988). Table I shows that trehalose depresses T_m of dry DPPC to even under that of the hydrated control. Also T_m of PLs extracted from *Typha* pollen does not rise appreciably during drying in the presence of sucrose, the major endogenous sugar in this species (23% of the DW). When isolated membranes of *Typha* pollen are dried without sucrose, a rise of T_m of approximately 60-70°C is

Table I. Effects of moisture level and disaccharides on transition temperatures (T_m) of dipalmitoylphosphatidylcholine (DPPC, determined by DSC), and phospholipid (PL) and membranes isolated from *Typha* pollen (FTIR). T_m values of intact pollen are also given.

Sources[a]	T_m		
	hydrated	dry	dry + disacch.
DPPC	42	108	24[b]
Pollen PL	-13	45	-8
Pollen membranes	-6	58	31
Intact pollen	-9		32[c]

[a]DPPC data from Crowe and Crowe (1988b); the other data from Hoekstra et al. (1991). [b]trehalose, the other data are with sucrose. [c]endogenous sucrose.

observed, which figure is also characteristic for the pure PLs of Fig. 1. Drying in the presence of sucrose reduces T_m of the membranes by about 30°C to a minimum value of 31°C. This value is strikingly similar to the T_m measured in intact dry pollen, suggesting that the relatively limited rise of T_m in intact pollen is due to the high endogenous sucrose content (Hoekstra et al., 1991).

Disaccharides are effective through their direct interaction with the PL polar head groups (reviewed in Crowe et al., 1987). They replace the water during dehydration, thereby preventing too close a packing of the PL molecules and gel phase formation. Monosaccharides are less effective in that respect (Crowe et al., 1985). The effect of an escape from gel phase formation is demonstrated in Fig. 2: When unilamellar liposomes produced from purified *Typha* PLs are dried and rehydrated, they leak entrapped solutes and fuse to an amorphic multilamellar mass. In the presence of increasing amounts of sucrose they remain intact, *i.e.* they do not leak and fusion is reduced. We believe that sucrose is one of the factors that enables pollen to escape from desiccation stress. This opinion is strengthened by the correlations in Fig. 3: with increasing sucrose contents during *Papaver* pollen development, survival from drying improves, whereas leakage is reduced.

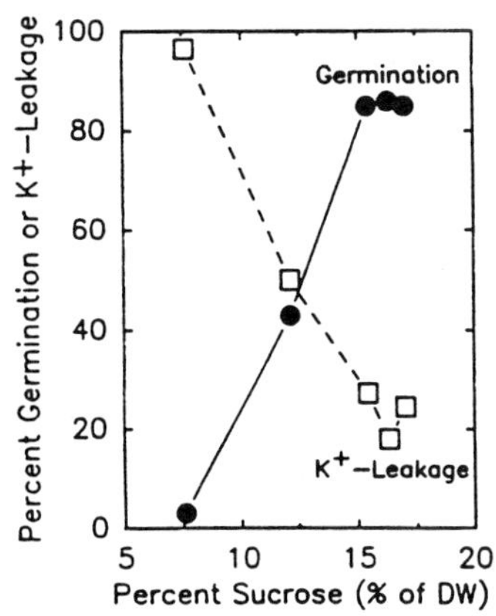

Fig. 3. Correlation between endogenous sucrose content of *Papaver* pollen during its development, and survival from drying, expressed as percentage germination or K^+-leakage (redrawn from Hoekstra and van Roekel, 1988).

Another factor that determines T_m of PLs is the extent of unsaturation of the acyl chains. Figure 1 clearly shows that the four double bonds of di-18:2-PC cause a reduction in T_m of more than 80°C as compared to the T_m of di-16:0-PC. Figure 1 further shows that *Rosa* pollen had lower T_m values than *Typha* pollen,

Table II. Content of phosphatidylethanolamines (PE) as a percentage of total phospholipids (PL) of 8 pollen species. The number of double bonds of esterified fatty acids per molecule PE and PL is also given.

Species	Desiccation tolerance	Mol% PE	Double bonds	
			per PE	per PL
Typha latifolia	yes	18.4	1.40	1.45
Quercus ilex	yes	19.8	1.52	1.45
Rosa moyesii	yes	25.4	1.92	1.99
Narcissus poeticus	yes	29.2	1.70	1.98
Impatiens glandulifera	yes	42.4	1.68	2.07
Zea mays	no	30.3	1.63	1.76
Secale cereale	no	31.9	1.97	2.26
Alopecurus pratensis	no	34.7	1.64	1.81

most likely caused by the higher degree of unsaturation of *Rosa* pollen fatty acids (Table II). If avoidance of a phase change would be a prerequisite for desiccation tolerance, an alternative for high levels of disaccharides would then be a high degree of unsaturation. However, extremely high levels of unsaturation are unfavourable for long term survival of pollen (Hoekstra, 1986), most likely due to the increased risk of lipid peroxidation (Schaich, 1980). There is no indication of extensive changes in PL composition and degree of unsaturation during the developmental shift from desiccation sensitivity to tolerance (Hoekstra and Van Roekel, 1988). Because of the considerable variation in degree of fatty acid unsaturation among pollen species, we could analyse whether the "saturated" species have more sucrose than the "unsaturated" species (Fig. 4). In all 15 species of Fig. 4 sucrose is the major soluble carbohydrate (>95%). There appears to be a tendency that less sucrose is available (or may be required) for protection of membranes if the extent of unsaturation is higher. In that respect it is still puzzling why, generally, *Gramineae* pollens are not desiccation tolerant. Their high degree of fatty acid unsaturation (Table II) would make a high sucrose content superfluous. It was just the low sucrose content of maize pollen that was held

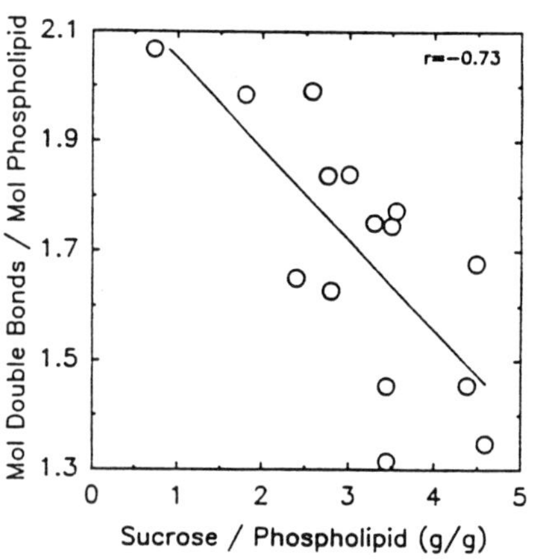

Fig. 4. Correlation between the extent of unsaturation of esterified fatty acids in phospholipids of 15 different pollen species, and the amount of endogenous sucrose that is available per unit of membrane phospholipid.

responsible for its lack of desiccation tolerance (Hoekstra et al.,1989a). This idea was based on the comparison with desiccation tolerant *Pennisetum* pollen (*Gramineae*) having a very similar lipid composition, but a much higher sucrose level. In the light of data from other sensitive *Gramineae* species such as *Secale* it is even less likely that lack of sucrose is the cause of the sensitivity: *Secale* pollen has sucrose contents of over 20% of the dry weight. However, the extremely high degree of fatty acid unsaturation in this species (Table II) is a risk in relation to lipid peroxidation, the more so as protection by lipid soluble antioxidants is rather marginal. It is evident from the above that a clear cause for the lack of desiccation tolerance has not been elucidated yet. That, nevertheless, phase transitions play an important role in the injury imposed by drying is explained in the next section on imbibitional damage.

Imbibitional Injury

Figure 1 shows that extremely dry *Typha* pollen (and also many other pollen species) has membranes, the PLs of which are in gel phase at room temperature. However, this pollen is nevertheless still viable, as can be observed when rehydration is carried out properly. We believe that the temporary loss of membrane integrity that accompanies the phase change to gel phase, does not cause injury, simply because the necessary free water is lacking for the transport of ions and other solutes. This is in contrast to rehydrating conditions, when the melting occurs in the presence of ample water. In this respect, two factors have been recognized as being important for maintenance of viability: the temperature of imbibition and the moisture content of the pollen at imbibition (Hoekstra and Van der Wal, 1988). When dry *Typha pollen* (0.06 g H_2O/g DW) is imbibed in germination medium at low temperature (and further incubated at 21°C), it does not germinate, and all K^+ ions leak within 5 minutes after the start of imbibition (Fig. 5; Hoekstra et al.,1989b). At higher temperatures of imbibition, germination is improved and leakage reduced. When pollen is brought to a higher moisture content by treatment with humid air prior to the imbibition, high germination scores and reduced leakage can be obtained, also in the cold.

Strikingly both factors also determine whether PLs are in gel or liquid crystalline phase. For instance, when extremely dry di-18:2-PC from Fig. 1 is kept at 0°C, it is in gel phase. Heating to 40°C will cause it to assume the liquid crystalline state. Melting also can be brought about at 0°C by elevating the moisture content to 0.1 g H_2O/g lipid. Similar phenomena occur in pollen PLs when the pollen is heated (or exposed to humid air): an upward shift (with temperature) of the frequency of the symmetric CH_2 stretch vibration band (Fig. 5), which is characteristic for the melting of lipids. Details of the *in situ* FTIR technique, that was employed here, are described elsewhere (Crowe et al., 1989b). What the coinciding curves of germination score and frequency may

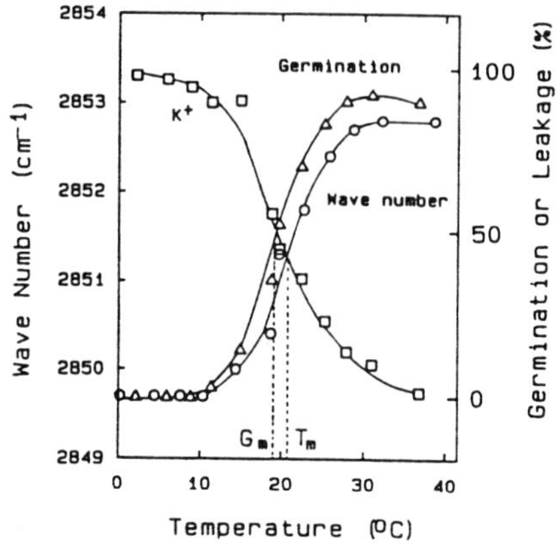

Fig. 5. Plot of the frequency of the CH_2 symm. stretch vibration band (FTIR) versus temperature for dry *Typha* pollen. Also plotted are curves of germination and K^+-leakage versus temperature of imbibition.

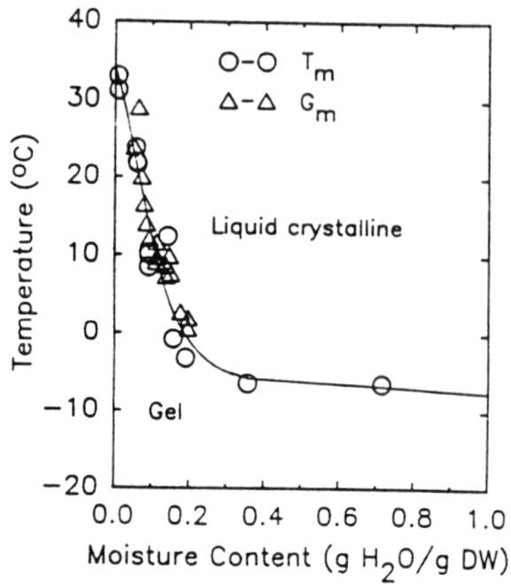

Fig. 6. Effect of moisture content of *Typha* pollen on T_m of PLs *in situ* and the temperature (G_m) at which half of the original germination capacity is expressed (from Hoekstra et al., 1989b).

indicate is that PLs must be in liquid crystalline phase at the time of imbibition. A phase change from gel to liquid crystalline during the massive entry of liquid medium is detrimental. From the germination and frequency curves of Fig. 5 the temperature can be determined at which half of the original germination capacity is expressed (G_m) and half of the membrane phospholipid has melted (T_m).

Figure 6 shows these values for large number of initial moisture contents of

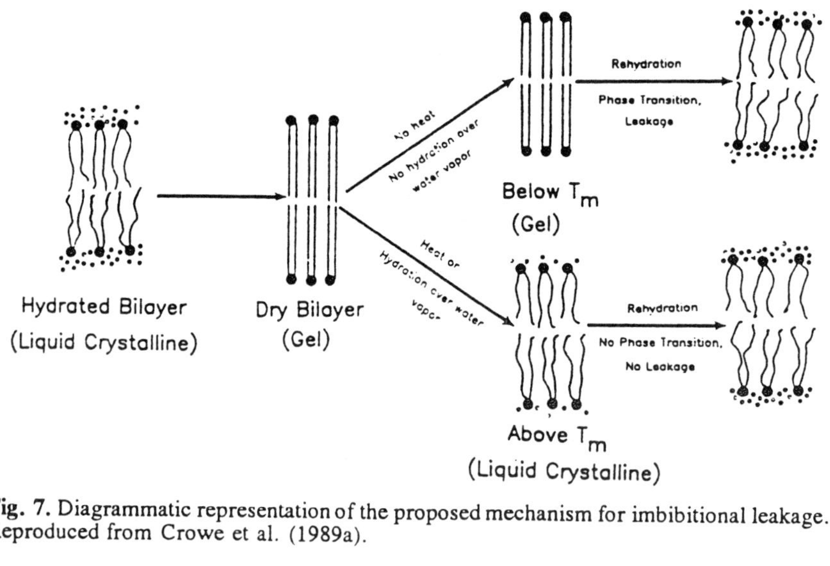

Fig. 7. Diagrammatic representation of the proposed mechanism for imbibitional leakage. Reproduced from Crowe et al. (1989a).

Typha pollen. The curve of Fig. 6 has predictive value as to which set of initial moisture content and temperature of imbibition should be selected to obtain maximal germination: above the curve PLs are in liquid crystalline phase and germination is maximal; below the curve PLs are in gel phase and germination is poor. That the rehydration in humid air does not cause imbibitional injury, in contrast to rehydration in liquid media, can be explained as follows: Uptake of a small amount of the water vapour will depress T_m which will lead to a phase change at room temperature. There is, however, not enough free water to enable large scale transport and leakage. As soon as free water is present, T_m is far below 0°C, and PLs are in liquid crystalline phase above 0°C. A diagrammatic representation of the above proposed mechanism for imbibitional leakage is given in Fig. 7. Strong support for the hypothesis came from dry *Impatiens* pollen that is tolerant to imbibition at low temperatures. This pollen species has very low T_m values, so that gel phase hardly occurs at 0°C, even when it is dry, and no phase changes take place during imbibition.

At low water contents germination capacity is not always restored by imbibition at elevated temperatures. *Typha* pollen below 0.03 g H_2O/g DW never reaches high germination and low leakage, although FTIR data indicate that the bulk of PLs had become fluid at elevated temperature. Another good example of such behaviour is given in Fig. 8A for dry *Quercus* pollen. After an initial improvement of germination with temperature, a decline of the germination capacity can be seen at temperatures of over 25°C, coinciding with an increased leakage during imbibition. Figure 8B shows that it is not thermal denaturation that is involved, since the prehydrated control is not sensitive to temperatures up to 40°C. This behaviour is discussed in the next section in the light of the possible involvement of non-bilayer phospholipid in imbibitional injury.

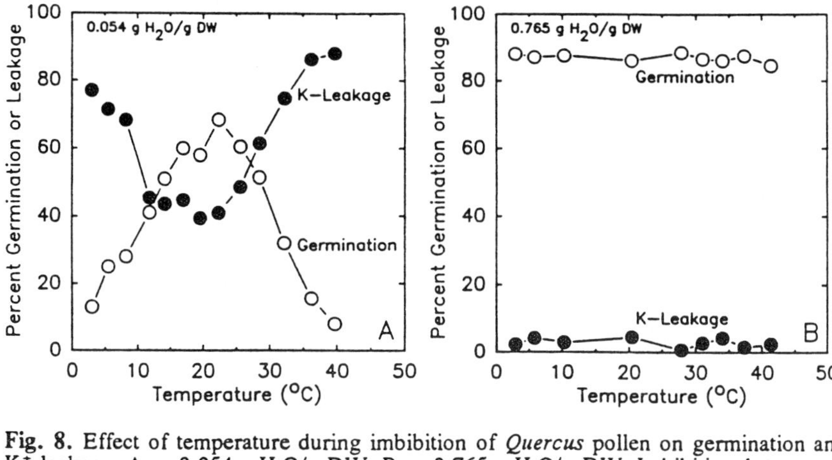

Fig. 8. Effect of temperature during imbibition of *Quercus* pollen on germination and K^+-leakage. A = 0.054 g H_2O/g DW; B = 0.765 g H_2O/g DW. Imbibitional treatment took 5 min; Incubation temperature was 21°C.

Non-bilayer Phase Phospholipid and Desiccation

Because PLs such as the phosphatidylethanolamines (PEs) can form bonds of strong ionic character between adjacent polar head groups, they are particularly sensitive to adopt the inverted hexagonal (H_{II}) phase (Cullis and De Kruijff, 1979). This tube like structure has the polar head groups directed to the interior, and the acyl chains oriented outwards. There are several conditions that favour the formation of H_{II} phase, among them heating and dehydration. Figure 9 shows the phase diagram of didodecyl-PE (DDPE): At 0.2 g H_2O/g, DDPE is in gel phase at 20°C; heating causes it to adopt the liquid crystalline phase at 35°C; upon further heating it will pass into H_{II} phase at 90°C when the DDPE molecules are forced to come out of the plane of the membrane, due the larger volume that the acyl chains occupy. A larger volume is also occupied when acyl chains are more cis-unsaturated: unsaturated PEs adopt H_{II} phase at lower temperatures than their saturated controls. Dehydration promotes bonding between polar head groups, causing the temperature at which H_{II} phase occurs, to decrease (Fig. 9). Liposomes having H_{II} phase domains leak solutes (Cullis and De Kruijff, 1979).

How does this knowledge apply to dry pollen? The improvement of germination and decrease of leakage observed when dry pollen is imbibed at elevated temperatures (Fig. 5) already excludes that H_{II} phase is the major cause of the imbibitional problems. If H_{II} domains were involved, injury would have aggrevated at elevated temperatures (cf Fig. 9). But how does pollen escape from H_{II} phase formation? Table II shows that it is not by the lack of PEs, in both desiccation tolerant and intolerant species. PE is the major PL class in pollen which could contribute to H_{II} formation. May be, here again, sucrose acts by intercalating between the polar head groups as dehydration proceeds, thereby keeping them at a distance of each other and preventing them to come out of the plane of the membrane and form H_{II} domains.

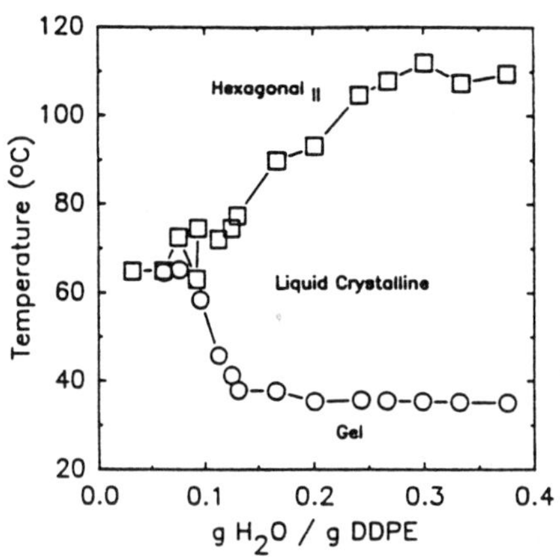

Fig. 9. Phase diagram of 12:0/12:0-phosphatidylethanolamine(DDPE) at various water contents. The configuration is indicated in the figure. Redrawn from Seddon et al. (1983).

At low water content the two curves of Fig. 9 merge. Translated to very dry pollen, this may mean that heating causes PEs in the membranes to go directly from gel into Hex_{II} phase. It might explain why very dry *Typha* pollen does not germinate even after imbibition at high temperature and why the dry *Quercus* pollen in

Fig. 8 starts leaking again at the elevated temperature range. In short, we are suggesting that Simon's hypothesis on the cause of imbibitional leakage may in principle be correct under special conditions of extreme desiccation.

Ageing and Membrane Behaviour

Similarly as for seeds, longevity of pollen is controlled by temperature, moisture content of the cells and specific endogenous biochemical parameters. Ageing is much faster at high moisture levels. There are indications that under conditions of high humidity, for instance when pollen shedding is retarded due to rainy weather, decline of viability starts already in the anther (Hoekstra and Bruinsma, 1975). Furthermore, high temperatures considerably accelerate the decline of viability, whereas the best conservation is achieved at subzero temperatures with pollen having low water contents.

When vigour starts to decline an increased leakage of solutes during imbibition usually is observed. This leakage can not be prevented by prehydration in humid air prior to the germination test. Because leakage points to reduced membrane integrity, we analyzed several parameters of membrane degradation. During both dry ageing and accelerated ageing at higher moisture contents free fatty acids (FFAs) and lyso-PLs are produced. Lyso-PLs are PLs with one acyl chain missing. In contrast to liposomes prepared from fresh pollen PLs, those from aged pollen PLs were considerably more leaky (Van Bilsen and Hoekstra, 1989). This leakiness could be attributed to the FFAs and lyso-PLs. The composition of one of the lyso-PCs, found, was 16:0-lyso-PC. Its thermotropic phase behaviour is different from 16:0/18:2-PC from which it originally was derived. In dry liposomes 16:0-lyso-PC phase separates, and melts at a temperature of far over 80°C. In aged pollen this will cause extensive leakage upon imbibition.

Pollen species have differential sensitivity to ageing conditions. Species that survive well under dry conditions, also have comparatively better longevity under more humid conditions. Hoekstra (1986) found a good correlation between longevity and an endogenous chemical parameter. Pollen PLs are either rich in linoleic acid (18:2) or linolenic acid (18:3), usually making up for 70% of the total of fatty acids. Long life tends to go together with reduced levels of linolenic acid. Indeed, the risk of lipid peroxidation is more severe in linolenic acid rich PLs than in those with 18:2 fatty acids (Schaich, 1980). We would expect the "unsaturated" species to be better protected by lipid soluble antioxidants. However, this is not the case. Particularly the desiccation sensitive *Gramineae* species contain small amounts. Apparently longevity is not of vital interest for those species, probably because they usually produce fresh pollen every day.

Acknowledgement: We acknowledge the support by grant 88-37264-4068 from the U.S. Department of Agriculture Competitive Grants Program.

Conclusion: We have shown that the behaviour of membranes in pollen is important for viability under various conditions of stress. Pollen defends itself against such stresses by accumulation of disaccharides, anti-oxidants and extent of fatty acid unsaturation. However, this only partly explains how desiccation tolerance is accomplished. In *Gramineae* pollen, particularly, the reason for the lack of desiccation tolerance is still speculative.

References

Chapman, D.; Williams, R. M.; Ladbrooke, B. D., Chem. Phys. Lipids 1:445-475; 1967.

Crowe, J. H.; Crowe, L. M.; Chapman, D., Science 223:701-703; 1984.

Crowe, J. H.; Crowe, L. M.; Carpenter, J. F.; Aurell Wistrom, C., Biochem. J. 242:1-10; 1987.

Crowe, J. H.; Hoekstra, F. A.; Crowe, L. M., Proc. Natl. Acad. Sci. USA 86:520-523; 1989a.

Crowe, J. H.; Hoekstra, F. A.; Crowe, L. M.; Anchordoguy, T. J.; Drobnis, E., Cryobiol. 26:76-84; 1989b.

Crowe, L.M.; Crowe, J.H.; Chapman, D., Arch. Biochem. Biophys. 236:289-296; 1985.

Crowe L.M.; Crowe J.H., In: Aloia, R.C.; Curtain, C.C.; Gordon, L.M., eds. Physiological regulation of membrane fluidity. New York: Alan R. Liss Inc., 1988a:75-99.

Crowe, L.M.; Crowe, J.H., Biochim. Biophys. Acta 946:193-201; 1988b.

Cullis, P. R.; De Kruijff, B., Biochim. Biophys. Acta 559:399-420; 1979.

Hammoudah, M.M.; Nir, S.; Bentz, J.; Mayhew, E.; Stewart, T.P.; Hui, S.W.; Kurland, R.J., Biochim. Biophys. Acta 645:102-114; 1981.

Hoekstra, F. A., In: Leopold, A.C., ed. Membranes metabolism and dry organisms. Ithaca/London: Comstock Publ. Ass.; 1986:p. 102-122.

Hoekstra, F.A.; Bruinsma, J., Z. Pflanzenphysiol. 76:36-43; 1975.

Hoekstra, F. A.; Van der Wal, E. G., J. Plant Physiol. 133:257-262; 1988.

Hoekstra, F. A.; Van Roekel, T., Plant Physiol. 88:626-632; 1988.

Hoekstra, F. A.; Crowe, L. M.; Crowe, J.H., Plant Cell Environm. 12:83-91; 1989a.

Hoekstra, F.A.; Crowe, J.H.; Crowe, L.M., In: Taylorson, R.B., ed. Recent advances in the development and germination of seeds. New York and London: Plenum Press; 1989b:p. 71-88.

Hoekstra, F.A.; Crowe, J.H.; Crowe, L.M., Plant Physiol. In press; 1991.

Koster, K. L.; Leopold, A. C., Plant Physiol. 88:829-832; 1988.

Luzzati, V.; Husson, F., J. Cell Biol. 12:207-219; 1962.

Lynch, D.V.; Steponkus, P.L., Biochim. Biophys. Acta 984:267-272; 1989.

McKersie, B. D.; Stinson, R. H., Plant Physiol. 66:316-320; 1980.

Priestley, D. A.; DeKruijff, B., Plant Physiol. 70:1075-1078; 1982.

Schaich, K.M., CRC Critical Rev. Food Sci. Nutr. 13:189-244; 1980.

Seddon, J. M.; Cevc, G.; Marsh, D., Biochem. 22:1280-1289; 1983.

Simon, E. W., New Phytol. 73:377-420; 1974.

Van Bilsen, D.G.J.L.; Hoekstra, F.A., In: Biacs, P.A.; Gruiz, K.; Kremmer, T., eds. Biological role of plant lipids. NY and London: Plenum Publ. Corp.; 1989:p. 595-596.

29
Recent Aspects of Sporopollenin Biosynthesis

S. Gubatz, S. Arendt, M. Rittscher, and R. Wiermann

I. Introduction

In the recent years tracer experiments were carried out to elucidate the biosynthesis and structure of sporopollenin. ^{14}C-labelled substances such as mevalonate, glucose, acetate, malonic acid, phenylalanine, tyrosine and p-coumaric acid were applied to anthers of Tulipa cv. Apeldoorn on the basis of an optimized application technique (RITTSCHER and WIERMANN, 1988 a). Regardless of the isolation and purification procedure used, a high level of incorporation into the sporopollenin fraction was always observed when ^{14}C-phenylalanine was applied. In addition radioactivity was incorporated to such an extent that the sporopollenin fraction could be degraded and the breakdown products could still be analysed. After the degradation of sporopollenin by potash-fusion, p-hydroxybenzoic acid was most strongly labelled among the breakdown products (RITTSCHER and WIERMANN, 1988 b). However, since uniformly labelled phenylalanine was used as a precursor, the results did not unambiguously prove the integral incorporation of the aromatic ring system. In order to clarify this question subsequent experiments were carried out using ring-labelled phenylalanine. Moreover, it had to be tested whether the incorporation of phenylalanine into sporopollenin of Tulipa is an unique phenomenon or whether this is true for other systems, too. Thus, for comparison, further tracer experiments have been performed with anthers of Cucurbita maxima.

II. Material and Methods

1. Material:

Anthers of Tulipa cv. Apeldoorn were prepared from bulbs purchased from Nebelung (Münster); anthers of Cucurbita maxima were obtained from flowers of plants cultivated in a greenhouse of the Botanical Garden Münster.

2. Methods:

Labelled precursors were applied to anthers of Tulipa according to RITTSCHER and WIERMANN (1988 a).

The androecia of Cucurbita were incubated directly in a medium with the respective precursor.

The sporopollenin fractions were isolated and purified by a gentle method comprising exhausting extractions, treatments with hydrolytic enzymes and alkaline hydrolyses (RITTSCHER and WIERMANN, 1988 b; GUBATZ and WIERMANN, in preparation).

The sporopollenin fractions were degraded by potash-fusion as described elsewhere (RITTSCHER and WIERMANN, 1988 b).

III. Results and Discussion

1. The results of tracer experiments with Tulipa anthers using ring-labelled phenylalanine
The results summarized in fig. 1 show that, in comparison with ^{14}C-glucose, ring-labelled phenylalanine was strongly incorporated into the sporopollenin fraction. The sporopollenin was labelled to an extent that it could be degraded. p-Hydroxybenzoic acid is the major labelled compound released during the potash-fusion (fig. 2). This proofs that the aromatic ring system of phenylalanine is incorporated into sporopollenin.

Fig. 1

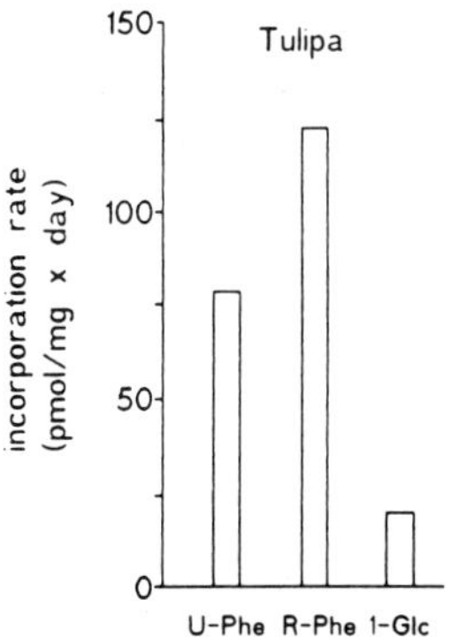

Fig. 2

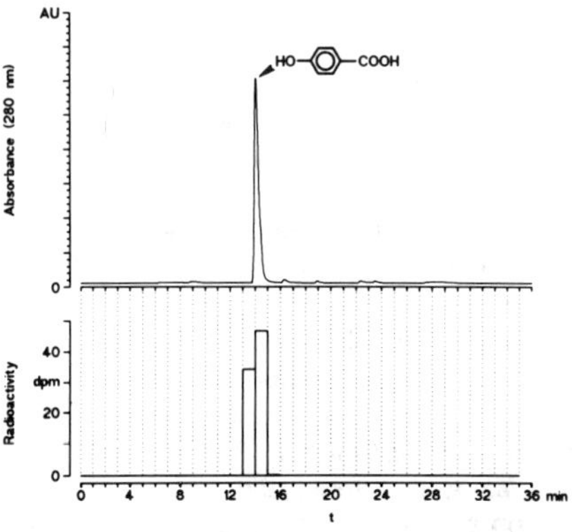

Fig. 1: The incorporation of [U–^{14}C]- and [R-^{14}C]-phenylalanine into the sporopollenin fraction of Tulipa cv. Apeldoorn in comparison to the incorporation of ^{14}C-glucose. (U-Phe: uniformly labelled phenylalanine, R-Phe: ring-labelled phenylalanine; 1-Glc: [1-^{14}C]-Glucose)

Fig. 2: HPLC-analysis of an acidified ether extract of degraded ^{14}C-labelled sporopollenin. Sporopollenin had been labelled by incubation with [R-^{14}C]-phenylalanine and degraded by potash-fusion. The fractions of the HPLC-run had been tested for radioactivity. Conditions according to GUBATZ and WIERMANN (in preparation).

2. The incorporation of ^{14}C-phenylalanine into the sporopollen infraction of Cucurbita

In these experiments [U-^{14}C]-phenylalanine was applied to anthers of Cucurbita maxima; for comparison, labelled glucose was used as a marker for the metabolic activity of the anthers. Fig. 3 shows the incorporation rates for phenylalanine as precursor. The results clearly demonstrate that phenylalanine is substantially incorporated in Cucurbita sporopollenin. Degradation of this sporopollenin fraction yielded p-hydroxybenzoic acid as the main labelled compound (fig. 4).

However, due to the forcing conditions during the degradation of sporopollenin it is unlikely that p-hydroxybenzoic acid is a real element of the sporopollenin skeleton. It is rather supposed that a phenylpropane unit such as p-coumaric acid forms the monomeric unit. This assumption was clearly confirmed by results of pyrolysis-GC-MS-analysis (WEHLING et al., 1989).

Fig. 3

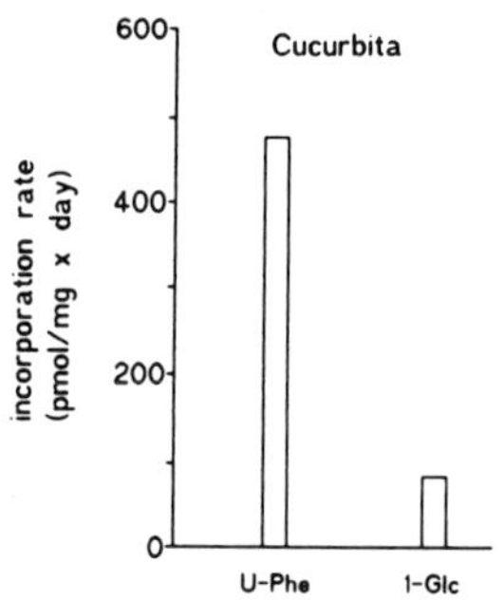

Fig. 4

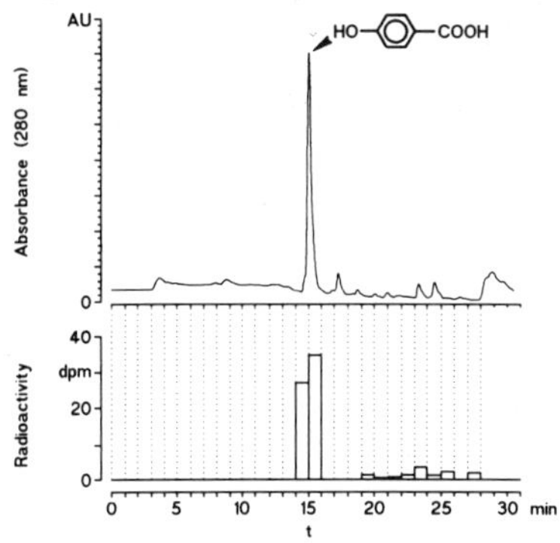

Fig. 3: The incorporation of ^{14}C–phenylalanine into the sporopollenin fraction of Cucurbita maxima in comparison to the incorporation of ^{14}C–glucose (U-Phe: uniformly labelled phenylalanine, 1-Glc: [1-^{14}C]-glucose).

Fig. 4: HPLC-analysis of an acidified ether extract of degraded ^{14}C-labelled sporopollenin. Sporopollenin had been labelled by incubation with [U-^{14}C]-phenylalanine and degraded by potash-fusion. The fractions of the HPLC-run had been tested for radioactivity. Conditions according to GUBATZ and WIERMANN (in preparation).

In summary, the substantial incorporation of labelled phenylalanine into sporopollenin has been shown. From these and other results (not shown) the occurrence of phenolics in the sporopollenin skeleton can be concluded. However, it should be stressed that this is only one aspect. According to the results of the ^{13}C-NMR-spectroscopy, sporopollenin is mainly an aliphatic polymer. Presumably sporopollenin is a mixed polymer consisting of long-chained aliphatics to a large degree and aromatic compounds such as phenols to varying degrees.

IV. Acknowledgements

These studies were financially supported by the Deutsche Forschungsgemeinschaft and the Fonds der Chemischen Industrie.

V. References

Rittscher M, Wiermann R (1988a) Studies on sporopollenin biosynthesis in Tulipa anthers. I. The development of an appropriate application method for radiolabelled precursors. Sex Plant Reprod 1:125-131

Rittscher M, Wiermann R (1988b) Studies on sporopollenin biosynthesis in Tulipa anthers. II. Incorporation of precursors and degradation of the radiolabeled polymer. Sex Plant Reprod 1:132-139

Wehling K, Niester Ch, Boon JJ, Willemse MTM, Wiermann R (1989) p-Coumaric acid - a monomer in the sporopollenin skeleton. Planta 179: 376-380

30
Developmental Synthesis of HSPs in the Male Gametophytic Phase of Heat Tolerant and Heat Sensitive Species

C. Frova, G. Taramino, M. Sari Gorla, and E. Ottaviano

Introduction

The fundamental requisite for a positive outcome of male gametophytic selection (MGS) is that the genes controlling the character to be selected are expressed in both phases of the plant life cycle. If this condition is met, MGS can be particularly efficient because of the large male gametophytic population size, its haploidy and the possibility to handle large experiments in limited space and controlled conditions. Moreover, detailed knowledge of the genetical basis of the character under selection is not an absolute need. This makes MGS an appealing selection strategy for traits with a complex and largely unknown genetical basis. One such case is thermotolerance, a character which is the result of several components acting at different levels of the plant organization. At the cellular level, the involvement of heat shock proteins (HSPs) in thermoprotection is indicated by a large body of correlative evidence. HSPs are transcriptionally induced by a marked temperature upshift in all sporophytic tissues, which concomitantly acquire the ability to withstand a further, otherwise lethal, temperature increase (Sachs and Ho, 1986; Lin et al., 1984; Krishnan et al., 1989). With regard to the male gametophytic phase, there are indications that HSP synthesis may be developmentally regulated: for instance in maize, HSP synthesis has been detected in immature, but not in mature, germinating pollen (Frova et al., 1986). Whether this feature is common among plants is not known: the results reported in the literature (Cooper et al., 1984; Mascarenhas and Altschuler, 1983; van Herpen et al., 1989) indicate a general lack of HSP synthesis in mature pollen, but the data refer mainly to species and genetic material adapted to temperate climates. The situation in species systematically exposed to high temperature stress has not been explored. It might be, as in the case of other environmental stresses (see Ottaviano and Mulcahy, 1989 for review), that recurrent selective pressures in both pre and post shedding phases have led to the evolution of adaptability to temperature stress in the sense of expression of the heat shock genes in those stages.

On the assumption that HSPs play an important role in cellular thermoprotection and with a view to considering MGS as a selection strategy, it is important to ascertain: 1) if and at what stage they are expressed during the male gametophytic phase, and 2) whether their synthesis is controlled by the same genes active in the sporophyte.

In this chapter we analyze the developmental expression of HSPs at different pre and post shedding stages in 2 plant species: i) maize, typical of temperate climates, and ii) sorghum, a thermotolerant species well adapted to tropical environments. The data we present have allowed the identification, in each case, of the developmental stages best suited for application of selective pressures. Accordingly we conducted a male gametophytic selection experiment in maize. The resulting progeny was evaluated at sporophytic level for overall ability to synthesize HSPs and for cellular degree of thermotolerance by means of an appropriate physiological test.

Materials and methods

HSP analysis. For pollen HSP studies in maize hybrid NI72 x OH43 was used in the analysis of developing microspores and germinating pollen. The response to heat shock treatment of sorghum gametophyte was studied using a genotype well adapted to tropical environments, line 13441 kindly supplied by the International Crop Research Institute for Semi-Arid Tropics (ICRISAT). Incubation conditions, extraction and HSP analysis were as described by Frova et al. (1989 and 1991)

In vivo gametophytic selection. A preliminary screening of different maize lines was performed in order to detect maximum differences for HSP synthesis and degree of thermotolerance. Two different genotypes, B37 (thermosensitive, HSP 17-) and Pa33 (thermotolerant, HSP 17+), were finally chosen and their F1 was used for the gametophytic selection experiment. Immature tassels (about 9 days before pollen shedding) of field grown plants were treated at 50° C for two hours using a portable battery powered thermostat. Pollen grains from the few surviving tassels were used to backcross the thermosensitive parental line B37. The gametophytic selection effect was assayed on the backcross progeny by means of quantitative analysis of the HSP profiles of individual root tips and by measuring cellular membrane stability (CMS) on leaf disks from the same individual seedlings. The quantitative evaluation of HSP expression was performed with a CAMAG densitometric scanner. CMS was tested by the electrolytic leakage assay, performed as described previously (Ottaviano et al., 1991)

Results and discussion

HSP synthesis during male gametophyte development in maize and sorghum

STAGE :	Tetrad	Uninuc	Binuc	Trinuc	Immgrain	Matgrain	Anthesis	Germination
HSPs : (kDa)		102,84 74,72 56,46 18	84,82 72	84 72	72 64	none	72 64 (const)	n.d.

Fig.1: HSP synthesis in relation to male gametophyte developmental stages in maize

As shown in Fig.1, five pre- and one post-shedding stage were analyzed in maize pollen, thus covering a significant part of its development. These stages correspond to important differentiation steps of pollen formation, i.e. mitotic division I and II, pollen grain maturation, germination and tube growth. The results clearly indicate that HSP synthesis in the male gametophyte is a developmentally regulated event: an almost complete set of HSPs is expressed in the undifferentiated uninucleate microspore, while in later developmental stages HSP expression progressively declines until at maturation it is completely absent. During germination and early tube growth no HSP synthesis is induced. Two HSP like (having the same molecular weight as) polypeptides are evident, but their synthesis is constitutive in that they are detectable also at control temperature. Later post shedding stages could not be analyzed because maize pollen shows only a very limited growth in vitro, and in vivo separation of elongating tubes from the stylar tissues they are growing through is not feasible. The picture outlined above is not unique to one genotype. Analysis of binucleate, mature and germinating pollen from other genotypes confirmed the general trend of reduction or lack of HSP expression as maturation proceeds and at shedding time..

In sorghum, however, the situation is quite different: in addition to confirming HSP synthesis in immature pollen, the analysis revealed that the ability to synthesize HSPs is maintained by sorghum pollen also in post germinative stages, during which the synthesis of several polypeptides is induced or greately increased under heat shock conditions (Table 1). This behaviour may reflect physiological differences between the two pollen species at maturation (maize is trinucleate while sorghum completes its second mitotic division during germination) and/or indicate a different evolutionary trend for adaptability to heat stress. The second hypothesis is supported by two lines of evidence: i) in Tradescantia, a binucleate pollen species, no HSPs are induced during germination (Mascarenhas and Altschuler,1983), and ii) also at the sporophytic level sorghum exhibits a more vigorous heat shock response than maize. In fact, in sorghum HSP synthesis is sustained over a prolonged (over 20 hours) heat stress period, and is still present at extremely high temperatures (50°-55° C) (Frova et al., 1991).

From a qualitative point of view, in both species two classes of HSPs are synthesized in the male gametophytic phase: those expressed also in sporophytic tissues (haplo-diploid transcription) and a few which appear to be gametophyte specific. Accordingly, a 2-fold effect of gametophytic selection can be expected:

Table 1: HSP synthesis in immature and germinating pollen in sorghum.

	HSPs (kDa) gametophytic/sporophytic	gametophytic only
immature (6 days pre-shed)	94; 84; 74; 72; 18	58; 54
germinating (in vitro)	89; 84; 74 72; 18	58; 54; 42; 14

improvement of sporophytic tolerance and increase of pollen fitness at high temperatures.

Effect of male gametophytic selection on HSP synthesis and thermotolerance in maize

To verify these expectations and on the basis of the data presented above (haplo-diploid transcription of heat shock genes, identification of the developmental stage at which they are expressed), we conducted a MGS experiment in maize. For experimental details see Materials and Methods. The general experimental design includes the evaluation of the effect of MGS at both sporophytic and gametophytic level. For this reason the selected and control progenies were divided in two groups: a first group was used for sporophytic analysis of HSP synthesis and cellular membrane stability (CMS), a major component of cellular thermotolerance. A second group, now in the field, will be used to evaluate increase in pollen fitness at high temperatures.

With regard to the sporophytic response, the relative intensity of 8 HSPs of individual seedlings was quantitatively determined. The same seedlings were then tested for CMS. This kind of analysis should allow the evaluation of: a) the effect of MGS on the molecular (HSPs) and physiological (CMS) components of cellular thermotolerance, and b) the correlation, if any, between the two components.

The results show an effect of MGS on some HSPs, in particular HSP 102 and 84 (Table 2). In these cases the mean value of the band in the selected progeny was significantly higher than in the controls, while for other HSPs the effect was not so clear. With regard to CMS, the data confirm that the physiological test used in this study is able to detect differences in thermotolerance between genotypes. In fact, the

Table 2. Mean value and standard error of HSP band intensity in control and selected backcross progenies.

HSP (kDa)	Selected	Control
102	1,49 + 0,09	1,08 + 0,10
84	4,83 + 0,12	4,42 + 0,12

Table 3. Correlation between CMS and HSP band intensity in parental lines

HSP (kDa)	r	P
102	0,31	0,01
18	0,27	0,03
17	0,31	0,01

values relating to inbreds Pa33 (thermotolerant) and B37 (thermosensitive) differed significantly. In addition, a positive correlation was found between the expression of some HSPs (102, 18, 17 kD) and CMS, even though for this latter character MGS seems not to be effective (Table 3). These results may indicate that MGS acts on other components of thermotolerance, or that CMS itself is controlled by many genes, only few of which respond to MGS. To clarify this point a more detailed statistical analysis of the above data and of those relating to the response at gametophytic level is required.

Acknowledgements

The authors wish to thank Marzio Villa for skillful technical assistance. Research supported by National Research Council of italy, Special Grant RAISA, Sub-project N.2.

Literature Cited

Cooper, P.; Ho, T.D.; Hauptmann, R.M. Tissue specificity of the heat shock response in maize. Plant Physiol. 75:431-441; 1984.

Frova, C.; Binelli, G.; Ottaviano, E. Male gametophyte response to high temperature in maize. In: Mulcahy, D.; Mulcahy, G.B.; Ottaviano, E., eds. Biotechnology and Ecology of Pollen. New York: Springer-Verlag; 1986:p.33-38.

Frova, C.; Taramino, G.; Binelli, G. Heat-shock proteins during pollen development in maize. Dev. Genet. 10:324-332; 1989.

Frova, C.; Taramino, G.; Ottaviano, E. Sporophytic and gametophytic heat shock protein synthesis in Sorghum bicolor. Plant Sci. 73:35-44; 1991.

Krishnan, M.; Nguyen, H.T.; Burke, J.J. Heat-shock protein synthesis and thermal tolerance in wheat. Plant Physiol. 90:140-145; 1989.

Lin, C.Y.; Roberts, J.K.; key, J.L. Acquisition of thermotolerance in soybean seedlings: synthesis and accumulation of heat shock proteins and their cellular localization. Plant Physiol. 74:152-160; 1989.

Mascarenhas, J.P.; Altschuler, M. The response of pollen to high temperatures and its potential applications. In: Mulcahy, D.; Ottaviano, E., eds. Pollen: Biology and Implication for Plant Breeding. New York: Elsevier; 1983:p.3-8.

Ottaviano, E.; Mulcahy, D.L. Genetics of angiosperm pollen. Adv. Genet. 26:1-64; 1989.

Ottaviano, E.; Sari Gorla, M.; Pè, E.; Frova, C. Molecular markers (RFLPs and HSPs) for the genetic dissection of thermotolerance in maize. Theor. Appl. Genet. (in press).

Herpen van, M.M.A.; Reijnen, W.H.; Schrauwen, J.A.M.; De Groot, P.F.M.; Jager, J.W.H.; Wullems, G.J. Heat shock proteins and survival of germinating pollen of Lilium longiflorum and Nicotiana tabacum. J. Plant Physiol. 134:345-351; 1989.

31

Electrophoretic Analysis of the Proteins Newly Formed as a Consequence of Pollen–Pistil Interaction

Gabriella Bergamini-Mulcahy, D.L. Mulcahy, A. Kalinowski, and W.A. Smith

The present research is aimed at investigating if the stylar activation phenomenon varies according to the type of pollen applied. Through electrophoretic analysis of the newly formed proteins of the style ahead of the pollen tubes we determine the specificity of the stylar response to different (but congruous) pollens. Similarly through the analysis of semivitro pollen tubes, grown through different styles, we can learn if there is a specific stylar input on the growing tubes.

Investigating metabolic changes occurring in the ovary during the progamic phase, Deurenberg (1976) was able to document a steady increase in protein synthesis. The maximum of activity in the ovary was observed well before the pollen tubes reach the ovary. Deurenberg was interested in quantitative and not qualitative analysis of the newly formed proteins. In 1979 Herrero and Dickinson observed a striking aggregation of ribosomes into polysomes in the cells of the transmitting tissues ahead of the pollen tubes. The focus of the interest then, as probably now, was to document the phenomenon of the activation of the pistil as a consequence of a signal triggered by pollination. Many hypotheses had been advanced about the nature of such a signal from an electric one (Sinyukhin and Britikov 1967; Linsken and Spanjers 1973) to an enzymatic (Roggen 1976) or hormonal one (Jensen 1983). The activated pistil, in turn, facilitates the growth of the pollen tube through the style into the ovary, with "the most impressive spring in biology" (J.D. Thomson, oral communcation). In previous investigations (Mulcahy, G.B., Mulcahy, D.L. 1985a, 1986b) we demonstrated, in Nicotiana alata, the effect of an "activated ovary" in the growth rates of compatible pollen. Yi-quin Li and Linskens (1983), working on Lilium longiflorum, observed that the quantity of wall-bound proteins of pollen tubes growing in vivo is 5-7 times those found in tubes grown in vitro. Furthermore, the protein pattern of wall-bound proteins is different between in vitro and in vivo grown pollen. Also well-established is the contribution of the pistil to the growth of the pollen tube but, so far, there is no evidence for the specificity of such contribution, except for the case of S-locus glycoprotein gene implicated in the pollen-style (or stigma) interaction in self-incompatibility (Nasrallah, 1985).

Two questions we try to address with the present work:
I. How specific is the action of a pollen on the activation of the pistil?
II. How specific is the contribution of a style to the growth of pollen tubes?

To carry out this research we chose two members of the Solanaceae: the gametophytic self-incompatible Nicotiana alata in two interfertile varieties; Beltsville (obtained from USDA) and Sensation (Burpee). [Each of the two varieties have been subjected to 10 generations of sib crosses by me] and the self-compatible (and congruous with N. alata) Nicotiana sylvestris (the product of several generations of spontaneous selfing)]. Grown in greenhouse 14 hr day temp. approx. 20-24oC, we employed Hurkman et al. (1986) protein extraction method to prepare the material for two-dimension electrophoretic analysis.

In a preliminary study, the choice of two different species of the genus (Nicotiana sylvestris and Nicotiana alata) exhibiting extensive electrophoretic differences of their ungerminated pollen analysis, was made to enhance the probability of finding differences in the newly formed proteins (both in styles ahead of tubes and in the tubes grown in semivitro). At 21oC (temperature used for incubation after pollination), the pollen of the Nicotiana alata variety Sensation - commonly used as pollen donor - and pollen of N. sylvestris grow, in the sylvestris style, at approximately the same rate - suggesting comparable grade of interaction. Using, separately, the two sources of pollen, pollination was made on emasculated styles of Nicotiana sylvestris. Emasculation was done one day before anthesis, pollination always done in "excess". Twenty-two hrs after pollination, when the pollen tubes have traveled the first 2.12 cm (X = 2.12 S.E. 0.07) of the style, the sections between 3.2-4.2 cm below the stigma of each of 15 styles, for each type of cross, were collected (approximately 0.04 gm fresh weight), immediately ground with Hurkman extraction buffer; and, at the completion of extraction protocol, stored at -70oC.

The part of the style above the section collected for protein extraction was, each time, fixed in FAA, 70% ethanol, cleared in 8N NaOH, and stained in decolorated aniline blue for epifluorescence observation to ensure that no pollen tubes have been included in the extracted material. At the same time the analogous sections wereextracted of 15 emasculated unpollinated styles of N. sylvestris. Fifteen μl aliquots of protein extract (prepared always in duplicate) have been used for each IEF column gel.

Proteins were separated according to Hochstrasser et al. (1988) method. The silver (Heukeskoven, 1985) stained second dimension gels, from styles pollinated with pollen of N. alata Sensation, those pollinated with N. sylvestris and those unpollinated have been compared. While the pattern of proteins of the unpollinated style shows unquestionable and extensive differences when compared with the similar section, ahead of the pollen tubes in the case of pollinated styles, the differences following pollination with different pollens are surprisingly few (Figure 1).

To investigate the contribution of the style to the growing pollen tube, I pollinated two different pistils (Nicotiana sylvestris and Nicotiana alata Beltsville) with the same pollen source (Nicotiana alata Sensation). To analyze the pollen tubes, grown through the styles, we employed the semivitro technique (Mulcahy, G.B. and Mulcahy, D.L. 1985). Twenty hrs after pollination, the pollinated styles were cut just ahead of the front of the pollen tubes, and were dipped in Brewbaker and Kwack (1983) medium (1.62 mM H_3BO_3, 1.27 mM $Ca(NO_3)_2$, 0.3 M sucrose) and incubated at 21o in the dark for 24 hrs. To avoid fungal or bacterial contamination, the media is filter sterilized, the styles surface sterilized with ethanol, and the glassware autoclaved. The pollen tubes extruding from the cut stylar end of 60 styles, from each type of pollination, collected with a razor blade, under dissecting scope on a microscope slide, are ground in Hurkman extraction buffer and, at the completion of the extraction procedure, frozen at -70° C. The analysis of the silver stained second dimension slab of the _in vitro_ grown _N. alata_ cv. Sensation pollen and the semivitro tubes grown out of the two different styles should have given us an estimate of the stylar influence on the newly formed proteins. But, so far, given the difficulty of collecting a consistent quantity of material makes comparison difficult. And even an apparently equal quantity of pollen tubes collected from _N. alata_ styles gives much poorer results on a silver stained second dimension gel, than tubes collected from _N. sylvestris_ styles. Pollen-style interaction has been for several decades one of the still unsolved puzzles on the reproductive biology of plants. Each new observation will bring us closer to the understanding of such paramount phenomenon.

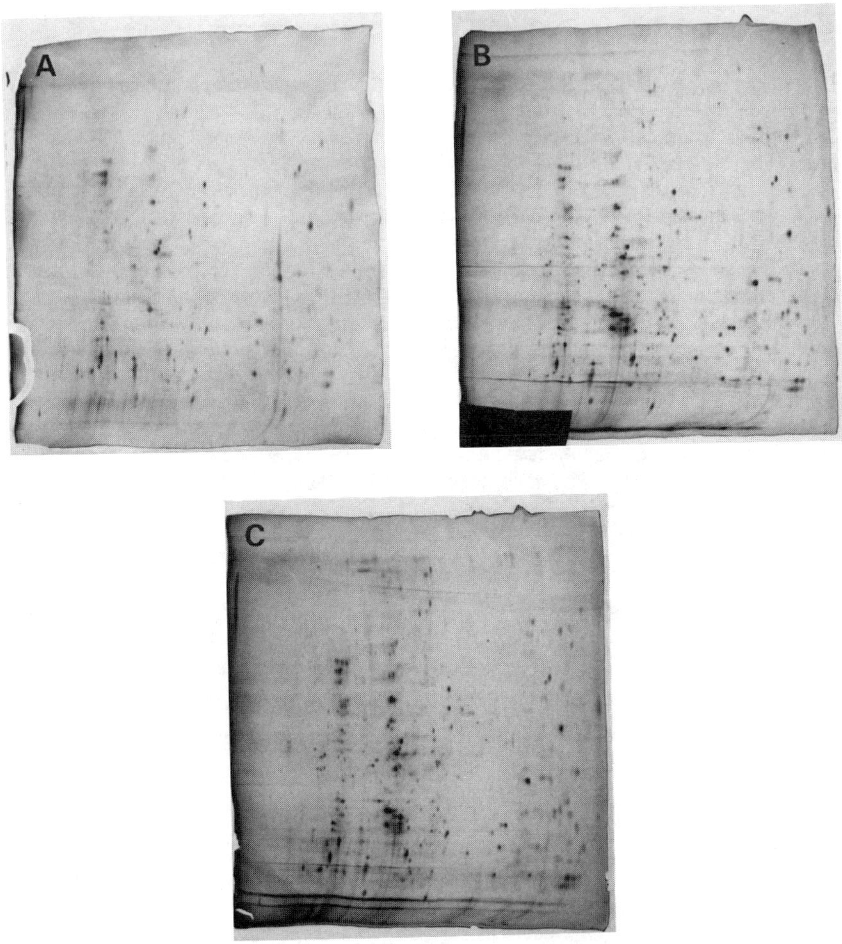

Figure 1. Two dimensional separation of stylar proteins.
A. *Nicotiana sylvestris* style, unpollinated.
B. *N. sylvestris* style, pollinated by *N. alata*
C. *N. sylvestris* style, pollinated by *N. sylvestris*.

References

Brewbaker, J.L., Kwack, B.H. 1963. The essential role of calcium in pollen germination and pollen tube growth. Am. J. Bot. 50: 859-865.

Deurenberg, J.J.M. 1976. In vitro protein synthesis with polysomes from unpollinated, cross- and self-pollinated Petunia ovaries. Planta 128: 29-33.

Herrero, M., Dickinson, H.G. 1979. Pollen-pistil incompatibility in Petunia hybrida: changes in the pistil following compatible and incompatible crosses. J. Cell Sci. 36: 1-18.

Heukeskoven, J., Dernicls, R. 1985. Simplified method for silver staining of proteins in polyacrylamide gels and the mechanism of silver staining. Electrophoresis 6: 103-112.

Hochstrasser, D.F., Harrington, M.G., Hochstrasser, A.C., Miller, M.J., Merril, C.R. 1988. Methods for increasing the resolution of two-dimensional protein electrophoresis. Analytical Biochem. 173: 424-435.

Hurkman, W.J., Tanaka, C.K. 1986. Solubilization of plant membrane proteins for analysis by two-dimensional gel electrophoresis. Plant Physiol. 81: 802-806.

Jensen, W.A., Ashton, M.E., Beasley, C.A. 1983. Pollen tube-embryo sac interaction in cotton. In: Mulcahy, D.L., Ottaviano, E., eds. Pollen: Biology and Implication for Plant Breeding 67-72. Elsevier.

Linskens, H.F., Spanjers, A.W. 1973. Changes of the electrical potential in the transmitting tissue of Petunia styles after cross- and self-pollination. Incompatibility Newsletter 3: 81-85.

Mulcahy, G.B., Mulcahy, D.L. 1985. Ovarian influence on pollen tube growth as indicated by the semivitro technique. J. Amer. Bot. 72(7): 1078-1080.

Mulcahy, G.B., Mulcahy, D.L. 1985. More evidence on the preponderant influence of the pistil on pollen tube growth. In: Biology of Reproduction and Cell Motility in Plants and Animals. Cresti, M. and Dallai, R., eds. pp. 139-144.

Mulcahy, G.B., Mulcahy, D.L. 1986. Use of vital dyes in conjunction with the semivitro technique for the detection of pollen tube sperm nuclei. Stain Technology, Vol. 61 : 382-383.

Nasrallah, J.B., Kao, T.H., Goldberg, M.L., Nasrallah, M.E. 1985. A cDNA encoding an S-locus specific glycoprotein from Brassica oleracea. Nature 318: 263-267.

Roggen, H.P.J.R. 1967. Changes in enzyme activities during the progame phase in Petunia hybrida. Thesis Nijmegen, The Netherlands. Sinyukhin, A.M., Britikov, E.A. 1967. Action potentials in the reproductive system of plants. Nature 215: 1278-1280.

Yi-quin, Li and Linskens, H.F. 1983. Wall-bound proteins of pollen tubes after self- and cross-pollination in Lilium longiflorum. Theor. Appl. Genet. 67: 11-16.

32

Adaptive Germination Responses by Walnut (*Juglans*) Pollen to Temperature during Pollen Development

Vito S. Polito and Steven A. Weinbaum

In winter-deciduous tree species, temporal variation in budbreak varies widely within species and is presumably related to requirements for winter chilling, level of chilling experienced and the post-dormant heat requirement (Couvillon and Erez, 1985; Spiegel-Roy and Alston, 1979). Temperature ranges and optima for pollen germination and pollen tube growth vary widely among species. Intraspecific variation in pollen responses to temperature is less well documented, and we have been interested in determining the extent to which variation in sporophyte responses to temperature (as in time of budbreak) is expressed in the gametophyte. In an investigation of two closely related *Prunus* species, *P. dulcis* (almond) and *P. persica* (peach), with different bloom dates, we (Weinbaum, et al., 1984) determined that there were no differences in pollen tube growth responses to temperature, but that temperatures at which pollen germination occurred maximally differed between the two species (16 C in the earlier blooming almond vs. 23 C in peach). No differences were evident within either species. Budbreak and bloom occur considerably later in *Juglans* spp. (walnut) than in *Prunus*, and the temporal range for these phenomena are much greater. Historical data from extensive collections of almond and walnut clones available at the University of California show that the range in mean bloom date is 22 days for almond clones (Weinbaum et al, 1984) and 50 days among *J. regia* clones (Hendricks, et al, 1985). Therefore, we have been investigating the response to temperature by pollen among clones and cultivars of walnut. This paper summarizes some of our results (Luza, et al., 1987; Polito, et al., 1991).

Materials and Methods

Pollen was collected from ten cultivars and clones of *J. regia* L. (English or Persian walnut) and three of *J. nigra* (Eastern black walnut) selected to include the range of staminate bloom dates for the *Juglans* collection in Davis, California (Table 1).

Table 1. Juglans regia and J. nigra cultivars used to determine pollen germination responses to temperature. Mean staminate bloom dates and temperatures for that date are based on 6 to 19-year averages at Davis, California.

Species Cultivar	Mean Staminate Bloom Date	Mean Daily Temperature
J. regia		
Serr	3 April	14.2
Manregian	5 April	14.4
Early Erhardt	11 April	14.2
Idaho	13 April	14.7
Chico	15 April	14.4
Sharkey	17 April	14.2
Amigo	22 April	14.4
S. Franquette	29 April	15.6
Meylan	3 May	16.4
XXX Mayette	5 May	16.1
J. nigra		
Student	7 May	16.9
Thomas	17 May	19.7
Ohio	19 May	19.2

In a second experiment, three branch units from 'Serr,' the *J. regia* cultivar with the earliest staminate bloom date, were heated above ambient using an apparatus described in Polito, et al (1991). The first from the time the presumed chilling requirement was met, a second two weeks later and the third one more week later. During this period, development in the anthers had proceeded from the sporogenous tissue stage to bicellular gametophytes. In each case, pollen was germinated *in vitro* on a temperature gradient apparatus described in Weinbaum, et al. (1984) adjusted to provide ten temperature environments from ca. 11 to 35 C. Pollen germination and tube growth were determined at each temperature and the data fitted to polynomial equations. The first derivatives of these equations were used to compute optimum germination temperatures.

Results and Discussion

Pollen tube growth. Models based on least-square polynomial regressions revealed no differences in optimum temperatures for pollen tube growth either between or within species. For both species maximum tube growth occurred at 32 to 33 C, except for *J. regia* 'Manregian' where it occurred at 29 C; at higher temperatures growth rates fell off rapidly. Some trends were evident in the response of tube growth to temperature, however. Typically, minimum temperatures that would support elongation of *J. regia* pollen tubes were lower than those for the later-blooming *J. nigra* cultivars, except for two of the *J. regia* cultivars which had low-temperature responses more similar to *J. nigra* (see, Luza, et al., 1987).

Relationship between pollen germination temperature and staminate bloom date. No pollen germinated at temperatures below 14 C; at 14 to 15 C only pollen from the earliest blooming cultivars (Serr, Manregian, Early Erhardt) germinated. Pollen of other *J. regia* cultivars showed minimal germination at 16 to 18 C with the exception of the latest-blooming cultivar, XXX Mayette, which required temperatures of at least 20 C. The earliest of the *J. nigra* clones germinated at a minimum temperature of 21 C, the two later-blooming cultivars required a minimum temperature of 25 C. Minimum temperatures for pollen germination were positively correlated with mean staminate bloom dates for the cultivars (Fig. 1). Optimum temperatures for pollen germination were interpolated from the first derivatives of polynomial equations fitted to the data for each cultivar. These data also showed a positive logarithmic correlation with mean staminate bloom date (Fig. 2). Thus, a clear relationship exists within these two *Juglans* species between time of staminate bloom and adaptation to temperature by the pollen. A similar relationship was also found for pollen of different clones of *Pistachia vera* (Polito, et al., 1988). This raises the question of whether the relationships are manifestations of genetically fixed responses by the male gametophytes to temperature conditions likely to be prevalent at the time of pollen dispersal, or if these differences are due to plasticity of the gametophytic phenotype in response to prevailing temperatures during pollen development.

To answer this question we heated three branches with staminate inflorescence buds on a single tree of *J. regia* 'Serr', the earliest blooming cultivar and the one showing the lowest temperature optimum for pollen germination. Heating began on 2 February, the approximate date when the chilling requirement for rest was met. The second and third branches were heated beginning 16 and 23 February. On 2 February the anthers contained sporogenous tissue, on 16 February they had early microsporocytes, and on 23 February they had microspore tetrads or bicellular microgametophytes present. Night temperatures were maintained at 16 to 17 C; day temperatures ranged to approximately 6 C above ambient (Fig. 3). Daily degree-days (a measure of integrated daily

Table 2. Timing of developmental events for 'Serr' walnut branch units under ambient conditions and in heated enclosures. Data are presented as accumulated degree-days with Julian date in parentheses. Accumulated degree-days for all treatments are calculated from the time the first branch unit was placed in a heated enclosure.

	Accumulated degree-days (Julian Date)			
Event	Ambient	Treatment 1	Treatment 2	Treatment 3
Branch enclosed and heat started	---	0 (33)	90 (47)	152 (55)
Catkin bud swelling	152 (55)	155 (44)	161 (52)	152 (55)
Terminal bud break	229 (63)	232 (49)	222 (56)	232 (60)
Leaf emergence from terminal	283 (69)	278 (52)	255 (58)	272 (63)
Catkins expanded	283 (69)	278 (52)	270 (59)	259 (62)
First staminate flowers open	292 (70)	310 (54)	286 (60)	273 (63)
First anther dehiscence	299 (71)	340 (56)	286 (60)	288 (64)
First pistillate flowers visible	359 (78)	356 (57)	377 (70)	363 (72)

temperature) were calculated and the rate of accumulation of heat units is shown in Fig. 4. Accumulation of heat units and times to several readily identifiable developmental stages for each of the treatments are tabulated in Table 2. Pollen from each of the four treatments was again germinated on the temperature gradient apparatus and optimum temperatures calculated from fitted polynomials. A highly significant regression (r^2 = 0.996) exists between optimum germination temperature and the natural log of mean daily degree-days from the start of the experiments to anther dehiscence over the range of conditions for these experiments (Fig. 5). Fig. 6 compares optimum pollen germination temperatures for each of the ten *J. regia* cultivars where development proceeded under ambient conditions and for the 'Serr' pollen from the single tree in this experiment. In the first experiment the optimum temperature was found to be 25.8 C; this compares with 25.6 C for the pollen that developed under ambient conditions in the second experiment. Note that the temperature range between pollen that developed under ambient conditions and that which developed under high temperature conditions is approximately half that shown by the ten clones of the species.

These results partially answer the question of whether the pollen germination response to temperature within the species is a fixed genetic

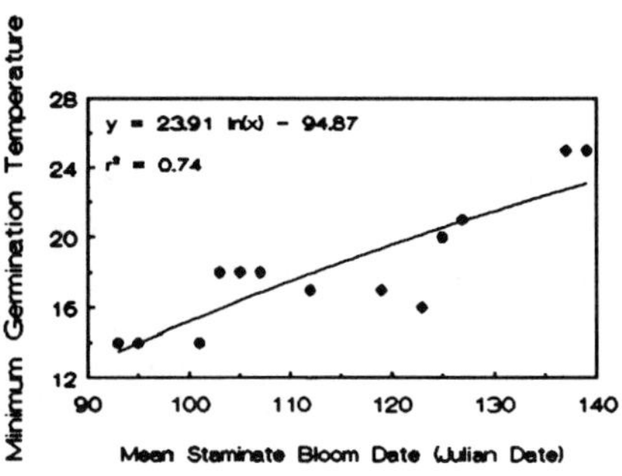

1

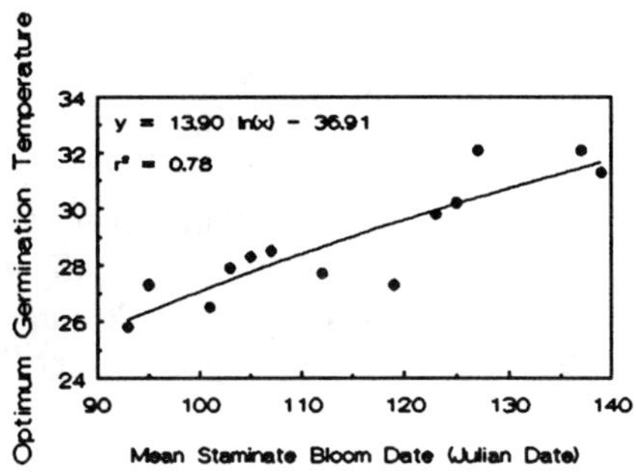

2

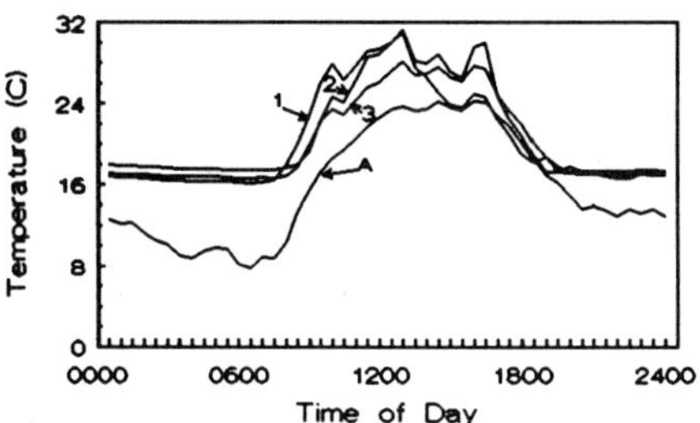

3

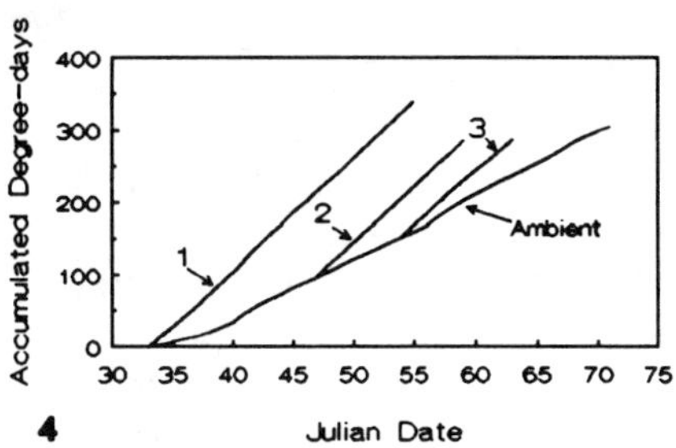

4

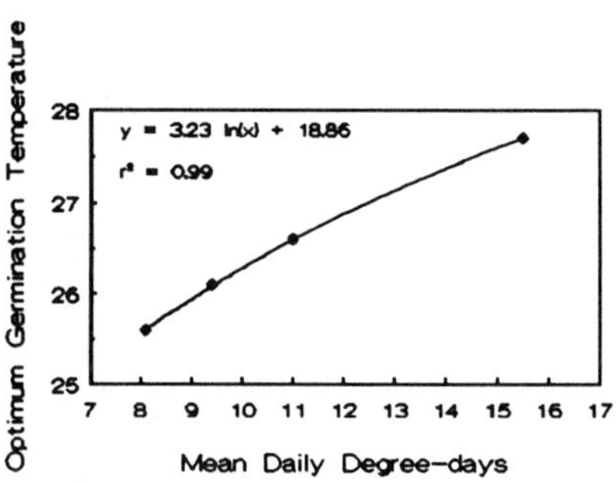

5

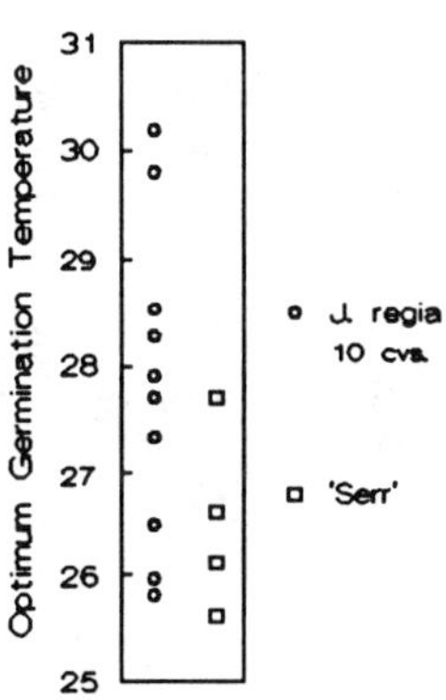

6

Fig. 1. Relationship between mean staminate bloom date and minimum temperature for pollen germination for *J. regia* and *J. nigra* cultivars. Fig. 2. Relationship between mean staminate bloom date and optimum temperature for pollen germination for *J. regia* and *J. nigra* cultivars. Fig. 3. Temperature readings from thermocouples in each of the three heated enclosures (1, 2, 3) and ambient conditions near an unheated staminate inflorescence bud. Fig. 4. Accumulated degree-days with time in each of the experimental treatments from the time heating was begun on the first branch. Fig. 5. Optimum temperature for pollen germination as a function of mean daily degree-days experienced during the period from February 2 to anther dehiscence. Fig. 6. Range of pollen germination temperature optima for ten *J. regia* clones collected after normal pollen development and the four pollen samples of *J.regia* cv. Serr described in Fig. 5.

adaptation of the male gametophyte that correlates with sporophyte phenology, or a plastic response to environmental conditions experienced during pollen development. Clearly, there is a plastic component of the *in vitro* pollen germination response to temperature that is at least partially affected by the environment during the pollen development period. The range of response covers about half of that exhibited by several clones in the species, however. Thus, it appears that the range of this plastic component of the phenotype may be limited within a given genotype.

References

Couvillon, G.A.; Erez, A. 1985. Influences of prolonged exposure to chilling temperatures on budbreak and heat requirement for bloom of several fruit species. J. Amer. Soc. Hortic. Sci. 100:47-50.

Spiegel-Roy, P.; Alston, F.H. 1979. Chilling and post-dormant heat requirement as selection criteria from late-flowering pears. J. Hortic. Sci. 54:115-120.

Luza, J.G.; Polito, V.S.; Weinbaum, S.A. 1987. Staminate bloom date and temperature responses of pollen germination and tube growth in two walnut (*Juglans*) species. Amer. J. Bot. 74:1898-1903.

Weinbaum, S.A.; Parfitt, D.E.; Polito, V.S. 1984. Differential cold sensitivity of pollen grain germination in two *Prunus* species. Euphytica 33:419-426.

Hendricks, L.C.; McGranahan, G.H.; Ramos, D.E.; Iwakiri, G.; Forde, H.I. Selection of varieties. In: Ramos, D.E., ed. Walnut Orchard Management. Berkeley CA: Univ. of Calif. Div. Agric. Natural Resources; 1985:p. 46-51.

Polito, V.S.; Luza, J.G.; Weinbaum, S.A.; 1988. Differential low-temperature germination responses by pollen of *Pistacia vera* clones with different bloom dates. Scientia Horticulturae 35:269-265.

Polito, V.S.; Weinbaum, S.A.; Muraoka, T.T. 1991. Adaptive responses of walnut pollen germination to temperature during pollen development. J. Amer. Soc. Hortic. Sci. 116:552-554.

33

Effect of the Carboxylic Ionophore Monensin on the Growth and Protein Pattern of Pollen Tubes

Gian Lorenzo Calzoni and Anna Speranza

Tip growth is a process in which the relations between structure and function, as well as between the single organelles and the cell as an integral whole are perhaps more evident than in other examples of cell morphogenesis (Sievers and Schnepfs, 1981).

Apical growth of pollen tubes is associated with the presence of large numbers of vesicles in the tip region: within the whole population, a larger vesicle class is likely to be concerned with the transport of cell wall constituents to the extending tip. In some cases, however, there is no membranous packaging nor are the materials dictyosomically-derived (Heslop-Harrison, 1987). In apple, the tube tip was found to be very rich in two types of secretory vesicles: large ones containing fibrillar substances and small ones containing highly electron-dense material (Speranza et al., 1982). According to Heslop-Harrison (1990) the smaller vesicles are likely to be concerned with enzyme emission by the pollen tube: this is an important function at every stage of growth , expecially in the case of tubes making their way through a solid style (as in Malus domestica).

The ability of eucariotic cells to conduct intracellular vesicular traffic is highly sensitive to prevailing monovalent cation levels (Tartakoff, 1983). Monensin is a well-characterized metabolite of Streptomyces cinnamonensis that binds Na^+, K^+ and protons, acting as a diffusional carrier mediating one-for-one cation exchange. Perturbing the internal ionic environment -that cells maintain upon continuous energy cost - slows down or arrests intracellular transport; in all cases, the principal site of arrest imposed by monensin has proven to be within the Golgi complex (Tartakoff, 1983).

The effect of monensin in pollen germination is the topic of the present chapter. The aim is to show the importance of intracellular cation homeostasis for vesicular traffic in a system such as pollen tubes, where morphogenesis and other related events greatly rely on secretion and transport.

Effect on tube emergence and growth

Apple pollen (Malus domestica Borkh. cv Starkrimson) was germinated in liquidmedium as described by Calzoni et al. (1979). Monensin was added from a stock solution in ethanol; identical ethanol concentrations were also present in the control cultures without monensin. Table 1 shows the range of monensin concentrations within which growth was inhibited. The system was very sensitive, since growth declined exponentiallly with increasing monensin concentration. Absolute values of tube emergence and elongation showed highly significant differences ($P < 0.01$) with respect

Table 1 - Germination and mean tube length of apple pollen incubated for 60 or 120 min in the presence or absence of monensin, expressed as % of controls.

Monensin (nM)	Germination 60 min	(% of control) 120 min	Tube length 60 min	(% of control) 120 min
5	96	91	108	90
50	83	85	87	75
100	72	43	81	33
200	41	15	59	26
500	12	4	38	18
1000	2	0	1	0

to controls (Student's t-test) already with 50 nM monensin, either after 60 or 120 min incubation. No many other plant systems are known to be so sensitive to this ionophore. Fungal hyphae, for example, which are also characterized by tip growth, were inhibited by 10 to 100 uM monensin (Pancaldi et al. , 1985).

Effect on protein secretion into the medium

After a rapid release of cell wall-stored proteins at the start of incubation , a second period of protein emission begins as tubular growth proceeds: the source of these latter proteins is the pollen tube cytoplasm (Kamboj et al. , 1985; Pettitt, 1985). Protein emission by pollen tubes is an energy-requiring process (Kamboj et al. , 1985); proteins are packaged into secretory vesicles and exported to the external environment by fusion of the membrane vesicle with the plasma membrane (Heslop-Harrison, 1987).

Four monensin concentrations were used to test a possible ionophore effect on protein release by pollen (fig. 1); the same amount of ethanol present in each monenesin treatment was given to untreated pollen as controls. A clear reduction in protein titer in the culture medium can be observed at 5 nM monensin . A reciprocal relationship linked this reduced amount in the medium with protein accumulation within the cytoplasm . Kinetics of protein release revealed that the monensin inhibition was evident in the later stage of incubation (fig. 2).

Numerous studies on animal cells show that protein secretion is greatly affected by monensin, which prevents their transfer from the Golgi complex to secretory vesicles (Tartakoff, 1983). In plants, secretion of a-amylase is markedly inhibited in monensin -treated barley aleurone and maize scutellum (Heupke and Robinson, 1985; Melroy and Jones, 1986; Sticher and Jones, 1988). In bean cotyledons monensin blocks the chase-out of phytohemagglutinin from the Golgi into the protein bodies (Chrispeels, 1983). In contrast, no effect was detected on polysaccharide slime secretion by maize root tips (Sticher and Jones, 1988); nor on the secretion of acid hydrolases by sycamore maple suspension culture (Driouich et al., 1989).

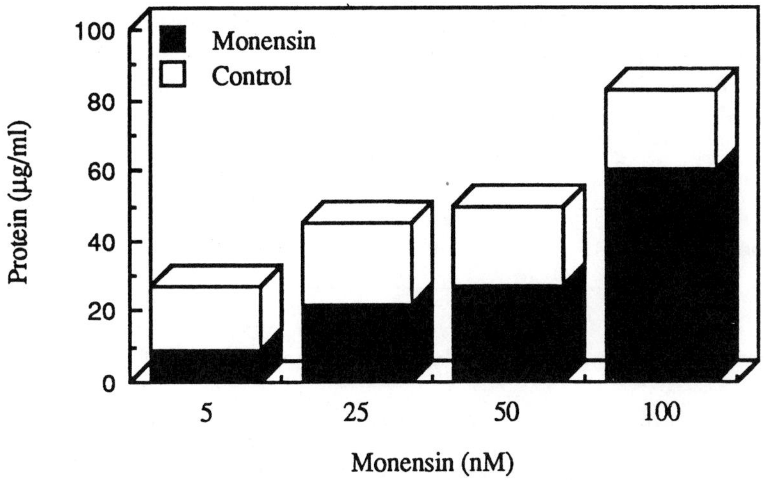

Figure 1. Protein release by apple pollen after 120 min incubation in the presence or absence of different concentrations of monensin. Ethanol in the cultures was 0.03 Σ in the case of 5 nM-treated pollen and the corresponding control; 0.02 % in the case of 25 nM-treated pollen and the corresponding control; 0.03 % in the case of 50 nM-treated pollen and the corresponding control; 0.07 % in the case of 100 nM-treated pollen and the corresponding control.

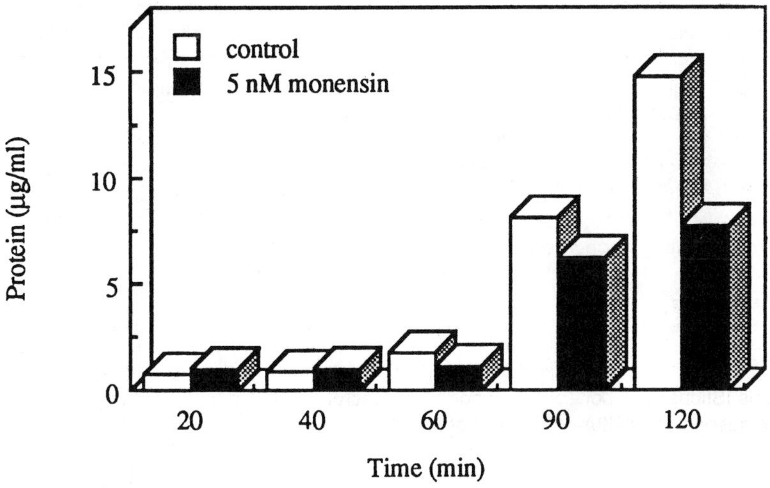

Figure 2. Time-course of protein release by apple pollen in the presence or absence of 5 nM monensin.

Table 2 . 3H- protein in the soluble (s) and membranous (m) fractions of cytoplasm of apple pollen germinated 120 min in the presence or absence of 5 nM monensin. The m/s percent ratio was calculated considering the P100K fraction.

	^{3}H- protein Soluble	(Bq mg^{-1} pollen) P 9K	P 100K	m/s (%)
Control	223	225	9	4.0
Monensin	184	275	12	6.5

Effect on transport of newly-synthesized proteins

We have already shown in apple that protein levels rise progressively in the germination medium while they decrease in the pollen (Speranza et al., 1984; Speranza and Calzoni, 1989). We also know that germinating apple pollen takes up increasing amounts of exogenous leucine which is incorporated into de novo -synthesized proteins starting from the 30 min lag phase of germination (Calzoni and Speranza, 1986).

In order to follow the changes in the movement of the newly-synthesized proteins, apple pollen was fed with L- [4, 5 - 3 H] leucine during 120 min incubation in the presence or absence of 5 nM monensin. Thereafter the cultures were filtered-off and washed with cold leucine. The walls, separated from the cytoplasm, were purified and extracted according to Calzoni and Speranza (1989). The cytoplasm was centrifuged at 9,000 g and 100,000 g : the correspondent pellets were referred to as P 9K and P 100K respectively, and the final supernatant as the soluble fraction. Incorporation of 3 H -leucine was measured after TCA precipitation on Whatmann glass filters.

Uptake of radioactive leucine was very similar either in the monensin-treated pollen or in the control. Table 2 shows the distribution of the newly-synthesized proteins in the different fractions of apple pollen cytoplasm. The 3 H - incorporation ratio in the membrane versus soluble fraction appeared to increase significantly in the presence of monensin, suggesting an interference on the routes of protein sorting from the membranous compartment. This seems to agree with Chrispeel's finding (1984) in bean cotyledons, where a 2 hrs pre-treatment with monensin had relatively little effect on the uptake of radioactive precursors, while labeling of proteins in the soluble fraction was lowered with respect to that in the membrane fraction.

Conclusions

Different developmental patterns during maturation may yield different models of pollen germination which may or may not rely on post-maturation syntheses. In pollen such as that of Malus domestica, if a pivotal role of the Golgi apparatus for morphogenetic events during germination can be hypothesized, the use of monensin seems to be promising. It is known that the Golgi apparatus in plant cells is not only the site of glycoprotein and proteoglycan processing, but also the site of synthesis and processing of complex matrix polysaccharides of the cell wall (Moore et al ., 1991).

In the present chapter, a very sensitive growth response of apple pollen to increasing monensin has been described (tab. 1). This may suggest that, if dictyosomes could represent the main source of membrane and wall components for the growing tube tip, at least a great part of these macromolecules should exit from the Golgi complex distal to the monensin site of action.

Monensin was not used before to probe protein secretion paths in pollen tubes. A very low monensin concentration was found to markedly inhibit protein export into the medium (figs. 1 and 2). The present finding on apple pollen at least indicates that these macromolecules, addressed to the exterior of the cell, may also exit from the Golgi apparatus, since the sensitivity of the intracellular transport of a given compound to monensin is diagnostic of the path followed by that compound.

The effect on transport of newly-synthesized proteins (tab. 2) has suggested to us further work on the biochemical consequences following the monensin block. Work is in progress to investigate late Golgi activities of glycoprotein processing, which could occur distal to the monensin site of inhibition.

In the light of these results, it is clear that the impact of cellular cation levels on secretion and vesicular traffic must be seriously considered. To quote Tartakoff (1983), although the gross features of vesicular traffic are well established, the use of monensin has introduced additional elements to the knowledge of the mechanism of intracellular transport, suggesting that secretory compartments can be also viewed as definite regions of the cell having a distinction composition.

Acknowledgements

Work supported by M.P.I. funds. The authors are very grateful to Dr. Stephanie Bond for kindly revising the English of the text.

References

Calzoni, G. L.; Speranza, A. ; Bagni, N. In vitro germination of apple pollen. Scientia Hortic. 10: 49-55 ;1979.

Calzoni, G.L.; Speranza, A. Developmental stage of mature apple pollen: study on RNA and protein synthesis during in vitro germination. Physiol. VÄg. 24:53-62 ;1986.

Calzoni, G. L.; Speranza, A . Glycoproteins in germinating apple pollen. Plant Physiol. Biochem . 27: 219-225; 1989.

Chrispeels, . The Golgi apparatus mediates the transport of phytohemagglutinin to the protein bodies in bean cotyledons. Planta 158: 140- 151; 1983.

Driouich, A.; Gonnet, P ; Makkie, M.; Laine, A. C.; Faye, L. The role of high- mannose and complex asparagine-linked glycans in the secretion and stability of glycoproteins. Planta 180: 96-104 ; 1989.

Heslop-Harrison, J. Pollen germination and pollen tube growth.. Int . Rev . Cytol. 107: 1-78 ; 1987.

Heslop-Harrison, J. Dynamic aspects of apical zonation in the angiosperm pollen tube. Sex. Plant Reprod . 3: 187-194; 1990.

Kamboj, R. K.; Linskens, H. F.; Jackson, J. F. Energy-driven protein release from germinating pollen of Petunia hybrida. Ann. Bot . 54: 647-652; 1984.

Melroy, D.; Jones, R. L. The effect of monensin on intracellular transport and secretion of a- amylase isoenzymes in barley. Planta 167: 252-259; 1986.

Moore, P. J.; Swords, K.M.M.; Lynch, M.A.; Stahelin, L. A. Spatial organization of the assembly pathways of glycoproteins and complex polysaccharides in the Golgi apparatus of plants. J . Cell Biol .112: 589-602; 1991.

Pancaldi, S.; Poli, F.; Vannini, G.L. Effetti della monensina sulla forma a lievito miceliare di Candida albicans (Robin) Berkhout. Giorn .Bot. It. 119: 113-114; 1985.

Pettitt, J. M. Pollen tube development and characteristics of the protein emission in Conifers. Ann . Bot . 56: 379-397; 1985.

Sievers, A.; Schnepf, E . Morphogenesis and polarity of tubular cells with tip growth. Cell Biol .Monogr. 8: 265-299; 1981.

Speranza, A.; Calzoni, G.L.; Cresti, M.; Ciampolini, F. Effects of gamma irradiation on in vitro germination and ultrastructure of apple pollen. Env. Exp. Bot . 22: 399-347; 1982.

Speranza, A.; Calzoni, G.L.; Bagni, N. Evidence for a polyamine-mediated control of ribonuclease activity in germinating apple pollen. Physiol. VÄg. 22: 323-321; 1984.

Speranza, A.; Calzoni, G. L . Cytochalasin B affects protein secretion in germinating pollen of Malus domestica Borkh. J. Plant Physiol. 133: 719-726; 1989.

Sticher, L.; Jones, R. L . Monensin inhibits the secretion of a- amylase but not polysaccharide slime from seedling tissues of Zea mays. Protoplasma 142: 36-45; 1988.

Tartakoff, A. M. Perturbation of vesicular traffic with the carboxylic ionophore monensin. Cell 32: 1026-1028; 1983

34
Localization of the Proteins Contained in the Stigma Exudate on the Lily Pistil Tissue

Hisako Miki-Hirosige, Sumio Nakamura, Masao Watanabe, and Kokichi Hinata

The pistil of Lilium longiflorum secretes stigmatic exudate and canal exudate. In the previous paper (Miki-Hirosige et al. 1987), electrophoretic studies of these exudates have revealed quantitative and qualitative differences in protein profiles, but these exudates contain the quantitatively different same 5 proteins, and also it is assumed that the exudatic components which are transferred to the cell wall with endoplasmic reticulum and Golgi vesicles, are stored within the cell wall of the secretive tissues and secreted from the cell walls directly. In this paper, an immunocytochemical study was done to probe the secretive route of the two exudates in the young and matured lily pistils, by means of immunofluorescence and EM immunogold staining methods using a polyclonal antibody against the proteins contained in the stigmatic exudate.

Materials and Methods

Pistils of Lilium longiflorum on two different stages were prepared for the material, one was the unpollinated matured pistil 1 day after the anthesis, and the other was the young unpollinated pistil from a bud of 10 cm length about one week before the anthesis. Production of antistigma exudate antiserum: For immunization the stigma exudate (containing 75 ug proteins) sampled from the unpollinated pistils of Lilium longiflorum, was injected intraperitoneally into an eight-week-old female mouse (Balb/c). Freund's complete adjuvant was used in the first and second injections. Three days after the third injection, the serum was collected and used for the polyclonal antibody.

Immunoflourescence staining procedure: Matured and young fresh pistils were fixed with 2% paraformaldehyde and 1% glutalardehyde in 0.05 M phosphate buffer pH 7.5, and kept in 2.3 M sucrose solution. These materials were cut with Reichert-Jung ultramicrotome FC4E at -90 C about 1 um thickness, and attached to a poly-L-lysin coated slide glass and used for usual immunofluorescence method.

Fixation for the EM immunocytochemistry was carried out according to the method of Maldonado et al. (1986). Application of the post-embedding, immunogold staining technique followed the method of Bloch et al. (1986).

Results

Distribution of anti-polyclonal antibody against the proteins contained in the stigmatic exudate of Lilium longiflorum is shown by the immunoelectron micrographs. As we have used polyclonal natibody, we can not decide what protein is indicated by the gold grains.

The numerous gold grains are observed over the stigmatic exudate surrounded by the matured stigma papillae (Fig. 1), which have also the grains over the cell wall, inside of the cell membrane, rER, Golgi bodies and Golgi vesicles (Figs. 1 & 2). Outside of the matured canal stigmatoid cell, the gold grains are present over the canal exudate, but the amount of the grains is less than those over the stigmatic exudate (Fig. 3). Inside of the stigmatoid cell, the gold grains are observed over Golgi vesicles, polysome areas, attaching vesicles to the cell membrane, electrondense areas of ingrowth and cell wall (Fig. 3). On young pistils, although the gold grains are seen over the stigmatic exudate (Fig. 4), but rare over the canal exudate, both of their numbers are less than those over the matured exudates (Fig. 5b). Over attaching vesicles to the cell membrane (Fig. 4a), inside of the cell membrane, Golgi bodies and vesicles (Fig. 4b), the gold grains are observed, but rare over the cell wall (Fig. 4). The grains are also shown over polysome areas, Golgi vesicles and attaching vesicles to the cell wall, but rare over the cell wall (Fig. 5). From the fluorescence results (data are not shown here), the matured papilla and the stigmatoid cells have fluorescence over the entire cell, but the papilla's brightness is stronger than that of the stigmatoid. The young tissues have not yet any fluorescence.

Discussion

From the result of the previous paper (Miki-Hirosige et al. 1987), the canal exudate yielded 10 bands in the densitometer scans, five more than stigmatic exudate. These exudates contain the same proteins but quantitatively these proteins differ. As the stigmatic exudate has more simple proteins and a larger amount can be more easily collected than from the canal exudate, the proteins which are contained in the stigmatic exudate were used for preparing the polyclonal antibody to probe the secretive route of two kinds of the exudates in a lily pistil tissue.

We suggested in the previous paper (Miki-Hirosige et al. 1987) that the exudatic substances transported by rER and Golgi vesicles were stored initially in the cell wall of the secretive cells and secreted later from here. Now we can recognize that the numerous gold grains on the immunoelectron micrographs are observed over the Golgi vesicles of the mature side of Golgi bodies, rER, attaching vesicles to the cell membranes, cell walls and the secreted exudates in the papilla and stigmatoid cells of the mature pistil. These results clearly support the previous suggestion, that is, that the proteins are synthesized in Golgi body, polysomes and rER, transferred to the cell wall directly or with the vesicles (the attaching vesicle shows this method); stored here; and then, secreted later to be contained in the secreted exudates. In the young pistil which does not secrete much exudate yet, but in which the proteins are synthesized in golgi body, rER and polysomes, transferred to the cell walls and stored there until the secretion time. The matured papilla and stigmatoid cells have fluorescence over the whole cells, but the tissues of the young bud have not yet any fluorescence, because the amount of the protein synthesis may yet be less in the young pistil.

Polyclonal antiproteins (from the stigmatic exudate) immunogold staining on the style tissues of Lilium longiflorum. CW:cell wall. E:exudate. G:Golgi body. M:mitochondrion.

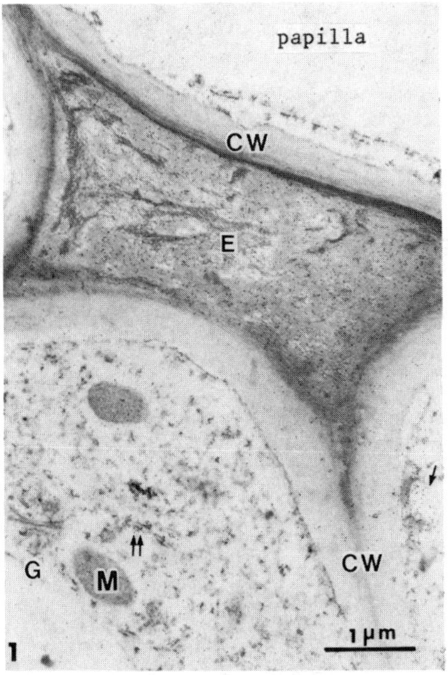
papilla
CW
E
G
M
CW
1 μm
1

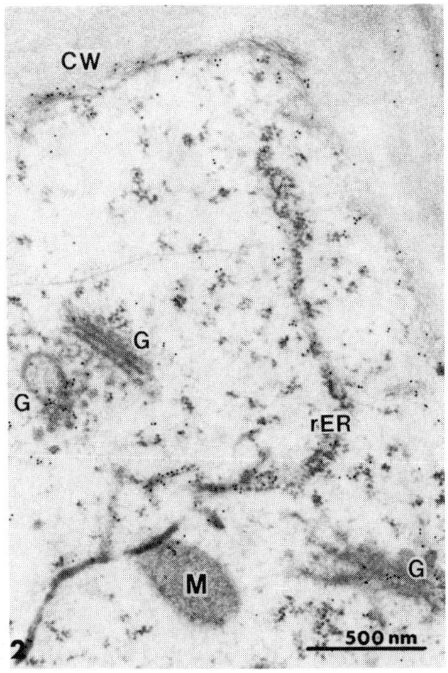
CW
G
G
rER
M
G
500 nm
2

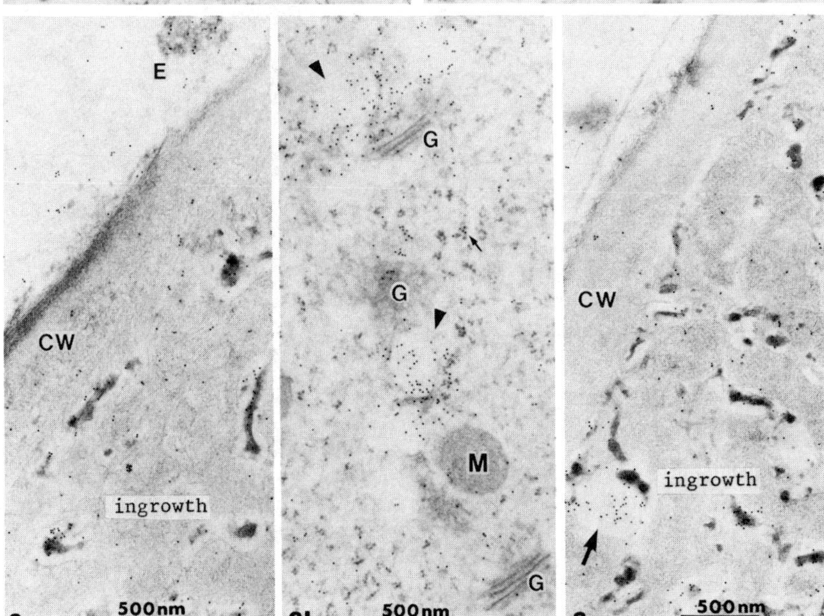
E
CW
ingrowth
500nm
3a
G
G
M
G
500nm
3b
CW
ingrowth
500nm
3c

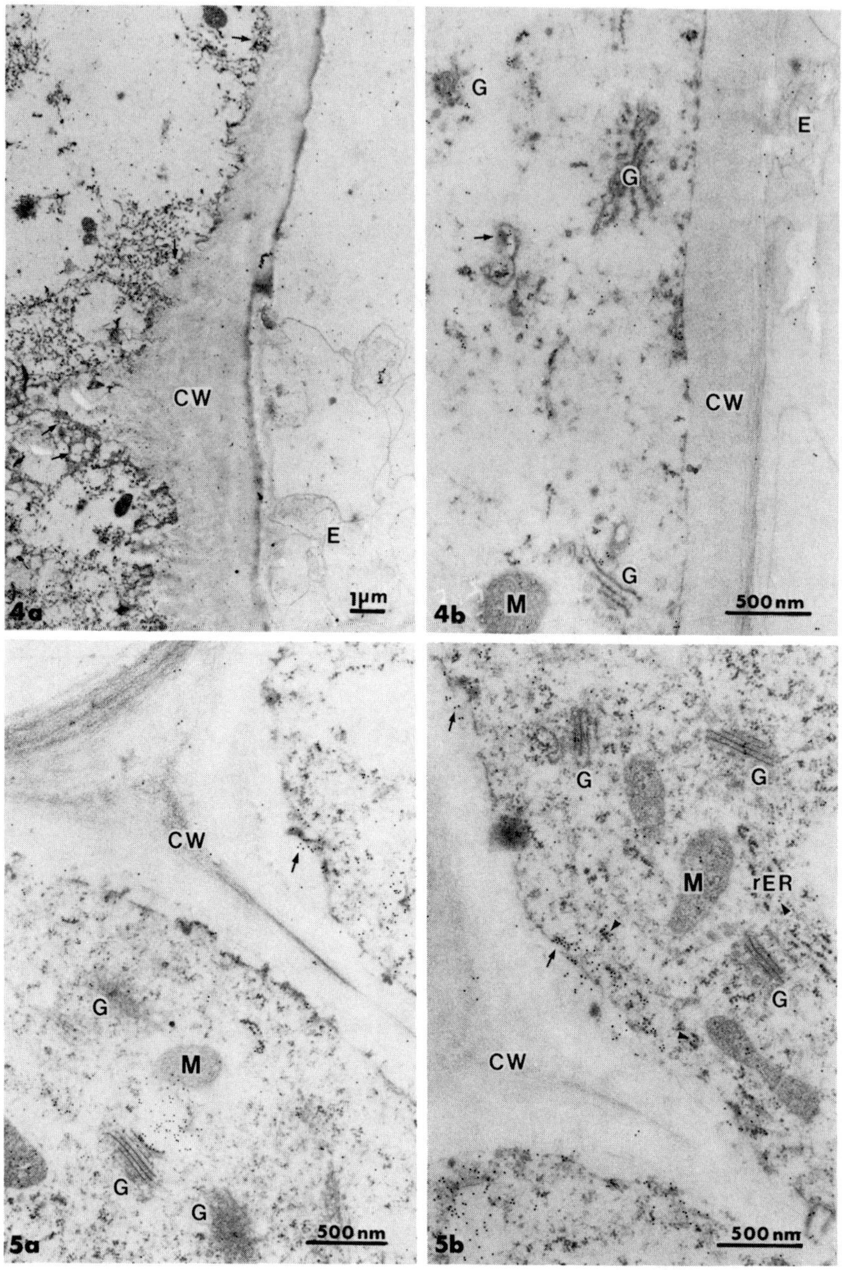
G
E
G
CW
CW
E
1µm
4a
G
500 nm
4b
M
CW
G
G
M
rER
G
CW
M
G
G
500 nm
5a
500 nm
5b

Fig. 1. The matured stigmatic papillae: The stigmatic exudate surrounded by the papillae has the numerous gold grains. The grains are also observed over the cell walls, a vesicle (arrow), rER (double arrow) and Golgi body.

Fig. 2. The matured papilla: The gold grains are shown over the cell wall, inside of the cell membrane, rER, Golgi bodies, and Golgi vesicles.

Fig. 3. The matured canal stigmatoid cell: a. The gold grains are shown over electrondense areas of the ingrowth, cell wall and exudate. b. The gold grains are shown over Golgi vesicles (arrow heads) and polysome area (arrow). c. Electrondense areas of the ingrowth and attaching vesicle (arrow) have the gold grains.

Fig. 4. The young stigmatic papilla: a. The gold grains are present over attaching vesicles (arrows) to the cell wall sectioned tangentially, cell wall and exudate. b. The gold grains are shown over the exudate, cell wall, inside of the cell membrane, Golgi bodies and namely Golgi vesicles (arrow).

Fig. 5. The young canal stigmatoid cell: a. The gold grains are seen over polysome areas, attaching vesicle (arrow) and Golgi vesicle, but rare over the cell wall and canal exudate. b. The gold grains are observed over attaching vesicles (arrows), polysome areas and rER (arrow heads).

References

1. Bloch KD, Seidman JG, Naftilan JD, et al. (1986) Neonatal atria and ventricles secrete atrial natriuretic factor via tissue-specific secretory pathways. Cell 47:695-702.

2. Maldonado CA, Saggau W, Forssman WG (1986) Cardiodilatin-immuno reactivity in specific atrial granules of human heart revealed by the immunogold stain. Anet Embryol 173:295-298.

3. Miki-Hirosige H, Hoek IHS, Nakamura S (1987) Secretions from the pistil of Lilium longiflorum. Amer J Bot 74:1709-1715.

35

Microtubular Motors in the Pollen Tube of *Nicotiana tabacum*

Giampiero Cai, Alessandro Bartalesi, Alessandra Moscatelli, Cecilia Del Casino, Antonio Tiezzi, and Mauro Cresti

Introduction

The Angiosperm pollen tube is specialized in delivering the sperm cells during the process of fertilization (Cresti and Tiezzi, 1990). Microtubules and actin filaments are the main components of the vegetative cell cytoskeleton, which study has been extensively carried out mainly by electron and fluorescence microscopy (reviewed by Tiezzi, 1991). Many types of motility characterize the cytoplasm of the pollen tube, including the movement of generative and sperm cells and the translocation of organelles and vesicles (Pierson et al., 1990). The dependence of this movement on the cytoskeletal elements is not well understood, even though some recent observations seem to support the actin filament system as an active component (Tang et al., 1989). On the other hand the role of microtubules in the cytoplasmic movement and consequently in the tube growth is not yet clear. Some recent remarks (Åström et al., 1991) have suggested possible correlations between the microtubule cytoskeleton and the movement of generative cell and vegetative nucleus.

In animal cells the movement of organelles and vesicles occurs along microtubules and is associated to motor proteins with ATPase activity. Two of these, kinesin and cytoplasmic dynein, have been identified in a wide variety of cell types. These proteins have been biochemically characterized and their role has been partially defined by using video-enhanced-differential-interference-contrast-microscopy to examine the in vitro movement of plastic beads or isolated organelles along microtubules (reviewed by Vallee and Shpetner, 1990).

We have already reported some indications on the presence of kinesin in the pollen tube of Nicotiana tabacum (Moscatelli et al. 1988) and in this chapter new immunocytochemical and biochemical analyses are presented. Additionally, some preliminary data on the identification of a pollen tube dynein-like protein are reported.

Materials and methods

1. Antibody preparation and specificity

A monoclonal antibody (named k71s23) against the heavy chain of bovine brain kinesin was prepared as already described (Moscatelli et al., 1988). The specificity of the antibody was verified by SDS-polyacrylamide-gel-electrophoresis (SDS-PAGE) (Laemmli, 1970) and subsequent immunoblotting (Towbin et al., 1979) on bovine brain kinesin purified by anion exchange high pressure liquid chromatography (AE-HPLC) on a BioRad MA7Q column after binding to microtubules and subsequent release.

Extracts of Nicotiana tabacum pollen tube after 4 hrs of germination in BK medium containing 15% sucrose (Brewbaker and Kwack, 1963) were tested by immunoblotting with the k71s23 anti-kinesin antibody.

2. Immunofluorescence staining

Immunofluorescence procedure was performed as previously described (Tiezzi et al., 1986). The k71s23 antibody was used undiluted. Confocal images were taken using a BioRad MRC 500 Confocal Laser Scanning Microscopy.

3. AMPPNP binding

Bovine brain taxol-stabilized MAPs-free microtubules (Fellous et al., 1977) were added to Nicotiana pollen tube extracts and incubated in presence of AMPPNP, a non hydrolizable analog of ATP. Microtubules were then pelleted and analyzed by SDS-PAGE and immunoblotting.

4. Sucrose gradient centrifugation

Extracts from Nicotiana pollen tube were sedimented through a 5-25% sucrose gradient and fractions were examined by SDS-PAGE and immunoblotting with the k71s23 antibody.

5. ATPase activity

ATPase activity of sedimentation fractions was evaluated by the colorimetric method of Ames (1966) both in presence and absence of bovine brain microtubules.

6. Isolation of the pollen tube kinesin homolog

A combination of AE-HPLC, sucrose gradient centrifugation and gel filtration on a BioGel A5m was used to separate the kinesin homolog. The k71s23 antibody was used to follow the protein during the purification steps.

Results

Using the k71s23 monoclonal antibody to the heavy chain of bovine brain kinesin, we moved to the search for immunologically related polypeptides in Nicotiana tabacum pollen tube. A 100 kD polypeptide was identified. By immunofluorescence analyses the cytoplasmic distribution of the 100 kD polypeptide was localized with a spot-like staining pattern in proximity of the pollen tube apex, where organelles and secretory vesicles are abundantly present. By optical sectioning the specimen with a Confocal Laser Scanning Microscopy, the localization of the 100 kD polypeptide was determined to be mainly cortical (fig.1). High magnifications revealed the presence of circular structures with the size of secretory vesicles.

The AMPPNP-stimulated binding to microtubules is a characteristic property of kinesins (Vallee and Shpetner, 1990) and was used to determine whether the pollen tube labelled polypeptide(s) could really interact with the microtubular surface. Immunoblotting analyses revealed that the polypeptide was bound to microtubules, even though a labelling in the unbound polypeptides-containing supernatant was also observed.

The 100 kD polypeptide was purified in association with a small amount of a 116 kd. Lower molecular weight polypeptides were not detected.

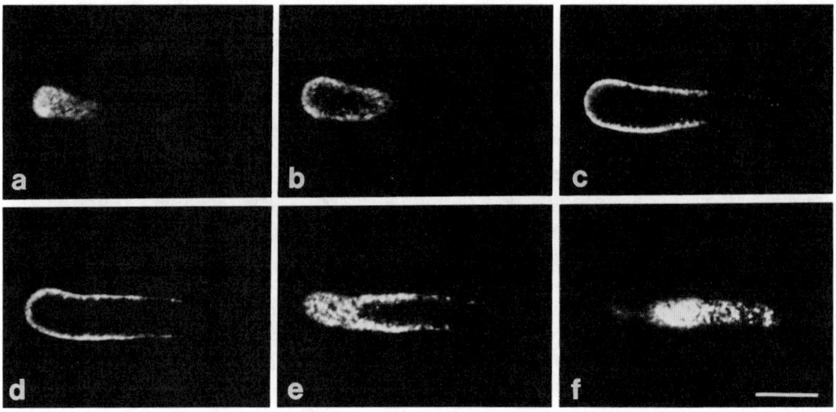

Fig.1 a,b,c,d,e,f) Confocal laser scanning micrographs at different focus levels of a fixed pollen tube of Nicotiana tabacum stained with the k71s23 antibody. The antigen is mainly detected in the cortical part of the apex region (bar: 20 μm).

When the cytosolic extract from Nicotiana pollen tube was centrifuged through a sucrose gradient and fractions assayed by immunoblotting analyses, the 100 kD polypeptide was found to sediment at 9 s, a sedimentation coefficient reported for kinesins (Vallee and Shpetner, 1990).

Sedimentation fractions were then tested for the ATPase activity in presence and absence of microtubules (fig.2). There were two main peaks of stimulation. One peak was found around fractions 11-13, containing the 100 kD polypeptide recognized by the k71s23 antibody. The other was around fractions 3-4, in which silver-stained gels revealed the presence of polypeptides with a molecular weight of about 300 kD (data not shown). The sedimentation coefficient of these fractions was around 19 s and they also showed a microtubule-stimulated CTPase activity.

Discussion

Current opinions are generally in agreement about the existence of a kinesin family with a similar aminoacidic sequence in the heavy chain, especially for the microtubule-dependent ATPase domain, that is accountable for the movement of the whole protein along microtubules (Yang et al., 1990). The rest of the polypeptide is susceptible of variations according to the particular function to play. The eventual presence and size of

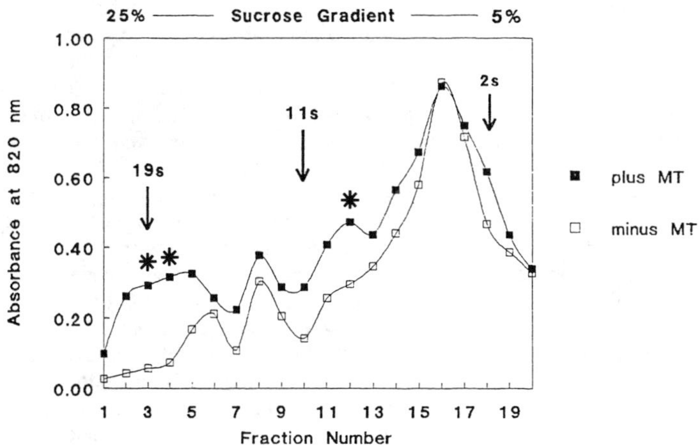

Fig.2 ATPase activity with and without bovine brain microtubules of fractions from the sucrose gradient centrifugation. Single asterisk indicates the stimulation of enzymatic activity in fractions containing the 100 kD polypeptide. Double asterisk marks a second activity peak corresponding to fractions containing the 300 kD polypeptides. Standards of sedimentation coefficient (19 s, 11 s and 2 s) are also reported.

light chains is in relation to the docking of membrane-bound organelles to be moved. Immunoblotting analyses establish the presence in Nicotiana tabacum pollen tube of a polypeptide immunologically related to the heavy chain of bovine brain kinesin. Biochemical investigations show that the 100 kD polypeptide co-purificates with a minor band at 116 kD. Presently we do not know if this polypeptide represents an associated protein or is part of the kinesin molecule. It is interesting to note that the molecular weight of the labelled polypeptide is almost similar to that of kinesin and kinesin-like protein heavy chains (105-130 kD) isolated from different organisms (Vallee and Shpetner, 1990). Gel filtration analysis suggests a lower native molecular weight for the pollen tube kinesin homolog in comparison to bovine brain kinesin.

AMPPNP is successfully used in the isolation of squid brain, bovine brain, sea urchin egg, pig brain and Drosophila kinesin (Vallee and Shpetner, 1990), but it shows no results in the isolation of the kinesin-like protein from Dictyostelium discoideum (McCaffrey and Vale, 1989). The lower AMPPNP-mediated binding to microtubules indicates that the pollen tube kinesin homolog is structurally different from those previously studied.

In animal cells the staining pattern of kinesin is punctate and enhanced in microtubule-containing areas, but it does not conform to microtubules (Hollembeck, 1989). Probably animal kinesin is mainly bound to organelles or vesicles and only little is found in association with microtubules or free in the cytoplasm. The vegetative microtubules in Nicotiana pollen tube are helical arranged along the longitudinal axis and there are few indications about their presence in the apex (Tiezzi, 1991), where the staining pattern of the kinesin homolog is higher. Nevertheless the apex is rich in secretory vesicles to which the kinesin homolog could be bound.

Analysis of the sedimentation coefficient indicates a value of 9 s, normally reported for kinesins. This means that the pollen tube kinesin possesses the appropriate shape to connect the microtubular skeleton to organelles or other cellular compartments.

The microtubule-mediated stimulation of ATPase activity is a distinctive property of microtubule-based motor proteins. The pollen tube kinesin co-sediments in fractions exhibiting an increase in the ATPase activity when microtubules are added. The low degree of stimulation could be partially related to the weak interaction observed between bovine brain microtubules and pollen tube kinesin in the AMPPNP experiments.

Fractions 3-4 of the sucrose gradient centrifugation show a considerable stimulation of ATPase activity in presence of bovine brain microtubules. High molecular weight polypeptides (300 kD) are also observed in those fractions. There are some affinities between these characteristics and those of cytoplasmic dyneins. These proteins sediment at 18-22 s depending on the cell types, their ATPase activity is stimulated by microtubules and the molecular size of the ATPase domain-containing polypeptide is normally around 300 kD (Vallee and Shpetner, 1990). In addition a microtubule-stimulated CTPase activity is usually observed. When all these data are considered, they clearly suggest the presence of dynein-like polypeptides in Nicotiana pollen tubes.

References

1. Ames, B.N. Assay of inorganic phosphate, total phosphate and phosphatases. In: Colowick, S.P.; Kaplan, N.O., eds. Methods in enzymology, vol. III, Academic Press, New York; 115-118; 1966.

2. Åström, H.; Virtanen, I.; Raudaskoski, M. Cold-stability in the pollen tube cytoskeleton. Protoplasma. 160:99-107; 1991.

3. Brewbaker, J.L.; Kwack, B.H. The essential role of calcium ion in pollen germination and pollen tube growth. Amer. J. Bot. 50:859-865; 1963.

4. Cresti, M.; Tiezzi, A. Germination and pollen tube formation. In: Blackmore, S.; Knox, R.B., eds. Microspores. Academic Press Limited; 239-264; 1990.

5. Fellous, A.; Francon, J.; Lennon, A.M.; Nunez, J. Microtubule assembly in vitro. Eur. J. Biochem. 78:167-174; 1977.

6. Hollembeck, P.J. The distribution, abundance and subcellular localization of kinesin. J. Cell Biol. 108:2335-2342; 1989.

7. Laemmli, U.K. Cleavage of structural proteins during the assembly of the head of bacteriophage T4. Nature. 227:680-685; 1970.

8. McCaffrey, G.; Vale, R.D. Identification of a kinesin-like microtubule-based motor protein in Dictyostelium discoideum. EMBO J. 8:3229-3234; 1989.

9. Moscatelli, A.; Tiezzi, A.; Vignani, R.; Cai, G.; Bartalesi, A.; Cresti, M. Presence of kinesin in tobacco pollen tube. In: Cresti M.; Pacini E.; Gori P., eds. Sexual reproduction in higher plants, Springer-Verlag, Berlin, Heidelberg, N.Y., London, Paris; 205-209; 1988.

10. Pierson, E.S.; Lichtscheidl, I.K.; Derksen, J. Structure and behavior of organelles in living pollen tubes of Lilium longiflorum. J. Exp. Bot. 41:1461-1468; 1990.
11. Tang, X.; Hepler, P.K.; Scordilis, S.P. Immunochemical and immunocytochemical identification of a myosin heavy chain polypeptide in Nicotiana pollen tubes. J. Cell Sci. 92:569-574; 1989.
12. Tiezzi, A.; Cresti, M.; Ciampolini, F. Microtubules in Nicotiana pollen tubes: ultrastructural, immunofluorescence and biochemical data. In: Cresti, M.; Dallai, R., eds. Biology of reproduction and cell motility in plants and animals. University of Siena Press, Siena; 87-94; 1986.
13. Tiezzi, A. The pollen tube cytoskeleton. In: Electron Microscopy Reviews. Pergamon Press, Vol.4, N.2; 1991. In press.
14. Towbin, H.T.; Staehelin, T.; Gordon, J. Electrophoretic transfer of proteins from polyacrylamide gels to nitrocellulose sheets: procedure and some applications. Proc. Natl. Acad. Sci. USA. 76:4350-4354; 1979.
15. Vallee, R.B.; Shpetner, H.S. Motor proteins of cytoplasmic microtubules. Ann. Rev. Biochem. 59:909-932; 1990.
16. Yang, J.T.; Saxton, W.M.; Stewart, R.J.; Raff, E.C.; Goldstein, L.S.B. Evidence that the head of kinesin is sufficient for force generation and motility in vitro. Science. 249:42-47; 1990.

36

The Microtubular Cytoskeleton during Pollen Development and Pollen Tube Growth in *Nicotiana tabacum*

Antonio Tiezzi, Joseph Bednara, Cecilia Del Casino, Alessandro Bartalesi, Giampiero Cai, and Alessandra Moscatelli

Introduction

The role of the male partner in the fertilization process in higher plants has been defined from long time (Stanley and Linskens, 1972). As shown by optical and ultrastructural investigations, some closely related steps such as pollen development, pollen release from the anther, pollen-stigma interaction, pollen tube emission and growth occur and each step is characterized by own morphological and physiological states.

According to the different steps, the array of cytoskeletal structures have been also studied. A large number of investigations have been carried out on pollen and pollen tube cytoskeleton contributing for a better evaluation and understanding of the cytoplasmic mechanisms occurring in the male partner during fertilization (for reviews see Tiezzi, 1991; Pierson and Cresti. 1992, in press).

Less is known on the role of cytoskeleton during pollen formation. Here a meiotic division occurs and in the few investigated plants, the cytoskeletal apparatus seems to play a prominent role in assuring the appropriate conditions for both chromosomes segregation and cell plate formation (Van Lammeren et al., 1985; Traas et al., 1989).

In order to contribute at the extention of the knowledge on the role of cytoskeleton in the male partner, here we report on the organization of the microtubular cytoskeleton during microsporogenesis and pollen tube growth in Nicotiana tabacum. The present contribution is based on immunofluorescence observations by both conventional and confocal laser scanning microscope.

Materials and Methods

Anthers and pollen grains were collected from plants of Nicotiana tabacum L. cultured at the Botanical garden of the University of Siena.

Microspore mother cells and young microspores were isolated from the anthers (Traas et al., 1989) and processed for immunofluorescence investigations and DAPI staining.

Pollen grains were germinated for 4h in BK medium (Brewbaker and Quack, 1963) containing 12% sucrose and processed for immunofluorescence investigations.

For immunofluorescence investigations we followed the procedure of Tiezzi et al. (1986), using a monoclonal to the α-tubulin subunit as primary antibody (Amersham) and a fluorescein isothiocyanate (FITC) labelled rabbit anti mouse as secondary antibody (Cappel).

Observations were done with both an epifluorescent Zeiss Axiophot microscope and with a BioRad MRC500 Confocal Laser Scanning Microscope. Photographs were taken using Kodak TRI-X films (400 ASA).

Results

During the process of pollen development, each microspore mother cell undergoes a meiotic division to the formation of a tetrad consisting of four microspores. Continuous rearrangements of cytoplamic and cytoskeletal arrays are related to the meiotic division. The present results, obtained by double staining with a monoclonal antibody specific to the α-tubulin subunit and the DAPI dye, show the simultaneous visualization of the array of both microtubules and chromosomes during the meiotic division.

At interphase (Fig.1a,b) microtubules form a network that surrounding the nucleus, extends to the microspores mother cell plasma membrane. The DAPI staining is diffuse and chromosomes are unorganized. At prometaphase I (fig.2a,b) the number of cytoplasmic microtubules decreases. Microtubules do not organize in any structure resembling the preprophase band. The chromatin condenses to form chromosomes. At metaphase I (Fig. 3a,b), opposite groups of microtubules radiate from the poles and form the spindle. The chromosomes, interacting specifically with the microtubules, characteristically locate in the central zone of the spindle. During anaphase I (Fig.4a,b) the microtubules consisting half spindles and the chromosomes migrate to the two poles. Some bundles of microtubules interconnect the two half spindles and the chromosomes organize into two daugther nuclei.

At telophase I (fig.5) the number of microtubules interconnecting the two daugther nuclei increases substantially. The use of the confocal laser scanning microscope improved for a better definition of the microtubular array. As shown by arrows (5d and 5f), microtubules are connected to the nuclear membrane. Microtubules are not organized to support the formation of the phragmoplast. The DAPI staining is again diffuse and chromatin is not organized in chromosomes.

The second division occur with the same modalities as the first division. At late prophase-early metaphase II (Fig.6a,b) microtubular connections between the daugther nuclei disappear and microtubules organize in two new spindles. The chromatin is organized in chromosomes. During anaphase II (Fig.7a,b) microtubules interconnect the new formed nuclei and subsequently in telophase II (Fig8a,b) they establish connections among all the four nuclei. Confocal immages offer a better visualization of the microtubular connections (Fig.8c,d,e,f). In telophase II the diffuse staining of DAPI is the result of chromosomes disaggregation to chromatin.

At the end of the division there is formation of a phragmoplast among the four microspores (not shown). After cytokinesis in each microspore, microtubules form a network that surrounds the nucleus and extends into the cytoplasm to the plasma membrane. Chromosomes are not aggregated (Fig.9a, b).

The microspores have then an asymmetric division (not shown) with the formation of a vegetative and a generative cell. The two cells together form the pollen grain.

A microtubular cytoskeleton is present inside the generative cell and shows to have similar distribution both when the generative cell is inside the pollen grain (fig.10) and in the growing pollen tube (fig.12). Also the pollen tube contains a large number of microtubules organized as bundles (fig.11).

Along the pollen tube the generative cell divides in two sperm cells and microtubules are involved in the process of generative cell division. After formation of a set between the new formed

sperm cells, microtubules reorganize inside the sperm (fig.13).

Discussion

The data presented here contribute to establish on the distribution of microtubules during microsporogenesis, pollen tube growth and generative cell division in Nicotiana tabacum. As in all higher plant cells, microtubules actively partecipate during the processes of cellular division occuring till to the formation of sperm cells. However, as already shown also in other plants (Van Lammeren et al., 1985; Traas et al., 1989), during microsporogenesis they do not organize to form the preprophase band as somatic plant cells do. This could reasonably mean that different mechanisms of cellular division occur in the same plant. In addition, as preprophase band does not form also during Nicotiana tabacun generative cell division (Bartalesi et al., 1991), we can conclude that both in microsporogenesis and the formation of the sperm cell pair, the mechanisms of cellular division occurs with similar modalities.

Many microtubules are present inside pollen tube. They have been studyed also by electron microscopy (Lancelle et.al., 1987). and although their cytoplasmic distribution is now well known (Tiezzi, 1991; Pierson and Cresti, 1992, in press), their role remains substantially unclear. Perhaps they could be involved in the movement of vesicles. playing an active role in the pollen tube growth (see Cai et al., this issue).

In summary, on the basis of the observations reported here, it is clear that the microtubular cytoskeleton plays fundamental roles during microsporogenesis and pollen tube growth and more in general in the fertilization process. A new basic question is now open concerning the regulation of the microtubular machinary. As in other cellular systems it has been shown the presence of some microtubule cooperating polypeptides (Dustin, 1984), perhaps further research will contribute to clarify on similar polypeptides in the above described microtubular system.

References

1. Bartalesi, A.; Del Casino, C.; Moscatelli, A.; Cai, G.; Tiezzi, A. Confocal laser scanning microscopy of the microtubular system of dividing generative cell in Nicotiana tabacum pollen tube. Giornale Botanico Italiano. 125; 1991. In press.
2. Brewbaker, J.L.; Kwack, B.H. The essential role of calcium ion in pollen germination and pollen tube growth. Amer. J. Bot. 50: 859-865; 1963.
3. Dustin. P. Microtubules. Springer, Berlin, Heidelberg, New York, Tokyo; 1984.
4. Lancelle, S.A.; Cresti, M.; Hepler, P.K. Ultrastructure of cytoskeleton in freese-substituted pololen tube of Nicotiana alata. Protoplasma. 140: 141-150; 1987.
5. Pierson, E.S.; Cresti, M. Pollen germination, pollen tube growth and cytoskeletal organization. Int. Rev. Cytol.; 1992. In press.
6. Stanley, R.G,; Linskens, H.F. Pollen: biology, biochemistry, management.

Springer-Verlag, Berlin, Heidelberg, New York; 1972.

7. Tiezzi, A.; Cresti, M.; Ciampolini, F. Microtubules in Nicotiana pollen tubes: ultrastructure, immunofluorescence and biochemical data. In: Biology of reproduction and cell motility in plants and animals (Cresti M, Dallai R, eds). University of Siena. Siena. 87-94; 1986.
8. Tiezzi, A. The pollen tube cytoskeleton. Electron microscopy review. Pergamon Press. Vol.4. N.2; 1991. In press.
9. Traas, J.A.; Burgain, S.; Dumas De Vaulx, R. The organization of the cytoskeleton during meiosis in egg plant (Solanum melongena (L.)): microtubules and F-actin are both necessary for coordinated meiotic division. J. Cell Sci. 92: 541-550; 1989.
10. Van Lammeren, A.A.M.; Keijzer, C.J.; Willemse, M.T.M.; Kieft, H. Structure and function of the microtubular cytoskeleton during pollen development in Gasteria verrucosa (Mill.) H. Duval. Planta. 165: 1-11; 1985.

List of Figures

Figs.1-9. Microtubular organization during microsporogenesis. Images collected with the optical microscope are at the same magnification (x120). Images collected with the confocal microscope are at the same magnification (x200). **1.** View of an interphase cell. **a**, anti tubulin staining; **b**, DAPI staining.

2. Prometaphase. **a**, antitubulin staining; **b**, DAPI staining.

3. Metaphase I. **a**, antitubulin staining; **b**, DAPI staining.

4. Anaphase I. **a**, antitubulin staining; **b**, DAPI staining.

5. Telophase I. **a**, **b**, two focal planes of the dividing cell, antitubulin staining. **c**, DAPI staining. **d,e,f**, confocal immages showing the different spatial distribution of the two nuclei. Some microtubule interacting points are located on the nuclear membranes (arrows). Antitubulin staining.

6. Late prophase II. **a**, antitulin staining; **b**, DAPI staining.

7. Anaphase II. **a**, antitubulin staining; **b**, DAPI staining.

8. Telophase II. **a**, antitubulin staining; **b**, DAPI staining. **c,d,e,f**, confocal images, antitubulin staining.

9. Microspore at interphase. **a**, antitubulin staining; **b**, DAPI staining.

Fig.10. Generative cell (GC) inside the pollen grain (PG). Antitubulin staining. Confocal image. Visualization of the microtubular cytoskeleton (MT) into the generative cell. (x520).

Fig.11. Pollen grain (PG) and pollen tube (PT). Antitubulin staining. Confocal image. Microtubules (MT) are present in the pollen tube (x430).

Fig.12. Pollen grain (PG) and pollen tube (PT). Antitubulin staining. Confocal image. The generative cell (GC) in the pollen tube. A consistent microtubular cytoskeleton (MT) is present into the generative cell (x430).

Fig.13. Dividing generative cell. Antitubulin staining. Confocal image. A set is forming and microtubules (MT) reorganize in the cytoplasm of the new formed sperm cells. SN, sperm nuclei (x430).

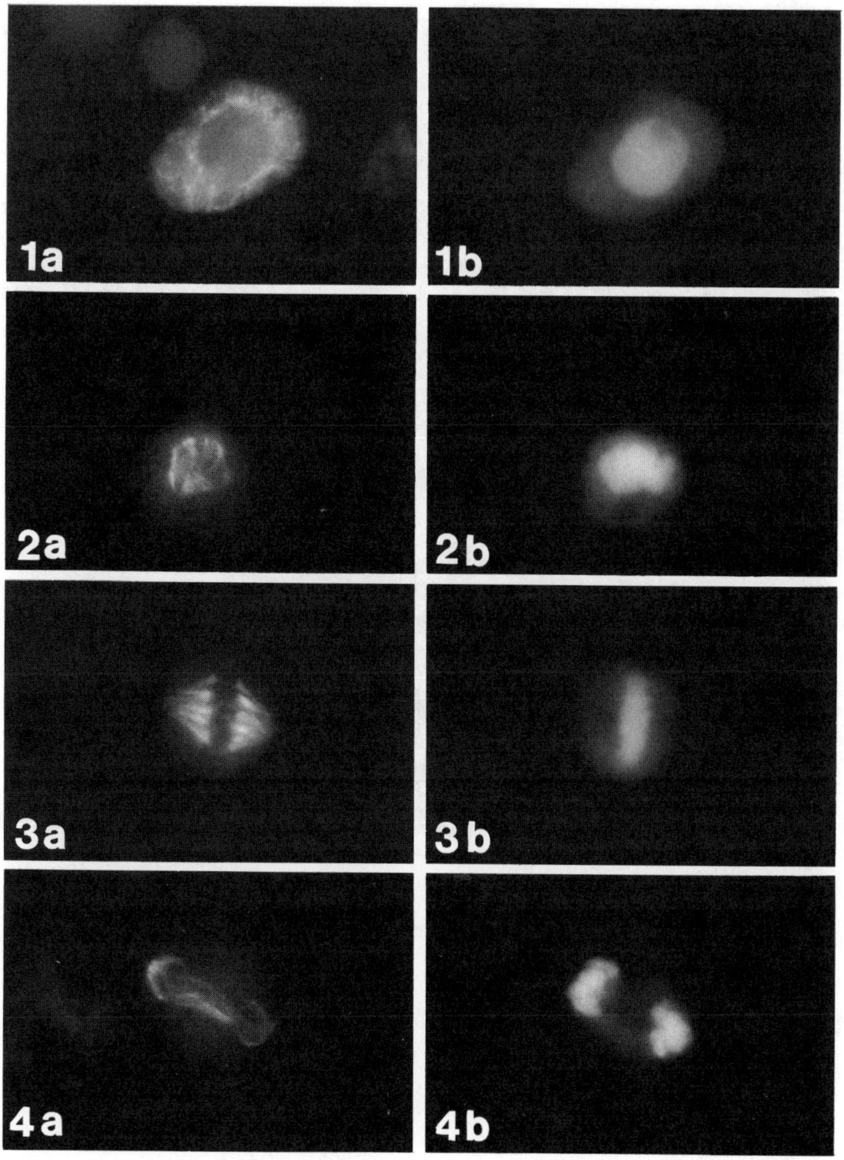
1a
1b
2a
2b
3a
3b
4a
4b

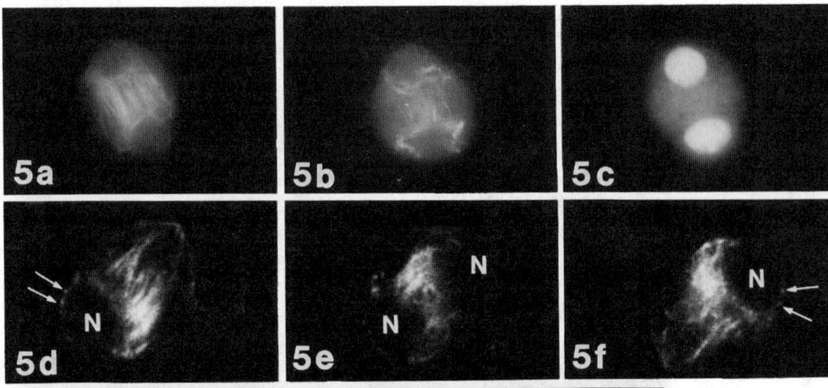
5a
5b
5c
N
5d
N
N
5e
N
5f

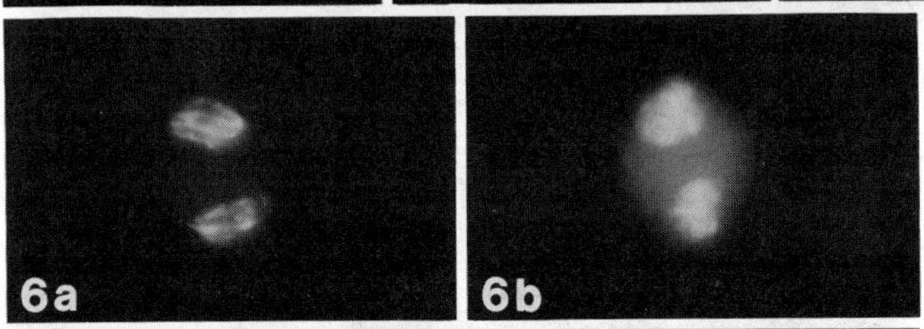
6a
6b

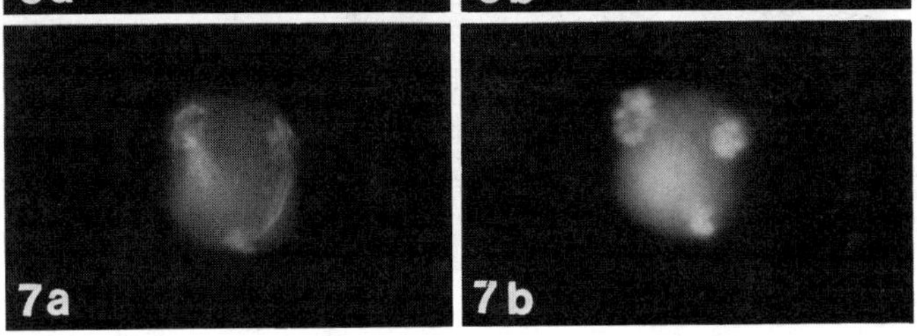
7a
7b

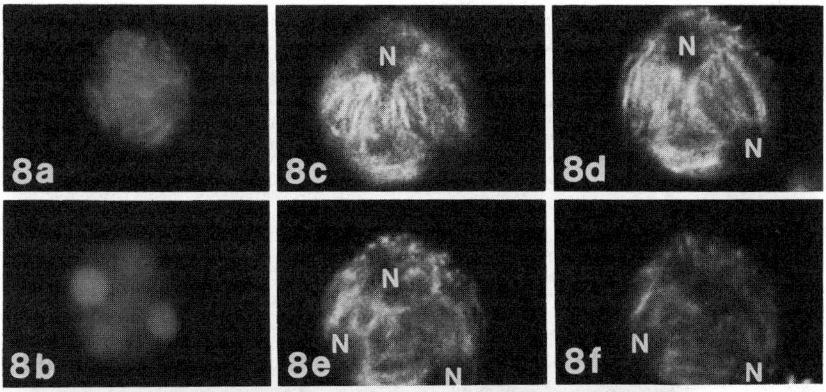
8a
8c
N
8d
N
N
8b
8e
N
N
N
8f
N
N

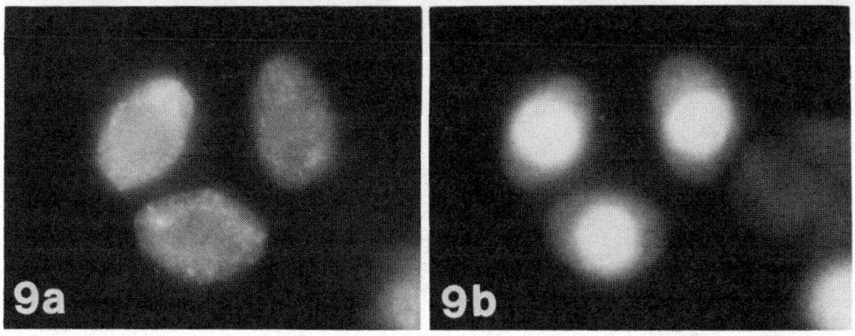
9a
9b

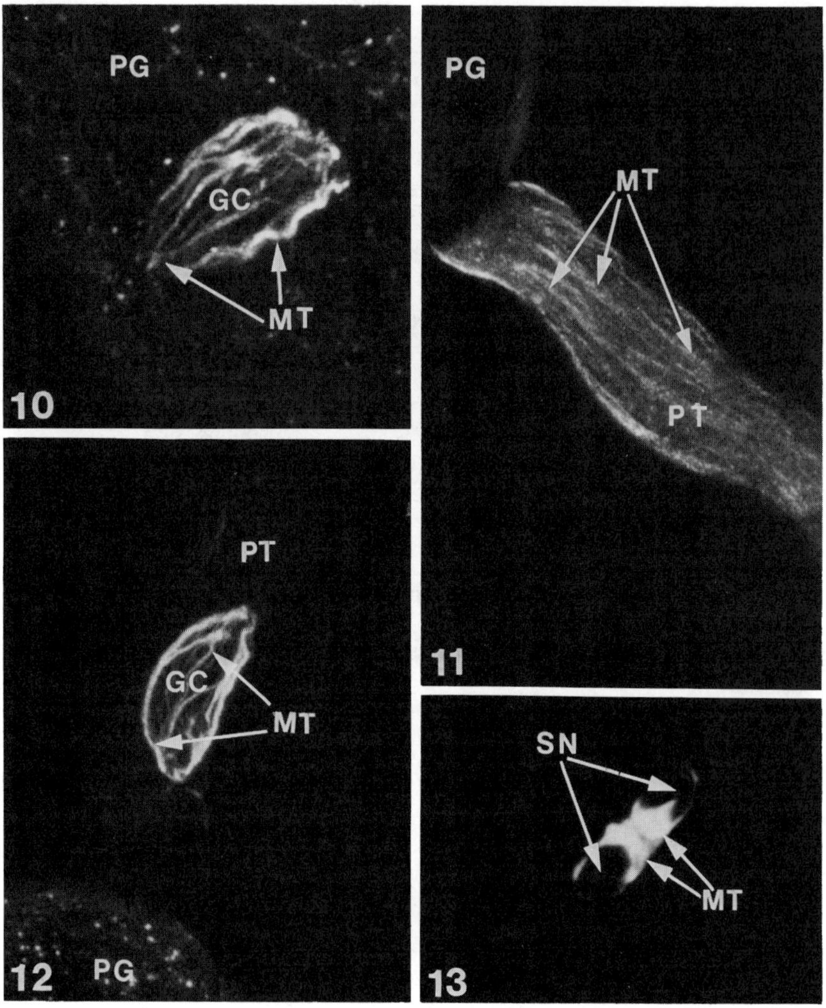
PG
GC
MT
10
PG
MT
PT
11
PT
GC
MT
PG
12
SN
MT
13

37

Mitosis and Colchicine Effects on Microtubule Organization in Isolated Generative Cells of *Allamanda neriifolia*

S.Y. Zee and Aziz-Un-Nisa

The generative cells of Allamanda neriifolia divide and form sperms within the growing pollen tubes. Recently immunocytochemical and ultrastructural studies have shown that these cells contain an extremely complex network of microtubules which is capable of changing its organizational pattern during mitosis (Zee & Aziz-Un-Nisa, 1991; Xu, 1991a). Furthermore, the organizational changes in the microtubules seen in the dividing generative cells are in many ways different from the changes seen in somatic cells (Gunning, 1982; Baskin & Cande, 1990). For example the generative cells do not form preprophase bands of microtubules and they also do not form cell plates.

Microtubules of generative cells in pollen tubes

The pattern of changes of the microtubules in the generative cell is very complex consisting of 8 different developmental stages (Xu, 1991a) (Fig. 1): Stage 1. At this stage the shape of the generative cell is like a tadpole and the microtubules inside the cell are tightly woven into a reticulate network. Stage 2. The cell becomes ellipsoidal in shape and the microtubule network appears much loosened. Stage 3. The cell becomes spindle shaped and the reticulate network of microtubules disappears. Inside the cell a new array of longitudinally oriented microtubules, parallel to the long axis of the cell, apppears. Stage 4. The shape of the cell does not seem to change much at this stage but the longitudinally oriented microtubules become more wavy. Stage 5. The cell now enters metaphase and the microtubules reorganize into a spindle consisting of kinetochore and non-kinetochore microtubules. Stage 6. After metaphase the cell enters anaphase A. At this phase the kinetochore microtubules disappear but in the interzonal region the non-kinetochore microtubules still remain. At each pole a new set of cap-like microtubules forms. Stage 7. At anaphase B the microtubules in the interzonal region elongate increasing the distance between sister nuclei. Stage 8. Telophase - at this final phase, the mid plane in the interzone forms a narrow junction. Within this junction microtubules do not form. This junction has been referred to as 'junction in phragmoplast' by Taylor et al. (1989). But in our view it is more appropriate to call this the 'junction between sperm cells' or simply the 'sperm junction' because the generative cells do not possess a proper phragmoplast of the type normally seen in somatic cells. Furthermore, in the generative cell of A. neriifolia the microtubules in the interzonal region tend to extend to the nuclear regions and

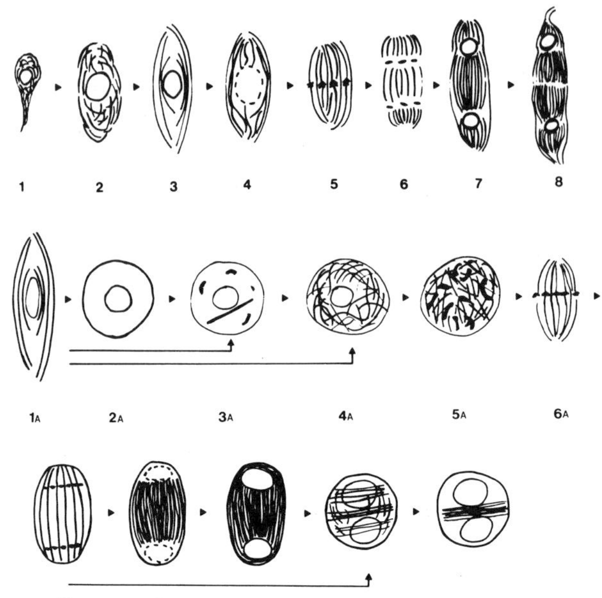

Fig. 1. Diagram showing the pattern of changes in the microtubules of *in situ* (1-8) and isolated (1A - 11A) generative cells of *A. neriifolia*. 1. Microtubules are tightly woven into a network; 2. Microtubules start to loosen up; 3. Spindle cells with longitudinally oriented microtubule bundles; 4. Microtubule bundles become wavy; 5. Spindle formation (metaphase); 6. Anaphase A; 7. Anaphase B; 8. Telophase. 1A. Spindle shaped cells possessing longitudinally oriented microtubules; 2A. Absence of microtubule cytoskeleton; 3A. Formation of dots, patches and thick bundles of microtubules; 4A. Reticulate microtubule network; 5A. Scattered chromatids mixing with microtubules; 6A. Spindle formation (metaphase); 7A. Anaphase A; 8A. Anaphase B; 9A. Elongated cells (telophase); 10A & 11A. Spherical cells (telophase).

beyond whereas the phragmoplast microtubules in the somatic cells seldom extend to the nuclear regions.

The general pattern of microtubule organizational changes in the generative cells of A. neriifolia outlined above has been seen in a number of other plants, including Zantedeschia aethiopica (Xu et al., 1992) and Alocasia macrorrhiza (Xu, 1991b). But this pattern is different from that seen in Tradescantia virginiana (Palevitz & Cresti, 1989).

Microtubules in Isolated Cells

After the generative cells have been isolated from the growing pollen tubes (using the osmotic shock method, Xu, 1991c) the shape of the cell rapidly changes from spindle to spheroidal. The microtubule organizational patterns in the spherical cells are quite varied. Based on the microtubule organizational patterns (as revealed by immunofluorescence microscopy and electron microscopy, Zee & Aziz-Un-Nisa, 1991; Zee, 1991) 3 different types of cells are present. Type 1. Cells do not possess a proper microtubule cytoskeleton. Type 2. Cells develop dots, patches and thick microtubule bundles. Type 3. Cells form a very complex reticulate network of microtubules. According to our interpretation these 3 different types of cells can change from one to the other in the following way (Fig. 1): Immediately after isolation the microtubules in the generative cells rapidly depolymerize (result of osmotic shock?). Then after a short period of time in the incubation medium the microtubules inside the cells begin to reassemble resulting in the formation of microtubule aggregates (Zee, 1991) (and/or microtubule organization centers?) in the form of dots, patches and thick bundles. Then at the final stage the microtubules reorganize themselves into a reticulate network. It is possible that some of the spindle-shaped cells may proceed directly to stage (3A) or (4A) without passing through stage(s) (2A) and/or (3A) (Fig. 1).

If these spherical cells are cultured for a longer period of time (about 1-3 hours) then the cells will start to divide. The first sign of cell division begins with the break up of the nuclear envelope and the scattering of chromatids. After the nuclear envelope has broken up the microtubules become mixed up with the scattered chromatids. Afterwards the cell enters metaphase with the chromatids become aligned at the equator and the microtubules forming a spindle consisting of kinetochore and non-kinetochore microtubules. Following metaphase the cell enters Anaphase A, sister chromatids separate and then migrate toward their respective poles. Inside the cell the kinetochore microtubules disappear. But the non-kinetochore microtubules still remain in the interzonal region. In addition to the interzonal (or interpolar) microtubules some cage microtubules also form. Following Anaphase A the cell enters anaphase B [N.B. this stage has been named Early Telophase in an earlier paper (Zee & Aziz-Un-Nisa, 1991)]. In the anaphase B cells the longitudinally oriented interzonal microtubules increase in number. But the polar microtubules (normally seen in cells under the in situ condition) do not form. Finally, the cells enter telophase. The telophase cells are still quite elongated. Interzonal microtubules are still present and some even extend to surround the newly formed nuclei. In these cells no 'sperm junctions' form. For some cells this may represent their last stage of development. But for some cells they may revert back to the spherical form. Inside these spherical cells the longitudinally oriented interzonal microtubules disappear. At the same time two new sets of microtubules form ----- an outer reticulate network and a more internally placed microtubule bands (or sheets). In some spherical cells the reticulate network of microtubules soon disappears and the cells are left with a thick band of microtubules in the equatorial plane between nuclei. In the spherical cells 'sperm junctions' also do not develop. The possibility that some of the elongated anaphase A cells may change to the spherical cells without passing through the elongated telophase cell stage cannot be ruled out (Fig. 1).

Effect of Colchicine

If the generative cells of A. neriifolia after isolation are treated with 10^{-3}M colchicine, cell divisions at all phases will stop. Higher concentrations of colchicine treatment (viz. 2×10^{-3}, 4×10^{-3} and 8×10^{-3}M) give the same result. Immunofluorescence microscopy reveals that the microtubules become thickened (Figs. 2 and 3) but not destroyed. These observations are in agreement with the current view that 'the mode of action of colchicine is to complex the polymerizing end of the microtubules thus inhibiting subsequent addition of subunits on the microtubules' but not destroying (necessarily) existing microtubules (Seagull, 1989). Furthermore, electron microscopic examination indicates that after colchicine treatment two new structures are produced inside the cell: an amorphous and a fine fribrillar structure (Fig. 4). We do not know whether these two new structures correspond to the thickened microtubules (seen by immunofluorescence microscopy) or not.

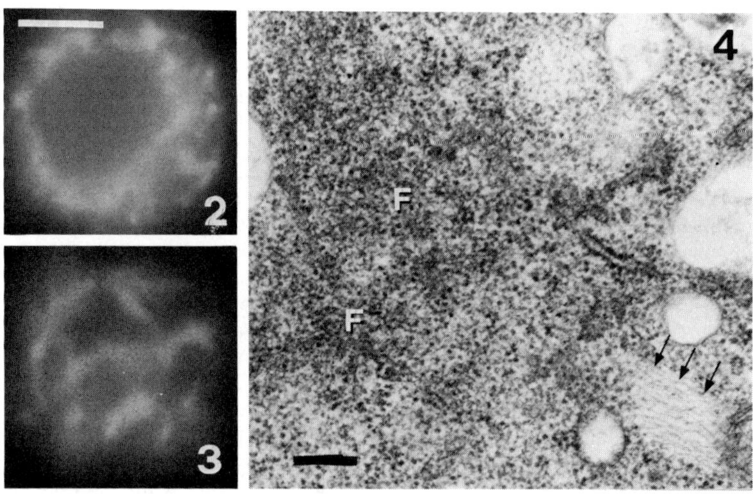

Figs. 2 and 3. Immunofluorescence micrographs of isolated generative cells of *A. neriifolia* after colchicine (10^{-3}M) treatment for 2 hrs at 28°C. Note that microtubules become thickened after colchicine treatment. Scale bar = 5um. Fig. 2. Prophase; Fig. 3. Metaphase.

Fig. 4. Electron micrograph of portion of an isolated generative cell treated with 10^{-3}M of colchicine for 2 hours. F = the amorphous structure; Arrow = the fine fibrillar structure. Scale bar = 0.25um.

References

Baskin, T.I.; Cande, W.Z. The structure and function of the mitotic spindle in flowering plants. Annu. Rev. Plant Physiol. Plant. Mol. Biol. 41:277-315; 1990.

Gunning, B.E.S. The cytokinetic apparatus: Its development and spatial regulation. In: Lloyd, C.W., ed. The cytoskeleton in plant growth and development.Academic Press; 1982: p230-290.

Palevitz, B.A.; Cresti, M. Cytoskeletal changes during generative cell division and sperm formation in Tradescantia virginiana. Protoplasma 150:54-71; 1989.

Seagull, R.W. The plant cytoskeleton. Critical Review in Plant Sciences 8:131-167; 1989.

Taylor, P.; Kenrick, J.; Li, Y.; Kaul, V.; Gunning, BES; Knox, R.B. The male germ unit of Rhododendron: quantitative cytology, three-dimensional reconstruction, isolation and detection using fluorescent probes. Sex Plant Reprod. 2:254-264; 1989.

Xu, S.X. (Zee, S.Y.) Microtubule cytoskeleton dynamics in the generative cells of Allamanda neriifolia Hook during mitosis. Acta Biologiae Experimentalis Sinica 24:79-87; 1991a.

Xu, S.X. (Zee, S.Y.) Changes in microtubule cytoskeleton of generative cells during mitosis and cytokinesis. Chinese Journal of Botany 1991b (In press).

Xu, S.X. (Zee, S.Y.) Division in isolated generative cells of Allamanda neriifolia. Acta Botanica Sinica 1991c (In press).

Xu, S.X. (Zee, S.Y.); Zhu, C.; Hu, S.H. Microtubule-cytoskeleton changes during generative cell division and sperm formation in the pollen tubes of Zantedeschia aethiopica. Chinese Journal of Botany 1992 (In press).

Zee, S.Y. Ultrastructure and microtubule organization of isolated generative cells of Allamanda neriifolia at interphase and prophase. Sexual Plant Reprod. 1991 (In press).

Zee, S.Y.; Aziz-Un-Nisa. Mitosis and microtubule organizational changes in isolated generative cells of Allamanda neriifolia. Sexual Plant Reprod. 4:132-137; 1991.

38
The Extracellular Matrix in Pollen Tube Growth

Luraynne C. Sanders and Elizabeth M. Lord

The role that the gynoecium plays in pollination has been considered a passive one, much like a nutrient conduit or vessel for the active growth of the pollen tube (Sanders & Lord, 1989). Most research has been on the pollen tube itself in this process of conveying the sperm to the egg, and on the phenomenon of self-incompatibility, or how this process can be blocked. This is in spite of the fact that self-compatibility is the predominant condition in the Angiosperms and no doubt the primitive one from which incompatibility arose (Nasrallah, 1989; Sanders & Lord, 1989). A more thorough knowledge of how compatible pollen tubes grow and interact with the gynoecium can only enhance our understanding of how the complexities of the derived condition of self-incompatibility arose evolutionarily, and suggest a mechanism for its function, especially in gametophytic systems.

The ability to grow pollen tubes in a defined medium greatly facilitated study of tip growth in pollen (Steer & Steer, 1989), but such an approach only strengthened the assumption that the transmitting tract of the gynoecium functioned much as a culture dish, providing a path of nutrient secretions which "guided" the pollen tube to the ovule. The difficulties with this view of pollination are that pollen tubes never grow as fast or as long in vitro as in vivo, and the appearance of tubes grown in vitro differs from those which are seen in vivo. They exhibit random wanderings in culture and generative cell division in bicellular species rarely occurs in vitro. One explanation may be that the ingredients of the medium are not complete, that some unknown factors are missing.

The stylar secretions of many species have been analyzed and found to contain a variety of components including sugars, amino acids, peptides, phenolic compounds, fatty acids, lipids, and glycolipids. The high molecular weight compounds are polysaccharides and complex proteins (Heslop-Harrison, 1983; Clarke, 1981). Arabinogalactans appear to be widespread and abundant (Clarke et al., 1979). The transmitting tract cells of the style secrete these components either onto the epidermis of a hollow style or between cells in the solid style (Knox, 1984). Pollen tubes track these secretions from the stigma to the micropyle of the ovule and rarely, if ever, cross free space. It is agreed that the secretions guide the tubes to the ovary but the idea that a chemotropic guidance is involved (via a calcium gradient) has been discredited for some time (Heslop-Harrison, 1987). The general assumption now is that the tube travels in the secretion of the style much as a fungal hypha would be guided by a nutrient medium; that the tissues of the gynoecium offer the path of least resistance for the pollen tube (Heslop-Harrison, 1987).

The question of rates of growth is not addressed by this argument. Why do pollen tubes grow so rapidly in vivo and not in vitro? In fact, there are species-specific rates of tube growth and we now recognize that the rate of travel is a critical component for a pollen tube's success in "siring" individuals of the next generation (Mulcahy, 1975).

Population biologists studying this phenomenon have provided data that pollen tube growth rates are determined not solely by the pollen, but by the female as well; thus, there exists a yet to be accounted for component of "female choice" (Stephenson & Bertin, 1983; Willson & Burley, 1983). Variation in the tube growth rates in vivo is a function of both maternal and paternal genotype. This component of female choice is unknown, but one could postulate that the secretions themselves play an active role in selecting certain pollen tubes for the "fast track" to the ovules (Hill & Lord, 1986; Malti & Shivanna, 1985). This implies recognition between the gynoecium and pollen tube in self-compatible crosses, something not yet explored by cell biologists (Sanders & Lord, 1989). The following discussion presents data that suggest a more active role for the gynoecium in pollen extension. We make a case for pollination being a unique example of cell migration in plants (Sanders & Lord, 1989), and then summarize our findings on the presence of a substrate adhesion molecule, vitronectin, in the style (Sanders et al., 1991). The glycoprotein vitronectin occurs in animal tissues and has been implicated as an agent of cell migration and tip growth in neurites (Underwood & Bennett, 1985; Neugebauer et al., 1991). We propose that it may play a role in actively guiding the pollen tube tip to the ovule.

Bead Movement in the Stylar Secretions

All cells secrete an extracellular matrix (ECM) that surrounds them and mediates their contacts with the environment (Adair & Mecham, 1990). In plants, this matrix is called the cell wall and has many roles in addition to that of support (Darvill et al., 1980; Roberts, 1990). More is known about the active biological role the ECM plays in animal systems, especially with regard to processes such as morphogenesis, recognition, cell division, polarity and cell migration (Hay, 1981). One set of specific ECM molecules that are agents of such a diverse array of activities are called substrate adhesion molecules (SAMs) (Edelman, 1988). These glycoproteins interact with the substrate and the cell plasma membrane via a class of receptors in the membrane called integrins (Hynes, 1987). Each SAM has one or a few specific integrins to which it attaches. These integrins are transmembrane proteins to which a SAM can attach on the exterior of the cell and cytoskeletal molecules on the interior of the cell (Dufour et al., 1988) (Fig. 1). In this manner, a cell establishes close physical contact with its environment. The known SAMs include fibronectin, laminin, vitronectin and fibrinogen. All of these molecules were isolated in various animals and only recently has one, vitronectin (VN) been found in plants (Sanders et al., 1991). Fibronectin (FN) has been implicated in neural crest cell migration during animal embryogenesis (Hynes, 1981). During early stages of development, cells move in animal embryos to defined regions where they then differentiate into complex tissues and organs. It was once thought that these cells move of their own accord in an amoeboid fashion over tracts of ECM molecules. Recent work by Bronner-Fraser (1982) and others has demonstrated that movement may also be controlled by external agents such as the ECM molecules themselves. This was shown by applying latex particles the size of neural crest cells onto specific areas of the ECM and tracking them to their destinations. When the beads were coated with FN, no movement occurred (Bronner-Fraser, 1985). Unidirectional migration of both chick embryo cells and latex particles occurred in vitro on artificial matrices formed using FN and collagen (Newmann et al., 1985). These results have opened a field of study which is providing a wholly new way of looking at cell migration (Fig. 1).

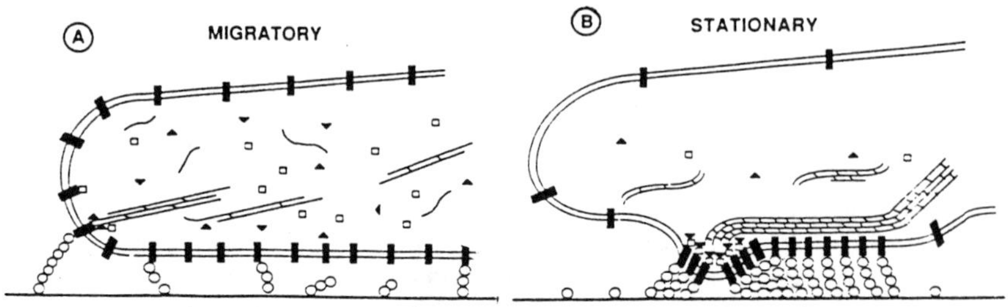

Figure 1. Speculative models for modes of cell-substratum adhesion of migratory and stationery cells. (A) Migratory cells have a disorganized cytoskeleton; the FN receptors are diffuse in the plane of the membrane and may be mobile. A number of receptors are not occupied by FN, and may constitute a reserve pool for new adhesion. (B) In stationary cells, the cytoskeleton is well organized and the receptors are localized mostly in clusters close to focal contact sites and in fibrillar streaks aligned with microfilament bundles. These receptors have restricted lateral mobility in the membrane. They are linked directly in fibrillar streaks both to t he cytoskeleton and to FN fibers. In focal contacts, FN receptors interact indirectly with actin microfilaments through talin and vinculin. |, FN receptors; = FN; l, talin; D, vinculin; alpha-actinin is symbolized by cross-links between actin microfilaments. (Figure adapted from Dufour et al., 1988, with permission from TIG.)

The structure of the ECM in animals is fibrillar and comprised of a variety of glycoproteins, not unlike the matrices of plant cells, especially that of the stylar ECM (Sanders et al., 1991). A phenomenon in plants that is parallel to cell migration in animal systems, is pollination. Here two or three cells are carried through the stylar matrix to the ovary. These are the vegetative cell and the generative cell, or the two sperm cells, within the pollen tube. Since no new cytoplasm is produced as the pollen tube grows, one can envision pollination as a case of cell migration contained within the elongating cell wall of the tube.. The cells themselves move forward leaving a trail of wall behind them (Sanders & Lord, 1989). Looking at pollination in this way, one can ask, does the stylar ECM play a role in this movement? To address this question, we applied 6 lm latex beads to the stigmas or exposed styles of three species of flowering plants representing the three style types (open, Hemerocallis flava; closed, Raphanus raphanistrum; intermediate, Vicia faba) (Sanders & Lord, 1989). The beads were translocated to the ovules in the secretions of the ECM and at the rates of movement of the respective pollen tubes. The most striking fact was that the beads moved only on the cells which produced the secretions and, in some cases, even entered micropyles (Fig. 2). This finding suggested that the ECM of the style was more than a nutrient bath for the

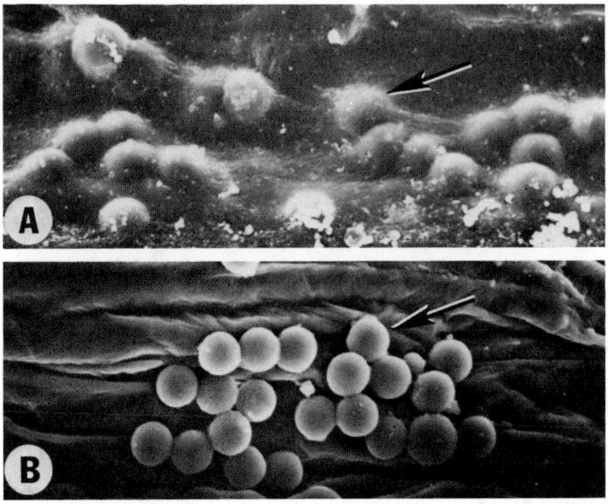

Figure 2. Cyro-scanning electron micrographs of latex beads in the transmitting tract. x 1150. a, *Hemerocallis flava*. Beads (arrow) beneath the cuticle which overlies the extracellular matric of secretions. b, *Vicia faba*. Beads on surface of transmitting tract cells. (Figure adapted from Sanders and Lord, 1990, with permission from Science.)

extension of pollen tubes and that a direct interaction was occurring between the stylar ECM and the pollen tube vegetative cell, much as occurs between the ECM and the moving animal cell (Fig. 1). On the basis of these results, we proposed a new model for pollination which incorporated both the model of cell attachment and that of cell movement in animals (Dufour et al., 1988; Sanders et al., 1991; Sanders & Lord, in press) (Fig. 1). The model is as follows: the pollen tube has both a tip which is moving forward (Fig. 1b) and a posterior part which is firmly attached to the ECM of the style (Fig. 1a). The tip of the pollen tube has transient interactions with the stylar ECM via integrins in the tube membrane which are exposed due to a patchy wall. The actin cytoskeleton is disorganized in the pollen tube tip (Lancelle et al., 1987) much as one sees in that of a moving cell (Fig. 1). Back from the tip, the pollen tube wall is thicker and attachments of the integrins in the tube membrane with the SAM in the ECM have become stabilized and form focal adhesions. These patches where adhesion of the cell and the substrate occur can be visualized by staining for either the microfilaments inside the cell or the ECM components to the exterior (Burridge et al., 1988). Indeed, studies of the cytoskeleton of lily pollen tubes grown in vivo show what look to be focal adhesions (Pierson et al., 1986). These are not evident in pollen tubes grown in vitro, as would be expected if our model is correct. In some manner, the tip of the pollen tube is pulled or

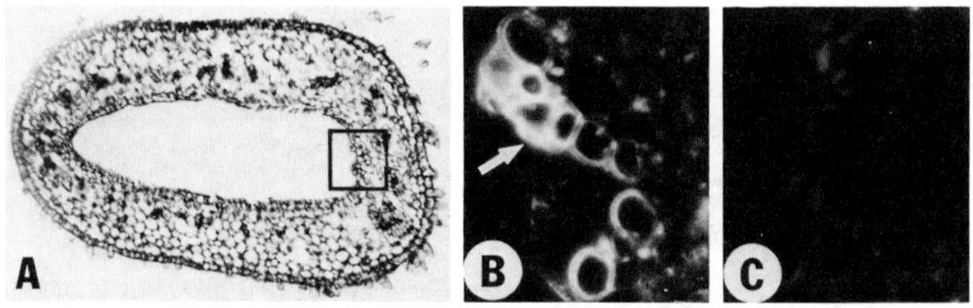

Figure 3. Localization of a vitronectin-like protein on frozen sections of *Vicia faba* gynoecium. (A) Cross-section of gynoecium stained with toluidine blue O, boxed area indicates transmitting tract. x 45. (B) Cross-section of transmitting tract incubated with human vitronectin antiserum. x 420. (C) Section of transmitting tract incubated with nonimmune serum. x 420. E, epidermis; M, mesophyll; C, cuticle; white arrows indicate areas of immunoreactivity. (Figure adapted from Sanders et al., 1991, with permission of the ASPP).

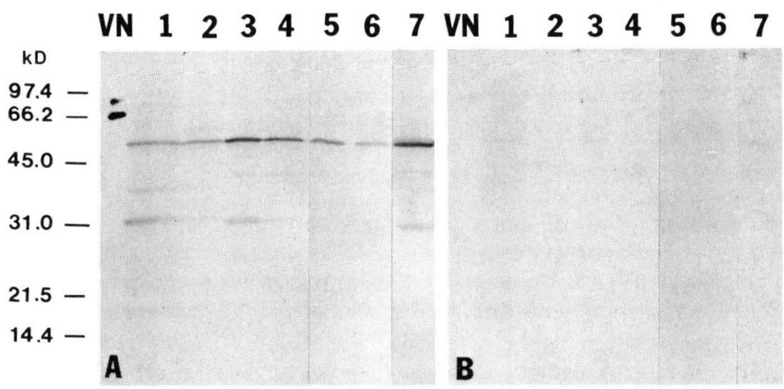

Figure 4. Detection of a 55 kD Plant Protein by Human Vitronectin Antiserum. Lanes 1-7 are protein extractions from plant tissue; lane 1, *L. longiflorum* leaf; lane 2, *L. longiflorum* root; lane 3, *V. faba* leaf; lane 4, *V. faba* l root; lane 5, *G. max* leaf; lane 6, *G. max* root; lane 7, *L. esculentum* leaf. The lane labeled VN contains 100 ng purified human vitronectin. (A) Protein blot incubated with human vitronectin antiserum. (B) Protein blot incubated with nonimmume serum. (Figure reproduced from Sanders et al., 1991, with permission of the ASPP).

pushed forward along the ECM of the style and adherence in the posterior part of the tube allows for a form of traction. The motive force could be generated at the tube tip by the SAM itself (Wolff & Lai, 1989; Newman et al., 1985), or by the contractile properties of the cytoskeletal elements in the pollen tube (actin and myosin) (Derksen et al., 1990), or both (Sanders & Lord, 1991). Instead of postulating a chemical, ionic gradient like that of CA+2, we are proposing a gradient of a bound substance, a SAM, that would induce a contact stimulated guidance; a haptotactic rather than a chemotactic phenomnon (Basara et al., 1985).

To test this model, we looked for the presence of SAMs in plants. Since both FN and VN were implicated in cell movement in animal systems, we began with these. We incubated plant tissues of Lilium longiflorum and Vicia faba with antibodies to human FN and VN and saw cross reactivity of the VN antibody with molecules in the wall of the plant cell, including the ECM of the style of Vicia faba (Sanders et al., 1991) (Fig. 3). We extracted proteins from leaves and roots of Lilium longiflorum, Vicia faba, Glycine max, and Lycopersicon esculentum, and showed cross reactivity of the labeled VN antibody to a 55 kD plant protein (Fig. 4). We were also able to show plant DNA hybridizations with a labeled human VN cDNA probe, suggestive of a family of genes in plants that have homology with human VN. We are reasonably sure that a VN-like glycoprotein occurs abundantly on the transmitting tract of Vicia faba. Whether it is the factor that moves the beads

is as yet untested. The next proposal is to look for the VN receptor or integrin in the pollen tube tip. There is already some evidence that a VN receptor occurs in plant plasma membranes (Schindler et al., 1989). We also must demonstrate, in vitro, that VN supports pollen tube extension and bead movement. As our knowledge of the cytoskeleton of the pollen tube emerges, a mechanism for movement of the component cells will no doubt be clarified. The pollen tube can grow in vitro in the absence of the stylar matrix (Steer & Steer, 1989), but a more complex system is operating in vivo, and a possible role for the gynoecium in the movement of the male germ unit must now be considered.

References

Adair, W. S.; Mecham, R.P. In: Adair, W. S.; Mecham, R. P., eds. Biology of Extracellular Matrix. New York: Academic Press; 1990.

Basara, M. L.; McCarthy, J. B.; Barnes, D. W.; Furcht, L. T. Cancer Research 45:2487-2494; 1985.

Bronner-Fraser, M. Dev. Biol. 91:50-63; 1982.

Bronner-Fraser, M. Dev. Biol. 108:131-145; 1985.

Burridge, K.; Fath, K.; Nuckolls, G.; Turner C. Annu. Rev. Cell Biol. 4:487-525; 1988.

Clarke, A.E. In: Tanner, W.; Loewus, F.A., eds. Plant Carbohydrates II, Extracellular Carbohydrates. New York: Springer-Verlag; 1981: p. 577-582.

Clarke, A. E.; Gleason, P.; Harrison, S.; Knox, R. B. Proc. Natl Acad. Sci. USA 76:3358-3362. 1979.

Darvill, A.; McNeil, M.; Albersheim, P.; Delmer, D. P. In: Tolberg, N. E., ed. The Plant Cell. New York: Academic Press; 1980: p. 91-162. (Tolbert, N.E. The Biochemistry of Plants; vol. 1).

Derksen, J.; Wilms, F. H. A.; Pierson, E. S. Acta Bot. Neerl. 39:1-18; 1990.

Dufour, S.; Duband, J-L.; Kornblihtt, A. R.; Thiery, J. P. TIG 4:198-203; 1988.
Edelman, G. M. New York: Basic Books, Inc. 1988.
Hay, E. D. In: Hay, E. D., ed. Cell Biology of Extracellular Matrix. New York: Plenum Press; 1981: p. 379-409.
Heslop-Harrison, J. Proc. R. Soc. Lond. B. 218:371-395; 1983.
Heslop-Harrison, J. Inter. Rev. Cytology 107:1-78; 1987.
Hill, J. P.; Lord, E. M. Evol. 40:1328-1333; 1986.
Hynes, R. O. In: Hay, E. D., ed. Cell Biology of Extracellular Matrix. New York: Plenum Press; 1981; p. 295-334.
Hynes, R. O. Cell 48:549-554; 1987.
Knox, R. B. Annu. Rev. Plant Physiol. 17:508-608; 1984.
Lancelle, S. A.; Cresti, M.; Hepler, P.K. Protoplasma 140:141-150; 1987.
Malti; Shivanna, K. R. Theor. Appl. Genet. 70:684-686; 1985.
Mulcahy, D. L. In: Mulcahy, D. L., ed. Gamete Competition in Plants and Animals. Amsterdam: North Holland; 1975: p. 1-4.
Nasrallah, J. In: Lord, E.M.; Bernier, G., eds. Plant Reproduction: From Floral Induction to Pollination. Rockville, MD: American Society of Plant Physiologists; 1989: p. 156-164. (Lord, E.M.; Bernier, G.: Current Topics in Plant Physiology, vol. 1).
Neugebauer, K. M.; Emmett, C. J.; Venstrom, K. A.; Reichardt, I. F. Neuron. 6:345-358; 1991.
Newman, S. A.; Frenz, D. A.; Tomasek, J. J.; Rabussi, D. D. Science 228:885-888; 1985.
Pierson, E. S.; Dirkson, J.; Traas, J. A. Eur. J. Cell Biol. 41:14-18; 1986.
Roberts, K. Curr. Op. Cell Biol. 2:920-928; 1990.
Sanders, L. C.; Lord, E. M. Science 243, 1606-1608; 1989.
Sanders, L. C.; Lord, E. M. Int. Rev. of Cytol. (in press)
Sanders, L. C.; Wang, C.-S.; Walling, L. L.; Lord, E. M. The Plant Cell 3:629-635; 1991.
Schindler, M.; Meiners, S.; Cheresh, D. A. J. Cell Biol. 108:1955-1965; 1989.
Steer, M. W.; Steer, J. New Phytol. 111:323-358; 1989.
Stephenson, A. G.; Bertin, R. L. Real, L., ed. Pollination Biology. Orlando, FL: Academic Press; 1983: p. 109-149.
Underwood, P. A.; Bennett, F. A. J. Cell Sci. 93:641-649; 1989.
Willson, M. F.; Burley, N. Princeton, NJ: Princeton Univ. Press; 1983.
Wolff, C.; Lai, C.-S. Arch. Biochem. Biophys. 268:536-545; 1989.

39
Analogies of Chromosome Reducing Events in Somatic Cultured Cells and in Microsporogenesis

V. Nuti Ronchi, L. Giorgetti, M. Tonelli, P. Belloni, and G. Martini

INTRODUCTION

Carrot somatic cells are considered totipotent,i.e committed to a determinate fate of development, since they can differentiate somatic embryos similar to zygotic embryos when the growth factors are withdrawn from the culture medium.

Beside pattern of development (through similar globular, heart and torpedo stages), somatic and zygotic embryos appear to share another important feature: the origin from a single cell, which has a specific cell wall composition (Pennell and Roberts, 1990) and some structural characteristic (nuclei, mitochondria) similar to those observed in gametic cells during maturation (Bonnelle et al., 1990).

The gametic-like condition of committed carrot cells has been recently suggested to be required for the development of somatic embryos from carrot hypocotyl cell culture (Nuti Ronchi et al., 1990; Nuti Ronchi, 1990). Genome reprogramming events have been shown to occur by means of meiotic-like divisions and prophase reduction mechanisms of chromosome segregation (Nuti Ronchi, 1991). These chromosome reducing events were detected in carrot hypocotyles during the first days of culture; single pollen-like cells, very similar to immature mononucleate pollen grains were released in the liquid medium. The embryogenic competence was acquired with subsequent medium changes, when embryogenic masses (PEMs) were formed from further divisions of these pollen-like cells.

The objective of this research was to induce embryogenic competence, similar to the carrot system, to other plant species known to be recalcitrant to regenerate in culture.

MATERIALS AND METHODS

Seeds of tomato (Lycopersicon esculentum) cv. Roma, sunflower (Helyanthus annuus) cv.Gloriasol and Vicia faba cv.Simonia were sterilized and germinated in agarized Gamborg 's B5 medium . Hypocotyles (of tomato and sunflower) or epicotyles (of Vicia) entire (15-20 mm) or cut in small pieces (2-4 mm) were cultured for 20 days in liquid B5 medium additioned with 2,2μM 2,4-D (2,4-Dichloro-phenossyacetic acid). Cytological analysis was performed on hypocotyl explants and released cells, fixed at different times of culture, stained with Feulgen and squashed on slides. Meiosis-like divisions and prophase-reduction events were recorded as percentage of all analyzed dividing cells.

RESULTS AND DISCUSSION

Tomato and sunflower hypocotyles and Vicia epicotyles started to swell and/or to elongate already during the first hours of culture, depending on the length of the explant. At 20 hours from the explant, cell divisions were numerous, chromosome reducing events being present in groups of cells along vascular strands. Cytological analysis showed that, as in carrot, the meiotic-like events were fairly similar to the ones occurring during microsporocytes meiosis in tomato, sunflower or Vicia anthers, but only up to anaphase I stage. At this point, the two reduced nuclei were separated by cytokinesis in two cells, the precocious centromere separation of the sister chromatid eventually restoring the diploid number but at the 2C DNA level. Already at the third day of culture differentiated structures were visible around the vascular strands within endodermis. During the following days of culture, these structures enlarged and differentiated further, resembling strikingly anthers, either isolated and/or accompanied by swollen structures similar to pistils. Inside the anther-like structures as the result of the reducing events, all species formed mononucleate immature pollen-like cells, which were released in the medium at different times of culture, depending on the species. In tomato, at the moment of the release into the medium (10-12 days of culture), the pollen-like cells showed a thick cell wall and were unable to divide further to give PEMs. An histological analysis is now being carried on, in the different species, to study the anatomy and the time course of the differentiation process of these structures which recall primitive ancestral sporangia and ovaries; these last ones containing ovules at different stages of growth, or freely attached to both ends of the hypocotyles. Further studies will clarify the origin and the true nature of these epigenetic phenotypes arising in culture.

An interesting indirect confirmation of our data is given by the recently reported action of auxin on soybean seedling: genes have been showed(by tissue print and in situ hybridization) to be expressed by auxin in the seedling hypocotyles and furthermore in developing reproducing structures (Guilfoyle et al ., 1991).

Particularly interesting will be the experiments, already programmed, to test the eventual expression, on these structures, of anther and pistil specific genes (Ursin et al., 1990; Gasser et al., 1988). Our data offer strategies alternative to androgenesis, to select haploid cultures and differentiate haploid embryos from plant species recalcitrant to regenerate in culture, beside giving new possibilities to isolate genes related to chromosomal reductional mechanisms and reproductive organs differentiation. Moreover, the somatic chromosome segregation in secondary polyploid plants could allow the regeneration of ancestral phenotypes and the transfer of genes from higher to lower polyploid. The occurrence of similar events has been already reported in somaclonal variants regenerated in vitro (Evens and Sharp, 1983; Evans et al., 1984).

ACKNOWLEDGEMENTS

The research was supported by the Ministry of Agriculture and Forestry program: "Sviluppo tecnologie alternative".

REFERENCES

Bonnelle,C., Lejenne,F., Fournier,B., et Tourté,Y., (1990) Remainiements infrastructuraux des cellules cotyledonaires en culture chez deux Legumineuses: relation avec l'acquisition de potentialitè embryogenese. C.R. Acad.Sci.Paris t 310 Serie III, 657-664.

Evans, D.A. and Sharp,W.R. (1983) Single gene mutation in tomato plants regenerated from tissue cultures. Science 221, 949-951.

Evans, D.A., Sharp,W.R. and Medina-Filho, H.P. (1984) Somaclonal and gametoclonal variation. Amer.J.Bot.71(6), 759-774.

Gasser,G..S., Budelier,K.A., Smith,A.G., Shah,D.M., and Fraley,R.T.,(1988) Isolation of tissue-specific cDNAs from tomato pistils.Plant cell 1,15-24.

Guilfoyle,G.,Hagen,M., Gee,Y.L., and Martin,G., (1991) Auxin-regulated gene expression in plants.In vitro 27.Part.II,93A.

Nuti Ronchi,V., (1990) Cytogenetic of plant cell cultures.In:Plant Tissue culture: Applications and Limitations (Bhojwani,S.S.,and Novak,F.J., Eds.).Elsevier, Amsterdam.

Nuti Ronchi,V., Giorgetti,L. and Tonelli,M., (1990) The commitment to embryogenesis, cytological approach. In: Progress in Plant Cellular and Molecular Biology(Nijkamp,Van der Plas, Van Martrijk, Eds.). Kluver Academic, Amsterdam.

Nuti Ronchi,V.(1991) Biological and genetic features of cell cultures in relation to plant morfogenesis.IAPTC Newsletter (in press).

Pennel,R.L. and Roberts,K. (1990) Sexual development in the pea is presaged by altered expression of arabinogalactan protein. Nature 344,347-349.

Ursin,V.M.,Yamaguchi,J. and McCormick,S. (1989) Gametophytic and sporophytic expression of anther-specific genes indevelopping tomato anthers .Plant Cell 1,727-736.

40

Cytological Studies on Plastid Inheritance in Alfalfa (*Medicago sativa*)

H. Lloyd Mogensen, Tong Zhu, and Steven E. Smith

Strict maternal cytoplasmic inheritance is the general rule among plants and animals (Gillham, 1978; Kirk and Tilney-Bassett, 1978; Sears, 1980; Whatley, 1982; Gyllensten, et al., 1985; Hagemann and Schroder, 1989; Smith, 1989a). Paternal plastid inheritance among the gymnosperms, particularly in the Coniferales, is a noted exception (Stine and Keathley, 1990), but paternal mitochondrial inheritance in plants is rare indeed (Soliman et al., 1987; Neale et al., 1989; Harrison and Doyle, 1990). Among the flowering plants, approximately one third of the species investigated show some degree of biparental plastid inheritance (Corriveau and Coleman, 1988); however, even in these cases, there is typically a strong maternal bias (Boblenz et al., 1990). Alfalfa is rather unique in this regard because, in this species, plastid inheritance is predominately from the male parent; and in controlled crosses using certain genotypes, the proportion of offspring containing only male plastids can be over 90 percent (Smith et al., 1986; Smith, 1989b).

The fate of the male plastids in those flowering plants displaying strict maternal plastid inheritance is variable and has been traced to one or more of the following pathways (for review see Chiu et al., 1988; Zhu et al., 1991): (1) the exclusion of plastids from the generative cell during the first pollen mitosis, (2) loss of plastids or plastid DNA from the generative or sperm cells, (3) exclusion of plastids from one sperm cell of a pair, (4) exclusion of plastids from the egg at syngamy, (5) preferential female plastid DNA replication or plastid multiplication rates within the zygote and embryo, and (6) sequestering of plastids into embryonic cell lines which do not become part of the viable embryo.

In alfalfa, a major question is: what is the mechanism(s) leading to a predominance of male plastid inheritance? In other words, what happens to the female plastids in this species? Since Smith et al. (1986) first demonstrated biparental plastid inheritance in *Medicago sativa* using chlorophyll-deficient mutants, several investigations have verified the occurrence of the phenomenon in this species utilizing plastid DNA restriction fragment length polymorphism analysis (Lee et al., 1988; Schumann and Hancock, 1989; Masoud et al., 1990). Schumann and Hancock (1989) suggest that in alfalfa, "The bias in favor of the paternal plastids may have occurred because there was an unequal input of plastids from the gametes of each parent. The zygotes may have initially contained more paternal plastids than maternal ones, and the maternal plastids may have been diluted out in subsequent cell divisions." Schumann and Hancock (1989) go on to say that "It is

Plastid Inheritance in Alfalfa

also possible that paternal plastids were preferentially replicated, and that they ultimately out-competed maternal ones." In their recent discussion on alternative hypotheses for the mechanisms determining differences in plastid inheritance patterns in *Pelargonium, Oenothera*, and *Medicago*, Tilney-Bassett and Almouslem (1989) point out that plastid input, along with the spatial disposition of male and female plastids and their propensity to mix within the zygote, may be of utmost importance.

Smith (1989b) has demonstrated that plastid inheritance is under genetic control in alfalfa, and he has characterized certain genotypes as being strong, intermediate, or weak males or females, based upon their plastid transmission strength. The availability of these genotypes, combined with the powerful techniques of serial ultrathin sectioning and computer-generated quantitation and three-dimensional reconstruction, provide a unique opportunity for a meaningful study of the cytological basis of plastid transmission in alfalfa.

Materials and Methods

A description of the genotypes used and techniques employed in our cytological studies have been published previously (Zhu et al., 1990, 1991; Shi et al., 1991).

Results and Conclusions to Date: Generative and Sperm Cells

Our initial approach has been to ask whether a positive correlation exists between the number of plastids per generative cell and known plastid transmission patterns in certain genotypes. Using the techniques of serial ultrathin sectioning and three-dimensional reconstruction, we found that the generative cells are morphologically similar among genotypes in that they all contain numerous plastids distributed equally at each end of the spindle shaped cell (Fig. 1A; Zhu et al., 1990a). Quantitative data show that the average number of plastids per generative cell varies among genotypes, but, unexpectedly, the intermediate male may have an equal or larger mean number of generative cell plastids than the stronger male (Table 1; Zhu et al., 1991).

Using a DNA-specific fluorochrome, the number of generative cell plastids containing nucleoids (DNA aggregates; Fig. IB) was also found to vary according to genotype, but was not significantly different in two genotypes that show markedly different male plastid transmission behavior (Table 1; Shi et al., 1991). Sperm cell analysis, likewise, does not appear to offer an explanation for the observed patterns of paternal plastid inheritance in

Plastid Inheritance in Alfalfa

alfalfa. Our data to date indicates that the two sperms of a pair within each pollen tube contain, collectively, approximately the same number of plastids as did the generative cell from which they were derived. Our results also indicate that each sperm of a pair is quite similar morphologically and that the two sperms contain an equal mean number of plastids, and plastid nucleoids (Zhu et al., 1990b; Zhu et al., in preparation).

Table 1. Comparison of plastid nucleoid number with plastid number in generative cells from three genotypes exhibiting different levels of male plastid transmission, i.e. MS-5 weakest, 7W intermediate, and 301 strongest. From Shi et al. 1991.

	Nucleoids/generative cell[a]			Plastids/generative cell[b]		
Genotype	Mean	SD[c]	Range	Mean	SD[c]	Range
MS-5	45.4d	14.4	20-80	62.0[d]	10.4	50-69
7W	59.9[e]	24.5	16-155	264.7[e]	107.6	190-388
301	54.4	21.5	18-113	174.7	26.3	151-203

[a]Counts from 150 generative cells (50 from each genotype).
[b]Data from Zhu et al. 1991; based on counts from serial ultrathin sections of nine generative cells (three from each genotype).
[c]SD = Standard deviation of the mean.
[d]Significantly less than genotype 301 or 7W ($p<0.05$).
[e]Not significantly different from 301 ($p>0.05$).

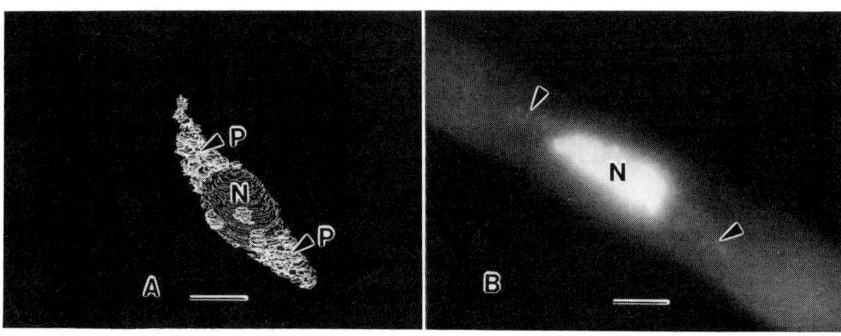

Figure 1. A. Reconstruction of the generative cell of alfalfa showing plastids (P) and the nucleus (N). From Zhu et al., (1990a). B. Fluorescence micrograph of a pollen tube of alfalfa showing the generative cell nucleoids (arrowheads) at each end of the nucleus (N) after DAPI staining. From Shi et al., (1991). Bar = 10um

Plastid Inheritance in Alfalfa

Preliminary Data From Egg Cells, and Future Research

Our next area of concentration in this study will be the examination of unfertilized eggs from strong and weak females to determine whether egg composition before fertilization is a determining factor in cytoplasmic inheritance in alfalfa. Preliminary data, derived from computer-generated reconstructions from serial ultrathin sections of unfertilized eggs, suggest that plastid distribution before fertilization may be an important determinant in plastid transmission. It is known from classical embryological studies on alfalfa (Cooper, 1935) that the first division of the zygote is transverse and unequal, resulting in the formation of a two celled embryo composed of a smaller apical cell and a larger basal cell. Subsequent development of the embryo is such that only those cells derived from the apical cell survive to form the new plant of the next generation. Derivatives of the basal cell form the suspensor, which eventually degenerates. Reconstructions of one egg each from genotype CUFB, a weak female, and genotype 6-4, a strong female, indicate that within the egg of the weak female, the majority of plastids are positioned such that after the first division of the zygote, they would be located within the basal cell and thereby eliminated from the path of inheritance. Conversely, plastids in the egg of the strong female are polarized such that they would most likely end up in the apical cell and thus be in a position for transmission to the next generation. We are currently obtaining additional examples to establish whether this is a consistent pattern for egg plastid distribution in these genotypes. In addition to unfertilized eggs, we plan to carry out detailed analyses of zygotes and early embryos to determine the behavior and fate of both male and female plastids after controlled crosses between strong and weak male and female genotypes.

Acknowledgements

Support for this work by the United States Department of Agriculture under grant 88-37234-3876, the Organized Research Fund of Northern Arizona University, and by the Arizona Agricultural Experiment Station is gratefully acknowledged.

References

1. Boblenz, K.; Nothnagel, T.; Metzlaff, M. Paternal inheritance of plastids in the genus *Daucus*. Mol. Gen. Genet. 220:489-491; 1990.

2. Chiu, W-L.; Stubbe, W.; Sears, B. B. Plastid inheritance in *Oenothera*: organelle genome modifies the extent of biparental plastid transmission. Curr. Genet. 13:181-189; 1988.

3. Cooper, D. C. Macrosporogenesis and embryology of *Medicago*. J. Agric. Res. 51:471-477; 1935.

4. Corriveau, J. L.; Coleman, A. W. Rapid method to detect potential biparental inheritance of plastid DNA and results for over 200 angiosperm species. Am. J. Bot. 75:1443-1458; 1988.

5. Gillham, N. W. Organelle Heredity. New York: Raven; 1978.

6. Gyllensten, U., D. Wharton and A.C. Wilson. 1985. Maternal inheritance of mitochondrial DNA during backcrossing of two species of mice. J. Heredity 76:321-324.
7. Hagemann, R.; Schroder, M. B. The cytological basis of the plastid inheritance in angiosperms. Protoplasma 152: 57-64; 1989.
8. Harrison, R. G.; Doyle, J. J. Redwoods break the rules. Nature 344:295-296; 1990.
9. Kirk, J. T. O.; Tilney-Bassett, R. A. E. The plastids: their chemistry, structure, growth and inheritance. 2nd ed. Amsterdam: Elsevier/North-Holland Biomedical Press; 1978.
10. Lee, D. J.; Blake, T. K.; Smith, S. E. Biparental inheritance of chloroplast DNA and the existence of heteroplasmic cells in alfalfa. Theor. Appl. Genet. 76;545-549; 1988.
11. Masoud, S. A.; Johnson, L. B.; Sorensen, E. L. High transmission of paternal plastid DNA in alfalfa plants demonstrated by restriction fragment polymorphic analysis. Theor. Appl. Genet. 79:49-55; 1990.
12. Neale, D. B.; Marshall, K. A.; Sederoff, R. R. Chloroplast and mitochondrial DNA are paternally inherited in *Sequoia sempervirens* D. Don Endl. Proc. Natl. Acad. Sci. USA 86:9347-9349; 1989.
13. Schumann, C. M.; Hancock, J. F. Paternal inheritance of plastids in *Medicago sativa*. Theor. Appl. Genet. 78:862-866; 1989.
14. Sears, B. B. Elimination of plastids during spermatogenesis and fertilization in the plant kingdom. Plasmid 4: 233-255; 1980.
15. Shi, L.; Zhu, T.; Mogensen, H. L.; Smith, S. E. Paternal plastid inheritance in alfalfa: Plastid nucleoid number within generative cells correlates poorly with plastid number and male plastid transmission strength. Curr. Genet. (in press); 1991.
16. Smith, S. E. Biparental inheritance of organelles and its implications in crop improvement. Plant Breed. Rev. 9:361-393; 1989a.
17. Smith, S. E. Evidence for genetic control of plastid inheritance in *Medicago sativa*. J. Hered. 80:214-217; 1989b.
18. Smith, S. E.; Bingham, E. T.; Fulton, R. W. Transmission of chlorophyll deficiencies in *Medicago sativa*, evidence for biparental inheritance of plastids. J. Hered. 77:35-38; 1986.
19. Soliman, K.; Fedak, G.; Allard, R. W. Inheritance of organelle DNA in barley and *Hordeum x Secale* intergeneric hybrids. Genome 29:867-872; 1987.
20. Stine, M.; Keathley, D. E. Paternal inheritance of plastids in engelmann spruce X blue spruce hybrids. Jour. Hered. 81: 443-446; 1990.
21. Tilney-Bassett, R. A. E.; Almouslem, A. B. Variation in plastid inheritance between pelargonium cultivars and their hybrids. Heredity 63:145-153; 1989.
22. Whatley, J. M. Ultrastructure of plastid inheritance: green algae to angiosperms. Biol. Rev. 57:527-569; 1982.
23. Zhu, T.; Mogensen, H. L.; Smith, S. E. Generative cell composition and its relation to male plastid inheritance patterns in *Medicago sativa*. Protoplasma 158:66-72; 1990a.
24. Zhu, T.; Mogensen, H. L.; Smith, S. E. Ultrastructural reconstruction and quantitative features of alfalfa sperms. Am. J. Bot. 77:s42; 1990b.
25. Zhu, T.; Mogensen, H. L., Smith, S. E. Quantitative cytology of the alfalfa generative cell and its relation to male plastid inheritance patterns in three genotypes. Theor. Appl. Genet. 81:21-26; 1991.

41
Modifications in the Interapertural Exine Following Pollinations of *Linum grandiflorum*

Rivka Dulberger

Due to inertness of sporopollenin, the outer exine is one of the most stable biological structures. Changes in the exine following pollination are therefore particularly interesting and thus far have been reported in only three plants: Pharbitis nil (Gherardini and Healey, 1969); Linum grandiflorum (Dickinson and Lewis, 1974) and Ulmus glabra (Rowley, 1983). In Linum grandiflorum degradation of sporopollenin was found to be restricted to the colpal areas.

In the course of a study of incompatibility in L. grandiflorum, modifications of the interapertural exine following pollination were incidentally revealed by the SEM. The present paper describes these changes, compares them with those previously reported in the same species and discusses their nature and significance.

METHODS

Pin and thrum stigmas were hand-pollinated with compatible or incompatible pollen and fixed in 3% glutaraldehyde in 0.05 M phosphate buffer, 30-60 min after pollination. Four samples of 4-6 stigmas each were examined per pollen-stigma combination. The samples were obtained during three different years. Preparation of the material for examination with SEM, TEM and light microscopy was as described previously (Dulberger, 1987, 1989).

RESULTS

Pollen of L. grandiflorum displays dimorphic exine sculpturing (Dulberger, 1981). The exine of thrum grains has processes of uniform size and shape. In pin grains, the majority of exines processes are small, but there are also larger ones, of about the same size as in thrum grains (Figs. 1, 2).

The exine contains loosely packed sporopollenin globules (Dickinson and Lewis, 1974). In cryostat sections of undehisced anthers, the main exine layer gave positive response to staining with alcian blue or coomassie blue, thus indicating presence of pectins and proteins. No difference was found between the two morphs in structure or staining responses of their exine (Dulberger, 1989 and unpublished).

In the vast majority of grains normally rehydrated on stigmas, processes and spaces between them were distinguishable and the characteristic exine ornamentation was evident. This was true also for most germinated grains (Figs. 3-6, 10).

EXPLANATION OF FIGURES

Figs. 1-9. The exine in pollen of *Linum grandiflorum* before and after pollination.

1,2. Exines of dry pollen. 1. Thrum exine. 2. Pin exine. 1,2 x3000.

3-9. Exines of grains rehydrated or germinated on stigmas.

3-6. Exines of rehydrated pollen grains showing morph-specific sculpturing.

3. Thrum grains on pin stigma. x780.

4. Thrum grain germinated on thrum stigma.x1800.

5. Pin grain germinated on thrum stigma. x1300. 6. Pin grain rehydrated on pin stigma. x1600.

7-9. Exines of rehydrated grains showing modifications.
7,8. Exines of aberrant thrum grains on pin stigmas.
7. x2000. 8. x3600.

9. As 7 but on thrum stigma. x4800.

Figs. 10-15. Exines of pollen rehydrated after various pollinations.

10. Exine of a thrum grain normally rehydrated on pin stigma. x 10000.

11-15. Exines showing modifications. 11. Thrum grain on pin stigma. x10000.

13. Exines of pin grains on thrum stigmas. 12. x2000. 13. x3000.

14,15. Exines of pin grains on pin stigmas. 14. x3000. 15. x10000.

In a few aberrant grains, however, modifications were observed, that blurred or distorted the characteristic exine patterning. In aberrant thrum grains, the exine processes appeared swollen in both compatible pin x thrum pollinations and incompatible thrum x thrum pollinations, while the spaces between the processes were considerably reduced or absent altogether (Figs. 7-9, 11).

Aberrant pin grains, on the other hand, displayed two opposed phenomena. In compatible thrum x pin pollinations the exine was swollen, which obviated distinction

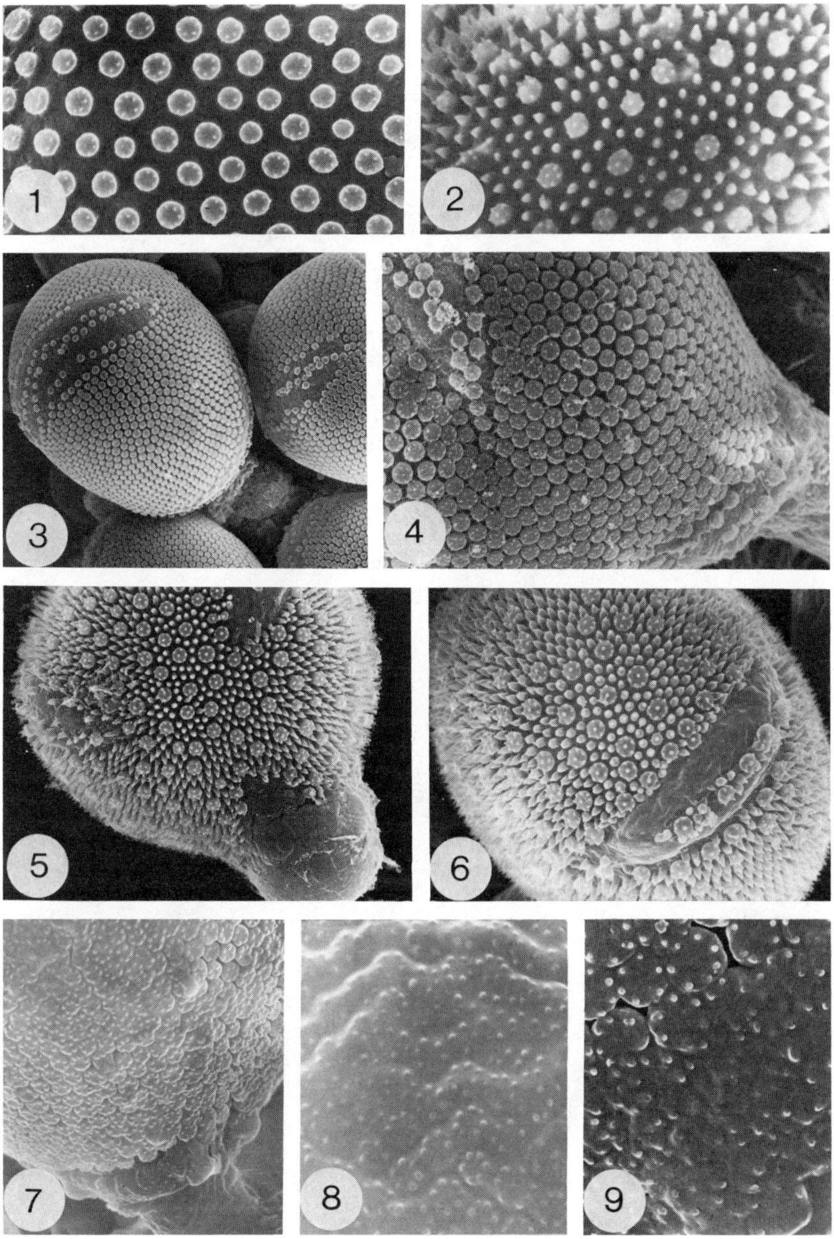
1
2
3
4
5
6
7
8
9

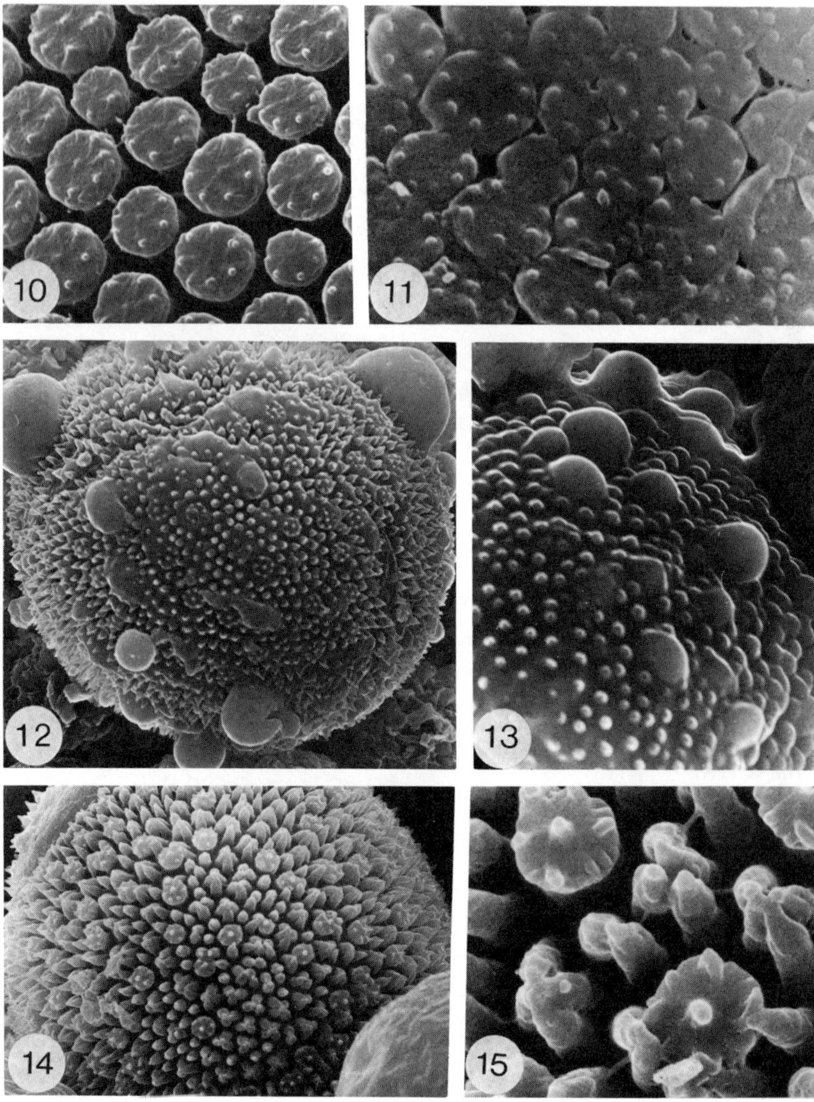
10
11
12
13
14
15

between large and small processes and enabled to see only the tips of some processes (Figs. 12, 13). By contrast, in pin x pin pollinations, aberrant grains had a shrunken exine, with the processes somewhat wrinkled and the smaller ones arranged in groups of three or four (Figs. 14, 15).

The described modifications pertained for a small fraction of pollen grains exposed on the stigmas, in two of the four samples examined.

Similar patterns of exine modification in L. pubescens will be reported separately.

DISCUSSION

Dickinson and Lewis (1974) reported rapid degradation of sporopollenin at the apertural areas after the two compatible pollinations and also after thrum x thrum pollinations, but not after pin x pin pollinations. Hydration of pollen grains by buffer solutions also stimulated degradation of the exine. These authors suggested that the degradation process is a consequence of hydration.

The phenomena reported here are fundamentally different from those described by Dickinson and Lewis (1974). They occur in the interapertural exine rather than in the colpal areas and are observable in a few grains only. Moreover, appearance of the modifications in SEM and the staining responses of the exine strongly suggest altered polysaccharides rather than degraded sporopollenin.

To explain the changes in the interapertural exine reported here, various processes may be invoked. For instance, at rehydration, expansion of the pollen wall at the apertural areas may entail a certain degree of contraction of the interapertural wall. However, despite contraction, most normally rehydrated grains displayed morph-specific exine sculpturing, as exemplified in Figures 3-6. Although in some grains patches of exine appeared with spaces between processes smaller than usual, contraction of the exine cannot explain the appearance of bubbles or the abrogation of exine patterning (Figs. 7-9, 11-15).

The observed changes are also attributable to incompatibility. It is known that in distylous L. grandiflorum intramorph pollinations are incompatible. In thrum x thrum pollinations pollen tubes are inhibited within the stigma. In pin x pin pollinations, on the other hand, most pollen grains fail to adhere to the stigmas; the few adhering grains fail to rehydrate, or their tubes are inhibited in the stigma (Ghosh and Shivanna, 1983).

Interestingly, the two opposing patterns of exine degradation, namely swelling vs. contraction, coincide with two contrasting categories of pollen behaviour on stigmas, namely rehydration of most grains deposited on stigmas vs. failure to rehydrate of a few grains in pin x pin pollinations. To what extent is this coincidence relevant for incompatibility?

The hypothesis that differences between the morphs in osmotic pressure of pollen and stigmas control incompatibility in L. grandiflorum (Lewis, 1943) is not supported by a recent study (Murray, 1986). It seems that low permeability of pin stigmatic papillae prevents the few adhering pin grains to rehydrate on pin stigmas (Dulberger, 1987). Thrum pollen grains apparently possess the ability to overcome the impermeability barrier. It should be pointed out that in L. grandiflorum the major barrier to self-fertilization of the pin morph is the failure of most pollen grains to adhere to the stigma rather than of a few grains to rehydrate.

Possibly, pin and thrum pollen differ in colloids found in the interstices of the sporopollenin globules. This assumption of a difference in colloids could explain the occurrence of two patterns of exine modification, but it fails to explain why modification of the exine is only displayed exceptionally by few grains, and not by all grains rehydrated on stigmas.

The present observations show that exines may change on stigmas. The ability of the exine to resist disintegration may vary among the pollen grains of a species. Deliberate rather than haphazard studies of exines deposited on stigmas should provide a better understanding of the rigorous control of exine integrity.

REFERENCES

Dickinson, H.G.; Lewis, D. Changes in the pollen grain wall of Linum grandiflorum following compatible and incompatible intraspecific pollinations. Ann. Bot. 38:23-29; 1974.

Dulberger, R. Dimorphic exine sculpturing in three distylous species of Linum (Linaceae). Pl. Syst. Evol. 139:113-119; 1981.

Dulberger, R. Fine structure and cytochemistry of the stigma surface and incompatibility in some distylous Linum species. Ann. Bot. 59:203-217; 1987.

Dulberger, R. The apertural wall in pollen of Linum grandiflorum. Ann. Bot. 63:421-431; 1989.

Gherardini, G.L. Healey, P.L. Dissolution of outer wall of pollen grain during pollination. Nature 224:718-719; 1969.

Ghosh, S; Shivanna, K.R. Studies in pollen pistil interaction in Linum grandiflorum. Phytomorphology 32:385-395; 1983.

Lewis, D. The physiology of incompatibility in plants. ii. Linum grandiflorum. Ann. Bot. 7:115-122; 1943.

Murray, B.G. Floral biology and self-incompatibility in Linum. Bot. Gaz. 147:327-333; 1986.

Rowley, J.R.; Rowley, J.S. Modification of the exine during germination of pollen grains. In Mulcahy, D.L.; Ottaviano, E. (eds) Pollen: Biology and Implications for plant breeding. Elsevier Biomedical. New York, Amsterdam, Oxford; 1983.

Biotechnology

42
Pollen Mediated Indirect Gene Transfer to Dicots and Monocots *

Dieter Hess, Markus Iser, Angela Schmid, Sabine Stegmaier, and Klaus Dressler

1. Introduction

Scientists will often be well advised to follow Nature as close as possible. Therefore, Hess proposed as early as 1974 to use Nature's masterly elaborated pollen transfer system in introducing exogenous gene material into higher plants. Beginning with the same year (Hess et al., 1974), in the following two decades experimental data on phenotypical, biochemical and formal genetic levels could be obtained, which could be interpreted with a pollen mediated gene transfer. In most of these investigations, however, no molecular proof of gene transfer could be achieved as suitable methods were not available. The only exception were attempts to transfer an E.coli transferase (E.C.2.7.7.12) gene to Petunia: Using Southern blot analysis, uptake and integration of transferase DNA into Petunia DNA could be demonstrated. The limited amount of DNA available from earlier experiments, however, allowed cleavage using only one endonuclease so that a correct restriction analysis was not possible (Hess, 1986, Hess and Dressler, unpublished data). To summarize: in all these early investigations convincing molecular evidence for a gene transfer could not be achieved (review Hess, 1987).

Usually, pollen preparations show a high nuclease activity, which is difficult or even impossible to overcome (Negrutiu et al., 1986, Van der Westhuizen et al., 1987, Body et al., 1989). Plasmid concentrations of 50 $\mu g \cdot ml^{-1}$ Petunia pollen suspension are degraded completely within a few minutes (Van der Westhuizen et al., 1987). Therefore, using "naked" DNA, no results interpretable as gene transfer could be obtained unless there was enough carrier DNA present, or - preferably - the DNA to be transferred protected within phage heads (Hess, 1978, 1979, Hess and Dressler, 1984, Hess et al., 1985, Hess, 1987).

Therefore, if one wants to profit from the advantages of plasmid techniques, it seems advisable to make use of indirect, Agrobacterium mediated transfer systems. The present chapter deals with the indirect pollen transfer of heterologous gene material to Petunia, a dicot, and to wheat, a monocot.

2. Indirect Pollen Mediated Gene Transfer to Petunia

Our ultimate intention will be to transfer agronomically useful genes to cereals, our most important nutritive plants. Before using wheat or other cereals as receptors, we investigated the possibility of an indirect pollen transfer with the model plant Petunia hybrida. In basic experiments the

* In Memoriam Ercole Ottaviano

system used should be as simple as possible. Nothing could be easier than imitating the normal mode of working of Agrobacterium tumefaciens: transferring tumor genes. By the way the screening technique will be provided: growth of transformed tumorous cells on medium without growth substances. So we tried whether A. tumefaciens would be kind enough to transfer its onc-genes via pollen to Petunia progenies.

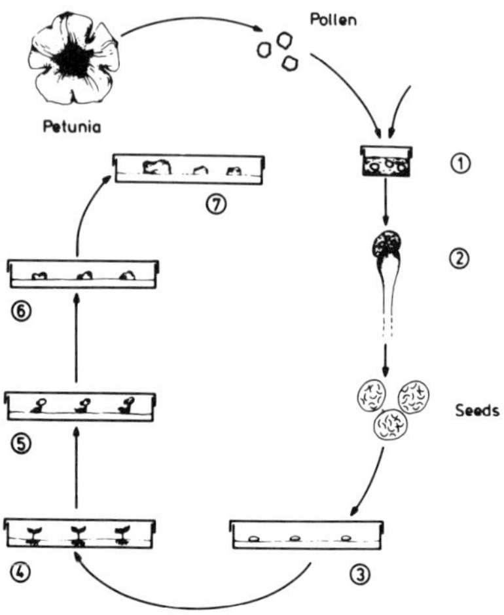

Fig.1. Indirect pollen mediated transfer of *onc*-genes to *Petunia hybrida.* 1 Coculture of *Petunia* pollen with *Agrobacterium tumefaciens* C58 (wild strain, addition indicated by an arrow). 2 "Wet" pollination using the pollen/*Agrobacterium* suspension. 3 Putting the seeds obtained on WH agar medium with hormones. 4 Development of seedlings. 5 Re-moval of one cotyledon from every seedling; the isolated cotyledons were put stalk upwards onto WH agar medium. 6 Transfer of the cotyledonar stalk calli onto Murashige-Skoog medium without hormones. 7 Growth of tumorous calli on medium without hormones (Hess, 1990).

A schedule of the procedure used is given in Fig. 1. Pollen of Petunia was cocultivated with an A. tumefaciens wild strain within a medium which allowed growth of both pollen and bacteria. Thereafter, a droplet of the pollen/Agrobacterium mixture was put on the stigmata of Petunia flowers. Following fertilization, the seeds obtained were screened for growth without hormones and assayed for nopaline synthase activity. Using Southern blot analysis, the presence of onc-genes in Petunia callus material could be demonstrated.

Step by step, the functioning of the pollen system could be proven:

2.1. Activation of the Plasmidial vir-Region

Before Agrobacterium will be able to transfer T-DNA into plant cells activation of its plasmidial vir-genes must occur. A vir/ lacZ fusion allows to monitor vir-activation by measuring ß-galactosidase activity. Using this assay, the group of Van Montagu had been able to identify vir-inducing plant factors, before all acetosyringone, the most potent inducer (Stachel et al., 1985). Acetosyringone has been detected so far from tobacco only. In other plant species it is replaced by chemically more or less related phenolic compounds.

Petunia stigma and pollen are excreting vir-inducing substances into aqueous medium. Using HPLC and the ß-galactosidase assay just mentioned, two vir-inducing fractions could be detected (Fig. 2.). In both cases, flavonol glycosides turned out to be the active principle (Zerback et al., 1989a,b). The activity of the main factor, a kaempferol glycoside, was comparatively low, but this was compensated by enormous quantities released from pollen (Zerback et al., 1989b). Therefore, we could use pollen/Agrobacterium mixtures without experimental addition of vir-inducing factors like acetosyringone.

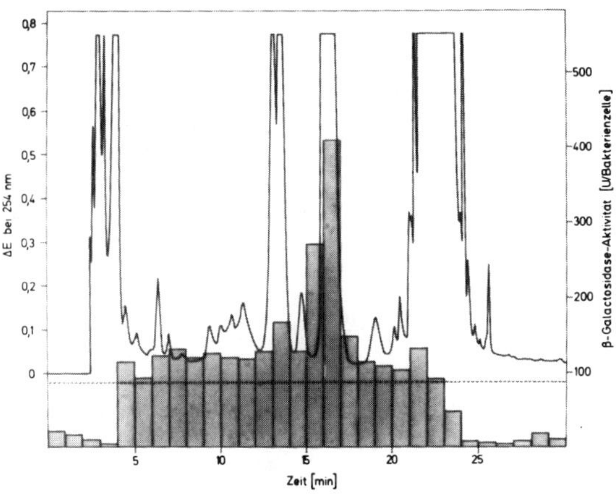

Fig.2. HPLC analysis of *vir*-activating factors from pollen extracts of *Petunia hybrida.* The solid line represents the elution profil at 254 nm. The HPLC fractions obtained were assayed for ß-galactosidase activity, which indicates *vir*-activation (shaded bar diagramm). The peak showing the highest *vir*-activation represents kaempferol-3-glucosyl-galactoside. The horizontal broken line represents ß-galactosidase activity from the pure medium control. On a lower level, essentially the same results were obtained from pollen diffusates (Zerback et al., 1989a).

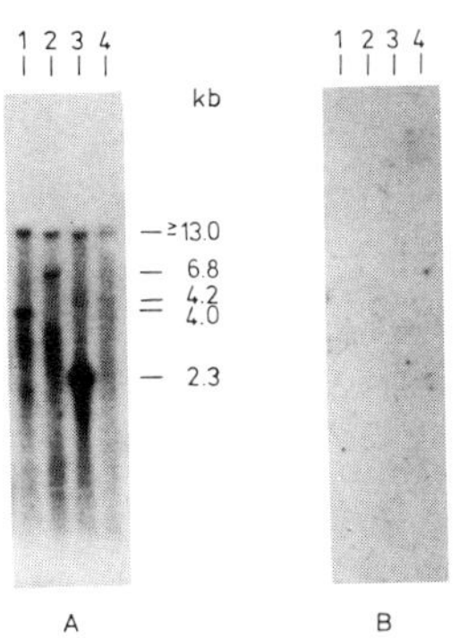

Fig.3. Southern blots of *Petunia* pollen DNA after coculture with *A.tumefaciens* LBA 4404ıpBI121 containing the GUS gene. Before DNA isolation, the bacteria had been removed by high stringency washes. A probed with the GUS gene. B probed with ^{32}P *vir*-DNA. Cleavage of pollen DNA: Lane 1 with *Sna*BI, lane 2 with *Sma*I, lane 3 with *Bam*H1, lane 4 with *Xba*I (Süssmuth et al., 1991).

2.2. Fixation of *Agrobacterium* to Pollen Tubes

Following *vir*-induction, *Agrobacterium* has to fix to the pollen tubes for the transfer of T-DNA. Fixation of the bacteria onto germinating and growing pollen tubes could easily be observed by light-microscopy (Hess and Dressler, 1989).

2.3. Transfer of T-DNA into Pollen Tubes

After fixation, the T-DNA has to be transferred. Using Southern blot analysis, transfer of GUS-DNA into pollen could be proven (Süssmuth et al., 1991, Fig. 3A). It should be stressed, that only the T-DNA transferred gives the signals detected by Southern blotting and no other DNA from *Agrobacterium* itself. There are several possibilities to exclude the presence of bacteria (cf. Hess et al., 1990). In our experiments all of these remained negative. A highly sensitive and reliable assay to detect *Agrobacterium* will be mentioned: *vir*-probing. The *vir*-region is localized on the Ti-plasmid outside the T-DNA. Assuming that a hybridization signal detected from a T-DNA probe was due to T-DNA within the bacteria, one has to expect a hybridization with *vir*-probes, too. All *vir*-probing remained negative (Fig. 3B). Therefore, the signals detected from pollen-DNA were due to T-DNA transferred into the pollen and not to *Agrobacterium*.

2.4. Expression of Transferred Genes within Pollen

The next question, whether the transferred gene material could be expressed in pollen itself is still not answered. The reason is a high intrinsic GUS-activity of *Petunia* pollen so that the activity of the transferred GUS-gene could not be studied (Süssmuth et al., 1991). New experiments using a luciferase reporter gene are initiated. Considering the intention of our experiments, however, it is of secondary importance whether a gene transferred will be expressed in pollen or not.

2.5. Screening for Tumor Characteristics

The seeds obtained from pollination and fertilization using pollen/*Agrobacterium* mixtures were screened for tumor formation. It is not to be expected that tumor formation of transformed plants occurs without any stimulation. Therefore, one cotyledon from

Fig.4. Tumorous growth of a *Petunia* callus transgenic for the *onc*-region transferred from *A. tumefaciens* C58 (wild strain). Above two smaller untransformed calli which had stopped growing (Original D. Hess, cf.Hess and Dressler, 1989).

seedlings developed on WH agar medium (Wagner and Hess, 1973) was removed and put stalk upwards onto agar (Fig. 1). Callus developing on the cut surface of the stalk was removed and put onto medium without hormones. In the experiment we started with 1853 and in the control (pollen treatment with medium only) with 1943 calli. Most of the calli died, but in the experiment four and in the control two calli showed growth on hormonfree medium (Fig. 4.), a characteristic of plant tumor cells - but of habituated cells as well (Hess and Dressler, 1989).

2.6. Southern Blot Analysis and NPT(II)-Assay

Using Southern blot analysis (Hess and Dressler, 1989), within two of the four growing experiment calli T-DNA covering the onc-genes for synthesis of tryptophan-2-mono-oxygenase and isopentenyl-transferase could be detected. In Fig. 5 blots from the best growing callus are shown. Vir-probing remained negative. No onc-DNA could be detected in the four other calli - two of experiment and two of the control - so that these must have been habituated.

None of the six calli showed nopaline synthase activity. This corresponds with the data from Southern blot analysis: in none of the calli, including the two onc-positive ones, the nos-gene could be detected. The explanation for this must be, that the nos-gene has not been transferred into the plant genome or has been lost during cultivation, an observation made with tobacco, too (Van Lijsebetten et al., 1986).

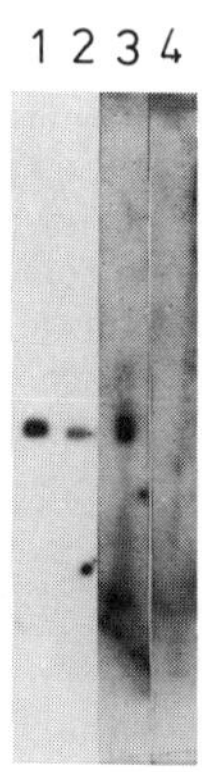

Fig.5. Southern blot analysis of DNA from a transgenic tumorous *Petunia* callus. The *Hin*d22 fragment codes for plant hormones or part of them, the *Hin*d23 fragment contains the nopaline synthase gene. Lane 1: *Hin*d22 fragment DNA probed with ^{32}P *Hin*d22 DNA. Lane 2: *Hin*d23 fragment DNA probed with ^{32}P *Hin*d23 DNA. Lane 3: *Hin*dIII cleaved DNA from a transgenic callus probed with ^{32}P *Hin*d22 DNA. Lane 4: *Hin*dIII cleaved DNA from the same transgenic callus probed with ^{32}P *Hin*d23 DNA. The region corresponding with the *Hin*d22 fragment can be detected from callus DNA, the *Hin*d23 region not (Original K. Dressler, cf. Hess and Dressler, 1989). The data just described give full evidence for the functioning of the indirect pollen system in *Petunia*: Agrobacterial *vir*-genes are activated (2.1.), bacteria attach to pollen tubes (2.2.), and T-DNA is transferred into pollen (2.3.). The T-DNA is integrated into receptor DNA (2.6.) and expressed in the progenies (2.5.).

3. Indirect Pollen Mediated Gene Transfer to Wheat

3.1. First Successful Transfers

The results obtained with the model plant Petunia hybrida encouraged to perform corresponding experiments with wheat, our most important nutritive plant (Hess et al., 1989).

3.1.1. "Cocultivation" by Pipetting Agrobacterium into Flowers

Using wheat as receptor the mode of cocultivation had to be adapted. As wheat pollen is produced only in minute quantities per flower it is difficult to collect and it shows, once collected, a rapid loss of viability. Therefore, when the spikelets in the middle of a given spike had opened, 5 µl of Agrobacterium suspension were pipetted into each flower of the spike. In the first experimental series, the Agrobacterium strain used contained the npt(II)-gene coding for kanamycin resistance within its T-DNA. The gene was regulated by the nos-promoter, a comparatively weak promoter.

3.1.2. Activation of the Plasmidial vir-Region

By pipetting Agrobacterium suspensions into flowers, the bacteria get into contact with pollen. Once more the problem of vir-activation had to be solved, and once more vir-activating factors are released from pollen. Using the methods mentioned above (cf. 2.1.), two vir-inducing flavonol-glycosides could be detected (Fig. 6) in the aqueous medium used for application of Agrobacterium. One of them, a kaempferol derivative, is diffusing from wheat pollen in quantities which are within the vir-inducing concentration range. Therefore, the basis for vir-activation, and subsequently T-DNA transfer to wheat pollen was given. The concentration of flavonol-glycosides released from wheat pollen, however, was lower than from Petunia pollen. So a further stimulation by the addition of acetosyringone could be useful. Experiments of this kind are currently performed.

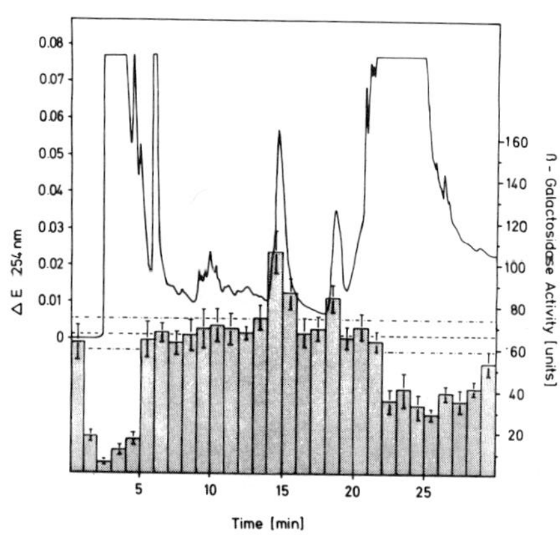

Fig.6. HPLC analysis of *vir*-inducing factors from pollen extracts of wheat. Broken lines: mean ß-galactosidase activity and standard deviations from pure medium controls. The highest shaded bar represents kaempferol-3-rhamnosyl-glucoside. Other details as in Fig. 2. On a lower level, essentially the same results could be obtained from pollen diffusates (Original R. Nimmrichter, cf. Hess et al., 1990).

3.1.3. Screening for Kanamycin Resistance

The grains obtained were first grown for three days in replica plates on kanamycin agar. This passage allows to remove infected seedlings, despite a rigid disinfection method (Hess and Kiefer, 1981) was applied. The best growing plants were transferred into 750 ml glass jars on agar medium with - in most cases - reduced kanamycin concentrations. If the seedlings in the replica plates showed no growth differences, they all were transferred into glass jars.

A wide range of kanamycin concentrations from 50 to 200 $mg\,l^{-1}$ in the replica plates and from 10 to 500 $mg\,l^{-1}$ in the glass jars were tested. Even with the highest kanamycin concentrations, kanamycin resistance proved to be no reliable marker: as judged from Southern blotting (cf. below), selection was possible only in a few experiments, in most cases it was impossible.

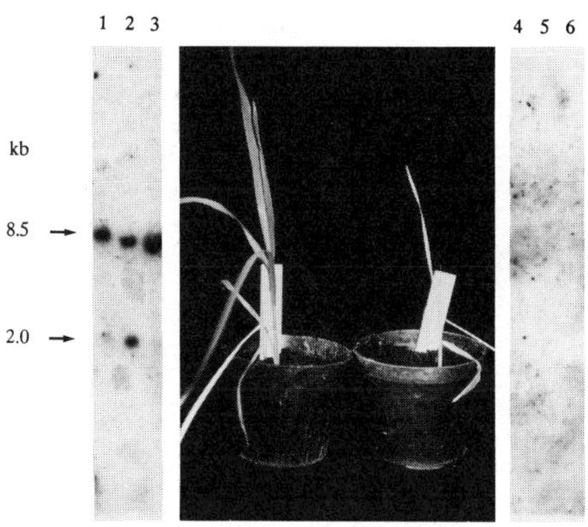

Fig.7. Transfer of the *npt*(II)-gene to wheat using the indirect pollen system. First generation plants - left a transgenic plant, right a control - are flanked by Southern blots. Lanes 1 - 3: DNA from three transgenic plants, cleaved with *Eco*RI and *Hin*dIII and probed with the ^{32}P *Eco*RI/*Sal*I fragment of plasmid pKM2, containing the *npt*(II) structural gene. Lanes 4 - 6: DNA from three control plants treated in the same way (Hess 1988, cf. Hess et al., 1990).

3.1.4. Southern Blotting in the First and Second Sexual Generation

Using Southern blotting, npt(II)-DNA could be detected from plants of the first generation which derived directly from Agrobacterium-treated spikelets. Controls showed no signals (Fig. 7). All transformants showed alterations concerning the transferred DNA sequences. Taking all experiments together, including some unsuccessful ones, transformation frequency was 1%, the highest frequency of one single experiment was 2.6%.

Some Southern positive plants of the first generation were selfed, and the second generation obtained was screened and used for Southern analysis. In some progenies, the transferred gene material had been lost, in
tamination with Agrobacterium was excluded by vir-probing (Fig. 8).

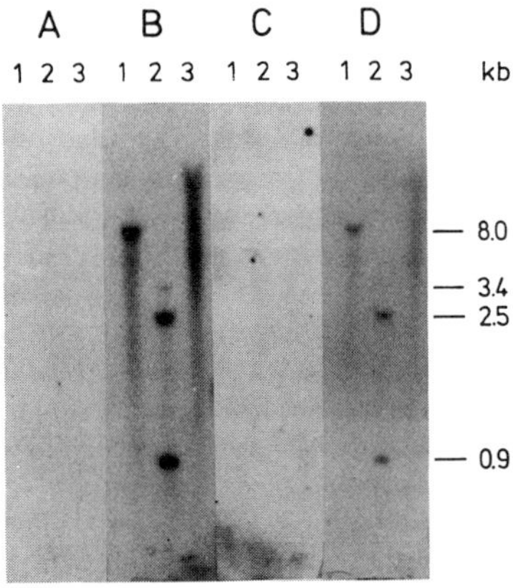

Fig.8. Transfer of the *npt*(II)-gene to wheat using the indirect pollen system: Southern blot analysis of the second generation. A DNA from a control plant; B DNA from a transformant; C Blot B reprobed with two ^{32}P *vir* fragments; D Blot C reprobed with ^{32}P *npt*(II)-DNA to demonstrate that the DNA was still present. Lane 1: DNA cleaved with *Eco*RI and *Hin*dIII. Lane 2: DNA cleaved with *Eco*RI, *Hin*dIII, and *Sph*I. Lane III: uncleaved wheat DNA. Hybridization of uncleaved DNA is unspecific (Hess et al., 1990).

3.1.5. NPT(II)-Assay

All Southern positive first and second generation transformants were assayed for NPT(II)-activity. Enzyme activity was absent or very low. This explains why selection on kanamycin was impossible in most cases. The reasons are DNA rearrangements or - more probably - DNA methylation (cf. below).

The data presented (3.1.1. - 3.1.5.) demonstrate for the first time the possibility to transfer genes to wheat.

3.2. Problems and Further Experimental Approaches

The main problem is the low expression of the genes transferred. For wheat breeders, this was not unexpected. In conventional wheat breeding the activity of a gene introduced from a diploid donor into a receptor with a higher level of ploidy is more or less repressed. A comparable situation was given in our experiments when we introduced a single gene into allohexaploid wheat. It had even been predicted that a gene transplanted into wheat should become inactivated (Ren and Lelley, 1989).

3.2.1. New Transfer Experiments: Southern Blot Screening

To overcome the difficulties just mentioned, we are testing other promoters (CaMV 35 S, 1',2'-(T_r)) combined once more with the marker genes npt(II) or hpt, so far without success. Neither using the npt(II)- nor the hpt-gene any selection was possible. What remains - at least at the present stage of our investigations - is to screen several hundreds of young wheat plants for the presence of the DNA under question. Using this "Southern Blot Screening" in appropriate experiments, once more npt(II)-DNA (Fig. 9.A), hpt-DNA (Fig. 9.B), and stilbene synthase-DNA (Fig.10.A) could be detected from wheat transformants. Transformation frequencies were about 1% (Iser, Stegmaier and Hess, unpublished data; Schmid and Hess, 1991).

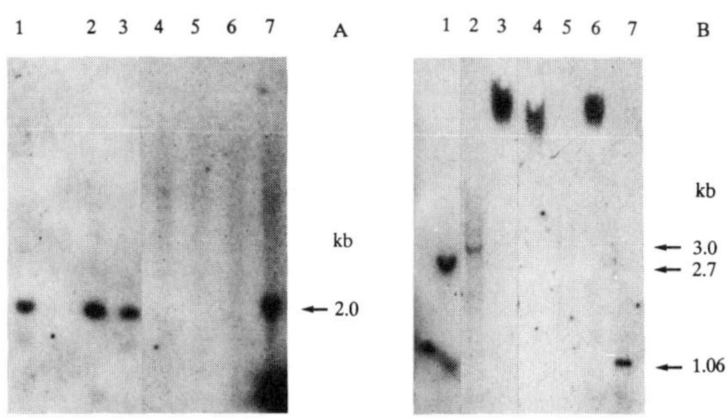

Fig.9. Transfer of *npt*(II)-DNA and *hpt*-DNA to wheat using the indirect pollen system: Southern blot analysis of first generation plants. A: *npt*(II)-DNA. Cleaved with *Pst*I and probed with ^{32}P *npt*(II)-DNA. Lane 1: DNA from *A.tumefaciens* pBILP 1, the donor used. Lanes 2 and 3: DNA from two transgenic tobacco plants containing the *npt*(II)-gene. Lanes 4 - 7: DNA from experiment wheat plants. Lane 7 represents a transgenic wheat plant (Original M. Iser, cf. Iser and Hess, unpublished results).

B: *hpt*-DNA. Probing with ^{32}P *hpt*-DNA. Lanes 1 - 3: DNA from a transgenic wheat plant, cleaved with *Bam*H1 and *Eco*RI (lane 1), with *Bam*HI (lane 2), and uncleaved (lane 3). Lanes 4 + 5: DNA from control wheat plants, uncleaved (lane 4), and cleaved with *Bam*HI (lane 5). Lanes 6 + 7: DNA from a transgenic tobacco plant containing the *hpt*-gene, uncelaved (lane 6), and cleaved with *Bam*H1 (lane 7). (Original S. Stegmaier, cf. Stegmaier and Hess, unpublished results).

3.2.2. Full NPT(II)-Activity from a Transgenic Wheat Plant

An irreversible inactivation by DNA rearrangements or methylation would not allow to make use of gene transfer. However, there is evidence that genes transferred to wheat must not necessarily be repressed. A transgenic wheat plant, which had been Southern positive for npt(II)-DNA, showed no NPT(II)-activity during screening for kanamycin resistance. In a later stage of development it exhibited a NPT(II)-activity as high as transgenic tobacco (Fig. 10.B, Schmid and Hess, 1991).

First these results show a transient inactivation of the npt(II)-gene, probably by methylation. Wheat is well known for its high DNA methylating capacity (Balzer, 1990). Secondly and more important: a reactivation of transferred and repressed genes could be expected in other cases, too.

4. Advantages of the Indirect Pollen Mediated Gene Transfer

In 1990 Vasil et al. succeeded in protoplast regeneration of wheat. In the publication there is no remark concerning the fertility of the regenerants. If they should be sterile, further intensive work would be necessary to get regenerants which are at least partly fertile. Thereafter, genes will be transferred into wheat protoplasts, and transgenic wheat plants will be regenerated.

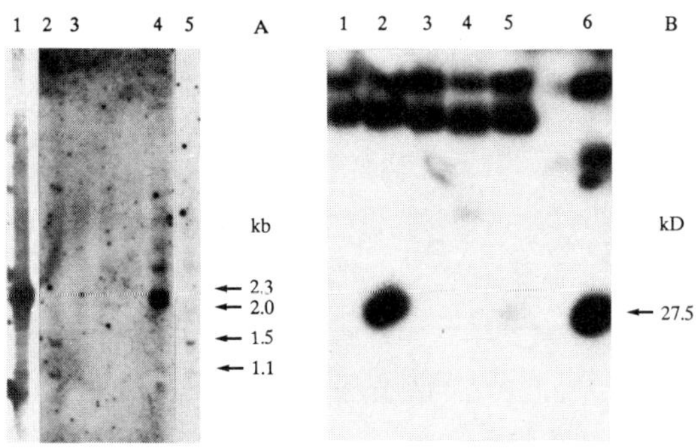

Fig.10.A Southern blot analysis of DNA restricted with *Pst*I and *Hind*III probed with stilbene synthase. Lane 1: Donor *Agrobacterium* DNA; Lanes 2+5: Transgenic wheat experiment plants; Lane 3: Wheat control plant; Lane 4: Tobacco positive control. **B** NPT(II)- assay of wheat and tobacco. Lanes 1 - 3: Wheat experiment plants. Lanes 4 and 5: Wheat control plants. Lane 6: Positive tobacco control. One wheat plant (lane 2) shows high NPT(II)-activity. (Original A. Schmid, cf. Schmid and Hess, 1991).

Transgenic maize plants derived from protoplasts (Rhodes et al., 1988; Donn et al., 1990), or from cell cultures which were microbombarded (Gordon-Kamm et al., 1990), showed morphological abnormalities and/or reduced fertility. With high probability the same will be found with protoplast-derived transgenic wheat. In contrast, all our pollen-derived wheat transformants were morphologically normal and showed a fertility as high as in controls.

These and other advantages of pollen mediated gene transfer and especially of its indirect version are shown in Table 1. Above all, the possibility for obtaining transgenic plants without in vitro-regeneration, and therefore the exclusion of somaclonal variation are convincing benefits of the pollen mediated gene transfer.

Table 1. Comparison of pollen based methods with other gene transfer technologies (d = direct, id = indirect). Each minus sign indicates an advantage of the method under question.

Method	Leaf disk	Protoplasts		Ballistic	Pollen	
		d	id		d	id
Sterile gene transfer necessary	+	+	+	+	+	-[a]
Degradation of donor DNA by DNase in the incubation medium	-	+	-	-	+	-
Sterile selection necessary	+	+	+	+	+	+
Somaclonal variation possible	+	+	+	+[b]	-	-
Formation of chimaeras possible	+	-[c]	-[c]	+	-	-
Regeneration needed to get plants	+	+	+	+	-	-

[a]Only cultivation of Agrobacterium under sterile conditions before gene transfer.
[b]Damaging of receptor DNA in microbombardements included.
[c]Regeneration from one single protoplast supposed.

Acknowledgements

Mrs. R. Frei and U. Glück-Behrens are thanked for skilful technical assistance. The work was supported by the Vater und Sohn Eiselen Stiftung, Ulm, and by the Bundesministerium für Forschung und Technologie, Bonn.

References

Balzer, H.-J. DNA-Methylierung in Weizen. Prien am Chiemsee: Intemann; 1990.

Donn, G.; Nilges, M.; Morocz, S. Stable transformation of maize with a chimaeric, modified phosphinothricin-acetyltransferase gene from Streptomyces viridochromogenes. Abstr. VIIth Int. Congress Plant Tissue and Cell Culture, Amsterdam. 1990: p. 53.

Gordon-Kamm, W.; Spencer, T.; Mangano, M.; Adams, T.; Daines, R.; Start, W.; O'Brien, J.; Chambers, C.; Adams, W.; Willets, N.; Rice, T.; Mackey, C.; Krueger, R.; Kausch, A.; Lemaux, P. Transformation of maize cells and regeneration of fertile transgenic plants. Plant Cell 2: 603-618; 1990

Hess, D. Neue Methoden der genetischen Rekombination. Biol. Rdsch. 12: 297-311; 1974.
Hess, D. Genetic effects in Petunia hybrida induced by pollination with pollen treated with lac transducing phages. Z. Pflanzenphysiol. 90: 119-132; 1978
Hess, D. Genetic effects in Petunia hybrida induced by pollination with pollen treated with gal transducing phages. Z. Pflanzenphysiol. 93: 429-436; 1979
Hess, D. The pollen system of gene transfer. In: Horn, W.; Jensen, C. J.; Odenbach, W.; Schieder, O. eds. Genetic Manipulation and Plant Breeding. Berlin: De Gruyter; 1986: p. 803-811.
Hess, D. Pollen based techniques in genetic manipulation. Int. Rev. Cytol. 107: 367-395; 1987
Hess, D. Pflanzenphysiologie. 8th ed. Stuttgart: Ulmer; 1988.
Hess, D. Gene technology and plant breeding. In: Sinha, S. K.; Sane, P. V.; Bhargava, S. C.; Agrawal, P. K. eds.: Proc. Int. Congress Plant Physiol., New Delhi 1990: p. 34-59.
Hess, D.; Dressler, K. Bacterial transferase activity expressed in Petunia progenies. J. Plant Physiol. 116: 61-272; 1984.
Hess, D.; Dressler, K. Tumor transformation of Petunia hybrida via pollen co-cultured with Agrobacterium tumefaciens. Bot. Acta 102: 202-207; 1989.
Hess, D.; Kiefer, S. Induction of bacterial nitrogenase activity in in vitro associations: a comparison of the inducing capabilities of Triticum aestivum and Sorghum nigricans. Z. Pflanzenphysiol. 101: 15-24; 1981.
Hess, D.; Lörz, H.; Weissert, E.-M. Uptake of bacterial DNA into swelling and germinating pollen grains of Petunia hybrida and Nicotiana tabacum. Z. Pflanzenphysiol. 74: 52-63; 1974.
Hess, D.; Dressler, K.; Konle, S. Gene transfer in higher plants using pollen as vectors: Bacterial transferase activity expressed in Petunia progenies. In: Chapman, G. P.; Mantell,S. H.; Daniels, R. W. eds. Experimental Manipulation of Ovule Tissues. London: Longman; 1986 p. 224-239.
Hess, D.; Dressler, K.; Nimmrichter, R. Transformation experiments by pipetting Agrobacterium into the spikelets of wheat (Triticum aestivum L.). Plant Sci. 72: 233-244; 1990.
Ren, Z. I.; Lelley, T. Hybrid necrosis in Triticale and the expression of necrosis genes in allopolyploids. Theor. Appl. Genet. 77: 742-748; 1989.
Rhodes, C.; Pierce, D; Mettler, I.; Mascarenhas, D.; Detmer, J. Genetically transformed maize plants from protoplasts. Science 240: 204-207; 1988.
Schmid, A.; Hess, D. Towards transgenic fungal resistant wheat: Transfer of stilbene synthase DNA to wheat using the indirect pollen system. Int. Conf. Genetic Engineering and Biotechnology Abstracts. Kathmandu: Nepal Biotechnol. Ass.; p. 36; 1991.
Stachel, S. E.; Messens, E.; Van Montagu, M.; Zambryski, P. Identification of the signal molecules produced by wounded plant cells that activate T-DNA transfer in Agrobacterium tumefaciens. Nature 318: 624-629; 1985.
Süssmuth, J.; Dressler, K.; Hess, D. Agrobacterium-mediated transfer of the GUS gene into pollen of Petunia. Bot. Acta 104: 72-76; 1991.
Van der Westhuizen, A.; Gliemeroth, A.; Wenzel, W.; Hess, D. Isolation and partial purification of an extracellular nuclease from pollen of Petunia hybrida. J. Plant Physiol. 131: 373-384; 1987.
Van Lijsebettens, M.; Inze, D.; Schell, J.; Van Montagu, M. Transformed cell clones as a tool to study T-DNA integration mediated by Agrobacterium tumefaciens. J. Mol. Biol. 188: 129-145; 1986.

Vasil, V.; Redway, F.; Vasil, I. K. Regeneration of plants from embryogenic suspension culture protoplasts of wheat (Triticum aestivum L.). Bio/Technol. 8: 429-434; 1990.

Wagner, G.; Hess, D. In vitro-Befruchtungen bei Petunia hybrida. Z. Pflanzenphysiol. 69: 262-269; 1973.

Zerback, R.; Bokel, M.; Geiger, H.; Hess, D. A kaempferol-3-glucosylgalactoside and further flavonoids from pollen of Petunia hybrida. Phytochemistry 28: 897-899; 1989a.

Zerback, R., Dressler, K.; Hess, D. Flavonoid compounds from pollen and stigma of Petunia hybrida inducers of the vir region of the Agrobacterium tumefaciens Ti plasmid. Plant Sci. 62: 83-91; 1989b

43
2n Gametes for True Seed Production (TPS) in Potato: Basic and Applied Aspects*

Luigi Frusciante, Amalia Barone, Donatella Consoli, Clara Conicella, and Luigi M. Monti

2n gametes occur spontaneously in Solanum and they are supposed to play a key role in the origin of polyploid species within the genus.

The genetics and cytology of 2n gamete formation have been extensively investigated, particularly for 2n pollen (Ramanna, 1974; Mok and Peloquin, 1975). Recessive genes with partial penetrance, pc1, pc2, ps determine 2n pollen phenotype through the following abnormalities in cytoskeleton behavior during microsporogenesis: premature cytokinesis after telophase I or parallel orientation of the spindles at anaphase II (Mok and Peloquin, 1975). In 2n egg formation, omission of the second meiotic division is the prevalent mechanism (Iwanaga and Peloquin, 1979; Stelly and Peloquin, 1986; Werner and Peloquin, 1987).

The genetic constitution of 2n gametes is dependent upon cytological mechanisms; if they carry sister or non-sister chromatids, they are potentially homozygous or heterozygous respectively and they are considered genetically equivalent to the products of a second division restitution (SDR) or a first division restitution (FDR).

2n gametes are widely used in potato breeding and their use is also proposed in the True Potato Seed (TPS) program aimed to reproduce potatoes from true seeds instead of conventional tuber seeds (Peloquin, 1983).

The main steps of a TPS program are:

constitution of TPS-functional parental lines through immission of wild germplasm

choice of appropriate breeding methods that maximize uniformity and heterozygosity.

Wild germplasm is utilized in potato breeding via 2n gametes when compatible species have different ploidy levels or via protoplast fusion when the species are sexually incompatible (Ehlenfeldt and Helgeson, 1987).

Within the framework of developing breeding schemes for TPS we are looking for mutants which modify meiosis so that 2n gamete production is high and genetic constitution can be predicted.

Selection of clones producing 2n gametes

The selection for 2n gametes was performed on hybrid diploid clones obtained by crossing wild Solanum species (chacoense, tarijense and phureja) with haploids of S. tuberosum (Table 1). All the clones that exhibit high frequency of 2n gametes from the

---

* Contribution no. 79 from Research Center for Vegetable Breeding

male or female side produce rare or null 2n gametes from the other side.

The clones with high frequency of 2n gametes were also evaluated for traits that are important in TPS production. Flower duration and flower intensity were taken into account in order to obtain a larger number of flowers; high frequency of 2n gametes together with high fruit set guaranteed high number of seeds per berry and per plant. The yield data, averages of several experimental trials carried out in the last five years in two locations of Southern Italy, were promising.

UP88P5 and T710, producing 2n pollen and 2n egg respectively at high frequency, were characterized by positive horticultural traits and gave high yields both in transplanting and seedling tuber techniques.

Table 1. Frequency of 2n gametes in hybrid diploid clones in Solanum spp.

Clone	Origin	Frequency of 2n pollen	Frequency of 2n egg
T710	S.Tub. x S. chc.	0	45-60
UP88C80	S.Tub. x S. chc.	1-5	1-2
UP88C75	S.Tub x S. chc.	1-5	14-29
UP88T101	S.Tub. x S. tar.	5-15	0
UP88T104	S.Tub. x S. tar.	1-10	0
UP88T164	S.Tub x S. tar.	1-10	14-25
UP88T163	S.Tub x S. tar.	1-2	0
I (W5295.7)	S.Tub. x S. phu.	35-60	0
UP88P5	S.Tub. x S. phu.	40-86	0
UP88P7	S.Tub. x S. phu.	70-90	0

T710 and I were selected by Prof. Peloquin (Madison, U.S.A.).

Cytological investigation

The mechanisms leading to 2n gametes were identified cytologically in micro- and macrosporogenesis.

In the clones producing 2n pollen, parallel spindles at anaphase II were evidenced (Fig. 1); this abnormality followed by a cleavage furrow at telophase II results in two 2n microspores carrying non-sister chromatids which are genetically equivalent to the products of a first division restitution.

In two clones producing 2n eggs, omission of meiosis II (Fig. 2), equivalent to SDR mechanisms, was observed (Table 2). In T710 the frequency of omission of meiosis II was lower than the frequency of dyads. In this clone additional meiotic abnormalities were found: abnormal orientation of the spindle axis at metaphase I (Fig. 3) and two nuclei at metaphase II in the chalazal cell (Fig. 4), the latter considered a consequence of the previous phenomenon. Such abnormalities might give rise to 2n egg formation by FDR mechanisms, and consequently T710 clone has a SDR-FDR mixture in the formation of 2n eggs (Conicella et al., 1991). Previous studies of Werner and Peloquin (1987) on clones producing 2n eggs including T710 revealed only the omission of meiosis II.

Table 2. Frequencies of meiotic abnormalities found in megasporogenesis of two diploid clones

Material	Metaphase I		2nd Meiotic division		Sporad stage	
	MMCs	abnormal spindle axis	MMCs	omission of 2nd division	Sporads	Dyads
	(no.)	(%)	(no.)	(%)	(no.)	(%)
T710	31	16.7	43	38.1	72	55.4
UP88C75	152	0	53	69.8	382	35.3

RFLPs identification

The use of RFLP markers is suggested to estimate the percentage of heterozygosity transmitted by 2n gametes. This determines which mechanism (SDR or FDR) is involved in 2n gamete production.

The segregation of 2/3 RFLP markers for each chromosome can also help to localize the centromere position on the RFLP map by means of recombination frequency. Moreover the selection of tetraploid plants with four different alleles in a locus could be achieved by RFLP markers.

For RFLP markers to be informative, they have to show polymorphisms between the parents and to reveal both alleles of heterozygous loci.

The screening of such markers is being performed on four diploid clones producing 2n gametes (T710, UP88, T710, I, UP88P5) by using as probes 54 cDNA or genomic potato clones from the RFLP potato map (Gebhardt et al., 1989). Up till now 4 out of 16 cDNA clones were informative and they are being tested on segregating populations from 2x x 2X crosses among the diploid clones T710, UP88C75, I and UP88P5.

Use of 2n gametes for TPS

Different breeding schemes were applied for TPS production (Table 3). Cultivars or tetraploid clones were crossed with the selected diploid clones forming 2n gametes (unilateral sexual polyploidization) and diploid clones were intercrossed (bilateral sexual polyploidization).

The 4x x 2x crosses gave excellent yields. The tetraploids obtained by means of these crosses displayed higher values in vigour and tuber uniformity than those coming from 4x x 4x crosses or open pollination (OP). Moreover the hybrid tetraploids showed positive characteristics crucial for TPS, such as seedling vigour and response to transplant failure.

The major limitation in these breeding schemes is the reduced number of available diploid parental lines forming 2n gametes.

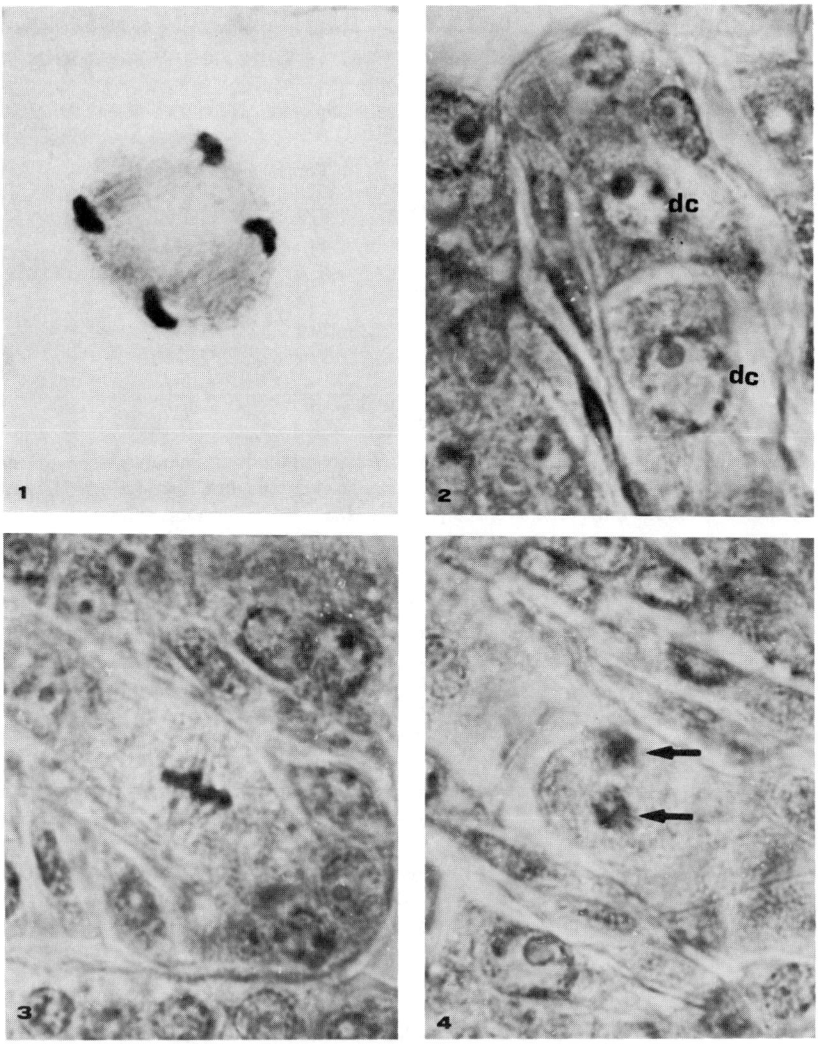

Fig. 1. Microsporogenesis in diploid clone UP88P5. Anaphase II showing parallel spindles (2000x).

Figs. 2-4. Macrosporogenesis in diploid clone T710. In fig. 2 two daughter cells (dc) after meiosis I at a resting stage (1800x). In fig. 3 spindle axis with abnormal orientation in metaphase I (1800x). In fig. 4 two nuclei (arrows) in one chalazal cell (1800x).

Table 3. Evaluation of plant vigour, tuber uniformity and yield in families coming from different breeding schemes (average of 4 years).

Type of cross	Plant* vigour	Tuber* uniformity	Yield (t/ha)	
			Transplant	Seedling
4x x 2x	2.41	2.48	34.1	52.7
2x x 4x	2.31	2.28	32.0	37.2
4x x 4x	2.20	2.13	32.2	43.6
2x x 2x	2.21	2.23	27.1	34.3
DT x 4x	1.91	2.10	29.5	42.5
OP	1.05	1.43	19.2	32.8*

1 min; 3 max

The use of 2n gametes in alternative breeding methods is valuable not only for developing TPS progenies with positive horticultural traits but also for transferring resistance to abiotic and biotic stresses (Iwanaga et al., 1989; See Watanabe in the text).

REFERENCES

1. Ramanna, M.S. The origin of unreduced microspores due to aberrant cytokinesis in the meiocytes of potato and its genetic significance. Euphytica 23: 20-30; 1974.

2. Mok, S.W.S.; Peloquin, S.J. The inheritance of three mechanisms of diplandroid (2n pollen) formation in diploid potatoes. Heredity 35: 295-302; 1975.

3. Iwanaga, M.; Peloquin, S.J. Synaptic mutant affecting only megasporogenesis in potatoes. J. Hered. 70: 385-389; 1979.

4. Stelly, D.M.; Peloquin, S.J. Diploid female gametophyte formation in 24-chromosome potatoes: genetic evidence for the prevalence of the second meiotic division restitution mode. Can. J. Genet. Cytol. 28: 101-108; 1986.

5. Werner, J.E.; Peloquin, S.J. Frequency and mechanisms of 2n egg formation in haploid Tuberosum-wild species F1 hybrids. Am. Potato J. 64: 641-654; 1987.

6. Peloquin, S.J. New approaches to breeding for the potato for the year 2000. In: Research for potato in the year 2000. International Potato Center (CIP), Lima, Peru. 1983: 32-34.

7. Ehlenfeldt, M.K.; Helgeson, J.P. Fertility of somatic hybrids from protoplast fusion of Solanum brevidens and S. tuberosum. Theor. Appl. Genet. 73: 395-492; 1987.

8. Conicella, C.; Barone, A.; Del Giudice, A.; Frusciante, L.; Monti, L.M. Cytological evidences of SDR-FDR mixture in the formation of 2n eggs in a potato diploid clone. Theor. Appl. Genet. 81: 59-63; 1991.

9. Gebhardt, C.; Ritter, E.; Debeuer T.; Schachtschabel, V.; Walkemuer, B.; Uhrig, H.; Salamini, F. RFLP analysis and linkage mapping in Solanum tuberosum. Theor. Appl. Genet. 78: 65-75; 1989.

10. Iwanaga, M.; Jatala, P.; Ortiz, R.; Guevara, E. Use of FDR 2n pollen to transfer resistance to root-knot nematodes into cultivated 4x potato. J. Amer. Soc. Hort. Sci 114 (6): 1008-1013; 1989.

44

Induction of Embryogenesis in Isolated Pollen Cultures of Tobacco

Oscar Vicente, Dolores Garrido, Viktor Zarsky, Norbert Eller, Ludmila Rihova, Maria Berenyi, Jaroslav Tupy, and Erwin Heberle-Bors

In vitro cultures of isolated tobacco pollen

The potential of immature pollen to deviate from its normal developmental pathway, giving rise to haploid embryos and plants, directly and asexually, has been known for a long time and is used in plant breeding for dihaploid production through anther culture (Heberle-Bors, 1989, and references therein). For basic and applied research, it would be very interesting to find out the molecular mechanisms underlying the transformation of the male gametophyte into an embryogenic cell. Anther culture, however, is not a suitable experimental system for these studies. The anther contains a heterogeneous cell population (a small percentage of embryogenic pollen, from which the embryos originate, together with normal gametophytic pollen and sporophytic cells from the anther tissues) which does not allow the use of biochemical and molecular techniques to study the process of induction of pollen embryogenesis. In addition, the anther wall can contribute substances, both stimulatory and inhibitory, affecting pollen embryo development in a way difficult to control.

In the last years we have established a highly efficient in vitro culture system of isolated tobacco pollen, that allows the strict control of pollen maturation and embryogenesis, and of the switch from gametophytic towards sporophytic development (Benito Moreno et al., 1988b, Garrido et al., 1991a). Immature pollen grains are isolated at the mid-binucleate stage, and cultured for 6-8 days in a medium without nutrients (MS macro minerals in 0.4 M manitol as osmoticum, pH 7.0). After this starvation treatment the young pollen grains develop into a characteristic type of cells, embryogenic pollen or P-grains, that can be distinguished cytologically from normal gametophytic pollen of any developmental stage. Transfer to a sugar-containing medium (0.25 M sucrose supplemented with macro and micro salts and Fe-EDTA) induces P-grains to divide and form embryos, that develop further into haploid plants. On the other hand, the same mid-binucleate pollen (as well as pollen isolated at earlier stages, including microspores), when cultured directly in a nutritionally rich medium, continues its normal developmental program, giving rise to mature, fully functional pollen grains (Benito Moreno et al., 1988a). The most important developmental decision during the life cycle of higher plants, the alternation of generations, is thus controlled in vitro by the nutritional status of the immature pollen grain.

These in vitro cultures of isolated tobacco pollen provide a unique tool for the study of different aspects of plant reproductive development, including the process of induction of pollen embryogenesis.

Starvation as the trigger of tobacco pollen embryogenesis

Since formation of P-grains is a prerequisite for subsequent pollen embryogenesis, and since they are formed from mid-binucleate pollen cultured in a sugar-free medium while culture in a sugar-containing medium leads to maturation, sugar starvation seems to be the only extracellular signal required to switch immature pollen, isolated at this specific stage, from gametophytic towards sporophytic development.

Old ultrastructural studies of embryogenic tobacco pollen induced in anther culture have shown the formation of large vacuole-like structures in the vegetative cell, regression of the cytoplasm and amyloplasts, decrease in polysome number and maintained integrity of the mitochondria (reviewed by Heberle-Bors, 1989). These are all characteristics of a starving cell that survives by converting stored nutrients into energy, suggesting that also in anther culture starvation is the trigger of pollen embryogenesis. The fact that embryogenic pollen formed in isolated pollen cultures under starvation conditions shows similar ultrastructural features (Garrido et al., 1991b) supports this idea. In this case, induction of embryogenic pollen is an almost quantitative process, whereas in anther culture only a small fraction of pollen grains is induced, in a population competing for a limited amount of nutrients (since the anther wall constitutes a barrier for the transfer of the components of the medium to the pollen).

Gene expression during induction of pollen embryogenesis

Induction of pollen embryogenesis is probably a quite complex phenomenon, that includes dedifferentiation of the male gametophyte, acquisition of embryogenic competence and derepression of cell division activity. These processes most likely involve a drastic change in the gene expression program of immature pollen, and starvation seems to be responsible for this reprogramming. Therefore, as a first step in our studies on the molecular mechanisms of induction of pollen embryogenesis, we have investigated changes in gene expression taking place in immature pollen cultured under starvation conditions.

In preliminary experiments we have observed a general decrease in total levels of RNA and proteins during starvation, as well as in their rate of synthesis, measured as incorporation of radioactivity in TCA-precipitable material after labelling with either (3H)Uridine or (35S)Met, in agreement with the cytological features of embryogenic pollen mentioned above.

In addition to this general degradative process, we assumed, as working hypothesis, that starvation activates the transcription of specific genes, whose products are required for the induction of cell division after transfer to the sugar-containing medium. Alternatively, starvation could simply provide the intracellular signal(s) necessary for activation of expression of those genes, but their actual transcription would take place only after transfer to the rich medium.

Inhibitor experiments gave preliminary evidence for the first hypothesis, i.e. that de novo transcription is required during the induction process. Actinomycin D was added to the cultures at different days and for different times during starvation, at concentrations previously established to inhibit RNA synthesis but not mRNA translation in pollen. These treatments did not affect pollen viability, nor, apparently, formation of embryogenic pollen since microscopic observations did not reveal cytological differences with P-grains formed in control cultures. However, cell division after transfer to the embryogenic medium was completely inhibited when actinomycin D had been included in the starvation medium.

Attempts to detect de novo synthesis of specific proteins, products of putative starvation-induced genes, by comparative analysis of silver-stained 2d gels of pollen extracts prepared at different times during the starvation treatment, did not show new protein spots. Similarly, we used 2d gel electrophoresis and autoradiography to analyse the patterns of in vivo (in pollen cultures) synthesised proteins, after labelling with pulses of (35S)Met. Labelling and preparation of pollen extracts were performed every day of the starvation treatment. Although the protein patterns obtained were different, in the sense that some proteins synthesised at the beginning of the culture were not detected towards the end of the starvation treatment, the synthesis of new, starvation-induced proteins was never observed. These results indicate that no new proteins are expressed (at the level of detection of these methods) during the starvation treatment.

In order to study gene expression at the mRNA level, total RNA was isolated from mid-binucleate and embryogenic pollen, and populations of mRNAs in both samples were analysed by 2d gel electrophoresis of their in vitro translated products (in rabbit reticulocyte lysates). In this case, specific mRNAs, not present in mid-binucleate pollen, were detected in embryogenic pollen. The differences included two major acidic proteins of 20 and 26 kDa that reproducibly appeared in all the experiments, carried out with RNA prepared from different cultures. These spots should correspond to two specific, abundant messengers that are induced during starvation and that accumulate in embryogenic pollen. Since these proteins were not detected after in vivo labelling, these results also suggest that the corresponding mRNAs (and probably other minor, starvation-induced mRNAs) are not translated in vivo.

Storage of preformed mRNAs as ribonucleoprotein particles not accessible to the polysomes is a common phenomenon in developmental systems arrested in a quiescent stage, before a period of active cell division. One well-known example in animals is the presence of maternal mRNAs in unfertilized eggs. Fertilization triggers the activation and translation of maternal messengers, whose protein products are used during the early stages of embryogenesis (Grainger and Winkler, 1987). Similarly, these starvation-induced mRNAs could be activated by transfer of embryogenic pollen to the sugar-containing medium. Experiments are in progress to confirm this hypothesis. In any case, these results indicate that de novo transcription of specific genes, required for subsequent embryogenesis, takes place already during the starvation period. Formation of embryogenic pollen, simply defined by its cytological characteristics, is a necessary, but not sufficient prerequisite for formation of pollen embryos.

Cell cycle progression during gametophytic and sporophytic pollen development

Pollen development is a highly ordered and often synchronous sequence of cell divisions, giving rise to the mature male gametophyte, one of the few examples of terminal differentiation in plants. Meiosis of the pollen mother cells is followed by microsporogenesis, and the unequal first pollen mitosis leads to the formation of the small generative and the large vegetative cell. The vegetative cell does not divide again, although it is not clear at which phase of the cell cycle it is arrested. The generative cell will divide once more, to form the two sperm cells, either already during pollen maturation or in the pollen tube, after germination.

During pollen embryogenesis this differentiation program breaks down. In tobacco, embryos develop from the (nearly) terminally differentiated and cell cycle arrested vegetative cell. Reactivation of cell cycle in the vegetative cell must be one of the crucial events in induction of pollen embryogenesis. To understand this process, it is therefore important to know at which phase of the cell cycle the vegetative cell is arrested during gametophytic development and whether the release of this arrest is a consequence of starvation or takes place after the already induced embryogenic pollen has been transferred to the rich medium.

Availability of the in vitro cultures of isolated pollen allowed us to use the technique of autoradiography after (3H)thymidine labelling, in addition to classical cytophotometry of Feulgen-stained nuclei, to follow DNA replication and study cell cycle progression in both pollen cells during normal pollen development and during starvation and early embryogenesis.

During gametophytic pollen development and soon after completion of first pollen mitosis, the generative nucleus enters S-phase, which is still in progress in mid-binucleate pollen. In mature pollen, DNA replication has finished and the generative cell is arrested in the G2 phase of the cell cycle, with a 2C level of DNA. The vegetative nucleus does not incorporate (3H)thymidine and remains with a 1C level of DNA throughout pollen development, that is, arrested in G1. While the results on the generative cell confirm what is known already, the results on the vegetative cell should end a controversy on the cell cycle state of this cell. Originally, the vegetative cell in tobacco was considered to be arrested in G2 (D'Amato et al., 1965) but later studies failed to detect DNA replication (Aruga et al., 1982).

When mid-binucleate pollen is cultured in the starvation medium, ongoing DNA replication in the generative nucleus is completed within the first two days of culture, and in embryogenic pollen at the end of the starvation period, the generative nucleus is in the G2 phase of the cell cycle, with a 2C level of DNA. In the vegetative nucleus, the first observable change during the starvation treatment is condensation of chromatin and a strong regression of the nucleolus. After this, in a fraction of pollen grains, the vegetative nucleus enters S-phase and replicates its DNA. Accordingly, cytophotometric analysis of embryogenic pollen showed two major populations of pollen, with 1C and 2C levels of DNA in the vegetative cell and pollen with intermediate DNA amounts still in S-phase. In another fraction of cells, DNA replication takes place only after transfer to the rich medium. Then the vegetative cell enters mitosis and continues dividing to form eventually an embryo.

Addition of hydroxyurea to the starvation medium inhibits DNA replication in the vegetative nucleus, but does not stop formation of embryos after transfer to hydroxyurea-free embryogenesis medium. The inhibitor simply postpones and synchronizes DNA replication in the cultured pollen.

These results indicate that one of the initial events in the process of induction of pollen embryogenesis under starvation conditions is the derepression of the cell cycle in the vegetative cell, which is arrested in G1 during gametophytic development. Release of the G1 arrest can be clearly separated from transition through the G1/S boundary, since actual DNA replication during starvation is not essential for embryogenic induction. We can conclude that starvation induces the vegetative cell to pass through a control point in the cell cycle that may be analogous to control points in late G1 defined in yeast and in mammalian tissue culture cells, that have been termed START and Restriction Point, respectively (Nasmyth, 1990; Pardee, 1989).

Conclusions and perspectives

Although the mechanisms of induction of pollen embryogenesis are still unknown, the experiments described here, making use of our in vitro cultures of isolated tobacco pollen, provide some insights into this phenomenon, as well as the basis for further experimental work. Induction of embryogenic pollen appears to be the consequence of different processes, all triggered by the starvation treatment. First, degradation of the cytoplasm of the immature pollen grains, isolated at a specific developmental stage (mid-binucleate pollen), probably necessary to inhibit and eliminate the products of gametophytic gene expression. Second, derepression of the cell cycle in the vegetative cell, arrested in G1 during normal pollen development. Third, activation of transcription of specific genes, required for subsequent cell division and embryogenesis.

Differential screening of cDNA libraries prepared from mid-binucleate and embryogenic pollen is already in progress, to isolate and characterise these starvation-induced genes. Expression studies of the plant cdc2 gene that has recently been isolated in our laboratory (Hirt et al., 1991) should give interesting information on the mechanisms involved in the derepression of the vegetative cell cycle. The product of this gene, the p34 protein kinase, is the major regulator of cell cycle in eukaryotic cells, controlling entry into both, S-phase and mitosis (see Lewin, 1990, for a review). Finally, for the study of the mechanisms mediating the effect of starvation on cell cycle and gene expression, comparison with other systems, in which a starvation treatment is responsible for similar developmental switches (e.g. in yeast), may be an interesting approach. Yeast genes involved in the signal transduction pathway of starvation can be used as heterologous probes to isolate homologous clones from tobacco cDNA libraries and then their expression during pollen development can be investigated. Since the products of many of these genes (as well as of cdc2 and other cell cycle genes) are protein kinases, it is likely that protein phosphorylation is involved in induction of pollen embryogenesis. The fact that we have observed changes in protein kinase activities in pollen extracts during starvation (unpublished results) is in agreement with this hypothesis. Biochemical characterisation of protein kinases during normal and embryogenic pollen development is under way, as a parallel approach that could also give valuable insights into the mechanisms of induction of pollen embryogenesis.

References

Aruga, K.; Yamamoto, K.; Nakajima, T. Nucleic acid synthesis required for embryogenic induction in pollen grains of Nicotiana tabacum L. In: Fujiwara, A., ed. Plant tissue culture. Tokyo: Japanese Association for Plant Tissue Culture; 1982:p.537-538.

Benito Moreno, R. M.; Macke, F.; Alwen, A.; Heberle-Bors, E. In situ seed production after pollination with in vitro matured, isolated pollen. Planta 176:145-148; 1988a.

Benito Moreno, R. M.; Macke, F.; Hauser, M.-T.; Alwen, A.; Heberle-Bors, E. Sporophytes and male gametophytes from in vitro cultured, immature tobacco pollen. In: Cresti, M.; Gori, P.; Pacini, E., eds. Sexual reproduction in higher plants. New York, Berlin, Heidelberg, Tokyo: Springer-Verlag; 1988b:p.137-142.

D'Amato, F.; Devreux, M.; Scarascia Mugnozza, G. T. The DNA content of the nuclei of the pollen grains in tobacco and barley. Caryologia 18:377-382; 1965.

Garrido, D.; Charvat, B.; Benito Moreno, R.M.; Alwen, A.; Vicente, O.; Heberle-Bors, E. Pollen culture for haploid formation in tobacco. In: Negrutiu, I., ed. Methods in plant molecular and cell biology. Klosterberg: Birkhäuser; 1991a, in press.

Garrido, D.; Vicente, O.; Heberle-Bors, E.; Rodríguez-García, M. I. Ultrastructural changes related to pollen embryogenesis. International Symposium on Angiosperm pollen and ovules. Basic and applied aspects. 1991b June 23-27. Como, Italy. Abstract.

Grainger, J. L.; Winkler, M. M. Fertilization triggers unmasking of maternal mRNAs in sea urchin eggs. Mol. Cell. Biol. 7:3947-3954; 1987.

Heberle-Bors, E. Isolated pollen culture in tobacco: plant reproductive development in a nutshell. Sex. Plant Reprod. 2:1-10; 1989.

Hirt, H.; Páy, A.; Györgyey, J.; Bakó, L.; Németh, K.; Bögre, L.; Schweyen, R. J.; Heberle-Bors, E.; Dudits, D. Complementation of a yeast cell cycle mutant by an alfalfa cDNA encoding a protein kinase homologous to p34cdc2. Proc. Natl. Acad. Sci. USA 88:1636-1640; 1991.

Lewin, B. Driving the cell cycle: M phase kinase, its partners, and substrates. Cell 61:743-752; 1990.

Nasmyth, K. A. FAR-reaching discoveries about the regulation of START. Cell 63:1117-1120; 1990.

Pardee, A. B. G1 events and regulation of cell proliferation. Science 246:603-608; 1989.

45

Genotype and Ploidy Effects on Maize Pollen and In Vitro Germination Characteristics

P.L. Pfahler, R.D. Barnett, and E.B. Blazey

In a number of species, marked in vivo transmission differences between 1x pollen grains from diploids and 2x pollen grains from autotetraploids have been reported (Zeven, 1980). In maize when 1x pollen grains compete with 2x pollen grains on either 2x or 4x silks, fertilization is almost exclusively by 1x pollen grains (Randolph, 1935). Many factors including rate of germination after silk attachment, rate of tube penetration into the silk and tube growth rate through the silk have been implicated in this differential transmission phenomenon (Green, 1946; Sarkar and Paria, 1980; Randolph, 1935).

Little information is available regarding ploidy effects on in vitro germination. This study was designed to determine the effect of genotype and ploidy level on various pollen characteristics (diameter, percent pollen fertility) and in vitro germination capacity (percent germination, tube length at various time periods after inoculation).

Materials and methods

Pollen grains from six populations were obtained on each of two collection dates using the detached tassel and screening method described earlier (Pfahler, 1965, 1968). The populations consisted of three single cross hybrids or genotypes (N6xW23=Nx3, W22xW23=Wx3, R75xW23=Rx3) with each genotype represented by two ploidy levels (2x, 4x).

Immediately after collection, pollen samples from each of the six populations at each collection date were placed in a killing and fixing solution (72 parts H_2O, 5 parts formaldehyde, 3 parts glacial acetic acid, 20 parts glycerin) and stored at 2°C for diameter and percent fertility measurements using microprojection equipment. The magnifications used were 400x and 67.5x for the diameter and percent fertility measurements, respectively. For each population at each collection date, the diameter of 50 randomly-selected grains was measured. For each population at each collection date, the percent fertility was determined from six 100-grain randomly-selected pollen samples. For each character, an analysis of variance in the form of a complete factorial with genotype and ploidy level as the main effects was performed (Table 1). Before statistical analysis, the arcsin transformation was applied to the percent fertility data (Snedecor, 1956).

Within 30 minutes after collection on each collection date, pollen grains from each of the six populations were inoculated on an in vitro germination medium [15% sucrose, 0.03% $Ca(NO_3)_2 \cdot 4H_2O$, 0.01% H_3BO_3] supplemented with four levels (0.6, 1.2, 1.8, 2.4%) of agar (bacto-agar). The medium preparation and procedure was given previously

(Pfahler, 1968). At 1, 2 and 3 hours after inoculation, the surface of the medium was flooded with the killing and fixing solution to stop activity and the petri dishes were stored at 2°C until classified using microprojection equipment. The magnifications used were 67.5x and 103.5x for percent germination and tube length, respectively. For each population at each collection date, the tube length of 40 randomly-selected germinated grains was measured for each agar level-hour after inoculation combination. For each character, an analysis of variance in the form of a complete factorial with genotype, ploidy level, agar level and hour after inoculation as the main effects was performed (Table 1). Before statistical analysis, the arcsin transformation was applied to the percent germination data (Snedecor, 1956).

The minimum difference for significance (MDS) values presented in the tables was obtained by means of the revised Duncan's ranges using for P only the maximum number of means to be compared (Harter, 1960).

Results

Diameter. The main effects (genotype, ploidy level) and their interaction were statistically significant (Table 1).

The expected relationship between ploidy levels was found with the 4x mean about 1.25 times (4x mean/2x mean) the 2x mean (Table 2). However, as indicated by the statistically significant genotype x ploidy level interaction (Table 1), this relationship was influenced by genotype.

Percent Fertility. The main effects (genotype, ploidy level) and their interaction were statistically significant (Table 1).

In each genotype, the 4x mean was considerably lower than the corresponding 2x mean (Table 2). However, as indicated by the statistically significant genotype x ploidy level interaction (Table 1), substantial differences among genotypes were present. In Nx3, the difference (2x mean-4x mean) was only 2% while in Wx3, this difference was 8% (Table 2).

Percent Germination. All main effects (genotype, ploidy level, agar level, hour after inoculation) and many of their interactions were statistically significant (Table 1).

At each genotype-agar level-hour after inoculation combination, the 4x mean was substantially less than the corresponding 2x mean (Table 2). The statistically significant genotype x ploidy level interaction (Table 1) indicated that genotype influenced this relationship. Over all agar levels and hours after inoculation, the differences (2x mean-4x mean) were 37, 38 and 23% for Nx3, Wx3 and Rx3, respectively (Table 2). Over all genotypes, ploidy levels and hours after inoculation, increasing agar levels reduced germination from 48% at 0.6% agar to 37% germination at 2.4% agar (Table 2). However, the magnitude of the decrease depended on ploidy level. For example, the decrease (0.6 agar mean-2.4 agar mean) was 8% for the 2x forms and 14% for the 4x forms. Differences resulting from hour after inoculation were relatively small but there was an indication that the 4x forms germinated more slowly than the corresponding 2x forms (Table 2). These germination rate differences between the ploidy levels were more pronounced at the higher agar levels.

Tube length. All main effects (genotype, ploidy level, agar level, hour after inoculation) and many of their interactions were statistically significant (Table 1).

Table 1. Mean squares and significance levels from the variance analyses of the various characters
(D = pollen diameter, %F = percent pollen fertility, %G = percent in vitro germination, TL = in vitro tube length) measured. The percent data was transformed (arcsin) before analysis.

Source[a]	df	D	%F	%G	TL
G	2	443.0**	92.2**	9035.7**	275411**
P	1	84520.1**	870.8**	60200.7**	21050987**
GxP	2	187.9**	106.6**	891.1**	485088**
A	3			1255.0**	4382765**
GxA	6			134.5**	55879
PxA	3			305.6**	387683**
GxPxA	6			109.4**	31740
H	2			305.7**	44362634**
GxH	4			23.9*	16309
PxH	2			133.4**	1540563**
GxPxH	4			2.6	63303
AxH	6			34.6**	439865**
GxAxH	12			10.6	32238
PxAxH	6			28.5**	80278*
GxPxAxH	12			7.5	41809
Error	b	26.1	12.8	7.7	31562

*,**F value significant at the 0.05 and 0.01 probability levels, respectively.
[a] The source abbreviations were G = single cross or genotype, P = ploidy level, A = agar level and H = time period or hour after inoculation.
[b] Error degrees of freedom were 594, 66, 504, and 5688 for D, %F, %G, and TL, respectively.

At each genotype-ploidy level-agar level-hour after inoculation combination, the 2x mean was considerably higher than the corresponding 4x mean (Table 2). The statistically significant genotype x ploidy level interaction (Table 1) indicated that genotype influenced this general relationship. In Nx3, the difference (2x mean-4x mean) was 150μ while in Wx3, this difference was 87μ (Table 2). The agar level also influenced tube length with increasing agar levels decreasing tube length. The statistically significant ploidy level x agar level interaction (Table 1) indicated that the two ploidy levels responded differently to increasing agar levels. At 0.6% agar, the difference (2x mean-4x mean) was 147μ while at 2.4% agar, this difference was 91μ (Table 2). The statistically significant ploidy level x hour after inoculation interaction (Table 1) indicated that the 4x forms grew at a slower rate than the 2x forms. For example, at 1 hour after inoculation, the difference (2x mean-4x mean) was 58μ while, at 3 hours after inoculation, this difference was 169μ (Table 2). Over all genotypes and ploidy levels, increasing agar levels also decreased tube growth rate as indicated by the statistically significant agar level x hour after

Table 2. Means of the three single cross hybrids or genotypes (Nx3 = N6xW23, Wx3 = W22xW23, Rx3 = R75xW23) at two ploidy levels (2x, 4x) for the various characters (D = pollen diameter in microns, %F = percent pollen fertility with the transformed mean in parentheses, %G = percent in vitro germination with the transformed mean in parentheses, TL = in vitro tube length) measured. For %G and TL, four agar percents (0.6, 1.2, 1.8, 2.4) were added to the medium with readings taken at three time periods (1, 2, 3 hours) after inoculation.

Character[a]	Agar percent	Time period	Nx3 2x	Nx3 4x	Wx3 2x	Wx3 4x	Rx3 2x	Rx3 4x
D			87.9	110.8	91.1	113.4	88.4	114.4
%F			97(80.9)	95(77.4)	97(82.1)	89(70.4)	96(78.3)	91(72.5)
%G	0.6	1	73(58.4)	39(35.7)	68(55.8)	33(34.7)	35(36.1)	22(27.2)
		2	79(62.7)	40(39.2)	72(58.3)	31(33.4)	42(40.6)	25(29.3)
		3	78(61.9)	42(40.1)	73(58.4)	39(38.3)	43(41.2)	28(31.7)
	1.2	1	65(53.9)	30(33.4)	68(55.3)	26(30.1)	34(35.4)	17(23.9)
		2	71(57.5)	33(35.2)	67(55.2)	32(34.3)	34(35.2)	18(24.0)
		3	66(54.3)	35(36.5)	67(55.0)	34(35.2)	36(36.6)	20(26.2)
	1.8	1	69(56.2)	21(27.3)	62(53.0)	20(25.8)	46(42.8)	14(21.5)
		2	68(55.6)	29(32.8)	62(51.8)	23(28.4)	42(40.4)	17(23.7)
		3	66(54.3)	27(31.3)	60(50.6)	24(28.9)	40(39.1)	17(23.4)
	2.4	1	56(51.4)	21(27.0)	56(48.2)	16(22.7)	45(42.0)	9(16.4)
		2	64(53.5)	26(30.7)	63(52.5)	20(26.2)	46(42.6)	11(19.1)
		3	62(51.9)	25(29.8)	62(52.1)	26(30.4)	44(41.6)	15(21.6)
TL	0.6	1	382	312	374	336	400	306
		2	673	483	634	511	629	474
		3	842	581	819	614	763	576
	1.2	1	393	319	392	336	379	322
		2	705	504	663	479	636	500
		3	876	595	771	640	820	574
	1.8	1	382	300	325	309	352	312
		2	588	446	529	450	557	435
		3	706	519	593	523	622	516
	2.4	1	362	279	303	272	313	255
		2	538	438	485	403	528	409
		3	605	474	539	509	656	468

[a]MDS values among means for each character at the 0.05 and 0.01 probability levels, respectively: D = 1.6 and 2.1; %F (transformed means in parentheses) = 3.3 and 4.3; %G (transformed means in parentheses) = 3.6 and 4.6; and TL = 72 and 92.

inoculation interaction (Table 1). At 0.6% agar, the difference (3 hour after inoculation mean-1 hour after inoculation mean) was 347μ while, at 2.4% agar, this difference was 245μ (Table 2).

Discussion

In this study, both pollen diameter and percent fertility were influenced by ploidy level and genotype. The results indicated that the pollen diameter of the 4x forms was about

1.25 times (or double the volume) of their corresponding 2x forms but genotypic differences were present. Previous studies (Lewis, 1980; Stebbins, 1947; Zeven, 1980) in a number of species have shown that the pollen volume of 4x forms is about twice that of the 2x forms. Genotypic differences in pollen diameter have been reported in maize (Pfahler and Wilcox, 1983). The reduction in pollen fertility found in this study has also been reported in maize (Randolph, 1935). A substantial reduction in fertility immediately after autotetraploidy induction presumably because of tetravalent pairing has been reported (Stebbins, 1947). After a number of generations of selection for seedset in the synthesized autotetraploids, the percent fertility increases so that 4x fertility approaches that in the 2x forms. This natural selection process for pairing affinity is termed allotetraploidization or diploidization and results in a higher frequency of bivalents and improved fertility (Doyle, 1979). Apparently, the 4x forms tested in this study had undergone extensive allotetraploidization and, therefore, the fertility of the 4x forms approached that of the 2x forms. The genotype x ploidy level interaction found in this study was probably a combination of the degree of allotetraploidization that was present in each genotype and the genotypic effects within the 2x and 4x forms.

The results of this study indicated that the in vitro germination characteristics were greatly influenced by genotype and ploidy level. Many morphological, physiological and biochemical differences resulting from genotype and ploidy level have been reported with genotypic differences in in vitro germination characteristics found among 2x genotypes (Mears, 1980; Pfahler, 1968; Tal, 1980; Stebbins, 1947). The 4x forms tested in this study had a slower germination rate and tube growth on the in vitro germination medium. These results are consistent with the in vivo studies (Green, 1946; Sarkar and Paria, 1980; Stebbins, 1947; Randolph, 1935) with maize which indicated that 1x pollen grains from 2x forms have a much greater transmission frequency on either 2x or 4x silks than 2x pollen grains from 4x forms. These transmission differences were so extreme that 4x forms are almost completely eliminated from interpollinating 2x-4x population mixtures in the next generation.

The effect of agar or a solidifying agent in the medium is not completely understood but the earliest study (Bair and Loomis, 1941) on in vitro maize pollen germination concluded that, during germination, the maize pollen grain could not be totally submerged in the medium but had to be partially exposed to the atmosphere because of oxygen exchange requirements. Since the 2x grains from the 4x forms were physically larger and presumably heavier than the 1x grains from the 2x forms, various agar levels were tested to determine if the degree of atmospheric exposure could be a factor. Since the germination rate and tube length of the 4x forms never equalled or exceeded the 2x forms at any agar level in this study, the degree of atmospheric exposure was not a major factor in the 2x-4x differences.

Acknowledgement

Appreciation is expressed to Laura Barry for her skill and diligence in typing and organizing this article.

References

Bair, R.A.; Loomis, W.E. The germination of maize pollen. Science 94:168-169; 1941.

Doyle, G.G. The allotetraploidization of maize I. the physical basis-differential pairing affinity. Theor. Appl. Genet. 54:103-112; 1979.

Green, J.M. Comparative rates of pollen tube establishment in diploid and tetraploid maize. J. Hered. 37:117-121; 1946.

Harter, H.L. Critical values for Duncan's multiple range test. Biometrics 16:671-685; 1960.

Lewis, W.H. Polyploidy in species populations. Lewis, W.H. ed. Polyploidy: biological relevance. New York: Plenum Press; 1980:p.103-147.

Mears, J.A. Chemistry of polyploids. Lewis, W.H. ed. Polyploids: biological relevance. New York: Plenum Press; 1980:p.77-101.

Pfahler, P.L. Fertilization ability of maize pollen grains I. pollen sources. Genetics 52:513-520; 1965.

Pfahler, P.L. In vitro germination and pollen tube growth of maize (Zea mays L.) pollen II. pollen source, calcium, and boron interactions. Can. J. Bot. 46:235-240; 1968.

Pfahler, P.L.; Wilcox, M. Genotype x ploidy level interactions for pollen geometry in maize. Erdelska O. ed. Fertilization and embryogenesis in ovulated plants. Bratislava, Czechoslovakia: VEDA; 1983:p.93-96.

Randolph, L.F. Cytogenetics of tetraploid maize. J. Agric. Res. 50:591-605; 1935.

Sarkar, K.R.; Paria, P. Pollen tube growth and fertilization in diploid and tetraploid maize. Indian J. Exp. Biol. 18:985-989; 1980.

Snedecor, G.W. Statistical methods. 5th ed. Ames, Iowa. Iowa State University Press; 1956.

Stebbins, G.L. Types of polyploids: their classification and significance. Adv. Genet. 1:403-427; 1947.

Tal, M. Physiology of polyploids. Lewis, W.H. ed. Polyploidy: biological relevance. New York: Plenum Press; 1980:p.61-75.

Zeven, A.C. Polyploidy and domestication: the origin and survival of polyploids in cytotype mixtures. Lewis, W.H. ed. Polyploidy: biological relevance. New York: Plenum Press; 1980:p.385-407.

46
Application of In Vitro Techniques in Cereal Pollen Biology

Beáta Barnabás and Géza Kovács

Improving the agronomically important characters (grain yield quality, quantity, disease resistance and tolerance to different abiotic stress conditions) of the main cereal crops is the most important task for the plant breeders. The above mentioned aims can be achieved by using traditional breeding techniques but already now some in vitro methods for haploid induction, mutant selection, somaclonal variation, etc. are well advanced and might also be used for applied aspects (Lörz et al. 1988). These new biotechnological techniques will probably not be competitive with the commercial breeding; rather they will contribute to creating more improved crops in an efficient manner by the combination of traditional technologies with the modern biotechnological methods.

Viewed in the short term plant in vitro techniques (cell, protoplast and tissue culture, genetic engineering) has had a rather minimal effect on plant breeding; but long term impact will be substantial and have a positive influence on agriculture and humankind (Tomes, 1990).

Recent progress in the field of in vitro techniques makes it possible to apply them more profoundly. The plant reproductive processes can serve as a suitable natural basis to carry out different manipulations. Since the male gametophyte has more active role in the process of fertilization, the pollen happened to be the primary research subject in our experimental work.

In vitro androgenesis in wheat and maize

The greatest interest in microspore-derived plants from microspore/anther cultures is based on their providing a rapid route to homozygosity and to expression of recessive genes. The main goal of research in this field is to provide the breeders with a practical tool applicable to all genotypes of specific interest for their breeding programmes. However, application of anther culture to wheat and corn breeding is hampered by low callus induction and green plant regeneration frequencies as well as prominent genotypic effects (Chu and Hill, 1988; Zhou and Konzak, 1989; Orshinsky et al., 1990; Genovesi, 1990).

In our lab a highly efficient anther culture technique was elaborated for wheat. The use of modified P2 induction medium and 190-2 regeneration medium (He and Ouyang, 1984) with the combination of suitable environmental conditions, enabled us to increase the haploid induction capacity of different genotypes to an advanced level. The processes of pollen callus/embryo induction and plant regeneration are of different genetic regulation, consequently these features can be improved jointly in a direct genetic way (Barnabás et al., 1988; Szakács et al., 1988; Barnabás et al., 1989).

Presently, culturing anthers in fluid media seems to be promising for certain genotypes (Table 1). The higher haploid induction in liquid medium can be explained by two possible reasons: 1, the better nutritional environment, and 2, the stimulating "conditioning factor" which can easily be transferred from one anther to another in the liquid environment He Guangyuan, personal information).

Table 1. Effect of different media on the haploid induction capacity of certain *T. aestivum* genotypes

Genotype	Medium							
	P2				PNC (Xue et al. 1987)			
	Solidified		Fluid		Solidified		Fluid	
	AR%	HI	AR%	HI	AR%	HI	AR%	HI
Benoist H77022	14.6	39.5	27.4**	60.5**	6.7	21.4	10.2	28.6
Orofen	25.2	46.3	28.9	59.2**	10.7	15.2	15.3	30.1**
Chinese Spring	9.1	9.3	8.8	11.7	12.8	14.6	15.2	12.2
Mv 8	5.2	7.1	6.1	14.8	12.1	17.3	15.3	30.1**

AR: Anther response; HI: Haploid induction (pollen embryos and calli)
** – significantly different at 0.01 probability level

From pollen calli induced in liquid media very fine fast growing suspension cultures can be initiated, that might serve as a suitable subject for haploid protoplast isolation (work in progress in our lab).

The wheat anther culture as a new biotechnological method will contribute to the faster improvement of the new varieties.

Nevertheless there is one important criterion to use anther culture derived haploids in a wide range: in plant breeding programs a large number of haploid plants that can easily and efficiently be transposed into genetically stable dihaploids is required. Comparing different chromosome reduplication techniques applied to pollen haploids of Triticum aestivum cv. Ciano¬ (Barnabás et al., 1991) obviously showed that the way of genome doubling produced by colchicine added to the induction medium before the first microspore mitosis, seems to be significantly more efficient than the conventionally used techniques. Generally the induction frequency of microspore-derived haploid structures was not considerably influenced by the necessary colchicine concentrations either in the other genotype examined (Table 2), which is important in this technique is to be used in practice.

Beyond the applied aspects of in vitro androgenesis the high pollen callus induction (up to 82%) and green plant regeneration (up to 60%) capacity of numerous wheat genotypes (T. aestivum cv. Benoist H77022, Diószegi 200, Mv 15, Mv 06-83, T. carthlicum, etc.) made it possible for us to study the in vitro morphogenetic processes on cellular and subcellular levels (Szakács and Barnabás, 1988; Barnabás et al., 1988) and to search for the genetic regulation of these processes (Szakács et al., 1988; Sági and Barnabás, 1989).

Table 2. Effect of colchicine on the anther induction and on the fertility of the plants regenerated from anther culture of cv. Benoist H77022

Treatment	No. of anthers cultured	Anther response %	Haploid induction %	No. of calli cultured	Plant regeneration Total	Plant regeneration Albino	Plant fertility %
Control	715	12.2	25.9	112	57.1	7.1	69.0
Colchicine							
0.01%	480	13.8	39.0	131	51.9	4.6	64.7
0.02%	498	13.1	24.9	107	60.8	3.7*	87.3
0.04%	462	6.3*	11.9*	50	52.0	0.0***	81.9

* : significantly differs from the control at p= 0.05
***: significantly differs from the control at p= 0.001

Since our first studies in the field of maize anther culture (Barnabás et al., 1987) a considerable progress has been achieved (Figure 1) on the basis of optimal medium-genotype interactions (unpublished results). Via crossing and subsequent selection the high androgenic capacity of some Chinese germplasms can be introduced into the local genotypes.

Good-quality, regenerable cell suspension cultures originated from haploid pollen calli or embryoids may be used for mutant selection, protoplast isolation and different ways of genetic manipulation.

Figure 1. Anther culture of a good-responding maize genotype (Chi 592)

Development of the male and female gametophytes in vitro provides an opportunity to study the regulation, nutritional, physiological and morphogenetical basis of their development (Pareddy et al., 1987; Benito-Moreno et al., 1988; Bommineni and Greyson, 1990).

To carry out gametophyte maturation in vitro in detached florets cultured from an early stage of development allows to apply selective environment continuously for both sexes and produce in vitro matured seeds in the controlled, stressed conditions.

Pollen maturation and selection in vitro would be very useful for those plant species where pollen is very short-lived in the nature (wheat for example) and carrying out in situ pollen selection is practically impossible.

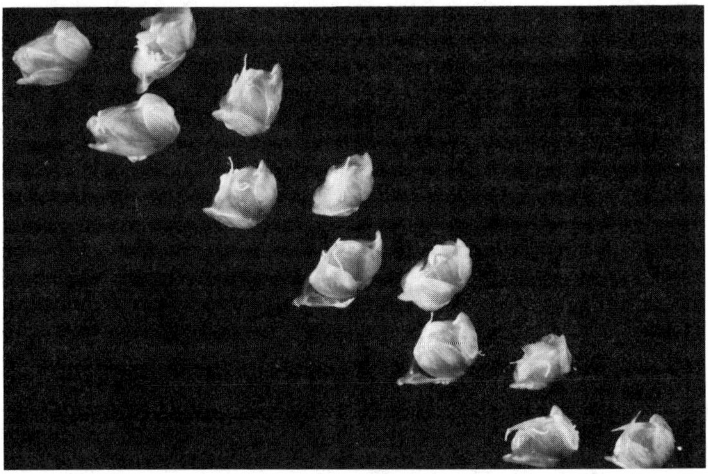

Figure 2. Floret culture of wheat cv. Chinese Spring

We have elaborated a method to produce functionable pollen in cultured wheat florets (Figure 2). The excised florets were inoculated in the late-uninucleate stage of microspore development onto a modified MS medium (supplemented with coconut water, different amino acids and cytokinins) containing 4% sucrose, and then incubated at 20°C with 8/16 hs photoperiod. After 7-9 days of incubation the anthesis occurred. There was a considerable high percentage (up to 66%) of normally developed trinucleate pollen grains in the in vitro matured pollen population (Figure 3), which did not differ morphologically from the control (in vivo developed) ones. Among the abnormally developed pollen grains sterile and immature forms could also be found and some of them had irregular wall (intine) development. Ultrastructural features of in vitro ripened pollen has also been studied.

Seed production of in vitro matured florets seems to be high enough in certain genotypes (Table 3) and fertile plants can either raised up directly from these seeds or via embryo culture (unpublished results).

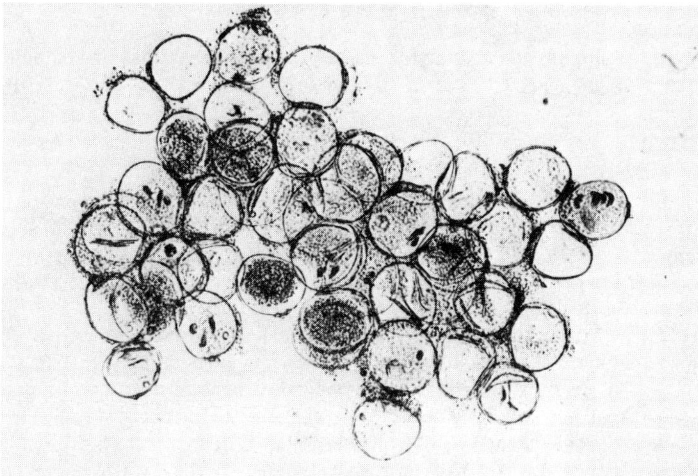

Figure 3. *In vitro* matured wheat pollen population M: 160x

Table 3. *In vitro* seed production of different wheat cultivars in detached floret cultures

Genotype	Seed set (%)
T. spelta	41.7
T. carthlicum	21.1
T. aestivum	
Mv 8	22.1
GK Kincso	16.7
Chinese Spring	9.6
Mv 15	5.0
Orofen	2.0

Pollen maturation in vitro "in anthero" or in isolated microspore cultures can give a lot of informations about the factors contributing to the pollen wall formation, cytoplasmic male sterility or self-incompatibility. Finally the premitotic microspore might be used as a target cell for gene transfer. In vitro pollen maturation appears to be rather important in this respect because using pollen as a transformation vector would provide an attractive alternative to methods that require prolonged tissue culture and in vitro regeneration (work has started in maize).

The other projects which have been carried out in our lab, such as:

- long term maintenance of the viability of maize and rye pollen, storage of young zygotic and pollen embryos of wheat in liquid nitrogen,
- improvement of early cold tolerance and herbicide tolerance in maize via in vivo pollen selection,

establishing favourable conditions for sperm cell manipulation in wheat,

beyond their importance in fundamental research all of them has real potential to be applied in practical fields.

References

1. Barnabás, B.; Fransz, P.L.; Schel, J.H.N. Ultrastructural studies on pollen embryogenesis in maize (Zea mays L.). Plant Cell Reports. 6: 212-215; 1987.
2. Barnabás, B.; Pfahler, P.L.; Kovács, G. Direct effect of colchicine on the microspore embryogenesis to produce dihaploid plants in wheat (Triticum aestivum L.). Theor. Appl. Genet. 81: 675-678; 1991.
3. Barnabás, B.; Szakács, É.; Kovács, G. Induction of haploid plants from wheat (Triticum aestivum L.) anther culture. Sveriges Utsadesförenings Tidskrift. 99: 125-130; 1989.
4. Barnabás, B.; Szakács, É.; Liszt, K. Cytological aspects of in vitro androgenesis in cereals. In: Cresti,M.; Gori,P.; Pacini,E. eds. Sexual Reproduction in Higher Plants. Springer-Verlag, Berlin Heidelberg. p. 113-118; 1988.
5. Barnabás, B.; Szakács, É.; Sági, L.; Kovács,G. Ways of the genetic improvement of haploid callus induction and plant regeneration in wheat (Triticum aestivum L.) anther cultures. Növénytermelés. 37: 289-292; 1988.
6. Benito-Moreno, R.M.; Macke, F.; Alven, A.; Heberle-Bors,E. In situ seed production after pollination with in vitro matured, isolated pollen. Planta. 176: 145-148; 1988.
7. Bommineni, V.R.; Greyson, R.I. Regulation of flower development in cultured ears of maize (Zea mays L.) Sex Plant Reprod. 3: 109-115; 1990.
8. Chu, C.C.; Hill, R.D. An improved anther culture method for obtaining higher frequency of pollen embryoids in Triticum aestivum L. Plant Science, 55: 175-181; 1988.
9. Genovesi, A.D. Maize (Zea mays L.): In vitro Production of Haploids. In: Bajaj,Y.P.S. ed. Biotechnology in Agriculture and Forestry, vol. 12. Haploids in Crop Improvement I.; Springer-Verlag, Berlin Heidelberg, 1990: p. 177-203.
10. He, D.G.; Ouyang, J.W. Callus and plantlet formation from cultured wheat anthers at different developmental stages. Plant Sci. Lett. 33: 71-79; 1984.
11. Lörz, H.; Göbel, E.; Brown, P. Advances in Tissue Culture and Progress Towards Genetic Transformation of Cereals. Plant Breeding. 100: 1-25; 1988.
12. Orshinsky, B.R.; McGregor, L.J.; Johnson, G.I.E.; Hucl, P.; Kartha, K.K. Improved embryoid induction and green shoot regeneration from wheat anthers cultured in medium with maltose. Plant Cell Reports. 9: 365-369; 1990.
13. Pareddy, D.R.; Greyson, R.I.; Walden, D.B. Fertilization and seed production with pollen from in vitro cultured maize tassels. Planta. 170: 441-443; 1987.
14. Sági, L.; Barnabás, B. Evidence for cytoplasmic control of in vitro microspore embryogenesis in the anther culture of wheat (Triticum aestivum L.). Theor. Appl. Genet. 78: 867-872; 1989.
15. Szakács, É.; Barnabás, B. Cytological aspects of in vitro androgenesis in wheat (Triticum aestivum L.) using fluorescent microscopy. Sex Plant Reprod. 1: 217-222; 1988.
16. Szakács, É.; Kovács, G.; Pauk, J.; Barnabás, B. Substitution analysis of callus induction and plant regeneration from anther culture in wheat (Triticum aestivum L.). Plant Cell Reports. 7: 127-129; 1988.

17. Tomes, D.T. Current Research in Biotechnology with Application to Plant Breeding. Nijkamp,H.J.J.; van Der Plas,L.H.W.; van Aartrijk,J. eds. Progress in Plant Cellular and Molecular Biology. Proceedings of the VIIth International Congress on Plant Tissue and Cell Culture, Amsterdam, 24-29 June 1990: 23-33.
18. Xue, X.; Hu, Y.; Bao, R. Studies of pollen plant transplantation and PNC medium in anther culture of maize. GMICN, 3: 1-14. 1987.
19. Zhou, H.; Konzak, C.F. Improvement of anther culture methods for haploid production in wheat. Crop Sci. 29: 817-821; 1989.

47
In Vitro Fertilization with Isolated Gametes of Maize and Its Application to Study Fertilization Processes and Early Events of Zygote Development

Erhard Kranz and Horst Lörz

In vitro fertilization with single gametes of maize was performed for studies of basic processes of differentiation, genetic transformation and incompatibility. Selected pairs of isolated egg and sperm cells as well as synergids or central cells and sperm cells were fused individually by an electrical pulse after dielectrophoretical alignment under microscopic observation (Kranz et al., 1990; 1991 a,b). High frequencies of microcallus formation from the fusion products of egg and sperm cells were observed. In addition, cytoplasmic organelles have been transmitted together with the sperm into the egg using individually selected cytoplasts from maize mesophyll. Present activities concentrate on the sustained growth of the fusion products toward regeneration of maize plants, the isolation and characterization of development-specific genes from early stages of in vitro-produced zygotes and genetic manipulation, including electroporation of DNA into isolated sperm cells and microinjection into the fusion product of egg cell and sperm cell.

Isolation and selection of single gametes and non-gametic cells, individual electrofusion and culture procedures

Maize (Zea mays L.) inbred line A 188 was used for sperm, egg cell, synergid, central cell, and cytoplast preparations. Line Seneca 60 was also used for sperm cell isolation. Pollen grains were collected from freshly exerted dehiscing anthers and used immediately or stored as described (Kranz and Lörz, 1990). Ears were bagged before silk emergence and collected after silk emergence. The outer leaves of the ears were sterilized with ethanol (70%).

Egg cells were isolated by cell wall degradating enzymes (Wagner et al., 1988; 1989) followed by a mechanical procedure (Kranz et al., 1990; 1991a,b). Pieces of tissue (containing the embryo sac) were dissected from the ovules under a dissecting microscope (Fig. 1a). After collecting the tissue pieces in mannitol solution they were incubated in enzyme solution (0.5% pectinase, Serva, Heidelberg, FRG; 0.2% pectolyase Y23, Seishin, Tokyo, Japan; 0.3% hemicellulase, Sigma, Deisenhofen, FRG and 0.3% cellulase "Onozuka" RS, Yakult Honsha, Tokyo, Japan, pH 5.0, adjusted to 570 mosmol/kg H_2O with mannitol). The ovular tissue was incubated in enzyme mixture at 24 ° +/- 0.5° C for about 10-20 min (depending on the "quality" of the plant material) without shaking. Egg cells and non-gametic cells were isolated manually after incubation, with very fine glass needles directly in the incubation dish under an inverted microscope (Fig. 1b).

Sperm cells were individually selected under microscopic observation after rupture of pollen grains in mannitol solution (0.5 M or 0.55 M mannitol). The isolated gametes and non-gametic cells of the embryo sac were transferred individually, using microcapillaries, into fusion droplets (2 μl each, consisting of 0.5 or 0.55 M mannitol) covered by mineral oil. Before fusion, the selected cells were transferred from one droplet into another for washing. The cells were selected by use of a computer-controlled dispenser/diluter. Electrofusion was performed under microscopic observation with platinium electrodes adjusted for fusion in the droplets (Koop et al., 1983; Koop and Schweiger 1985 a,b; Schweiger et al., 1987). In each droplet one egg cell or one non-gametic cell of the embryo sac was fused with one sperm cell (Fig. 1 c,d).
Fusion of egg and sperm cells was induced by single or multiple (2-3) negative DC-pulses (50 μs; 0.9-1.0 kV x cm^{-1}) after dielectrophoretic alignment of defined gametes (1 MHz; 71 V x cm^{-1}) on one of the electrodes for a few seconds with an electrofusion apparatus (CFA 400, Krüss, Hamburg, Germany). Fusion medium was 0.55 M mannitol. Fusion of sperm cells with central cells or with egg cells and cytoplasts respectively was induced by negative DC-pulses (50 μs; 0.7 kV x cm^{-1}) after dielectrophoretic alignment of defined gametes and cytoplasts (1 MHz; 48 V x cm^{-1} on one of the electrodes for a few seconds. Fusion medium was 0.5M mannitol.
A "Millicell-CM" dish (diameter 12 mm, Millipore) was inserted into a 3.5-cm plastic dish previously filled with 1.5 ml of a maize feeder suspension. The insert

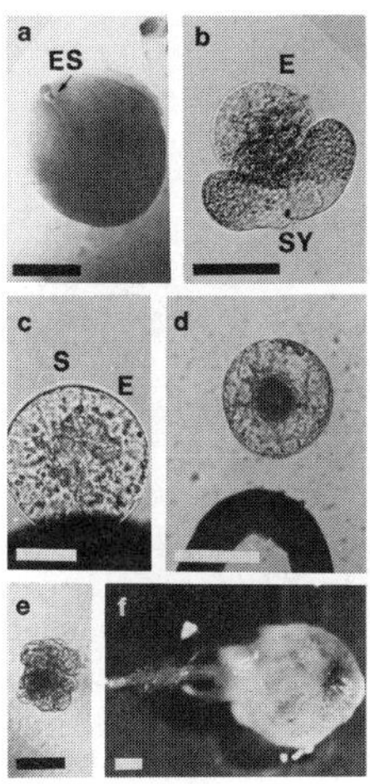

Fig. 1a-f. Electrofusion-mediated in vitro fertilization with single gametes of Zea mays L.
a. Piece of ovular tissue incubated 10 min in enzym solution for manual isolation of the cells of the embryo sac (ES). Bar = 1.0 mm.
b. Isolated egg cell (E) with adherant synergids (SY). Bar = 50 μm.
c. Alignment of a selected pair of an egg cell (E) and a sperm cell (S) on one of the electrodes for electrofusion. Bar = 25 μm.
d. Fusion product of a selected pair of a central cell and a sperm cell. Bar = 100 μm.
e. Multicellular structure 7 days after fusion of an egg cell and a sperm cell on the semipermeable membrane of a "Millicell" insert. Bar = 50 μm. f. Regenerated structure of a fusion product of an egg cell and a sperm cell 29 days after fusion, with a root and meristematic green spots. Bar = 1.0 mm.

was filled with 200 μl mannitol solution (0.55M) and the fusion products were transferred into the insert with a microcapillary. For the feeder culture system MS-medium (Murashige and Skoog, 1962) with 1.0 mg/l 2.4 D and 0.02 mg/l kinetin as well as N_{6ap}-medium (Rhodes et al., 1988) with 1.0 mg/l 2,4 D and additional L-lysine (100 mg/l),

L-serine (100 mg/l), L-glutamic acid (100 mg/l) and kinetin (0.02 mg/l) were used. Both media were adjusted to 600 mosmol/kg H_2O with glucose, pH 5.7. Microcalli, grown in modified N_{6ap}-medium for seven days, were cultivated after one week in the same medium but with reduced osmolarity (300 mosmol/kg H_2O) and 3 weeks after fusion without hormones and with sucrose (30 g/l). One week later, the calli were grown on solid N_6-medium without hormones. Culture conditions: 24° +/- 0.5°C in the dark or a light/dark cycle of 16/8h and a light intensity of approximately 50 $\mu Em^{-2}s^{-1}$, on a rotary shaker at 70 rpm.

In vitro fertilization mediated by electrofusion of single gametes - current status, prospects and limitations

In vitro fertilization was performed by individual electrical fusion of isolated egg cells and sperm cells of maize. The "bottleneck" of the method is to get a sufficient number of isolated egg cells for routine work. The yield of isolated embryo sacs and egg cells could be improved by reducing the enzyme concentration and by shortening the time of incubation of ovule-pieces in the enzyme mixture to 10 minutes. As found by Wagner et al. (1989) and Huang and Russel (1989) the timing of exposure of the embryo sac containing tissue in the enzyme solution was important to prevent unwanted fusions. Depending on the "quality" of the plant material, the enzyme concentrations and the incubation time must be determined exactly. A yield of up to 80% isolated egg cells could be obtained (mean isolation frequency 55%). The yield was finally determined by the manual isolation step. Using thin glass needles, up to 40 egg cells could be isolated per 3 hours.

The conditions for the electrical fusion of somatic cell protoplasts could be used also for the gametic protoplasts of maize. The sperm cells are protoplasts in a mature pollen grain of maize (Dupuis et al., 1987), Matthys-Rochon et al., 1988; Cass and Fabi, 1988). By an electrical pulse, they can be fused efficiently with the egg cell, synergids and central cells (mean fusion frequencies 79%, 75% and 55% respectively).

Our attempts to induce cell division of isolated egg cells of maize have so far failed. These cells show protoplasmic streaming during 22 days of culture, but they do not divide. After fusion of one sperm with an egg cell however, the fusion products started to divide within 2.5 - 3 days. Microcalli developed from the fusion products with a mean frequency of 83% in the modified Murashige and Skoog medium (Fig. 1e). Four to five weeks after fusion, calli derived from single fusion products, regenerated roots and developed structures with green meristematic spots (Fig. 1f).

Cytoplasmic organelles were transferred into the egg cells by their fusion with cytoplasts, selected from a mesophyll protoplast preparation of maize (Kranz et al., 1991b). Either cytoplasts were fused first with the sperm cell followed by fusion with the egg or cytoplast, or sperm and egg cell were fused in one step. Single or multiple (2-3) electrical pulses were applied and resulted in a mean fusion frequency of 73%. The transmission of cytoplasm through the in vitro fertilization process will facilitate studies on cytoplasmic inheritance. The fate of organelles (plastids and mitochondria) transmitted by sperm cells or additionally by cytoplasts from somatic cells can be studied with this system. In addition, cell reconstitutions were performed, using single cytoplasts from maize mesophyll cells for fusion with one or two sperm cells (Kranz et al., 1991b). With the aim of studying in vitro fertilization processes, fusions were performed with the following combinations: synergid with sperm cell, synergid and adherent second synergid and egg cell with sperm, central cell alone with sperm, central cell and adherent synergids and egg cell with sperm, and sperm with sperm cells (Kranz et al. 1991b). Recently the development and growth of fusion products of a central cell with a sperm cell in culture could be observed (E.Kranz, unpublished data). By fusion of these single cells, endosperm tissue could be produced in vitro for the first time. Such experiments may facilitate studies to perform double fertilization in vitro at the single cell level. The possibility of making such fusion combinations of single sperm cells with central cells or synergids will allow studies of fertilization mechanisms as well as the effects of non-gametic cells on the development of artificially-produced zygotes.

The transfer of foreign genes into egg and sperm cells via electroporation as well as into zygotes before the first division takes place via microinjection, seems to be promising, as synchronized single cells can be used. Fusion products have been used in recent experiments for transformation studies. Preliminary experiments reveal high survival rates (mean 83%) and division rates (mean 67%) of the fused gametes following microinjection of plasmid DNA (E.Kranz, unpublished data). Additionally, by using the polymerase chain reaction (PCR) method for amplification of mRNA's, it may be possible to detect development-specific genes from only a few in vitro-created zygotes of early stages.

The observation that unfertilized egg cells do not divide in vitro, but the fused gametes divide with high frequencies, may facilitate studies to determine cell division factors which are produced or activated after fusion of the cells and nuclei, and will allow studies of growth and differentiation of the zygotes in vitro.

A method has been established for in vitro fertilization using single plant gametes and subsequent growth of the fusion products, the zygotes. It has become possible to handle single gametes and to use them for fusion, microinjection and electroporation. First studied with maize as a model, in vitro fertilization using isolated gametes may also be applied to other economically important crops.

REFERENCES

1. Cass, D.D.; Fabi, G.C. Structure and properties of sperm cells isolated from the pollen of Zea mays. Can. J. Bot. 66:819-825; 1988.
2. Dupuis, I.; Roeckel, P.; Matthys-Rochon, E.; Dumas, C. Procedure to isolate viable sperm cells from corn (Zea mays L.) pollen grains. Plant. Physiol. 85:876-878; 1987.
3. Huang, B.Q.; Russel, S.D. Isolation of fixed and viable eggs, central cells and embryo sacs from ovules of Plumbago zeylanica. Plant. Physiol. 90:9-12; 1989.
4. Koop, H.U.; Dirk, J.; Wolff, D.; Schweiger, H.G. Somatic hybridization of two selected single cells. Cell. Biol. Int. Rep. 7:1123-1128; 1983.
5. Koop, H.U.; Schweiger, H.G. Regeneration of plants from individually cultivated protoplasts using an improved microculture system. J. Plant Physiol. 121:245-257; 1985a.
6. Koop, H.U.; Schweiger, H.G. Regeneration of plants after electrofusion of selected pairs of protoplasts. Eur. J. Cell Biol. 39:46-49; 1985b.
7. Kranz, E.; Lörz, H. Micromanipulation and in vitro fertilization with single pollen grains of maize. Sex. Plant Reprod. 3:160-169; 1990.
8. Kranz, E.; Bautor, J.; Lörz, H. In vitro fertilization of single, isolated gametes, transmission of cytoplasmic organelles and cell reconstitution of maize (Zea mays L.).

Nijkamp, H.J.J.; Van der Plas, L.H.W.; Van Aartrijk, J., eds. Progress in plant cellular and molecular biology: proceedings of the VIIth International Congress on Plant Tissue and Cell Culture; 1990 June 24-29; Amsterdam, The Netherlands: Kluwer Academic Publishers, Dordrecht, Boston, London; 1990:252-257.

9. Kranz, E.; Bautor, J.; Lörz, H. In vitro fertilization of single, isolated gametes of maize mediated by electrofusion. Sex. Plant Reprod. 4:12-16; 1991a.

10. Kranz, E.; Bautor, J.; Lörz, H. Electrofusion-mediated transmission of cytoplasmic organelles through the in vitro fertilization process, fusion of sperm cells with synergids and central cells, and cell reconstitution in maize. Sex. Plant Reprod. 4:17-21; 1991b.

11. Matthys-Rochon, E.; Vergne, P.; Detchepare, S.; Dumas, C. Male germ unit isolation from three tricellular pollen species: Brassica oleracea, Zea mays, and Triticum aestivum. Plant Physiol. 83:464-466; 1987.

12. Murashige, T.; Skoog, F. A revised medium for rapid growth and bioassays with tobacco tissue cultures. Physiol. Plant 15:473-497; 1962.

13. Rhodes, C.A.; Lowe, K.S.; Ruby, K.L. Plant regeneration from protoplasts isolated from embryogenic maize cell cultures. Bio/Technol. 6:56-60; 1988.

14. Schweiger, H.G.; Dirk, J.; Koop, H.U.; Kranz, E.; Neuhaus, G.; Spangenberg, G.; Wolff, D. Individual selection, culture and manipulation of higher plant cells. Theor. Appl. Genet. 73:769-783; 1987.

15. Wagner, V.T.; Song, Y.; Matthys-Rochon, E.; Dumas, C. The isolated embryo sac of Zea mays: structural and ultrastructural observations. Cresti, M.; Gori, P.; Pacini, E. eds. Sexual reproduction in higher plants: proceedings of the tenth international symposium on the sexual reproduction in higher plants; 1988 May 30-June 4; University of Siena, Siena, Italy: Springer, Berlin, Heidelberg, New York; 1988:125-130.

16. Wagner, V.T.; Song, Y.C.; Matthys-Rochon, E.; Dumas, C. Observations on the isolated embryo sac of Zea mays L. Plant Sci. 59:127-132; 1989.

48
In Vitro Maturation of Maize Pollen

D.R. Pareddy and J.F. Petolino

In vitro procedures for crop improvement have most typically involved dedifferentiation and abnormal proliferation of somatic cells followed by plant regeneration via either organogenesis or embryogenesis. Even when immature reproductive structures have been used, abnormal development, resulting in callus formation and plant regeneration, has been induced (i.e., androgenesis). Reports of normal reproductive development *in vitro*, culminating in the formation of mature gametophytes capable of reproductive function, are rare (Benito-Moreno et al., 1988).

Male and female inflorescences of maize have been cultured as a means of studying *in vitro* gametophyte development (Pareddy and Greyson, 1985; Bommineni and Greyson, 1987). Immature tassel meristems have been used for the *in vitro* production of mature, trinucleate pollen capable of germination and fertilization (Pareddy et al., 1987). While this technique has some utility for basic and applied research, *in vitro* gametophyte development, with little or no associated somatic tissue, would be even more desirable for certain types of genetic manipulation such as germ-line transformation and gamete selection. This chapter describes procedures for the production of maize pollen *in vitro* from immature detached spikelets, excised anthers, and isolated microspores.

Pollen Development in Maize

Male gametophytic development begins with meiosis in pollen mother cells and, following pollination, culminates with double fertilization of egg and polar nuclei in embryo sacs of ovaries within female inflorescences. Although a continuous process, pollen development can be conveniently sub-divided into several phases. Inside developing anthers, each pollen mother cell undergoes meiosis resulting in four haploid, uninucleate microspores. These microspores are characterized by a large nucleus which occupies one-third to one-half of the cell volume. A germ pore appears at this stage, just prior to enlargement of the microspore, while the cytoplasm is dispersed throughout the cell. As the microspore increases in size, the nucleus remains near the pore and becomes smaller. The formation of a large central vacuole pushes the nucleus to the side opposite the germ pore. The nucleus then enlarges and undergoes a first pollen grain mitosis. Resulting daughter nuclei, which are initially similar in size, begin to differentiate while still within the original cytoplasm. At this point, a vegetative nucleus is distinguishable by its larger size in contrast to a smaller generative nucleus.

The vegetative nucleus then migrates to the opposite side of the pollen grain. The generative nucleus, along with a small amount of cytoplasm, becomes separated from the vegetative nucleus by the formation of an arch-shaped cell wall attached to the inner microspore wall. Starch accumulation follows, progressing from the pore side, until the entire pollen grain is filled. The generative cell then detaches from the inner wall, becomes rounded, and migrates toward the vegetative cell where a second pollen grain mitosis occurs, prior to being shed from the anther, resulting in a trinucleate pollen grain.

In Vitro Culture of Immature Gametophytes

The *in vitro* culture of early-staged male gametophytes has been used as an experimental method to study various aspects of pollen development (Vasil, 1967). Methods differ based on the explant used for culture and the degree of association of the male gametophytes with surrounding somatic tissue. *In vitro* culture of "detached" gametophytes, involving whole inflorescences, floral buds, or immature flowers, resulting in functional pollen development, has been reported in a diverse number of species (Rastogi and Sawhney., 1989). In this system, complex reproductive tissues, containing anther primordia, are excised from the plant and cultured on nutrient medium containing growth regulators. Considerable variability has been observed with regard to the cultural requirements for male gametophyte development of different species. This variability is compounded by the fact that the external hormonal and nutritional requirements appear to be different at various stages of male gametophyte development and depend on the amount of accompanying somatic tissue.

Reports of successful pollen production from excised anther or isolated microspore cultures are rare. Identifying the necessary cultural conditions (to replace requirements that would normally be provided to the developing gametophytes by the surrounding somatic tissue) appears to be technically more demanding in these systems. Nonetheless, in certain binucleate species, isolated male gametophytes have been cultured to produce pollen from the uninucleate or binucleate phases of development. Uninucleate microspores of lily, mechanically isolated from anthers and cultured on the surface of agar-solidified medium containing mineral salts, vitamins, yeast extract, and sucrose, resulted in binucleate pollen capable of germination (Tanaka and Ito, 1981). Similarly, late-uninucleate and early-binucleate microspores isolated from tobacco have been used to produce mature, functional, binucleate pollen by culturing them in a series of liquid medium containing decreasing concentrations of salts and sugar (Benito-Moreno et al., 1988).

Upon detachment, male gametophytes of trinucleate species have proven to be difficult to manipulate (Knox et al., 1986). Detached gametophytes of a trinucleate species, rye, could only be cultured from the premeiotic to the uninucleate phase of development (de la Pena, 1986). This work involved mechanical isolation of premeiotic cells from anthers followed by culture in liquid medium containing mineral salts, amino acids, and sucrose. Although cytological analysis suggested that meiosis appeared to proceed normally, development was arrested at about the early-uninucleate phase. Recently, binucleate microspores of wheat were matured *in vitro* to produce functional pollen at low frequeny (Stauffer et al., 1991). In maize, whole inflorescences have been cultured to maturity (Pareddy et al., 1987). In this system, entire tassel primordia (containing premeiotic pollen mother cells) are excised and cultured in a liquid medium containing mineral salts, sucrose, and kinetin such that gametophytic development occurs entirely enclosed within numerous layers of somatic tissue. *In vitro* development of detached male gametophytes of maize (i.e., anther and microspore culture), culminating in the formation of functional pollen, has not been previously described.

Detached Male Gametophytes of Maize

Maize pollen has recently been produced from cultured immature spikelets, excised anthers, and isolated microspores. Tassels with anthers containing microspores at the mid-uninucleate to late-binucleate phase were harvested and surface sterilized in a 0.5% sodium hypochlorite solution for 20 min followed by a sterile water rinse before use. Immature spikelets or anthers were excised and cultured in 2 mL of liquid "pollen maturation" medium in a 60x15 mm Petri dish (10 spikelets or 60 anthers per dish). The "pollen maturation" medium consisted of Murashige and Skoog (1962) major and minor salts, White's (1943) vitamines, 100 mg/L myo-inositol, 120 g/L sucrose, 1 g/L casein hydrolysate, 25 ug/L kinetin, 2% coconut milk, 35 mg/L L-proline, and 45 mg/L L-glutamine adjusted to pH 5.8.

Microspores were isolated by blending 250-300 anthers in 40 mL of "pollen maturation" medium (180 g/L sucrose) at low speed (2000 rpm) for 10 sec in a 110 mL stainless steel attachment (MC-2, Waring Products Division, New Hartford, CT). The contents were passed through a 113 micron screen and centrifuged at 1000 rpm for 5 min. Microspores were then re-suspended in fresh "pollen maturation" medium. Approximately 100-200 uL of suspended microspores were cultured in a 60x15 mm Petri dish containing 10 mL of "pollen maturation" medium solidified with 1% Noble-agar (Difco Laboratories, Detroit, MI) covered with a piece of No. 1 Whatman Filter Paper.

Detached spikelets, excised anthers, and isolated microspores were incubated at 28°C as stationary cultures in the dark for 10-16 days. Nuclear divisions and starch accumulation were observed at different intervals of culture with mithramycin and acetocarmine stains, respectively. Viability was assessed with fluorescein diacetate (FDA). For spikelet culture, pollen was collected by placing anthers onto No. 1 Filter Paper and placing in a desiccator containing a mixture of dehydrated silica gel and calcium chloride for about 2-3 hours. During this time pollen shed onto the filter paper. Dehisced pollen was sprinkled onto the surface of 5 mL of "pollen germination" medium. For anther culture, matured anthers were opened with a scalpel in liquid "pollen germination" medium. For microspore culture, suspended pollen was placed directly onto "pollen germination" medium using a Pasteur pipet. "Pollen germination" medium consisted of 12% sucrose, 0.03% calcium chloride, and 0.01% boric acid solidified with 1.5% Bacto-agar (Cook and Walden, 1965). Pollen was considered germinated if the tube was at least as long as the grain diameter. Recovered pollen grains were applied to receptive silks (approximately 10^4 per ear), protected with a bag before and after pollination, of greenhouse grown plants.

In Vitro-Cultured Maize Pollen

Golden yellow in color and spheroidal in shape with a diameter between about 90 and 100 microns, *in vitro*-produced pollen is almost completely starch filled and contains three nuclei (two smaller, densely staining and one larger, less densely staining). Under the conditions described above, the amount of time required to reach maturity depends on the stage of development at the time of culture initiation. Generally, mid- to late-uninucleate microspores require between about 10 to 16 days to mature, early- to mid-binucleate microspores about 8 to 14 days, mid- to late-binucleate microspores about 6 to 12 days, and early- to mid-trinucleate mature in 4 to 10 days. Typically, 80-100% of the cultured spikelets (Figure 1) developed mature anthers, about 40-60% of which extruded anthers by filament elongation (Figure 2). Mature anthers from cultured spikelets dehisced, upon drying in a desiccator, liberating dry, mature pollen, which germinated at a frequency of about 40-70%. About 30-70% of anthers developed normally when cultured excised from spikelets and floated on "pollen maturation" medium (Figures 3 and 4). About 5% of the pollen from excised anthers germinated. Late-uninucleate to late-binucleate microspores could be isolated

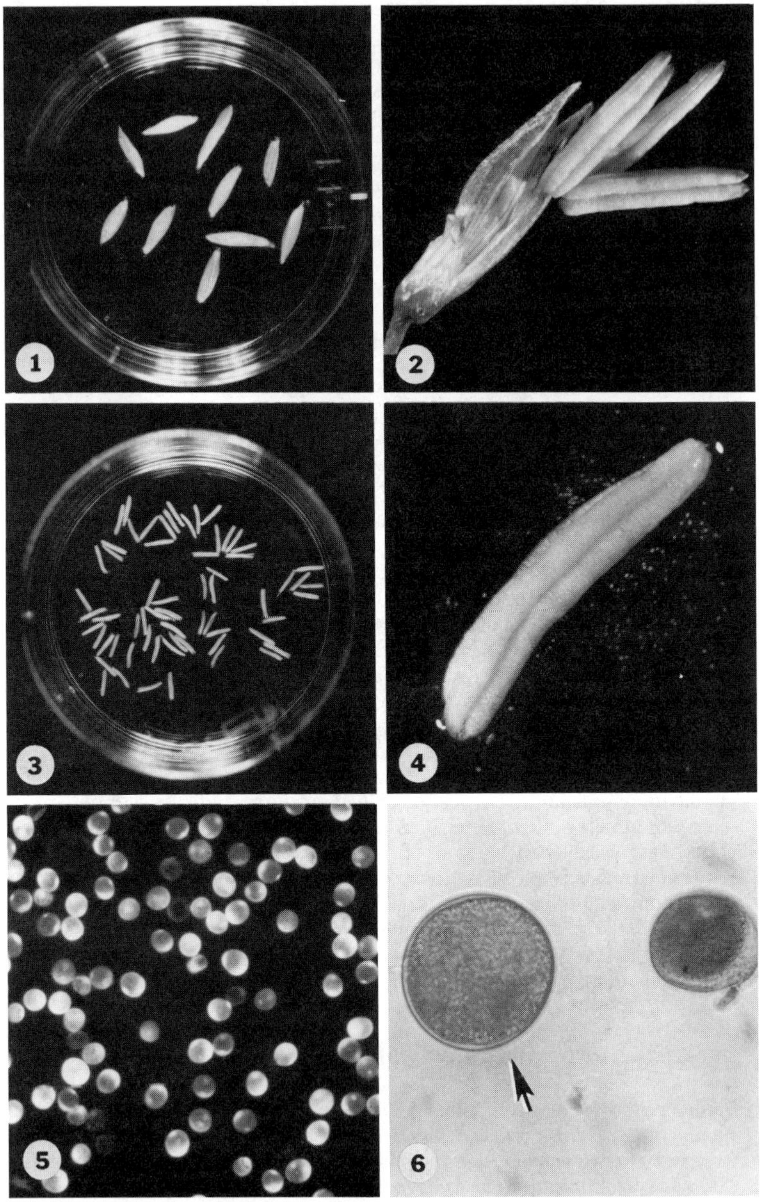

Figure 1. Immature spikelets explanted to liquid medium.
Figure 2. Mature spikelet showing extrusion of anthers after 10 days of culture.
Figure 3. Immature anthers at mid- to late-uninucleate stage in liquid medium.
Figure 4. Mature anther developed after 12 days of culture.
Figure 5. FDA stained microspores isolated by blending.
Figure 6. Mature pollen grain (arrow) developed after 13 days of culture.

from anthers by blending and maintain 70-80% viability (Figure 5). When cultured on a semi-solid "pollen maturation" medium,these microspores underwent normal development through 1 or 2 divisions resulting in mature, trinucleate pollen at a frequency of 5-20% (Figure 6). Although viable as measured by FDA staining, less than 1% of the pollen produced from isolated microspores germinated *in vitro*. Kernel formation, following fertilization with *in vitro* matured pollen, was variable (ranging from 0 to 100 kernels per ear).

These results demonstrate that maize pollen can be produced *in vitro* by culturing immature detached spikelets, excised anthers, and isolated microspores on a medium comprising mineral salts, sugars, amino acids, and kinetin. Although functional pollen could be produced in all culture systems, the frequency of maturation and germinability of pollen decreased as the surrounding somatic tissue was removed (i.e., isolated microspores < excised anthers < detached spikelets). This may have been a function of the extreme sensitivity of developing gametophytic cells to environmental perturbations or the removal of some essential component provided by the surrounding somatic tissue. Indeed, the cultural requirements for isolated microspores appeared to be more complex than for detached spikelets or excised anthers. For example, higher concentrations of sucrose, the addition of coconut milk, proline, and glutamine, and the necessity of culturing on the surface of a semi-solid medium were necessary for maturation of isolated microspores. Nonetheless, these procedures provide methods for the *in vitro* manipulation of a significant phase in the sexual cycle of maize, a notoriously difficult species to culture. The methods also make large numbers of developing male gametophytes, at precisely known stages of development, accessible for study. Up to two mitotic divisions occur under controlled laboratory conditions, thus, making the germ-line more available for genetic experimentation.

Acknowledgement

The authors would like to thank Kristin Zahaitis for her technical assistance.

Literature Cited

Benito-Moreno, R.M.; Macke, F.; Alwen, A.; Heberle-Bors, E. In-situ seed production after pollination with in-vitro-matured, isolated pollen. Planta 176:145-148; 1988.

Bommineni, V.R.; Greyson, R.I. *In vitro* culture of ear shoots of Zea mays and the effect of kinetin on sex expression. Amer. J. Bot. 74:883-890; 1987.

Cook, F.S.; Walden, D.B. The male gametophyte of Zea mays L. II. *In vitro* germination. Can. J. Bot. 43:779-786; 1965.

de la Pena, A. *In vitro* culture of isolated meiocytes of rye, *Secale cereale* L. Environ. Exp. Bot. 26:17-23; 1986.

Knox, R.B.; Williams, E.G.; Dumas, C. Pollen, pistil, and reproductive function in crop plants. Plant Breeding Rev. 4:9-79; 1986.

Murashige, T; Skoog, F. A revised medium for rapid growth and bioassays with tobacco tissue cultures. Physiol. Plant. 15:473-497; 1962.

Pareddy, D.R.; Greyson, R.I. *In vitro* culture of immature tassels of an inbred field variety of Zea mays cv. Oh43. Plant Cell Tissue Organ Culture 5:119-128; 1985.

Pareddy, D.R.; Greyson, R.I.; Walden, D.B. Fertilization and seed production with pollen from *in vitro* cultured maize tassels. Planta 170:141-143; 1987.

Rastogi, R.; Sawhney, V.K. *In vitro* development of angiosperm floral buds and organs. Plant Cell Tissue and Organ Culture 16:145-174; 1989.

Stauffer, C.; Benito Moreno, R.M.; Heberle-Bors, E. Seed set after pollination with *in vitro* matured, isolated pollen of Triticum aestivum. Theor. Appl. Genet. 81:576-580; 1991.

Tanaka, I.; Ito, M. Studies on microspore development in Liliaceous plants III. Pollen tube development in lily pollen cultured from the uninucleate stage. Plant Cell Physiol. 22:149-153; 1981.

Vasil, I.K. Physiology and cytology of anther development, Biol.. Rev. 42:327-373; 1967.

White, P.R. A handbook of plant tissue culture. Jacques Cattell Press, Lancaster; 1943.

49
Differentiation and Maturation of Tobacco Pollen In Situ and in Suspension Culture

Jaroslav Tupy, Ludmila Rihova, Vera Capkova, and Viktor Zarsky

Pollen development
1. Introduction and Terminology

In the cycle of sexual plant reproduction consisting in alternation of diploid and haploid generations, the diploid sporophyte produces spores, the mitotic division of which gives rise to the haploid gamete-producing gametophyte. According to this general scheme, the microsporocytes in the anther undergo meiosis to produce microspores through the process of microsporogenesis. The mitotic division of the unicellular microspore starts the ontogeny of pollen (microgametophyte) during which two gametes are produced by a process of gametogenesis. The pollen grain, the developmental equivalent of the embryo sac (macrogametophyte) should thus not be termed microspore, similarly as the embryo sac is not called macrospore, and the term microgametophyte is to be limited to the two- or three-cellular structures arising from microspore. Using this phylogeny-based terminology, the two postmeiotic mitotic divisions should be termed microsporemitosis and pollenmitosis, rather than 1st and 2nd pollen mitosis.
Nicotiana tabacum L. is a suitable model for pollen developmental studies mainly because of possibility for an efficient induction of pollen embryogenesis, continual flowering and large flowers and anthers. The relatively easy isolation of microspores and pollen grains and their culture in suspension offer a number of opportunities both for basic pollen cells studies and for the progress of pollen biotechnologies (rev. Heberle-Bors,1989). From the earlier and new data obtained in that way, the normal ontogeny of tobacco pollen is shown here to proceed in two phases which can be developmentally well defined as terminal cell differentiation resulting in functional gametophyte and maturation leading to pollen dormancy required for pollen dispersal. Towards the end of differentiation phase (mid- to late-binucleate stage) the pollen grains are already able to enter the functional phase, both in pollen germination medium and on the stigma, and pollination with immature pollen results in seed production. Application of this concept of two-phase pollen ontogeny in the anther led to the development of a technology for production of dormant fertile pollen from microspores invitro (Tupy et al.,1991). The method consists in gradual lowering of water potential of a medium for differentiation and it is shown here that starch accumulation to its maximum during the differentiation phase before the addition of osmotica appears important for the fertility of pollen produced in vitro.

2. Material and Methods

Plants of Nicotiana tabacum L. cv. Samsun were grown in free soil in a greenhouse. The developmental stages 1 to 6 (Tupy et al.,1983) correspond to the bud lengths 13-16, 17-21, 26-31, 36-45, 48-51 and 52-57 mm, respectively. The size of buds within one stage, however, varies according to the season and cultural conditions and the pollen stages were checked microscopically. The stages 1 and 6 correspond, respectively, to microspore mitosis and one day before anthesis. The advancement for one stage normally takes about one day.

All experiments and analyses were carried out on isolated microspores or pollen grains. Starch content was determined according to Lustinec et al.(1983) or cytologically after staining with IIK. The DNA content in Feulgen stained pollen nuclei was estimated cytophotometrically. The size of nucleolus was measured using differential interference contrast. The pattern of pollen glycoproteins was examined by ConA-peroxidase staining of polypeptides separated by 1D-SDS-PAGE and transferred to nitrocellulose.

The procedure for in vitro pollen production from microspore was essentially as described (Tupy et al.,1991) using the complex differentiation medium M1 and maturation medium P (1.17M sucrose, 100mM L-proline). The in vitro pollen germinated in medium G (O.3M sucrose, 1.6mM H3BO3).

3. Results and Discussion

Pollen differentiation and maturation in situ

The initial pollen development from microspore is characterized by a rapid activation of processes of gametophytic cytoplasm formation (Fig.1). The increase in size of the vegetative nucleolus indicates a rise of rRNA synthesis which is in accord with the results on uridine incorporation into RNA in isolated pollen (Tupy et al.,1983) and of density gradient analysis of the synthesized RNA (Tupy,1982). The synthetic activity reaches its maximum at the 3rd stage and falls steadily during the subsequent pollen development making at the beginning of germination only about 3% of the maximal rate (Tupy et al.,1986). Consistently with the rise of transcription activity, there is a dramatic increase of RNA. Total RNA and mRNA increase about 7times and 20times to reach the final levels of 230pg and 6.2pg per pollen grain and the mature pollen grain contains an amount of ribosomes ten times higher than the dividing microspore. Nearly all RNA present in mature pollen is thus made after microspore mitosis. Similar results were obtained recently and developmental mRNA diversity has been demonstrated (Schrauwen et al.,1990). The period of the 3rd stage is also characterized by an enhanced proteosynthesis and amylogenesis and by DNA replication in the generative nucleus (Figs1,2). The cycle of the vegetative cell is arrested at G1 (Zarsky et al.,submitted).

Some mRNAs synthesized during pollen development in the anther may not be translated before pollen germination (Capkova et al.,1988). The 69kDa glycoprotein, the most intensely synthesized protein in pollen tubes (Capkova et al.,1990) is absent in ungerminated pollen (Fig.3). The patterns of glycoproteins exhibit considerable changes also during pollen development in the anther. The most marked is the appearance of prominent species of about 110kDa at stage2 and their strong diminution or

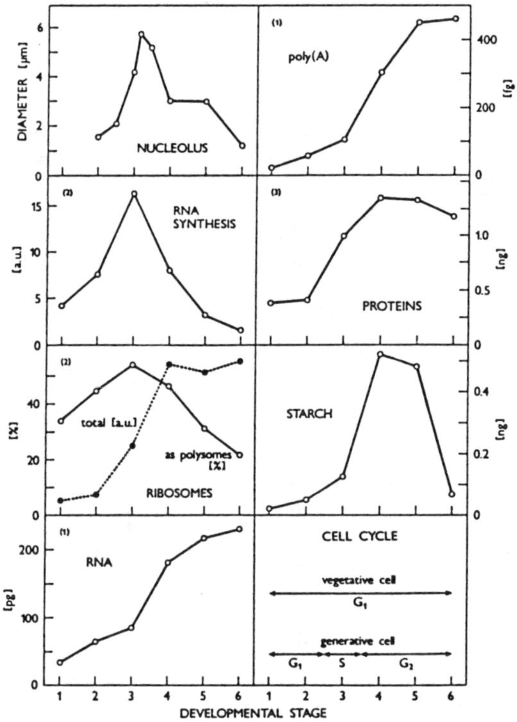

Fig.1. Outline of changes during *in situ* pollen differentiation and maturation.The levels in mass units are per pollen grain. The numbers in parenthesis are literature references (1-Tupy, 1982; 2-Tupy et al., 1983; 3-Zarsky et al., 1985).

disappearance in the next stages. Mature dry pollen is characterized by high levels of glycoproteins of about 60 and 90kDa.

The dramatic change after the 3rd stage from the rise to a rapid decline of the synthetic activities is accompanied by polysome disassembly, starch degradation (Fig.1) resulting in a rise of sucrose and by proline accumulation (data not given). High sucrose (Hrabetova and Tupy,1964) and proline (rev. Stanley and Linskens,1974) contents in mature pollen have been demonstrated in many species and were shown to play an important role in desiccation and heat tolerance (Hoekstra and Van Roekel,1988, Hoekstra et al.,1989, Mutters et al.,1989).

The developmental changes described indicate that the initial 3 stages of pollen ontogeny can be considered as gametophytic differentiation and that this phase proceeds between stages 3 and 4 to the phase of maturation in terms of transition to dormancy implying desiccation and protective changes against environmental stress factors. After the 2nd but before the 3rd stage the pollen population consists mainly of mid-binucleate grains and can undergo dedifferentiation and pollen embryogenesis (Kyo and Harada,1986).

Germination of in situ differentiated immature pollen on stigma and in vitro

Pollen from stages 2 to 6 was applied on stigmas of flowers on the day of anthesis or cultivated in sugar-mineral medium with MES-KOH buffer pH 5.9 (Tupy and Rihova,1984) and its germination was followed at 25 oC during 3 or 4 days. Except for stage 2,

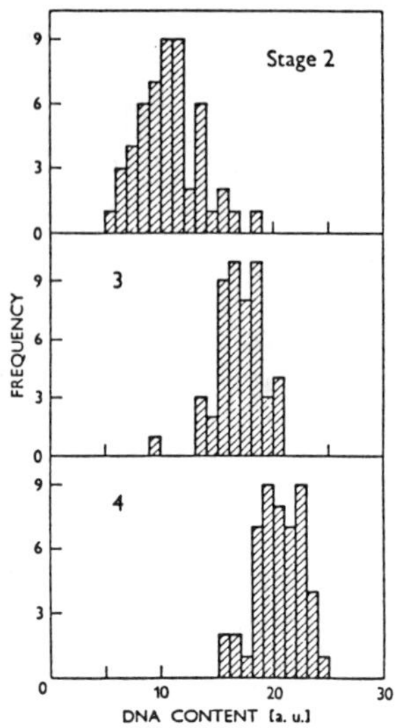

Fig.2. Cytophotometry of Feulgen stained nuclei in generative cells at pollen stages 2,3 and 4.

immature pollen was able to proceed on the stigma to the phase of function and produced normally growing tubes. The beginning of germination in pollen of stages 3,4,5 and 6 was delayed by about 3,2,1 and 0.8 days with respect to the mature dry pollen (Fig.4). In comparison with the normal course in the anther, pollen development on the stigma to the stage of germination was thus about one day shorter from stages 3,4 and 5, but nearly the same from stage 6. This further suggests that like in maize pollen (Lin and Dickinson,1984), tobacco pollen desiccation is not necessary for good germination. This could be confirmed by in vitro cultivation. Pollen of stages 3,4,5 and 6 started germination after 20,24,28 and 36 hours, that is with similar delay as on stigma in the case of stages 5 and 6, but about 1 and 1.5 days earlier in stages 4 and 3. The pollen tubes grew at similar rate and to similar length as in normally matured pollen (data not shown). Pollen of stage 2 germinated very poorly (3-5%) and produced only short tubes.

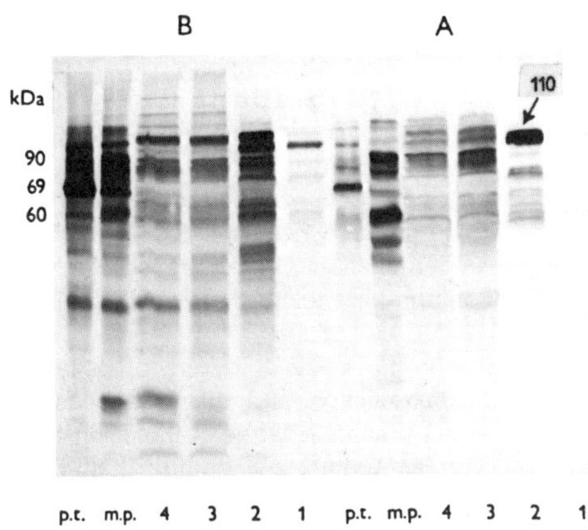

Fig. 3. SDS-PAGE pattern of soluble (A) and insoluble (B) glycoproteins extracted from immature pollen at stages 1,2,3,4, mature pollen (m.p.) and from pollen tubes (p.t.) grown for 22 hours.

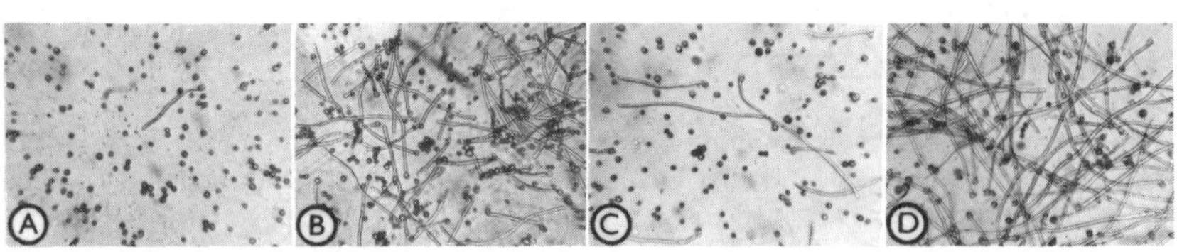

Fig. 4. Pollen tube lengths in styles 1,2 and 3 days after pollination with immature (stages 3-6) and mature (m.p.) pollen.

The seed set after pollination with immature pollen strongly depended on pollen stage. It was reduced, in comparison with normal pollination, by about 1/3 in stages 5 and 6 and by 75 and 97% in stages 4 and 3. One reason for this reduction could lie in the delay of pollen germination and in ovule abortion before fertilization.

Production of fertile dormant pollen from early microspores in vitro

In the earlier work (Tupy et al.,1991) pollen differentiation from microspore and pollen maturation to dormancy was achieved by successive culture of late microspores for 3-4 days in medium M1, 1 day in M1+medium P (1:1) and 1 day in M1+P (1:3). Mature pollen grains were stored on filters or in P at -20 oC, and the pollination with the stored pollen produced up to 300 viable seeds per pod. Using this protocol, early microspores liberated from tetrads were cultivated for various times in M1 and starch deposition in pollen grains was registered as possible marker of optimal phase for lowering water potential of the differentiation medium to induce maturation.

The frequency of pollen grains filled with starch increased to the 5th day in medium M1 and decreased during the next 3 days. Medium P stopped starch increase when added on the 4th day and slightly reduced its decrease after 5 and 6 days of culture in medium M1. The best germination and tube growth in the pollen produced from early microspores was achieved when the culture in M1 was supplied with medium P at the highest frequency of pollen grains with starch, and seeds were obtained after pollination with this pollen. Relatively high values for both, frequency of grains with starch and pollen germination, were obtained for late microspores cultivated in M1 for 4 days. Starch degradation during the phase of maturation is, however, not necessary for germination, since in pollen of stages 3 and 4 cultivated in vitro the starch content did not decrease prior to germination. The essential role of the medium P is in lowering of water potential as suggested by production of pollen with 19% germinability when using 50% PEG 4000 instead of medium P.

4. References

Capkova, V.; Hrabetova, E.; Tupy, J. Protein synthesis in pollen tubes: preferential formation of new species independent of transcription. Sex. Plant Reprod. 1:150-155; 1988.

Capkova, V.; Hrabetova, E.; Tupy, J. Glycoprotein occurrence and synthesis in pollen tubes. In: Proc. XIth Internat. Symp. on Embryology and Seed Production, Leningrad 1990 (in press).

Heberle-Bors, E. Isolated pollen culture in tobacco: plant reproductive development in a nutshell. Sex. Plant Reprod. 2:1-10; 1989.

Hoekstra, F.A.; Van Roekel, T. Desiccation tolerance of Papaver dubium L. pollen during its development in the anther: possible role of phospholipid composition and sucrose content. Plant Physiol. 88:626-632; 1988.

Hoekstra, F.A.; Crowe, L.M.; Crowe J.H. Differential desiccation sensitivity of corn and Pennisetum pollen linked to their sucrose contents. Plant, Cell Environ. 12:83-91; 1989.

Hrabetova, E.; Tupy, J. The growth effect of some sugars and their metabolism in pollen tubes. In: Linskens, H.F. (ed.) Pollen Physiology and Fertilization. Pp. 95-101. North-Holland Publ. Co., Amsterdam 1964.

Kyo, M.; Harada, H. Specific phosphoproteins in the initial period of tobacco pollen embryogenesis. Planta 182:58-63; 1990.

Lin, J.-J.; Dickinson, D.B. Ability of pollen to germinate prior to anthesis and effect of desiccation on germination. Plant Physiol. 74:746-748; 1984.

Lustinec, J.; Hadacova, V.; Kaminek, M.; Prochazka, Z. Quantitative determination of starch amylose and amylopectin in plant tissues using glass filter paper. Anal. Biochem. 132:265-271; 1983.

Mutters, R.G.; Ferreira, L.G.R.; Hall, A.E. Proline content of the anthers and pollen of heat-tolerant and heat-sensitive cowpea subjected to different temperatures. Crop Sci. 29:1497-1500; 1989.

Schrauwen, J.A.M.; de Groot, P.F.M.; van Herpen, M.M.A.; van der Lee, T.; Reynen, W.H.; Weterings, K.A.P.; Wullems, G.J. Stage-related expression of mRNAs during pollen development in lily and tobacco. Planta 182:298-304; 1990.

Stanley, R.G.; Linskens, H.F. Pollen - Biology, Biochemistry, Management. Springer-Verlag, Berlin-Heidelberg-New York 1974.

Tupy, J. Alterations in polyadenylated RNA during pollen maturation and germination. Biol. Plant. 24:331-340; 1982.

Tupy, J.; Rihova, L. Changes and growth effect of pH in pollen tube culture. J. Plant Physiol. 115:1-10; 1984.

Tupy, J.; Suss, J.; Hrabetova, E.; Rihova, L. Developmental changes in gene expression during pollen differentiation and maturation in Nicotiana tabacum L. Biol. Plant. 25:231-237; 1983.

Tupy, J.; Suss, J.; Rihova, L. RNA synthesis and ribosome status in pollen tube growth of Nicotiana tabacum L.; Effects of external pH. J. Plant Physiol. 123:467-476;, 1986.

Tupy, J.; Rihova, L.; Zarsky, V. Production of fertile tobacco pollen from microspores in suspension culture and its storage for in situ pollination. Sex. Plant Reprod. (in press).

Zarsky, V.; Capkova, V.; Hrabetova, E.; Tupy, J. Protein changes during pollen development in Nicotiana tabacum L. Biol. Plant. 27:438-444, 1985.

50
Pollen Tube Growth and Fertilization in *Lilium longiflorum* after Stigmatic and Placental Pollination

Juliette Janson

In Lilium longiflorum the pollen grains meet a large amount of exudate at the stigma when pollination is carried out. The pollen tubes grow through the exudate at the stigma, in the style and in the ovary. In L. longiflorum stigmatic exudate differs in composition from that produced in the style (Dickinson, Moriarty and Lawson, 1982; Miki-Hirosige, Hoek and Nakamura, 1987).
Although in L. longiflorum and L. regale the ultrastructure of the stigma papillae differs from that of the secretion cells in the style, similarities occur in the organelles observed and the presence of a transfer wall, which is in case of a papillar cell not as pronounced as in the secretory cells in the style (Dashek, Thomas and Rosen, 1971; Dickinson, Moriarty and Lawson, 1982; Miki- Hirosige, Hoek and Nakamura, 1987; Rosen, 1971; Rosen and Thomas, 1970; Vasil'ev, 1970). The relation between exudate and pollen tube growth has been a topic of study since many years. Pollen tubes take up substances from the exudate while growing through the style (Kroh, Miki-Hirosige, Rosen and Loewus, 1970; Labarca and Loewus, 1972 and 1973). Because of the hollow style of L.longiflorum the exudate serves not only as a source of nutrition but is also the medium for the pollen tubes to grow through.

Exudate production and pollen tube growth after stigmatic pollination

Exudate of L. longiflorum is largely rinsed away when using traditional fixation methods. To study the production of exudate low temperature scanning electron microscopy, CRYO-SEM, was used. Pieces of pistil material were frozen in solid nitrogen to preserve the position of the exudate. The material was subsequently cryo-fractured so that the observations were not made at the original cutting surface at which the exudate had probably been dislocated. The production of stigma and style exudate starts before the flower bud opens. In a bud of 80 mm, approximately 8 days before anthesis, some fluid is already covering the secretion cells in the style (Fig. 1). At least initially the secretion at the stigma and in the style is not evenly distributed over the surface of the exudate producing cells, leading to an exudate layer of varying thickness (Fig. 2). This is especially obvious at the secretion cells in the top of the style. Here the exudate produced is at first restricted to each secretion cell, and as production proceeds those accumulations merge (Fig.3). Exudate production proceeds from the top of the style towards the basis (compare Fig. 3 with Fig. 4). At two days after anthesis the exudate produced by the different cells in the top of the style has merged into an almost continuous layer. The stylar canal has three lobes and the exudate is preferentially gathering in the most abaxial parts. In between the lobes the layer of exudate is thinner and has an irregular surface (Fig. 5). This uneven distribution of exudate over the secretory cells is remarkable, because of the liquid nature of the substance. It might be that the cuticle is still restricting the movement. The fate of the cuticle could not be observed using this technique. In the ovary exudate is present at the time pollen tubes arrive. The ultrastructure of the secretory cells of the placenta (Fig. 6) is similar to those in the style, as described by several authors (see introduction). However, in the style the secretion cells are long in longitudinal direction compared with the spherical cells at the placenta. At anthesis a huge amount of exudate is covering the stigma. After pollination and germination the pollen tubes grow in all directions, but helped by the morphology of the stigma finally grow through one of the three slits into the style. Two days after pollination and in the top of the style the pollen tubes and secretory cells are separated by a layer of exudate (Fig. 7). Half way down the style there is less exudate with less pollen tubes at that moment resulting in a smaller distance between the two (Fig. 8). After at least two days the pollen tubes have reached the end of the style. In the ovary the pathway of pollen tube growth is restricted to exudate-covered areas. From the pollen tube bundles formed in between the two rows of ovules in each of the three cavities, individual pollen tubes bend to grow in between the ovules towards the micropylar side. Arriving at this side the secretory tissue is restricted close to the funiculus, being absent at the pericarp. The pollen tubes bend to stay in the proximity of these secretory cells, leading them along the inner integuments of the ovules. To reach the micropyles, however, growth from this zone is needed. The space between the micropyle, the secretory zone and the pericarp is filled with exudate, serving again as a growth medium for the pollen tube. In fact the complete pathway of the pollen tube is restricted to the presence of exudate. The contact between the pollen tubes on one hand and the stigma papillae and the secretory cells in the style and probably also in the ovary on the other seems to be limited. Communication between the two has to take place via the exudates, which seems the only way the pistil can interfere with the pollen tube growth. In flower buds in which the embryo sacs have not reached their ripe stage yet, the pollen tube growth is slowed down. A part of the embryo sacs happen to be ripe when the pollen tubes finally arrive in the ovary. The development of the embryo sacs is not influenced by the pollen tube growth. Before the entry into the embryo sac no degeneration of the synergid was observed. The polar nuclei have not fused yet. The fusion of the sperm nucleus with the nucleus of the egg cell probably takes a few days to complete. After 13 days the first division of the zygote was observed.

Pollen tube growth after placental pollination

Using placental pollination in vitro, fluid from the culture medium runs over the placenta and in between the ovule and the ovary wall. After germination at the placenta the pollen tube growth is initially still directed from the placenta to the micropylar side of the ovules. Although the inner integuments are often covered with pollen tubes, only a few enter the micropyle, after this placental pollination. Due to the accumulation of starch in the nucellus and in the outer integument, embryos are hard to detect. The development of the embryos originating from placental pollination is much slower compared with in vivo embryo formation.

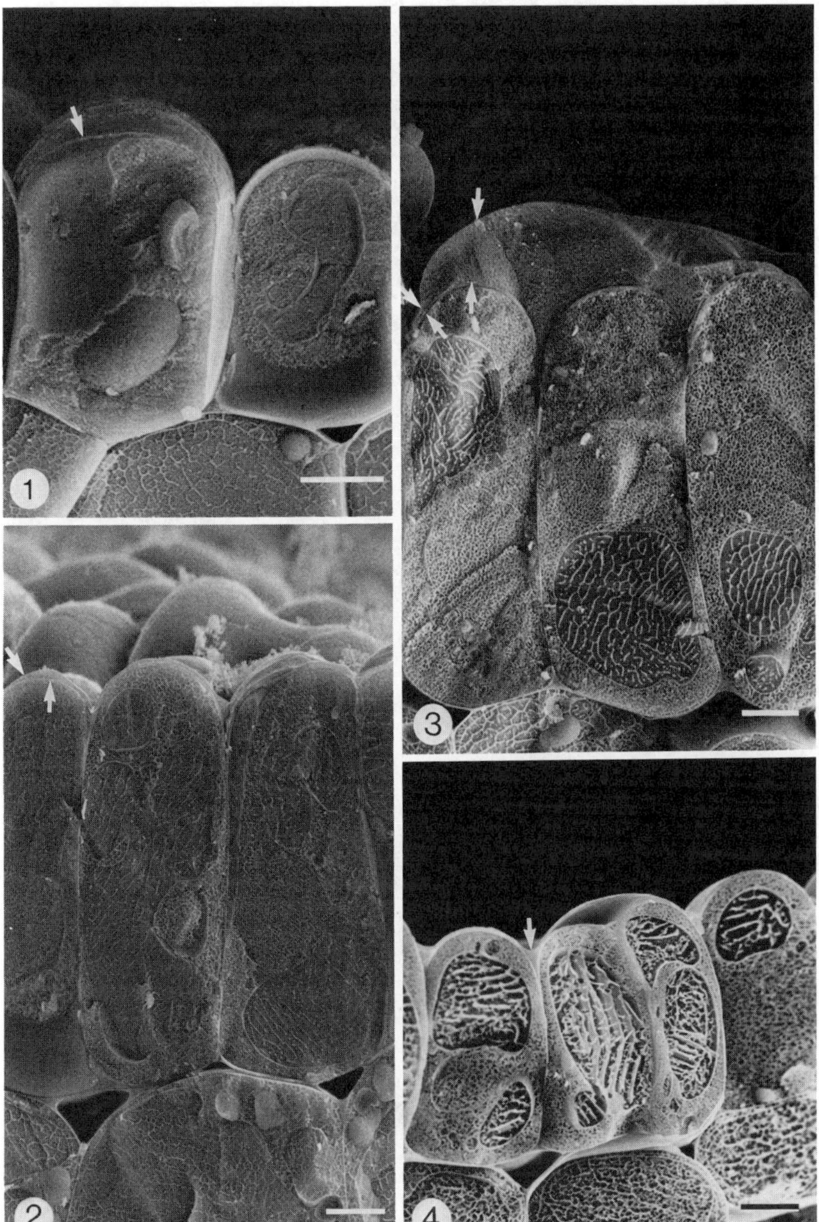
1
2
3
4

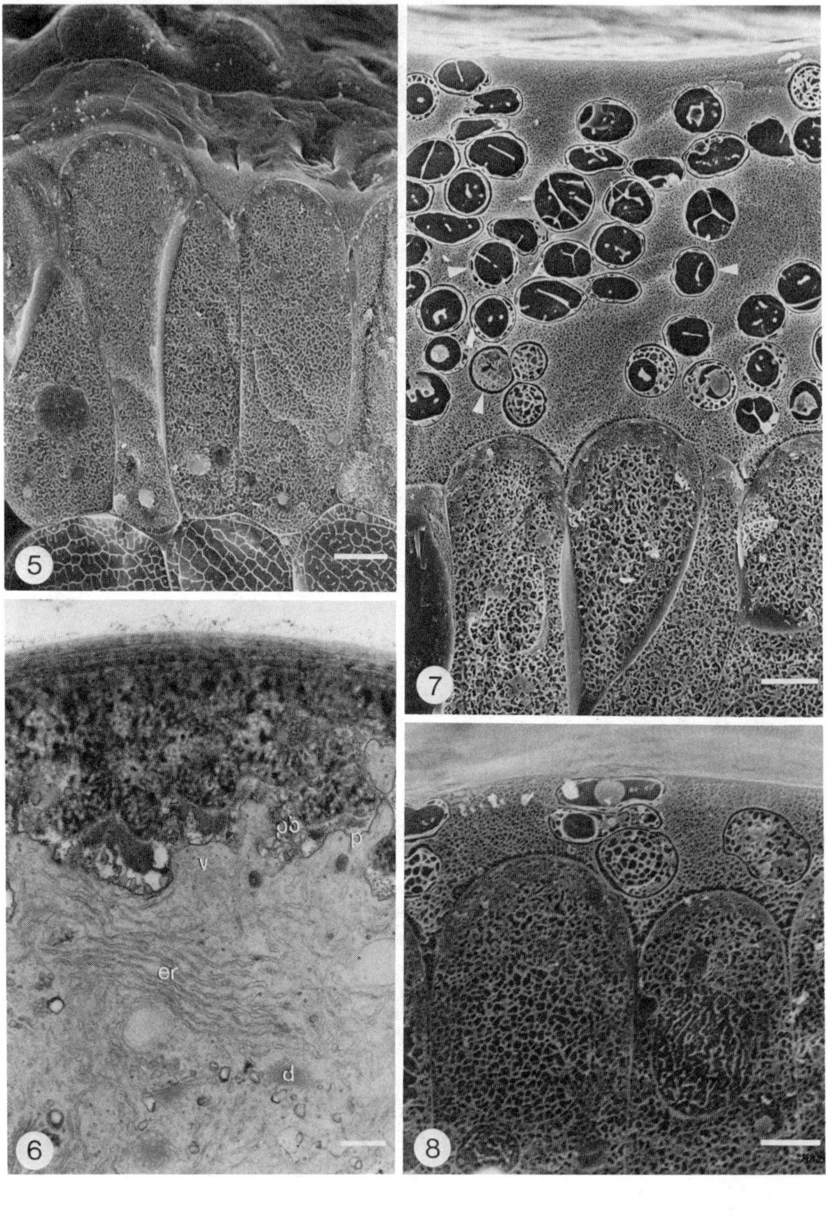
5
6
pb
p
v
er
d
7
8

Legends

Figs. 1 to 5, 7 and 8 are scanning electron micrographs of secretory cells of a transversely broken hollow style of *Lilium longiflorum* cv. Gelria. Pollen tube growth was observed two days after pollination, which was carried out two days after anthesis, with pollen of cv. White American. The bar represents 10 μm.

Fig. 1. Secretory cells in the style of a flower bud of 80 mm, approximately 8 days before anthesis. Some exudate is covering the cells (arrow).

Fig. 2. Secretory cells with exudate (arrows) in the top of the style of a flower bud of 100 mm, approximately 6 days before anthesis.

Fig. 3. Secretory cells in the top of the style of a bud of 120 mm, approximately 3 days before anthesis. Note the varying thickness of the exudate layer (arrows). The exudate accumulations of a few cells have merged.

Fig. 4. Secretory cells at the basis of the style of a flower bud of 120 mm. Exudate is gathering in between the cells (arrow). Compare the difference in the amount of produced exudate with Fig. 3.

Fig. 5. Secretory cells in between the lobes of a style of a flower two days after anthesis. The surface of the exudate layer is irregular.

Fig. 6. A secretory placental cell as observed with the transmission electron microscope. The bar represents 500 nm. Note the wall ingrowths against the more 'fibrillar' cell wall. Abbreviations : d dictyosome, er endoplasmic reticulum, p plasmalemma, pb paramural body, v vesicle.

Fig. 7. Pollen tubes (arrowheads) growing in the thick layer of exudate in the top of the style.

Fig. 8. Half way down the style there are less pollen tubes in a thinner layer of exudate when compared with Fig. 7.

Acknowledgements

The author wishes to thank Siep Massalt and A.B. Haasdijk for the artwork, dr.ir.C.J. Keijzer for critically reading the text and the 'LEB fonds' for the financial support.

References

Dashek, W.V.; Thomas, H.R.; Rosen, W.G. Secretory cells of lily pistils. II. Electron microscope cytochemistry of canal cells. Amer. J. Bot. 58: 909-920; 1971.

Dickinson, H.G.; Moriarty, J.; Lawson, J. Pollen-pistil interaction in Lilium longiflorum: the role of the pistil in controlling pollen tube growth following cross- and self-pollinations. Proc. R. Soc. Lond. B 215: 45-62; 1982.

Kroh, M.; Miki-Hirosige, H.; Rosen, W.; Loewus, F. Incorporation of label into pollen tube walls from myoinositol-labelled Lilium longiflorum pistils. Plant Physiol. 45: 92-94; 1970.

Labarca, C.; Loewus, F. The nutritional role of pistil exudate in pollen tube wall formation in Lilium longiflorum. I. Utilization of injected stigmatic exudate. Plant Physiol. 50: 7-14; 1972.

Labarca, C.; Loewus, F. The nutritional role of pistil exudate in pollen tube wall formation in Lilium longiflorum. II. Production and utilization of exudate from stigma and stylar canal. Plant Physiol. 52: 87-92; 1973.

Miki-Hirosige, H.; Hoek, I.H.S.; Nakamura, S. Secretions from the pistil of Lilium longiflorum. Amer. J. Bot. 74: 1709-1715; 1987.

Rosen, W.G. Pistil-pollen interactions in Lilium. In: Heslop-Harrison, J., ed. Pollen: Development and physiology. London: Butterworth; 1971:p. 239-254.

Rosen, W.G.; Thomas, H.R. Secretory cells of lily pistils. I. Fine structure and function. Amer. J. Bot. 57: 1108-1114; 1970.

Vasil'ev, A.E. Ultrastructure of stigmatoid cells in Lilium. Sov. Plant Physiol. 19: 1035-1044; 1970.

51
Isolation of Generative Cells and Sperm from Pollen Tubes of *Lilium*

D. Southworth, R. Copley, and S. Kwiatkowski

Following meiosis, haploid cells do not differentiate directly into sperm cells; rather they divide to form vegetative and generative cells. The latter divide into sperm cells which function in double fertilization.

The changes that occur in differentiation of generative cells to functional sperm are unknown. The minimum change would be a division of the cytoplasm and organelles. The value of this division would lie in the production of two cells essential for double fertilization. Because only one pollen tube penetrates an ovule, two sperm from one pollen tube are required. Maturation of generative cells would be equivalent to sperm development, except for nuclear division.

Alternatively, changes from generative cells to sperm might include differentiation into sperm with distinct surface receptors, different cellular organelles, and different proteins. Gene expression during sperm development could create sperm distinct from generative cells.

To compare generative cells and sperm, we need a system in which the two cell types can be isolated for protein analysis. In tricellular pollen, generative cells divide in anthers. Although sperm can be obtained from tricellular pollen at anthesis, isolation of generative cells requires separation of immature pollen from anthers.

In bicellular pollen, generative cells divide in pollen tubes. By managing the timing of division so that generative cells are isolated before division and sperm after division, both cell types can be obtained.

We report a method for isolation of generative cells and sperm from pollen tubes of Lilium longiflorum.

Materials and Methods

Pollen of L. longiflorum was collected on the day of anthesis from partially open buds. Sterile forceps were used to remove anthers to sterile Petri dishes for drying overnight. Dry anthers were stored frozen.

Pollen tubes were germinated in sterile medium containing 0.3 M sucrose, 0.16 mM H_3BO_3, 1.3 mM $Ca(NO_3)_2$ and 1.0 mM KNO_3 for 6 to 72 hr at 25° C with reciprocal shaking (Dickinson, 1965).

To discharge germ cells, the mass of pollen tubes was removed with forceps into fresh medium. An equal volume of distilled water was added as an osmotic shock (Yang and Zhou, 1989). To test the effect of variations in hypotonic media on quality of released cells, 10 mM ascorbic acid, 10 mM HEPES buffer, 0.3% polyvinylpyrrolidone, and 1% bovine serum albumin were added to hypotonic shock media. The germination medium without sucrose was also tested. The fraction of nuclei present in whole cells and of FDA-positive cells were scored.

The suspension was shaken on ice for 30 min, filtered through a 62 uM nylon mesh filter into an equal volume of germination medium containing 0.6 M sucrose for a final concentration of 0.38 M sucrose, and centrifuged at 2,000 x g for 10 min. Sperm and generative cells formed a thin white pellet. Cells were stained with 4',6-diamidino-2-phenyl-indole (DAPI) or fluorescein diacetate (FDA).

For gel electrophoresis, proteins were extracted from isolated generative cells, sperm, or whole pollen by boiling for 3 min with equal volumes of extraction buffer (0.125 M tris-Cl, pH 6.8, 4% SDS, 20% glycerol, and 10% mercaptoethanol) and isolation medium, further solubilized for 4 hr at 21° C, and centrifuged. Supernatants were run on a 12% acrylamide gel and stained with coomassie blue.

Results

Spherical germ cells were FDA-positive (Figs. 1, 2); cells that were DAPI-positive were FDA-negative (Figs. 3, 4). Sperm or generative cells appeared as free nuclei, nuclei with partially attached cytoplasm, or whole cells. Nuclei were oval to elongate, or curved. Vegetative nuclei (Fig. 5) were also observed, sometimes attached to one sperm or generative cell. Generative cells responded to changes in osmoticum. A decrease of solutes from 0.6 M to 0.3 M sucrose resulted in a doubling of volume in FDA+ cells (Figs. 6-8).

Addition of ascorbic acid, polyvinylpyrrolidone, and bovine serum albumin or HEPES buffer did not improve the fraction of germ cells released as intact cells.

Generative cells divided asynchronously. After 17 hr in culture, most nuclei had condensed chromatin in prophase (Fig. 9). At 18-22 hr, nuclei had elongated; chromosomes were visible (Fig. 10, 11). At 30 hr, chromosomes were in metaphase and anaphase with some cytokinesis. At 47 hr, nuclei were in telophase with shortened, condensed chromosomes in two separate nuclei (Fig. 12).

From 0 to 18 hr, generative cells could be isolated. From 18 to 48 hr, dividing cells ruptured and released the entire mitotic apparatus. After 48 hr, the number of mitotic figures declined. Few were seen after 57 hr; however, isolated germ cells would include undivided generative cells as well as sperm (Figs. 13-16).

Chromatin in both generative cells and sperm was more condensed than that of vegetative nuclei. No distinction could be made in the pattern of DAPI staining between generative cells and sperm. Nuclei of generative cells were larger than those of sperm.

Figs. 1-16. *Lilium longiflorum*. Bar = 20 μm. Figs. 1-4. Isolated generative cells, 7 hr. 1, 3. Phase contrast. 2. FDA. 4. DAPI. Fig. 5. Isolated vegetative nucleus, DAPI. Figs. 6-8. Isolated generative cells, 6 hr. phase contrast. 6. In 0.15 M sucrose medium. 7. In 0.3 M sucrose medium. 8. In 0.6 M sucrose medium. Figs. 9-12. Generative cell division, DAPI. 9. Prophase, 17 hr. 10. Late prophase, 18 hr. 11. Anaphase, 30 hr. 12. Telophase, 47 hr. Figs. 13-16. Isolated sperm cells, 72 hr. 13, 15. Phase contrast. 14, 16 DAPI. (Southworth, unpublished.)

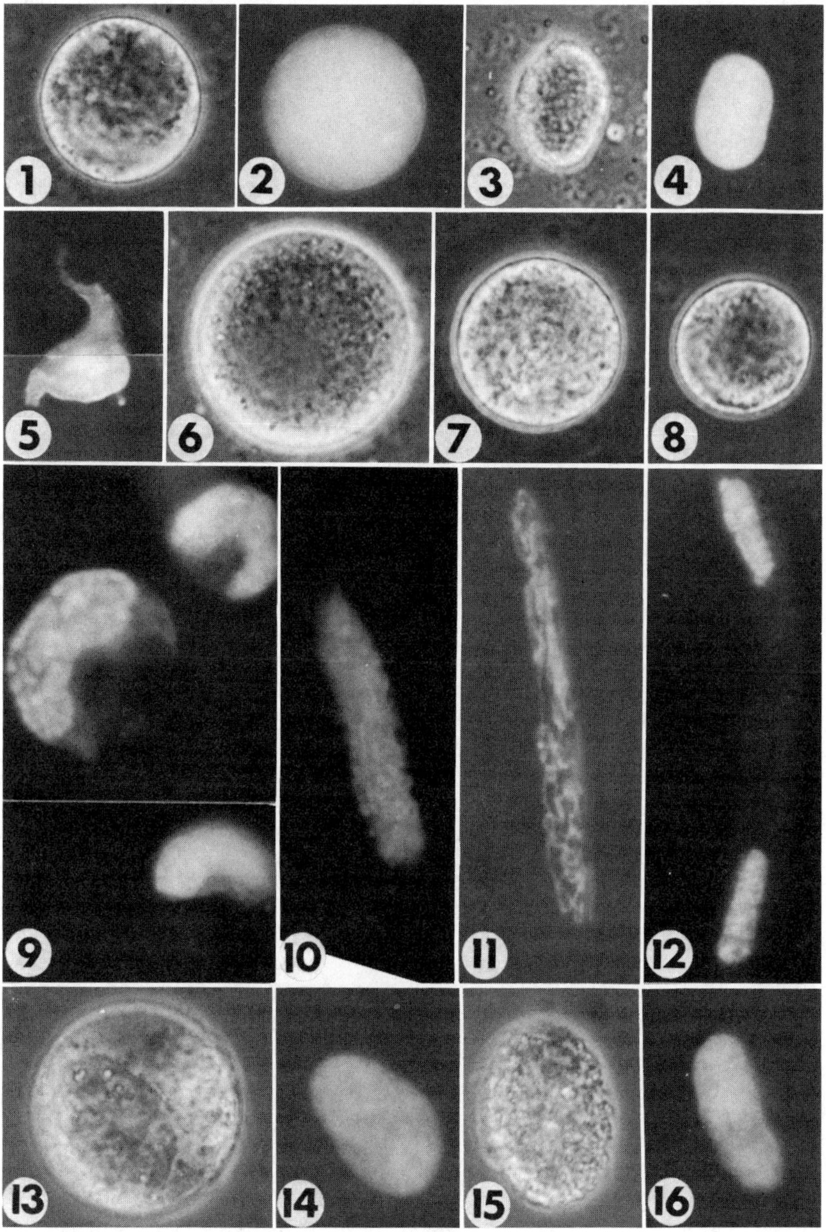
1
2
3
4
5
6
7
8
9
10
11
12
13
14
15
16

The major proteins of sperm and generative cell fractions were similar with predominant bands at 14.5, 18, 20, 31, and 59 kD, and numerous other bands. No difference between proteins of sperm and generative cells could be detected by one-dimensional SDS-PAGE.

Discussion

Generative cells of L. longiflorum will divide in shake culture. Generative cell division has been reported in other species of Lilium and in Tradescantia (O'Mara, 1933; Lafleur et al., 1981). Mitosis occurs with no definite metaphase plate (Palevitz, 1989). Paired chromosomes are arranged along the pollen tube. In anaphase, they slide past each other. No cell plate forms.

Tanaka (1988) isolated generative cells from mature pollen of L. longiflorum by enzymatic digestions, first of intines and then of the generative cell wall. We were unsuccessful in isolating generative cells with enzymes. Because generative cells reported here are in pollen tubes, they are further into the cell cycle and are no longer enclosed by a cell wall.

Proteins from sperm and generative cells of L. longiflorum were similar. In Triticum, there were few differences in proteins between anthers containing late bicellular pollen and those with early tricellular pollen (Vergne and Dumas, 1988). Protein differences were found between vegetative cells and sperm in Plumbago (Geltz and Russell, 1988).

Sperm and generative cells isolated by osmotic shock from pollen tubes showed similar structure by light microscopy, and similar proteins by SDS-PAGE.

Acknowledgements

This work was supported by National Science Foundation Grant DCG-8801902 through Research at Undergraduate Institutions.

References

1. Dickinson, D. B. Germination of lily pollen: respiration and tube growth. Science 150:1818-1819; 1965.
2. Geltz, N. R.; Russell, S. D. Two-dimensional electrophoretic studies of the proteins and polypeptides in mature pollen grains and the male germ unit of Plumbago zeylanica. Plant Physiol. 88:764-769; 1988.
3. Lafleur, G. J.; Gross, A. E.; Mascarenhas, J. P. Optimization of culture conditions for the formation of sperm cells in pollen tube of Tradescantia. Gamete Res. 4:35-40; 1981.
4. O'Mara, J. Division of the generative nucleus in the pollen tube of Lilium. Bot. Gaz. 94:567-578; 1933.
5. Palevitz, B. A.; Cresti, M. Cytoskeletal changes during generative cell division and sperm formation in Tradescantia virgianiana. Protoplasma 150:54-71; 1989.
6. Vergne, P.; Dumas, C. Isolation of viable wheat male gametophytes of different stages of development and variations in their protein patterns. Plant Physiol. 88:969-972; 1988.
7. Tanaka, I. Isolation of generative cells and their protoplasts from pollen of Lilium longiflorum. Protoplasma 142:68-73; 1988.
8. Yang, H-Y.; Zhou, C. Isolation of viable sperms from pollen of Brassica napus, Zea mays, and Secale cereale. Chinese J. Bot. 1:80-84; 1989.

52

New Interspecific Hybrids in the Genus *Medicago* through In Vitro Culture of Fertilized Ovules

Marco Piccirilli and Sergio Arcioni

In the genus *Medicago* interspecific crosses are often prevented by post-zygotic mechanisms of incompatibility, that interfere with the development of the hybrid embryos (Oldemeyer, 1956; Fridricksson and Bolton, 1963; Sandguen et al., 1983; McCoy, 1985). As a consequence, germplasm sources available for alfalfa improvement are limited to the *M. sativa* complex, which includes the species *M. falcata, M. media* and *M. glutinosa*, that are freely crossable with *M. sativa* (Lesins and Lesins, 1979). Out of this group, conventional sexual cross between *M. sativa* and other *Medicago* species has succeeded in only a few cases (Clement, 1963; Lesins, 1961a, 1961b, 1962, 1968, 1970, 1972; Lesins and Lesins, 1979; Sandguen et al., 1982).

Ovule-embryo culture (McCoy, 1985), a technique suitable for overcoming post-zygotic mechanisms of incompatibility existing in the genus *Medicago*, has allowed the recovery of new hybrids between *M. sativa* and other perennial *Medicago* species (McCoy, 1985; McCoy and Smith, 1986; McCoy and Quarisa, 1989).

Table 1 - Composition of the medium used for the ovule-embryo culture (mg l-1).

Macroelements		Microelements	
KNO3	2500	MnSO4x4H2O	10
MgSO4	400	H3BO3	5
CaCl2	200	ZnSO4x7H2O	1
KH2PO4	355	KI	1
(NH4)2SO4	991	CuSO4x5H2O	0.2
		NaMo2O4x2H2O	0.1
		CoCl2x6H2O	0.1
Organics		FeSO4x7H2O	15
Inositol	1000	Na2EDTA	20
Nicotinic Acid	5		
Thiamine HCl	5		
Piridoxine HCl	0.5		
Proline	11000		
Maltose	30000		

M. rugosa and *M. scutellata* are the only polyploid annual *Medicago* species known

(2n=30; Bauchan and Elgin, 1984); together with *M. noëana* (annual; 2n=16) they are highly resistant to alfalfa insect pests, due to the presence of dense glandular hairs on their stems and leaves (Barnes and Ratcliffe, 1969; Shade et al., 1975, 1979). *M. arborea*, (perennial; 2n=4x=32), is the only *Medicago* with a shrubby growth habit; it is resistant to environmental stresses and is characterized by an appreciable forage yield during winter. In this study, these four species were crossed with *M. sativa* in order to introduce their interesting traits into cultivated alfalfa germplasm. Since these species are totally incompatible with alfalfa, the ovule-embryo culture technique was used to recover hybrids, and in particular, between *M. rugosa* and *M. sativa*.

To our knowledge, this is the first report concerning fertile and non-chimeral hybrid plants between an annual and a perennial species of Medicago.

Materials and Methods

Normal tetraploid, male fertile plants of *M. sativa* (2n=32), cv. Adriana, were used as the female parent. Before crossing, alfalfa flowers were emasculated and to eliminate possible contaminant pollen grains present on the surface of the stigmas, they were treated with 50% (v/v) ethanol (30 sec), rinsed in distilled water (1 min) and wiped off. Flowers were then hand pollinated with the pollen taken from the wild species and covered with a paper bag.

Developing legumes were removed from the mother plant 10 days after pollination; they were sterilized in 70% (v/v) ethanol, rinsed in sterile distilled water and aseptically dissected at the stereoscope. Fertilized ovules, that could be distinguished due to their enlarged size, were transferred to 5 cm Petri dishes containing 7 ml of a hormone-free medium (table 1) derived from SH medium (Schenk and Hildebrandt, 1972). Cultures were incubated under fluorescent light (23°C, 27 µE s-1m-2, 12 h photoperiod). After one week, viable embryos were excised from ovules, subcultured onto the same medium and maintained under the above reported conditions. Two weeks later, unifoliate plantlets were transferred to vessels containing the same medium but without proline. When plantlets had 5-7 trifoliate leaves, they were transplanted into Jiffy pots and maintained under a plastic bag for about two weeks to protect them from dessication.

Cytological analysis of somatic chromosomes was conducted on root tips obtained from plants that were vegetatively propagated. Roots were pre-treated in a saturated aqueous solution of α-bromonaphthalene for 4 h at room temperature and fixed in ethanol:acetic acid (3:1). Root tips were then stained using the Feulgen method, squashed in 1.5% (v/v) acetic-orcein and observed at the microscope.

Electrophoretic assays were conducted on proteins extracted from vegetatively propagated plants. For peroxidase analysis, 200 mg of shoot apexes and 100 mg of rootlets were crushed with 3 drops of chilled gel buffer and the juice was absorbed in Whatman no. 3 filter paper wicks (0.5x1.0 cm). The gel (15x18 cm) consisted of 11% (w/v) hydrolized potato starch in Tris-citric acid buffer (1.52x10-2M and 3.6x10-3M, respectively; pH=7.8). The wicks were inserted across the gel 5 cm from the cathode; the buffer was a solution of 0.3 M H3BO3 and the pH was adjusted to 7.8 with NaOH. Electrophoresis was carried out at 5°C and 150 V for 30 min, thereafter the wicks were removed and the voltage was increased to 300 V. Gels were then sliced into three layers (2 mm thick), stained for peroxidases (Hoyle, 1978) and fixed in 50% (v/v) ethanol.

For the analysis of esterases, 100 mg (fresh weight) of leaves were taken from hybrid and parent plants. Leaves were frozen in liquid nitrogen, homogenized with 100 μl of 0.2 M Tris-glycine buffer (pH=8.9) and centrifuged (12.000xg at 5°C; 30 min). The supernatant was removed and its protein concentration was determined according to Bradford (1976), using bovine albumine as a standard. For electrophoresis, 400 μg of protein solubilized in 100 μl of 0.2 M Tris-glycine (pH=8.9) containing 5% (v/v) glycerol were used. Discontinuous polyacrylamide gels (2 mm thick; staking gel: 3.5% acrylamide, 1 cm long; separating gel: 7.5% acrylamide, 8 cm long) were allowed to run for 4 h at 150 V and 20 mA on a vertical electrophoresis unit maintained at 5°C. Gels were stained for esterases (Kahler and Allard, 1970) and fixed in 50% (v/v) ethanol.

Results

The results of the four interspecific crosses are reported in table 2. From the *M. sativa* x *M. noëana* and *M. sativa* x *M. scutellata* crosses it was not possible to obtain any plants, even if some viable embryos were recovered; in fact, all these embryos were very weak and died during the *in vitro* culture or when transferred to the soil. On the contrary, it was possible to sustain the growth of some embryos derived from the *M. sativa* x *M. rugosa* and *M. sativa* x *M. arborea* crosses and some plants were obtained.

Table 2 - Results of the crosses between *M. sativa* and other wild species of *Medicago*.

male parent	flowers pollinated	legumes set	viable embryos	hybrid plants
M. noeana	1160	36	24	0
M. rugosa	1088	24	56	16
M. scutellata	504	12	12	0
M. arborea	1232	52	56	3

M. sativa x M. rugosa

The plants derived from this cross had some characters in common with both parents; in fact, as *M. sativa*, they were perennial, had purple flowers and did not have glandular hairs, but at the same time, they had fibrous roots and reduced height, like *M. rugosa* and the other annual species. Some of their morphological traits were intermediate to those of the parents; this was particularly evident with the size and shape of the leaves and of the stipules, (fig. 1).

Cytological analysis confirmed the hybrid origin of the plants, since the somatic metaphases consisted of 31 chromosomes (fig. 2), i.e. the sum of the haploid complements of the two parental species. Of particular significance was the presence of three nucleolar chromosomes, two coming from *M. sativa*, that has four, and one coming from *M. rugosa*, that has only two. In fact, it was possible to recognize the nucleolar chromosome provided by *M. rugosa*, since in this species it is characterized by a marked secondary constriction and by a large satellite, whose length constitutes nearly one half of the whole chromosome (Mariani and Falistocco, 1991).

The analysis of peroxidase isoenzymes provided a further confirmation of the hybrid

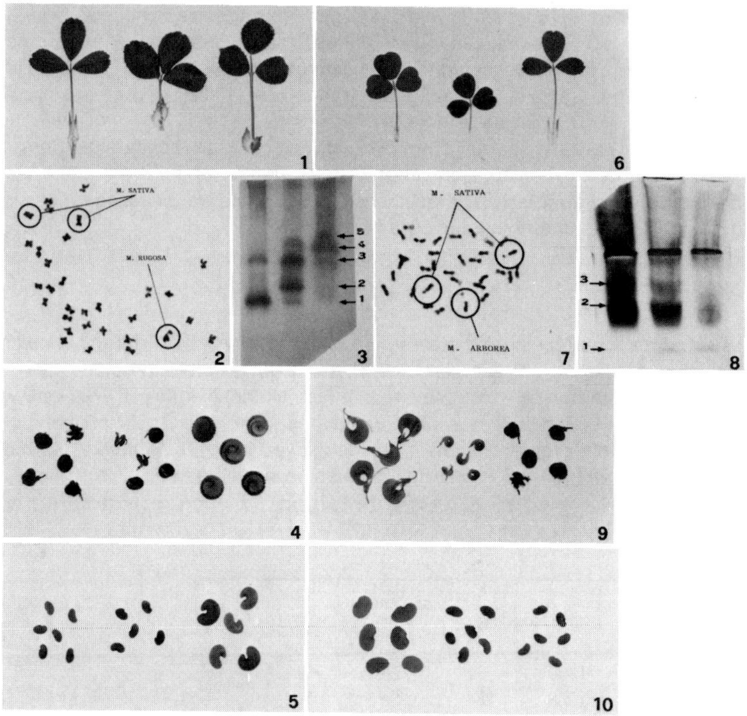

Fig. 1 - Leaves and stipules of *M. sativa* (left), *M. rugosa* (right) and their hybrid (center). Fig. 2 - Somatic metaphase of the hybrid *M. sativa x M. rugosa*; the nucleolar chromosomes provided by the two parent species are shown. Fig. 3 - Electrophoretic banding pattern of the peroxidases of *M. rugosa* (left), *M. sativa* (right) and their hybrid (center). Figs. 4 - 5 - Legumes (fig. 4) and seeds (fig. 5) of *M. sativa* (left), *M. rugosa* (right) and their hybrid (center). Fig. 6 - Leaves and stipules of *M. arborea* (left), *M. sativa* (right) and their hybrid (center). Fig. 7 - Somatic metaphase of the hybrid *M. sativa x M. arborea*; the nucleolar chromosomes provided by the two parent are shown. Fig. 8 - Electrophoretic banding pattern of the esterases of *M. arborea* (right), *M. sativa* (left) and their hybrid (center). Figs. 9 - 10 - Legumes (fig. 9) and seeds (fig. 10) of *M. arborea* (left), *M. sativa* (right) and their hybrid (center).nature of the plants (fig. 3); their electrophoretic pattern was characterized by the presence of a band typical of *M. rugosa* (band 1) and of a band typical of *M. sativa* (band 4). Band 2 was expressed only in the hybrids and its presence excluded the chemical nature of the plants.

All the plants obtained from this cross flowered and were male and female fertile. After selfing they produced legumes that were smaller than those of the parents, with 1-2 coils per pod but resembling in shape those of *M. sativa* (fig. 4). Each pod contained 1-2 seeds whose size and shape were intermediate to those of the parents (fig. 5). Seeds were viable and the F2 progeny is presently under evaluation.

M. sativa x M. arborea

Hybrids between *M. sativa* and *M. arborea* were more difficult to produce than those between *M. sativa* and *M. rugosa*. In fact, embryos were extremely weak and developed slowly, so that only 3 out of the 56 embryos obtained survived (tab. 2).

Morphology of the hybrids appeared to be a blend of characteristics from the two parental species and in particular, the size and shape of the leaves and stipules (fig. 6). Cytological analysis revealed that the plants were tetraploid (2n=32). In this case also three nucleolar chromosomes were detected, and since in the two parental species the nucleolar chromosomes are different as regards the extension of the nucleolar organizer region and the size of the satellite (Falistocco, 1987), it was possible to recognize those provided by *M. sativa* from the one provided by *M. arborea* (fig. 7).

Isoenzyme phenotypes of esterases confirmed the hybrid status of the plants, which showed all the parental bands (band 1 from *M. arborea* and band 3 from *M. sativa*; fig. 8).

The hybrid plants had purple flowers and were fertile. Their legumes were smaller than those of both parents, but they were very similar in color and shape to those of *M. arborea* (fig. 9). Each pod contained one seed only, whose morphology was intermediate to that of the parental species (fig. 10). From these seeds an F2 progeny was obtained, that is now under evaluation.

In conclusion, the above reported results indicate that in the genus *Medicago* the potential of ovule-embryo culture in bypassing post-zygotic mechanisms of incompatibility is not only limited to crosses between perennial species, but can also be extended to crosses between annual and perennial species. This could greatly enhance the role of the annual species of *Medicago* that have always been scarcely considered not withstanding their importance as a source of useful characters not expressed in the germplasm of *M. sativa*.

The hybrid plants obtained in this study, as well as their progeny, will be utilized for transferring the desired traits from the wild species to alfalfa; therefore, the next step will be the identification of the best plants to be backcrossed to *M. sativa*.

References

Barnes D.K., Ratcliffe R.H. Evaluation of annual species of *Medicago* as sources of alfalfa weevil resistance. Crop Sci. 24: 640-642; 1969.

Bauchan G.R., Elgin J.H. Jr. A new chromosome number for the genus *Medicago*. Crop Sci. 24: 193-195; 1984.

Bradford M.M. A rapid and sensitive method for the quantitation of micrograms of protein utilizing the principle of protein dye binding. Anal. Biochem. 12: 248-254; 1970.

Clement W.M. Chromosome relationships in a diploid hybrid between *Medicago sativa* L. and *M. dzhawakhetica* Bordz. Can. J. Genet. Cytol. 5: 427-432; 1963.

Falistocco E. Cytogenetic investigations and karyological relationships of two Medicago: *M. sativa L. (alfalfa)* and *M. arborea*. Caryologia 40: 339-346; 1987.
Fridriksson S., Bolton J.L. Development of the embryo of *Medicago sativa L.* after normal fertilization and after pollination by other species of Medicago. Can. J. Bot. 41: 23-33; 1963.
Hoyle H.C. Illustrated handbook for high resolution of IAA oxidase-peroxidase isoenzymes by isoelectric focusing in slabs of polyacrylamide gel. V.S. For. Ser. Gen. Tech. Rep. NE-37. USDA For. Exp. Sta. 22pp; 1978
Kahler A.L., Allard R.W. Genetics of isozyme variants in barley. Crop Sci. 10: 444-448; 1970.
Lesins K.A. Interspecific crosses involving alfalfa. II. Medicago cancellata *M.B. x M. sativa* L. Can. J. Genet. Cytol. 3: 316-324; 1961a.
Lesins K.A. Interspecific crosses involving alfalfa. I. *Medicago dzhawakhetica* (Bordz.) *Vass. x M. sativa* L. and its peculiarities. Can. J. Genet. Cytol. 3: 135-152; 1961b.
Lesins K.A. Interspecific crosses involving alfalfa. III. *Medicago sativa L. x M. prostrata L.* Jacq. Can. J. Genet. Cytol. 4: 14-23; 1962.
Lesins K.A. Interspecific crosses involving alfalfa. IV. *Medicago glomerata x M. sativa* with reference to *M. prostrata*. Can. J. Genet. Cytol. 10: 536-554; 1968.
Lesins K.A. Interspecific crosses involving alfalfa. *V. Medicago saxatilis x M. sativa* with reference to *M. cancellata* and *M. rhodopea*. Can. J. Genet. Cytol. 12: 80-86; 1970.
Lesins K.A. Interspecific crosses involving alfalfa. VII. *Medicago sativa x M. rhodopea*. Can. J. Genet. Cytol. 14: 221-226; 1972.
Lesins K.A., Lesins I. Genus *Medicago (Leguminosae)*: a taxogenetic study. W. Junk bv Publishers, The Hague, The Netherlands; 1979.
Mariani A., Falistocco E. Cytogenetic analysis of *Medicago rugosa* and *Medicago scutellata*. J. of Genet. and Breed. in press; 1991.
McCoy T.J. Interspecific hybridization of *Medicago sativa L.* and *Medicago rupestris* M.B. using ovule-embryo culture. Can. J. Genet. Cytol. 27: 238-245; 1985.
McCoy T.J., Quarisa G.L. Allotetraploid behavior of hybrids of *Medicago sativa L.* and *M. papillosa* Boiss. Genome 32: 6-11; 1989.
McCoy T.J., Smith L.Y. Interspecific hybrids of perennial *Medicago* species using ovule-embryo culture. Theor. Appl. Genet. Cytol. 71: 772-783; 1986.
Oldemeyer R.D. Interspecific hybridization in *Medicago*. Agron. J. 48: 584-585; 1956
Sandguen N., Sorensen E.L., Liang G.H. A perennial x annual *Medicago* cross. Can. J. Genet. Cytol. 24: 361-365; 1982.
Sandguen N., Kreitner G.L., Sorensen E.L. Light and electron microscopy of embryo development in an annual x perennial *Medicago* species cross. Can. J. Bot. 61: 1241-1257; 1983.
Schenk R.U., Hildebrandt A.C. Medium and techniques for induction and growth of monocotyledonous and dicotyledonous cell cultures. Can. J. Bot. 50: 199-204; 1972.
Shade R.E., Thompson T.E., Campbell W.R. An alfalfa weevil larval resistance mechanism detected in *Medicago*. J. Econ. Entomol. 8: 399-404; 1975.
Shade R.E., Doskocil M.J., Maxon N.D. Potato leafhopper resistance in glandular haired alfalfa species. Crop Sci. 19: 287-288; 1979.

53

In Vitro Fertilization: A Method Facilitating the Production of Hybrid Embryos and Plants

Maciej Zenkteler

In vitro pollination techniques facilitate detailed studies on fertilization, and embryo, and endosperm development in controlled environment including the effects of various physical and chemical factors on the processes. The term in vitro pollination includes: 1/ direct pollination of ovules followed by their culture together with the placenta and parts of the calyx, on a suitable medium; this is referred to as placental pollination, and 2/ pollination of stigmas of the pistils which are then cultured along with a part of the pedicel and calyx. There can be various modifications of the above mentioned procedures depending on the type of experimental material. For example, in Solanaceae, the whole ovary wall is removed, while in Brassicaceae only the upper part of the ovary is cut off in order to make an opening for the deposition of pollen grains.

The technique of direct pollination of ovules would enable to bypass all the prefertilization barriers, as a direct contact of the ovules with the environment may induce the zygote/embryo to develop further. Stigmatic pollination in vitro seems to be less important for bypassing prefertilization barriers. However, it provides a good system for analyzing the role of various factors present in the media and in the environment on self-incompatibility system (Douglas and Connolly, 1989) and on the growth of hybrid embryos which fail to develop further in vivo due to postzygotic barriers (Zenkteler, 1990).

The method of direct self-pollination of ovules has been successful in achieving fertilization in about 50 species, however, viable seeds capable of producing plants have been obtained in only 36 species. Interspecific, intergeneric and even interfamiliar crosses were tried in 53 combinations out of which in 19 combinations viable seeds were produced. In the remaining combinations hybrid embryos collapsed at different stages of embryogenesis (Stewart, 1981; Zenkteler, 1990).

This chapter reports the results of placental pollination in some species of Brassicaceae.

Material and Methods

Plants which constituted the experimental material were grown in the Botanical Garden or in the field of the Institute of Plant Genetics, Poznan. The details of the species used in crosses are given in Table 1.

Table 1: Cross-pollination of ovules attached to the placenta

Species	Number of ovaries cultured	Number of ovules with embryos	Number of hybrid seedlings
B. oleracea+ x B. cretica	60	18	12
B. oleracea x E.+ +pallidiflorum	54	12	-
B. oleracea x E. crepidifolium	46	8	-
B. oleracea x E. pulchellum	28	3	-
B. oleracea x E. virgatum	15	2	-
B. oleracea x Barbarea vulgaris	31	6	-
B. oleracea x B. intermedia	16	3	-
B. oleracea x Sinapsis alba	139	9	-
B. oleracea x S. arvensis	65	14	4
B. oleracea x A+ + + officinalis	22	2	-
B. napus x Barbarea vulgaris	62	4	-
B. napus x E. pallidiflorum	12	1	-
B. napus x S. arvensis	36	6	-
A. officinalis x Hesperis matronalis	124	3	-

\+ Brassica oleracea var. sabellica
\+ + Erysimum
\+ + + Alliaria

All the experiments were carried out during the period May-August 1990 and May-June 1991. Closed flower buds were harvested 24 hours before natural dehiscence. Others were dissected from the just opening buds and kept in sterile Petri dishes until they dehisced. Closed flower buds were surface sterilized for 30-40 sec in 70% ethanol followed by 6-8 min in chlorine water and rinsed several times in sterile water. In a laminar box the petals and some sepals were removed, and the stigma jointly with the style were cut off. Pollen grains were subsequently placed directly on the surface of the exposed ovules. The pedicels were then trimmed, and the entire ovary placed in the MS (Murashige and Skoog, 1962) salts supplemented with myoinositol 20 mg 1-1, nicotinic acid 0.5 mg 1-1, pyridoxine HCl 0.5 mg 1-1, thiamine HCl 1 mg 1-1, sucrose 3 or 6% and 8 g 1-1 agar with pH adjusted to 5.8. This medium was enriched with kinetin 1-2 mg 1-1, IAA 1-2 mg 1-1 individually and in combinations. Cultures were incubated at 22-24oC under continuous low level hours after in vitro pollination, sample ovaries were removed from culture, stained directly in aniline blue and examined under epifluorescence microscope. For embryological investigations the enlarged ovules were fixed in Carnoy's solution, passed through alcohol-xylene series and embedded in paraffin. Sections were stained with Heidenhain's hematoxyline and fast green.

Results and Discussion

Pollen germinated with various intensity and the pollen tubes did not follow any particular direction, but some entered the ovules through micropyle. The entrance of pollen tubes inside the embryo sac was visible in all the analyzed ovules representing various combinations of crosses. Ovules which failed to receive pollen tubes could be distinguished from the fertilized ones by their smaller size. After 5-8 days of pollination, some ovules were visibly enlarging (Fig. 1). In the fertilized ovules globular embryos were discernible 5-6 days after culture. In certain fertilized ovules the globular embryos developed concomitantly with an endosperm, but in some other ones endosperm was poor or not present. The best development of globular hybrid embryos occurred in ovule culture of B. oleracea var. sabellica pollinated with pollen grains of Barbarea vulgaris and Sinapis alba. In those crosses globular embryos reached nearly a heart stage (Fig. 2). In the cross B. napus x Sinapis arvensis well developed embryos at heart stage (Fig. 3) were noticed but they did not develop further and seedlings were not obtained. Very poor globular embryos were found in the cross Alliaria officinalis x Hesperis matronalis.

All stages of embryogenesis were observed after pollinating ovules of B. oleracea var. sabellica with B. cretica (Fig. 4). In this cross fully developed cotyledonary embryos were transferred on a fresh medium and later on they grew into seedlings. Those seedlings were kept on the medium during the winter 1990/91 and in April 1991 they were transferred into the soil. The hybrid plants are grown with a high intensity (Fig. 5) and in June they already started to flower. The well developed cotyledonary embryos were also obtained after pollinating in vitro ovules of B. oleracea var. sabellica with pollen of Sinapis arvensis. So far, ovules with embryos, or embryos alone are cultured on various media and the formation of small seedlings are already observed.

The results of the present investigation indicate that direct interspecific and intergeneric pollination of exposed ovules of various species of Brassicaceae result in the formation of embryos which grow to different stages. So far, only in one interspecific combination (B. oleracea var. sabellica x B. cretica) and in one intergeneric combination (B. oleracea var. sabellica x Sinapis arvensis) embryos have been realized, which produced plants after transferring into the medium. In all the remaining combination embryos developed up to the globular stage with the exception of B. oleracea var. sabellica x Barbarea vulgaris and B. napus x Sinapis arvensis in which embryos reached an early heart stage.

It is worth mentioning that 3 genera which were used as pollinators do not belong to the Brassiceae tribe. Erysimum is included to the Hesperidineae, Barbarea to Arabideae and Alliaria to Sisymbriae (Shulz, 1936). The genus Sinapsis belongs to the tribe Brassiceae which includes Brassica crops and some wild relatives. It can be presumed that some genetic homology exists between Brassica and Sinapis arvensis and, therefore a successful hybridization occurred after crossing these 2 species. Irrespective of this successful cross all the investigations with other species show that the barriers of incompatibility between those species are postzygotic. Globular hybrid embryos were always noticed but they did not develop further. It seems probable that an early transfer of ovules with globular hybrid embryos from the placenta on more suitable medium, may help promote further growth of those embryos.

Fig. 1. Enlarged ovules (arrows) inside the opened ovary of B. oleracea, 5 days after test tube pollination (ttp.) with pollen of Barbarea vulgaris. Fig. 2. B. oleracea x Barbarea vulgaris, 8 days after ttp., immature embryo at an early heart stage inside the ovule, endosperm not present. Fig. 3. B. napus x Sinapis arvensis, 8 days after ttp., embryo at the heart stage inside the ovule, endosperm not present. Fig. 4. B. oleracea x Brassica cretica, 6 days after ttp., immature globular embryo and endosperm inside the ovule. Fig. 5. Hybrid plant of B. oleracea x B. cretica obtained from in vitro pollinated ovules.

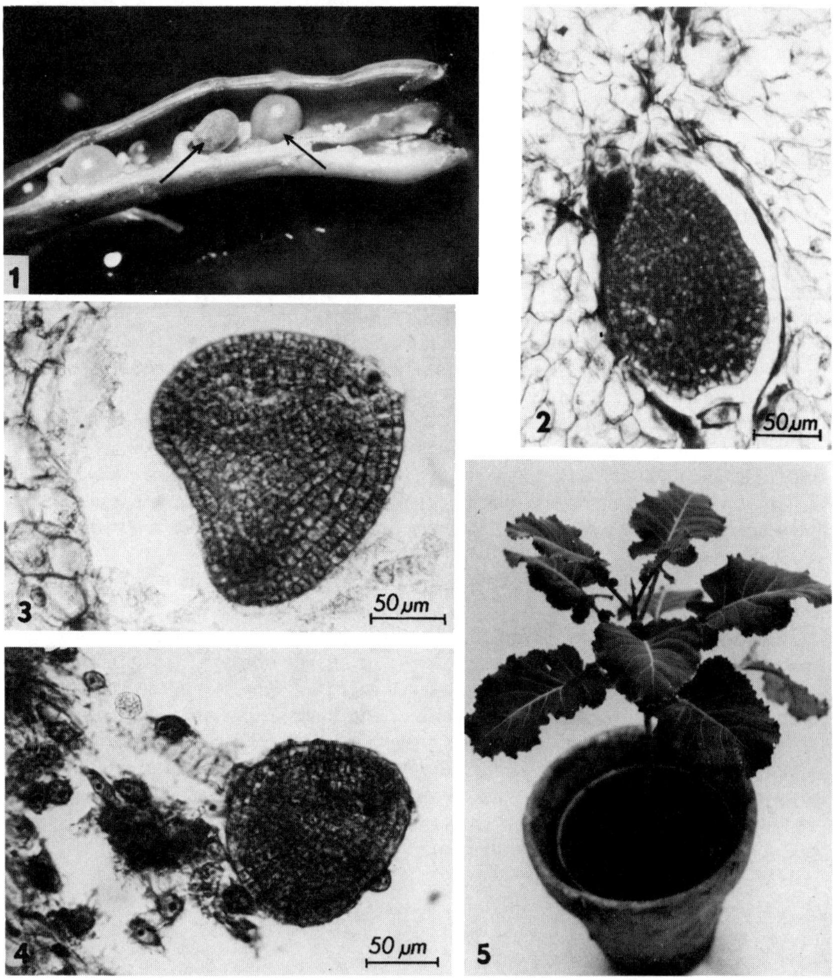

References

Douglas, G.C., Connolly, V. Self-fertilization and seed set in Trifolium repens L. by in situ and in vitro pollination. Theor. Appl. Genet. 77: 71-75; 1989.

Murashige, T., Skoog, F. A revised medium for rapid growth and bioassay with tobacco tissue cultures. Physiol. Plant. 15:473-497; 1962.

Shulz, O. Cruciferae. In: Engler A. (ed.). Die Naturlichen Pflanzenfamilien. Wilhelm Engelmann, Leipzig, 17b: 1-799; 1936.

Stewart, J. McD. In vitro fertilization and embryo rescue. Environmental and Experimental Botany 21: 301-315; 1981.

Zenkteler, M. In vitro fertilization and wide hybridization in higher plants. Crit. Rev. Plant Sci. 9: 267-279; 1990.

54

Introduction of Exogenous DNA into Plants after Pollination via the Pollen Tube Pathway

Guang Yu Zhou

The technique of introducing exogenous DNA into plant after pollination has been used successfully on cotton and rice during last thriteen years by five institutes in China. New varieties of agricultural improvement on diseases resistances and other better traits have been obtained [1-6]. with total donor DNA transfer. The technique was confirmed to be useful for breeding by a Specialists Committee which was organized by Chinese Academy of Scienses and Ministry of Agriculture, Animal husbandry and Fishery in 1986. . A workshop (1986) and a Symposium (1988) of Molecurlar Breeding were held. Since then, there are sixteen provinces including cities with twenty five laboratonies starting to apply this technique on wheat,soya bean, corn, forage, vegetables. as well as cotton, rice and etc. All produced transformed offsprings, some of which had disease resistance, quality or yield improvement.

The technique was designed specially to introduce exogenous DNA "after pollination" in order to have normal seed {s}. An other reason of emphysizing after pollination is that a pollen tube pathway forms to allow the pollen tube[s] carring two sperm nuclei to pass through the nucellus, reach the egg apparatus to complish the double fertilization. The exogenous DNA goes along this pollen tube pathway into the embryonic sac and egg, zygote or early embryonic cells. In general, these cells have no normal cell walls and could be considered as natural protoplasts to take foreign DNA.. The exogenous DNA which passed along the pollen tube pathway to reach the egg apparatus was demonstrated with tritium labeled 50 kb DNA by Gong et al. (1988) The DNA was injected into the cotton ovaries from their top centre 24 hours after pollination. Ovaries were then harvested 30 minutes to 8 hours after the DNA injection. It was shown that, within 30 minutes, the DNA was seen in the egg apparatus. It was clearly shown that the DNA is restricted to travel nowhere but along the pollen tube pathway and reaches the region of the egg apparatus. There was no DNA inside of the pollen tube.

Weng J. et al 1984 introduced a modified plasmid of Mp7 which had been recombined with a segment of cotton repetitive sequence into an upland cotton. Sourthern blot analysis showed the integration of the foreign DNA in D1 [the first generation after DNA transfered. Weng (1988) also studied the kanamycin resistant gene transfer on cotton. The positive data of the expression and integration of the kanamycin resistant gene were presented in his doctoral thesis.

With pNE0105, a plasmid containing a Kanamycim resistant gene provided by Dr. J. Sampson and pRR plasmids which were constructed by recombining pNE0105 with different rice repetitive sequences, Gong and Zhou have published (1987) the rice data of a second generation. It was shown that the seedlings which had been transformed with the plasmids pNE0105 or pRRs showed higher resistance to kanamycim then the controls which were treated only with the DNA solution. Among them the seedlings of second generation [D2]] by introducing pRR23 and pRR29 were growing normally in 200 ug/ml kanamycin solution, strong NPTII activity was found. The southern blot was shown in pRR23 transfered seedlings. No23 is a rice special repetitive sequence, while no29 is a cereal specific. The rice technique was adopted and some of the results were reported by Ray Wu's lab. (1988) This technique was also applied for transfering bacterial toxin protein gene into cotton by Xie et al (1991) .

Thus the technique we used for transfering DNA into plant could be applied with total donor DNA as well as recombinant DNA. It is the germline cells in situ of a living plant which are the DNA acceptor. The DNA transfered cell is maintained by the whole plant to develop to seed. There are no problems of protoplast preparation, cell and tissue culture or plant regeneration. The technique can be applied directly to the cultivars of dicots or monocots. There are no somatic variations. New varieties for agricultural application are easier to obtain than to have that from a laboratory transgenic plant by regeneration.

Construction of recombinant DNA by genetic engineering needs isolated selected genea. Which gene and how many genes determined an interesting trait for agricultural improvement is usually uncertain. It is rather difficult and it takes time to recognize and isolate a useful gene.

The technique of molecular breeding we used is very simple, one which ordinary breeders adopt easily. We used total donor DNA which contained the gene of a goal trait for transfering into plant. This was done not only because we have no valuable isolated gene interesting for transfer at that time [1978] or even now. Another reason was that I was attempting to prove a hypothesis of DNA segment hybridization among remotely related plants which was my working hypothesis for starting (1974) the study of plant DNA transfer (Zhou, 1981). The results of our using total donor DNA were very encouraging. We have not only proved the hypothesis of DNA segment hybridization but also selected the DNA transfer plants for which the inheritance is stable and useful for agriculture. Although there are still many qustions remained, e.g. how the DNA segments integrate, express and are inhertied. It is difficult to have a clear answer before we have proper probes. This approach has been used to search for the cotton gene resistant to Fusarium oxsporium. A promising way to recognize the goal gene in total donor DNA could be to try to compare first the gene product of the donor with that of the acceptor and the hybrid. A protein in donor as well as in the DNA transfered hybrid which responds directly to the spore germination of the pathgen was found. It is a pathogen strain specific, menbranous glycoprotein or a recognition molecule.

The technical details for introduction of exogenous DNA into plant differ according to the flower structure, the time and duration of the pollination and fertilization of different plant in different region and the climate of the operation day if there are no temperature and humdity controled facilities, etc. With the help of the gene transfer from total donor DNA, it could be easier to recognize a unknown gene and speed up the work of genetic engineering for agricultural improvement. Generally genetic engineering is regarded as a high technology, a precise gene and precise engineering to create new varieties which could be wholly directed by man. But I think a present strategy which was proposed in l987 for developing molecular breeding, especially in developing countries, is practical. That is: Recombinant DNA as well as total donor DNA should be both emphasized.

I was often qustioned that if without a definite and isolated gene to be transfered, was the new variety believable? For the moment who can tell exactly what is happening genetically in a new variety from somatic variation or cell fusion? I present here some facts about molecular breeding using the technique of pollen tube pathway. The mechanism, especially that of introducing total donor DNA, still needs further study.

References

Huang,J.,Qian S. and Liu,G.; Weng J., Zeng Y. and Zhou G. Y, Variation in the characters of upland cotton (G. hirsutum) induced by exotic DNA of sea island cotton (G. barbadense). Acta Genetica Sinica 8: 715-719; l981
Zhou G.Y., Weng J. and. Zeng Y.; Huang J., Quian S., and Liu G. Introduction of Exgenous DNA into cotton embryos. Methods in Enzymology. New York : Academic Press; 101: 433-481; 1983.

Duan X. and Chen S. Variation in rice induced by foreign DNA uptake. Acta Agronomica Sinica 3: 6-11.1985

Li D. Y. and Chen C. B.; Zhou G. Y. Yang W. X. and Gong Z. Z.. 1) Variation and Inheritance in the Chactracters of Rice(O. sativa) Induced by Exogenous wild Rice (O. offidinalis) DNA. 258-274.

Huang J., Quan S., Liu G., Xue D., Ying M., and Hong A.; Zhou G. Y.,Weng J., Zhen Y., Gong Z., Wang Z. and Yang W. Transfer of Fusarium wilt resistance between cotton cultivares via exogenous DNA introduction. Acta Agrinomica Sinica . 3: 32-36, l986.

Gong Z., Shen W.and Zhou G. Y.; Huang J.and Qian S.. Introducing exogenous DNA into pLants after pollination--DNA traverse pathway of pollen tube to reach embryonic sac. Scientia Sinica Seria B 31{5), 1080-1084.,1988.

Weng J., Shen W., Wang Z., Chen K. Yang W., Gong Z., Zhen Y., and Zhou G. Y.; Huang J., Quian S. and Liu G. A molecular demenstration of the introduction into cotton embryos of exogenous DNA. Acta Biochemica et Biophysica. l6: 325-327. l984.

Weng Jian. Transformation von Baumwolle und Reis mit chimaren Gene-- Injektion von DNA in die Baumwollblute und DNA Bchandlung von Reisbluten.. Inaugural-Dissertation zur Erlangung des Doctorgrades der Mathematisch- Naturwisscenschafthchen Fakuhat der Universitat zu Koln. Koln l988

Zhou G.Y, and Gong Z. The technique of introducing exogenous DNA into plant and it's appllication on Agriculture. in Proceedings of Advancing Biochemistry. Celebration for the eightieth birthday of Professor Y L.Wang. 139-152, l987.

Zhong-xun Luo and Ray Wu. A simple Method for the Transformation of Rice Via the Pollen-Tube Pathway Plant Molecular Biologgy Reporter. 6(3) 165-174. l988.

Xie B. Q. Fan Y.L.,Ni W. C. and Huang J Q. Transgeneic Plant Obtained by introducing the Gene of Bacillus thuringiases into Cotton. Science in China Series B Chemistry,Life Sciences & Earth Sciences. 34 (4) 359-366. 1991

Zhou G.Y.,Weng J., Gong Z., Shen W.,Wang.W. and Tao Q.; Huang J.,Quian S., .Liu G.,Lin M.,Xue D.,Hong A. and Xu Y.; Duan X. and Chen. Plant molecular Breeding . International Crops Research Institute for the semi-Arid Tropics. l988 Biotechnology in tropical crop improvement proceeding of the International Biotechnology Workshop, . 12-15 Jan. l987, ICRISAT Center p. 33-36 India. Patancheru, A,P. 502324, India: ICRISAT

Zhou G. Y.,Weng J., Gong z., Zhen Y., Yang W., Shen W., Wang Z and Rao Q.;Huang J., Qian S., and Liu G., Ying M., Xue D., Hong A and Xu Y.; Chen S. and Duan X. Molecular Breeding of Agriculture--Technique for introducing exogenous DNA into plants after self pollination. Scientia Agriculura Sinica. 21(3), 1-6. 1988.

55

Biotechnology in Reproductive Biology

S.K. Hahn, K.V. Bai, R. Asiedu, A.G.O. Dixon, S. Tavoletti, A. Mariani, F. Veronesi, J.M. van Tuyl, M.G.M. van Creji, M.P. van Dien, D.M.F.Y. Kerckhoffs, G. Laublin, M. Brault, M. Cappadocia, A. Makoveychuk, A. Semenova, T.B. Batygina, P. Taylor, J. Kenrick, C. Blomstedt, B. Knox, A. Agnihotri, A. Lakshmikumaran, K.R. Shivanna, V. Jagannathan, M. Ikeda, T. Nakanishi, M. Yoshida, R.W. van den Bulk, J.J.M. Dons, H. Veronneau, G. Lavoie, G. Kovács, I. Takács, Z. Kristof, R. Nagy, B. Barnabás, K. Dressler, M. Iser, A. Schmid, S. Stegmaier, D. Hess, L.M. Plegt, B.C.E. van de Ven, J. Franken, R. Bino, A.J. van Tunen, E. Stöger, R.M. Benito Moreno, B. Ylstra, O. Vicente, E. Heberle-Bors, A. Gaillard, M. Beckert, E. Matthys-Rochon, P. Vergne, C. Duitas, C. Digonnet, and C. Dumas

Plant reproduction is a core process in plant breeding. For decades, one can even say for centuries, plant reproductive biology has largely been a matter of observation, description, and philosophical speculation. Only in the last two decades has it entered the experimental phase, and it is in the last decade that tools are emerging that allow manipulation of plant reproduction for the sake of human needs and goals.

While vegetative reproduction has gone through this process more rapidly, sexual and asexual reproduction lags behind, and it is only now that generalized techniques appear that allow to manipulate plant sex and the multitude of its diversions, both natural and induced, for plant breeding. The methods in this area range (also historically) from induced polyploidization, selection and pollination with unreduced pollen, embryo culture after distant crosses, to the use of in vitro culture methods for gametophytes and in vitro fertilization of isolated gametes, as well as molecular methods such as pollen and embryosac transformation and molecularly engineered male sterility.

Intraspecific polyploids

The use of unreduced pollen and embryo sacs to produce polyploids by natural pollination has been investigated in great detail in potato. At this meeting, it has been reported that a similar approach is feasible also with alfalfa and cassava. In cassava, tetraploids and triploids have been detected (Hahn, Bai, Asiedu, Dixon, IITA, Ibadan, Nigeria) after crossing cultivated cassava, Manihot esculenta ($2n=36$) with its wild relatives M. glaziovii and M. epruinosa (both $2n=36$). Since unreduced pollen was found in the parent species in frequencies from 0.1 to 35.6 %, it was concluded that these hybrids were produced by zygote formation from unreduced gametes and were possibly maintained by facultative apomixis. With respect to agricultural performance, the polyploids appeared to be superior to the diploid parent species, with some advantages in terms of yield and plant architecture for the triploids.

In diploid alfalfa (Medicago sativa L.), a previously described mutation was studied cytologically in greater detail that showed irregularities in postmeiotic cytokinesis of megaspore and microspore mother cells (Tavoletti, Mariani, Veronesi, CNR Perugia and university of Udine, Italy). On the male side, only 4n pollen are produced while on the female side predominantly 2n eggs are formed. It is suggested that male sterility resulting from the 4n pollen formation and availability of unreduced eggs may be used in sexual polyploidization.

Interspecific hybrids, in vitro culture of female reproductive structures

Going from intraspecific or "auto"polyploids to interspecific or allopolyploids, the problem to overcome in many cases are crossing barriers. A group at CPO in Wageningen, The Netherlands, has succeeded in lily as a model crop to set up in vitro cut-style pollination and the in vitro grafted style technique to overcome pre-fertilization barriers and ovary culture and ovule culture to overcome post-fertilization barriers (Van Tuyl, van Creji, van Dien). A number of interspecific hybrids within the genus Lilium could be made more efficiently.

These techniques are now being transferred to Tulipa, Nerine, Amaryllis and Lycoris. First results in tulips with in vitro pollination to get seeds and with ovary-slice culture and ovule culture are very encouraging (Van Creji, Kerckhoffs, van Tuyl).

Ovary culture has also been used by Laublin, Brault, and Cappadocia from the Institut de Recherche en Biologie Végétale, from Montreal, Canada to rescue embryos from a cross of Iris versicolor and I. ensata, and a high number of plantlets were regenerated.

Apart for embryo rescue after distant crosses, embryo culture is also used to rescue embryos that require specific developmental and nutritional conditions. Today, this technique is also used to save rare plants from extinction. Makoveychuk, Semenova, and Batygina from the Komarov Botanical Institute, St. Petersburg (former Leningrad), USSR, have succeeded to rescue plantlets from protocorms of several terrestrial orchids, such as Gymnadenia conopsea, Dactylorhiza maculata and Epipactis palustris.

The real radical solution to overcome crossing barriers by a biotechnological approach (within sexual reproduction, excluding somatic protoplast fusion) is to fuse in vitro isolated sperms and eggs (see Kranz and Lörz in this volume). Isolation of sperms and eggs is a prerequisite for such attempts. Isolation of sperms has made considerable progress during the last years. Taylor, Kenrick, Blomstedt, and Knox from the university of Melbourne, Australia, have succeeded to isolate sperm cells by osomotic shock that appeared to remain intact after isolation when incubated in an osmotically balanced medium at low temperature. Apart from biotechnological application, isolated sperms can be used as material to study sperm cell development and function. One of the enigmas of sperm function in plants is how the sperm cell cytoplasm is excluded from being incorporated into the egg cytoplasm, to guarantee female inheritance of plasmonal genes. The finding by Taylor et al. that both, in vivo and in vitro, the sperm nucleus always remained attached to a portion of the sperm plasma membrane, suggests that this portion might be the site for membrane fusion, possibly excluding other parts of the sperm cell to be taken up by the egg cell.

An alternative to produce wide hybrids by deliberately trying to overcome crossing barriers through in vitro or other methods is to screen wild populations. This seems to be particularly appropriate for traits such as disease resistance which do not reveal themselves so easily as hybrid vigour. RFLP-mapping and DNA finger printing have shown their value to identify individuals and genotypes. Agnihotri, Lakshikumaran, Shivanna, and Jagannathan from the Delhi Tata Institute and the University of Delhi, India have shown that finger prints of a 177 bp tandem repeat unit can distinguish Brassica species and their hybrids opening the way for identifying hybrids with new properties.

Apomixis

Apomixis is an area of research that is still in its infancy but which offers both, exciting new possibilities to study plant reproduction and new strategies for plant breeding, particularly for hybrid seed production. Feared by some breeders and seed propagators and embraced by Third World enthusiasts, recent developments in the field of variety and plant property rights may end this controversy and open the way to channel more money into the research of apomixis and its uses to benefit both the breeders and the farmers. Although apomixis work was not presented at this meeting, it was intensively discussed. Ironically, in one contribution, apomixis was presented as a nuisance. In polyembryonic Citrus, nucellar embryogeny is so overwhelming that the zygotic embryo cannot be recovered. Attempts to selectively destroy the nucellar embryos by gamma irradiation have been presented but more work seems to be required to establish the validity of this method (Ikeda, Nakanishi, and Yoshida, Kobe University, Japan).

Haploids from pollen

The ability of somatic cells of a gametophyte to form an embryo is not confined to the embryo sac (apogamy). Haploid embryos are also produced when immature pollen is cultured in vitro, either in anther or isolated pollen culture. Apart from the use of doubled haploids to speed up production of pure lines, isolated pollen culture in particular should be an interesting tool for in vitro selection on the haploid level. It is, however, not much used until now, mainly because embryos in isolated pollen cultures are much more difficult to get than in anther cultures. The lily group from CPO in Wageningen, The Netherlands (Van den Bulk, van Tuyl, and Dons) are seemingly trying to set up for lilies all the reproductive biotechnologies one can think of, including pollen culture. After some unsuccessful work, they have decided to spend more work on the importance of the developmental stage of the pollen and on methods to classify and determine pollen stages. They also applied shock treatments known to induce pollen embryogenesis in other species, such as heat shock and starvation. A combination of precise staging and heat shock, followed by culture in a standard pollen culture medium resulted in the formation of pollen containing more than the usual number of nuclei. These results are promising to overcome recalcitrance of this and maybe other ornamental species.

The formation of pollen embryos in anther cultures is known to be strongly genotype-dependant. In addition, environmental conditions and season play an important role. Interestingly, regeneration of the plantlets after induction of cell division in the pollen grains, does not seem so much dependant on environment and season, similar to regeneration from sporophytic tissues (Veronneau, Lavoie, Cappadocia, University of Montreal, Canada). A detailled genetic analysis performed on reciprocal hybrids of diploid

Solanum chacoense clones revealed that in fact a significant positive correlation existed between anther culture and leaf disc response. This suggests that in Solanum, these two regeneration systems may be controlled by the same or similar genes. This opens the way to isolate these genes by RFLP mapping and chromosome walking.

In vitro maturation of gametophytes

Apart from forming embryos, isolated pollen can develop, in appropriate culture conditions, into mature pollen that is fertile (see Vicente et al., this volume). This in vitro pollen maturation can be used for in vitro selection on the haploid level. In spikelet cultures, not only pollen mature in vitro, but also embryosacs (Kovács, Takács, and Barnabás from the Agricultural Research Institute in Martonvásár, Hungary). Viable seeds are produced in these spikelets in a frequency of up to 20-25 %. Attempts to use this method for selection for low temperature tolerance seem to be very promising. In a more detailled study, this group has investigated pollen development in the in vitro cultured spikelets on the ultrastructural level (Kristof, Nagy, and Barnabás). It turned out that pollen development was very heterogeneous and that many had burst inside the anther. In those grains that seemed to develop to maturity, differences in the amount and structure of organelles were found. A particular feature of these grains was a thickened intine.

Pollen and embryosac transformation

One of the most attractive aspects of biotechnological applications of research in plant sexual reproduction could be the use of pollen as a natural vector for DNA. DNA transferred into pollen grains would be transported into the embryo sac, and transgenic plants would be formed after fertilization. If feasible this technique would revolutionarize plant genetic engineering. Such a general method for the production of transgenic plants would overcome host range limitations that exist for bacterial or viral vectors, as well as limitations for in vitro regeneration that exist when DNA is transferred in somatic cells. In addition, somaclonal variation that is inherent to in vitro culture would also be avoided. The same would apply if the embryosac could be transformed. However, the embryosac is much more difficult to target by DNA delivery devices making it a less attractive object of study.

In the last decade, a number of techniques have been developed to produce transgenic plants. From this work, a consensus has emerged with respect to the criteria that have to be fulfilled to prove the transgenic nature of a plant. The most important of these criteria are activity of marker genes, correspondance of transferred DNA and phenotype, physical presence of the transferred DNA integrated in the genome of the host and segregation of the (marker) gene in the offspring. Results being published on the production of putative transgenic plants have to be judged on this background.

Pollen transfomation has a history that goes back to 1974 when Hess published his original work on Petunia (for review see Hess, 1987). Since then Hess has published a number of reports claiming to provide complete evidence for successful pollen transformation. Unfortunately, the techniques to transfer DNA into pollen and the type of DNA being transferred were not the same in all these experiments, therefore the data published in the different reports cannot be compared. One thing is however clear: in not a single instance have data been presented on the segregation of the transferred gene(s) in the offspring, be it on the phenotypical or on the molecular level. A number of labs have tried to reproduce the claims of Hess, and a number of papers have appeared on the failure of these attempts (for review see Heberle-Bors et al., 1990). It is ironic that John Sanford, out of frustration with his experiments on pollen transformation (Sanford and Skubik, 1986), has developed the DNA particle gun (Klein et al., 1987) that is today the only proven method to transfer DNA into pollen (see also below).

The original method of transferring DNA into pollen was thought to be passive entry during imbibition of dry pollen in DNA solutions. Later, with the onset of molecular biology, Agrobacterium tumefaciens, microinjection, and electroporation have been tried. In no case have transgenic plants been produced that fulfill the criteria mentioned abouve, and in other reports, clear negative results have been obtained (see reviews by Heberle-Bors et al., 1990, Heberle-Bors, 1991).

The group of Hess (University of Hohenheim-Stuttgart, Germany) presented results at this meeting on the use of Agrobacterium tumefaciens for gene transfer into pollen in petunia (by cocultivation with germinating pollen) and wheat (by pipetting agrobacteria into flowers). In the latter case, to call this approach pollen mediated gene transfer is inappropriate. If there is transfer at all, it is as likely that embryosacs or even zygotes and embryos were transformed as pollen. No data have been presented to distinguish between these possibilities (e.g. segregation of marker genes after test crosses). Indeed, no evidence of DNA transfer into pollen or other cells, by transient expression experiments, was shown, although some of the constructs used included the GUS reporter gene that can be easily assayed histochemically.

Different constructs have been made to transfer nitrogen fixation genes (Dressler and Hess), the ß-glucuronidase gene (Iser and Hess), the stilbene synthase gene (Schmid and Hess), and the hygromycin resistance gene (Stegmaier and Hess) to wheat. Again, the results presented in these posters were unclear and inconclusive. Several, or all the controls that should be provided for Southern blot data, particularly when Agrobacterium is used as a vector, were missing. Experiments on the expression of the marker genes were not presented. Data on the segregation of the transferred genes in the offspring were also not provided. The claim of full evidence for the functioning of the their method to produce transgenic plants was therefore not supported by the data presented.

As mentioned above, the particle gun is the only device that can introduce DNA into pollen (Twell et al., 1989). Two groups have reported at this meeting that this approach can be reproduced. The group of van Tunen (Plegt, van de Ven, Franken, Bino, Van Tunen, CPRO, Wageningen, The Netherlands) showed that the PA2 promoter of the chi gene of Petunia and the TR2'promoter of the Agrobacterium mannopine synthase gene (using the GUS gene as a reporter gene) are expressed in mature tobacco and lily pollen while the 35S-promoter is not. No transgenic plants have been recovered after pollination with this transformed pollen. Mascarenhas reported in his lecture similar data on a pollen-specific promoter of maize.

Similar data have also been produced by Stöger, Benito Moreno, Ylstra, Vicente, Heberle-Bors (University of Vienna, Austria) comparing the expression of the PA2 and the 35S promoter (fused to the GUS-gene) in tobacco pollen. In addition they have shown that transfer of DNA by the particle gun into tobacco microspores gives expression of the chimeric gene in in vitro matured pollen (Stöger et al., 1991). It is expected that DNA transferred into microspores should be present not only in the vegetative cell but also in the generative cell of the in vitro matured pollen (Alwen et al., 1990). Selection during in vitro maturation should be a further advantage of this system.

The same group has also tested whether immature, mature, or germinating tobacco pollen can take up DNA after cocultivation with Agrobacterium tumefaciens (Stöger et al., 1991). In these experiments, the GUS gene was interrupted by a plant intron that prevented expression of GUS in agrobacteria. Transgenic plants produced by leaf-disk transformation, as well as bombarded pollen containing this construct, did show GUS activity in pollen demonstrating that the intron was spliced out. Cocultivation of immature, mature or germinating pollen with agrobacteria containing this construct did not lead to the formation of GUS-positive pollen although agrobacteria could be seen attached in large quantities to the surface of the pollen grains and tubes. Special care was taken to inhibit the activity of an endogenous ß-glucuronidase that is present in plants (Alwen et al., submitted) by carefully controlling pH and other GUS assay conditions.

The same construct was also used to incubate pollen from freshly dehisced anthers of tobacco into DNA solutions of varying concentrations and buffer conditions. Again, no GUS-positive pollen grain could be detected (Stöger et al., 1991). These experiments make DNA uptake by imbibition and Agrobacterium tumefaciens very unlikely and give further reason to be extremely cautious with respect to claims on successful pollen transformation.

As mentioned above, lack of efficient regeneration systems is the cause for the failure to produce transgenic plants in some species. This is particularly true for cereals. In these species, pollen embryogenesis is a quite efficient regeneration system. Transfer of DNA into microspores or young pollen grains followed by regeneration has been tried in some species already (group of Potrykus in Zürich, Switzerland), with no success however until now. Gaillard, Beckert, Matthys-Rochon, Vergne, Dumas (INRA Clermond-Ferrand and INRA Villeurbanne, France) are trying to follow Potrykus' approach in maize. Maize anther and pollen culture were very inefficient until now. They have developed an efficient microspore culture system from a highly androgenic maize line detected by anther culture, and are using microinjection to introduce DNA into young pollen embryos that have high regeneration capacity. Data were presented that indicate that dyes (Lucifer Yellow, FITC-dextran) can be injected into such early embryos without impairing embryogenesis.

The same group (Mattys-Rochon, Digonnet, Dumas) has used microinjection also to transfer dyes into the egg and central cell of isolated embryosacs of maize. If the same can be done with DNA and if the isolated and fertilized embryosacs (coinjection of the sperm nucleus) can form plants in an efficient way, this approach will be very promising.

References

Alwen, A.; Eller, N.; Kastler, M.; Benito Moreno, R.M.; Heberle-Bors, E. Potential of in vitro pollen maturation for gene transfer. Physiol. Plant. 79:194-196; 1990.

Heberle-Bors, E. Germ line transformation in higher plants. IAPTC Newsletter 64:2-10; 1991.

Heberle-Bors, E.; Benito Moreno, R.M.; Alwen, E.; Stöger, E.; Vicente, O. Transformation of pollen. In: Nijkamp, H.J.J.; Van der Plas, L.H.W.; Van Aartrijk, J., eds. Progress in plant cellular and molecular biology. Dordrecht Boston London: Kluwer; 1990:p. 244-251.

Hess, D. Pollen-based techniques in genetic manipulation. Int. Rev. Cytol. 107:169-190; 1987.

Klein, T.M.; Wolf, E.D.; Wu, R.; Sanford, J.C. High-velocity microprojectiles for delivering nucleic acids into living cells. Nature 327:70-73; 1987.

Sanford, J.C.; Skubik, K.A. Attempted pollen-mediated transformation using Ti-plasmids. In: Mulcahy, D.L.; Bergamini Mulcahy, G.; Ottaviano, E., eds. Biotechnology and ecology of pollen. New York Berlin Heidelberg Tokyo: Springer-Verlag; 1986:p. 71-76.
Stöger, E.; Benito Moreno, R.M.; Ylstra, B.; Vicente, O.; Heberle-Bors, E. Comparison of different techniques for gene transfer into mature and immature tobacco pollen. Transgenic Research, in press; 1991.
Twell, D.; Klein, T.M.; Fromm, M.E.; McCormick, S. Transient expression of chimeric genes delivered into pollen by microprojectile bombardment. Plant Physiol. 91:1270-1274; 1989.

Gametophytic Selection

56

Effects of High Temperature and Humidity Stresses on Tobacco Pollen and Their Progeny

Mandhu Bajaj, M. Cresti, and K.R. Shivanna

In recent years many investigations have shown that genes imparting resistance to many biotic and abiotic stresses such as pathotoxin, salinity, low temperature and herbicide are expressed in the pollen grain also (Mulcahy, 1979; Mascarenhas, 1989). This has opened up new avenues for application of section pressure on pollen grains to achieve preferential transmission of adaptive genes to the progeny (Ottaviano and Mulchay, 1989). In some systems, selection pressure on pollen in the form of ageing (Milcahy et al., 1982; Pallais et al., 1986) or high density pollinations (Mulcahy et. al., 1975, Lee and Hartgerink, 1986; Schlichting et. al., 1987) has been shown to increase the vigour and uniformity of the progeny. We have been investigating the feasibility of using high temperature and high humidity treatment to pollen as selection pressures to increase the vigour of the progeny. Our earlier studies have shown that in many systems, high temperature and/or RH significantly affects pollen vigour but not viability (Shivanna and Cresti 1989, Shivanna et al., 1991 a,b).

In their chapter the results on the ability of high temperature and/or RH-treated pollen of tobacco to set seeds, and on the vigour of the resulting seedlings are presented.

Material and methods

Pollen grains collected from freshly dehisced anthers of Nicotiana tabacum were subjected to high temperature (38 and 45 ± 20C), and/or high RH (>95%) for 4 h (Shivanna et al., 1991a). After the treatment, pollen samples were tested for viability (on the basis of fluorochromatic reaction (FCR) test, Heslop–Harrison & Heslop–Harrison, 1970) and vigour (on the basis of the speed of germination in vitro and pollen tube emergence in semi–vivo styles, Shivanna et al., 1991a), and were used to carry out pollinations of emasculated and bagged flowers. Pollen germination and pollen tube growth in pollinated pistils was studied through aniline blue flourescence (Shivanna and Johri 1985). Seeds were collected from mature fruits and stored under laboratory conditions in paper bags until use.

In vitro germinability of seeds obtained from different treatments was tested during July. Seeds were sown in soil under field conditions during October. Emergence of seedlings and their morphological details were recorded at regular intervals during their growth. The seedlings were transplanted to individual pots (30 cm diameter) after 40 days (from the day of sowing) and their further growth was monitored. Fresh and dry weights of the plantlets were measured at the time of transplantation, and eight weeks after transplantation. For this seedlings were carefully uprooted, and roots were gently and thoroughly washed to remove the soil; the seedlings were blotted to remove surface water and used to determine fresh for determining the day weight, weights. The seedlings were maintained at 80 ± 20C Results in treatment paper bags for seven days and weighed.

Table 1

Details of viability, in vitro germination and vigour of pollen samples used for pollinations

Pollen sample	FCR (%)	in vitro germination 2 h (%)	No. pollen tubes emerged in 1 cm semi-vivo styles 16 h	24 h
Fresh F	83.3	71.8	100	100
High RH Lab (L HH)	93.5	83.9	100	100
High RH 380C (38 HH)	90.7	53.0*	10.4	100
High RH 450C (45 HH)	79.7	0	0	0
Dry 450C (45 dry)	91.5	82.3	100	100

* 4th after culture

Table 2

Details of germination and tube growth in the pistils of treated pollen samples.

Time after pollination	Pollen treatment: F. Lab HH, 38 HH, 45 D	Pollen treatment: 45 HH
40 h	pollen germination profuse, pollen tubes entered ovary	pollengermination profuse,abnormalities in many pollen tubes grew for 1. 5-2 cm in the style.
64 h	stigma and style started senescing; corolla wilted	pollen tubes grew ca 3.5cm in the style, no pollen tubes in the ovary; stigma, style and corolla fresh.
84 h	stigma and upper part of style dried, ovary showed visible enlargement	pollen tubes entered ovary, pistil still fresh, no enlargement of the ovary.

Table 1 presents details viability and in vitro germination of pollen samples used for pollination.

There was no significant difference in the viability of different pollen samples. However, in vitro germination was significantly reduced and delayed in pollen samples treated at high Rs at 380C (38 HH). Pollen samples treated at high Rs at 450C (45 HH) failed to germinate in vitro. However, all the pollen samples showed good germination on the stigma although many pollen tubes in 45 HH-treated pollen samples showed abnormalities in the form of swelling, curling and excessive deposition of callose. There was no apparent difference in the details of pollen tube growth in the style in different treatments except in the sample treated to 45 HH; in the former pollen tubes reached the ovary by 40 h while in the latter they took 84 h.

Seeds collected from each treatment were germinated in petri plates during July at 25 ± 20C. More than 50% of the seeds in 38 HH and 450C dry (45 D) germinated by third day while in the other treatments only about 20-30% of the seeds germinated in 3 days. Most of the seeds germinated in all the treatments by 5 days (Fig. 1).

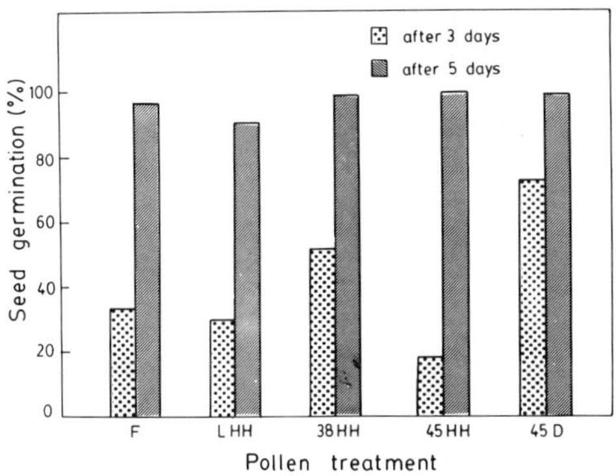

Fig. 1. Germination of seeds derived from different pollen treatments. F. fresh pollen, HH - laboratory, 38 HH - high RH at 380C, 45 HH - high RH at 450C, 45 D - 450C day.

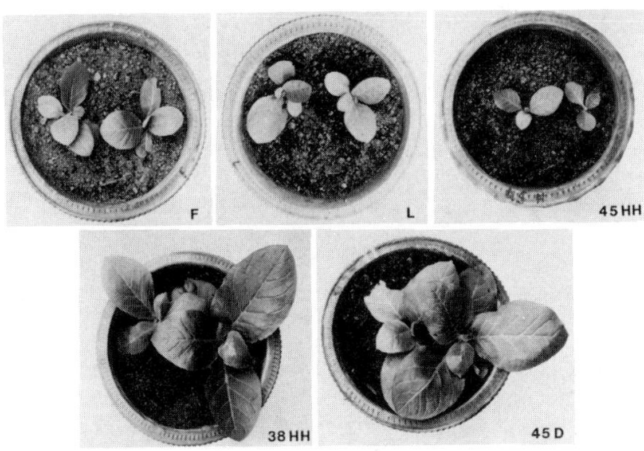

Fig. 2. Three representative seedlings derived from different pollen treatment at th time of transplantation. Abbreviations of treatments as in Fig. 1.

In seed samples sown in pots, cotyledons emerged above the level of the soil and turned green by eight days in 38 HH and 45 D treatments, while in others cotytedon emergence took 10-12 days. The size of the cotyledons was also larger in 38 HH and 45 D seeds.

In three weeks, most of the seedlings from 38 HH and 45 D showed expanded first and second leaves whereas in the other treatments, most of the seedlings showed the expansion of only the first leaf. The differences in seedling growth of the two sets remained obvious during their further growth even after transplantation (Figs. 2, 3).

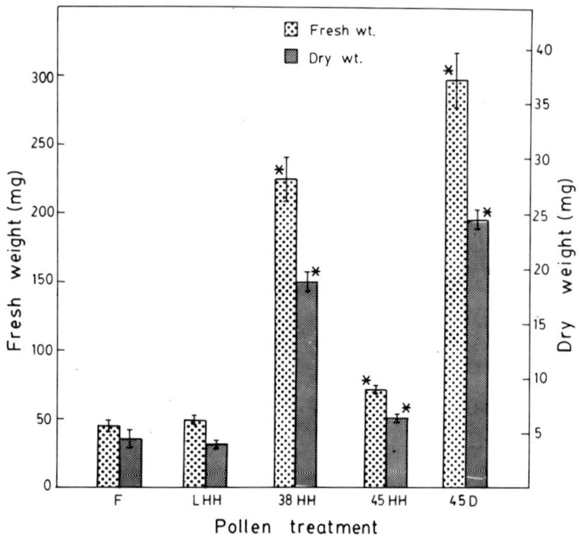

Fig. 3. Seedlings derived from different pollen treatments 8 weeks after transplantation. Abbreviations or treatments as in Fig.1.

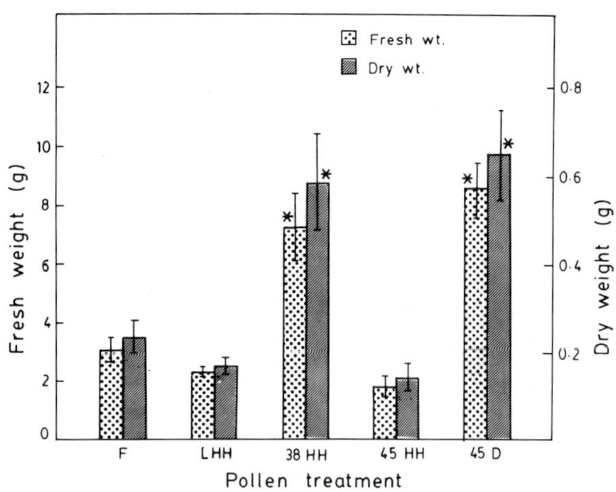

Fig.4. Fresh and dry weights of seedlings at the time of transplantation. Abbreviations as in Fig. 1.

Figure 4 and 5 presents fresh and dry weight determinations of plantlets at the time of transplantation, and eight weeks after transplantation. At the time of transplantation, plantlets derived from 38 HH, 45 D and 45 HH pollen samples, showed significantly higher fresh and dry weights when compared to the control (Fresh pollen); the values for plantlets derived from laboratory maintained. Pollen were not significantly different from those of fresh pollen (Fig. 4). Even after 8 weeks of transplantation, plantlets derived from 45 D and 38 HH pollen showed significantly higher fresh and day weight when compared to those from other treatments (Fig. 5).

The plants derived from 38 HH and 45 D showed early flowering when compared to those from other treatments. Plants from all the treatments produced functional pollen and set fruits and seeds. Pollen grains from these plants, particularly those derived from 38 HH and 45 D treatments, have again been subjected to temperature and RH stresses and used for pollination. The seeds would be collected and their performance would be studied in the coming year. The plants are still flowering and other details of the reproductive phase would be collected.

Discussion

Many investigations during the last decade have shown that pollen competition results in differential gene transmission. In Zea the speed of pollen tube growth is positively correlated with many features such as weight of the seedling, ear and kernel of the resulting sporophytes (Ottaviano et al., 1980, see also Ottaviano and Mulcahy 1989). Selection pressures which favour more vigorous pollen (In terms of speed of germination and/or rate of pollen tube growth) in effecting fertilization have resulted in vigorous and uniform sporophytes when compared to those from pollen not subjected to selection pressures.

Two of the selection pressures which are relevant to the present studies are high density pollination and storage of pollen before pollination. In Petunia (Mulcahy et al., 1982), Dianthus ( ), Cassia (Lee), and Lotus and Cucurbita (Schlichting), 1987, high density pollination, which increases gametophytic competition, resulted in sporophytes which germinated more rapidly and seedlings which grew more vigorously when compared to the controls. In Dianthus plants derived from intense competition were competitively superior to lolium perinne than the controls. Pollination with stored pollen has also been shown to improve the performance of the resulting sporophyte in Petunia (Mckenna and Mulcahy, 1983) and potato (Pallais 1986). In Zea, however, pollen storage did not affect the performance of the progeny (Pfahler 1986).

Conclusions

Results of the present investigation clearly show that application of high temperature stress alone or in combination with high humidity to pollen acts as an effective selection pressure, and increases the vigour of the resulting seedlings. The simple explanation for these results would be that the stress reduces the chances of less vigorous pollen effecting fertilization. However, more investigations are needed on the details of the biology of stressed pollen and of the progeny for a better understanding of the effects of high temperature and humidity stresses on pollen function. Similar studies are being carried out in other systems.

References

Heslop-Harrison, J.; Heslop-Harrison, Y. Evaluation of pollen viability by induced fluorescence; intracellular hydrolysis of fluorescein diacetate. Stain Technol. 45:115-120; 1970.

Lee, T.D.; Hartgerink, A.P. Pollination intensity, fruit naturation pattern and offspring quality in Cassia fasciculata (Leguminosae). Mulcahy, D.L.; Mulcahy, G.B.; Ottaviano, E., eds. Biotechnology and ecology of pollen. Springer-Verlag, Heidelberg. 1986:417-422.

Mascarenhas, J.P. The male gametophyte of flowering plants. Plant Cell 1:657-644; 1989.

Mulcahy, D.L. The rise of the angiosperns: a genecological factor. Science 206:20-23; 1979.

Mulcahy, D.L.; Mulcahy, G.B.; Ottaviano, E. Sporophytic expression of gametophytic competition is Petunia hybrida.

Mulcahy D.L. ed. Gamete competition is plants and animals:

Elsevier Press, Amsterdam; 1975:227.

Mulcahy, G.B.; Mulcahy, D.L.; Pfahler, P.L. The effect of delayed pollination in Peunia hybrida. Acta Bot. Neerl. 31:97:103; 1982.

Ottaviano, E.; Mulcahy, D.L. Genetics of anqiosperm pollen. Adv. Genet. 26:1-64; 1989.

Pallais, N.; Malagamba, P.; Fong, N.; Garcia, R.; Schmiediche, P. Pollen selection through storage: A tool for improving, true potato seed quality? Mulcahy, D.L.

Mulcahy, G.B.; Ottaviano, E., eds. Biotechnology and ecology of pollen: Springer - Verlag, Heidelberg; 1986:153-158.

Schlichting, C.D.; Stephanson, A.G.; Davis, L.E.; Winsor, J.A. Pollen competition and offspring variance. Evol. Trends Plants 1: 35-39; 1987.

Shivanna K.R.; Cresti, M. Effects of high humidity and temperature stress as pollen membrane and pollen vigour. Sex. Plant Repord. 2:137-141; 1989.

Shivanna K.R.; Johri, B. M. The angiosperm pollen: structure and function. Wiley Eastern, New Delhi; 1985.

Shivanna, K.R.; Linskens, H.F.; Cresti, M. Responses of tobacco pollen to high humidity and heat stress: siability, and germinability in vitro and in vivo. Sex. Plant Repord. 1991a (in press).

Shivanna, K.R,; Linskens, H.F.; Cresti, M. Pollen viability and pollen vigour. Theor. Appl. Genet. 81:38-42; 1991 b.

57
Gamete Selection for Heat Stress Tolerance in Maize

J.F. Petolino, N.M. Cowen, S.A. Thompson, and J.C. Mitchell

Grain yields in maize are historically lower in those years when above average temperature and below average rainfall conditions prevail during the months of July and August. This is an implicit expression of the relative sensitivity of the reproductive process to heat and drought stress. Delay in silk emergence, pollen sterility, reduced ear receptivity, and kernel abortion have been associated with stress during this period. Breeding plants for greater environmental resiliency is possible, since tolerance to heat and drought stress are heritable traits. However, conventional breeding for stress tolerance has been difficult due to the lack of effective, easily measured selection criteria.

The ability of maize pollen to develop and grow at higher temperatures has been suggested as a means of screening to identify genotypes less sensitive to heat stress during the reproductive stage of development (Herrero and Johnson, 1980)). This trait, when combined with improved drought tolerance and high productivity, would result in genotypes whose yields are less sensitive to unfavorably high temperatures. The development of a technique capable of identifying genotypic differences in gamete behavior is a prerequisite for the exploitation of this trait for plant breeding.

Pollen Selection

There has been an increasing body of evidence regarding the ability to shift allelic frequencies by allowing gametes to develop under selective conditions (Mulcahy et al., 1986). Selection at the pollen level has been proposed as a strategy to enrich the frequencies of genes associated with useful agronomic traits (Zamir, 1983). Exposure of pollen to thermal stress during its formation, germination, tube growth, and/or fertilization may represent a potentially powerful means of altering the genetic structure of the resulting progeny by skewing allelic frequencies in favor of greater tolerance. The utilization of pollen selection in maize breeding assumes a degree of independence on the part of the male gametes with respect to gene expression and the occurrence of considerable genetic overlap between gametophytic and sporophytic generations (Sari Gorla et al., 1986).

In an attempt to capture different segments of the gametic population, selection at the pollen level has been attempted using various procedures prior to or during pollination. Factors such as the length of pollen storage (Mulcahy et al., 1982), the amount of pollen applied to the stigmatic surface (Ter-Avanesian, 1978), the length of stigmatic tissue to traverse (Ottaviano et al., 1982), and the chemical environment inside the stigma (Pfahler et al., 1986b) have all been found to play a role in differential gamete transmission.

Promising results have emerged from those studies where a concerted effort was made to create an environment to selectively differentiate between developing gametes. Self-pollinating an inter-specific hybrid between the chilling-sensitive tomato (Lycopersicon esculentum) and a cold-tolerant wild relative (L. hirsutum) under low temperature conditions resulted in an F2 population that was more cold-tolerant than that produced by self-pollinating at higher temperatures (Zamir and Gaddish, 1987). Moreover, using isozyme analysis of back-crossed progeny, gametes containing two chromosomal regions of L. hirsutum were found to be favored over those with the corresponding L. esculentum alleles in crosses at low temperature (Zamir et al., 1982). Recent work in maize demonstrated that forced pollen shedding from detached tassels and prezygotic styler treatments altered the transmission frequency of various endosperm mutant loci to resulting progeny (Pfahler et al., 1986a; 1986b). These results established that selective fertilization can be directed such that allelic frequencies of specific traits are increased.

The development of a reliable technology capable of identifying differences in gamete vigor is a prerequisite for the application of pollen selection to plant breeding. The precise control of the gametophytic phase of the life cycle, made possible through *in vitro* pollination-fertilization, represents a powerful experimental system. Using this procedure, male and female gametes can be isolated, combined, and the resulting zygote grown to maturity under relatively controlled laboratory conditions.

Gamete Selection via *In Vitro* Pollination-Fertilization

In order to assess the feasibility of utilizing gamete selection as a means of developing heat stress tolerant maize germplasm, a series of selected and non-selected progenies were generated by exposing pollen to elevated (38°C) and normal (28°C) temperatures, respectively. The approach was to use a modification of a crossing scheme for improving inbred lines which involved crossing an adapted inbred line with a random sample of pollen from a variable pollen source, such as an open-pollinated variety (Stadler, 1944). Individual plants so obtained are then self-pollinated to generate progenies as well as out-crossed to an appropriate tester line. As originally conceived, genotypes whose test-cross performance indicate that they had received a desirable gamete from the variable pollen source are then maintained from the selfed seed. In the present application, rather than a random sample of gametes from a variable pollen source, attempts were made to capture only those gametes which can function under high temperature stress conditions using *in vitro* pollination-fertilization (Petolino et al., 1990).

The inbred lines used as females were B73 and LH74. These genotypes represent Corn Belt-adapted inbred lines, related to the Iowa Stiff Stalk Synthetic, which had been previously found to be relatively heat stress sensitive. The source of heat stress tolerance was a synthetic population known as "Arizona Arid Environment Maize" (AZ100). This material was derived from maize collections make on various southwestern U.S. Indian Reservations and maintained as an open-pollinated population, isolated from all other maize, for 23 years in southern Arizona. Differences among the genotypes were apparent with regard to heat stress-induced reductions in pollen germination. Pollen from the inbreds B73 and LH74 exhibited complete lack of germination at 38°C while AZ100 pollen maintained a germination frequency near 40% at this temperature.

Using *in vitro* pollination-fertilization procedures a series of selected and non-selected progenies were generated. Unpollinated ear shoots were harvested from the inbreds 4 days after first silk emergence. After a surface sterilization, the husks were removed to expose the unfertilized ovaries after which wedge-shaped sections were cut and placed onto medium which could support kernel development. Pollen was then collected from AZ100 plants and prepared for use by passing through a 0.5 mm filter.

Approximately 250 pollen grains were applied to the end of each silk and the entire system was incubated for 24 hours. Selected progenies were those incubated in 38°C during this period while non-selected progenies were incubated in 28°C. During this time the pollen germinates and fertilization occurs. The silks were then removed and all of the wedges were maintained at 28°C. After about 40 days, the fertilized ovaries displayed varying degrees of development ranging from swelling, with little or no embryo growth, to complete kernel formation. These kernels produced normal plants which were grown to maturity and self-pollinated in the greenhouse. Seed from individual plants were the progenies evaluated.

Four separate populations of progenies were generated using these procedures: B73xAZ100, B73x(B73xAZ100), LH74xAZ100, and LH74x(LH74xAZ100). Out of 9,914 ovaries pollinated, 1,250 kernels were formed from which 689 plants were obtained. A total of 112 progenies (14 per cross x temperature treatment), designated selected or non-selected depending on whether the plant was derived from the 38°C or 28°C treatment, respectively, were produced by self-pollinating the resulting in vitro-derived plants.

Performance of Selected vs. Non-Selected Progenies

In order to have fertilized an ovary and become a selected progeny, a pollen grain would have had to function at 38°C. Thus, the question arises as to the relative sensitivity of the pollen from the selected and non-selected progenies. In order to evaluate the relative pollen sensitivity of the progenies, tassels were harvested from representative plants on the day the first anthers began to shed. The cut portion of the stalk was immediately submerged in water and transported to incubators maintained at a temperature of either 28°C or 38°C. Pollen was collected the next day and sprinkled on the surface of a germination medium. Pollen was classified as germinated if, after a 2 h dark incubation, the tube was at least as long as the diameter of the grain. A total of 250 pollen grains were counted per tassel and 5 tassels were taken for each progeny x temperature treatment. No significant differences were observed for pollen gemination at 28°C. However, in all four crosses, pollen from the selected progenies was on average more tolerant to heat stress-induced reductions in germination than was pollen from non-selected progenies. This was particularly apparent in the single cross B73xAZ100, where the average percent pollen germination at 38°C of the selected progenies was more than three-times that of the non-selected progenies of the same cross (i.e., 11.5% vs. 3.7%). Thus, exposure to elevated temperature during gamete function appeared to influence the heat stress tolerance of pollen in the following generation.

In order to assess the relative field performance of these materials, all 112 progenies were grown in Champaign and various traits were measured. Seed was planted on 23 May and plots were harvested on 18 October. The planting date was later than recommended in order to ensure higher than average temperatures during reproductive development. Average maximum and minimum temperatures at the site were 25.7°C and 18.8°C during July and 31.9°C and 18.0°C during August. Temperature reached or exceeded 32°C on 18 and 16 days, respectively, during July and August. The experimental design was a split plot with the main plots consisting of the four crosses and subplots were single rows of either selected or non-selected progeny. Averaged over all crosses, significant differences between the selected and non-selected progenies were observed for seedling vigor, plant height, stalk lodging, root lodging, grain moisture, and grain yield. In most instances, the performance of the selected progenies was superior, as evidenced by higher grain yields, greater seedling vigor, and reduced stalk and root lodging. This was particularly apparent in the B73xAZ100 cross where the selected progenies exhibited 17% greater grain yield and 43% less root lodging than the non-selected progenies.

Heat stress selection during the generation of these progenies resulted in increased heat tolerance at both the gametophytic and sporophytic levels. The deliberate late planting of these materials in a year when above average temperatures and below average rainfall conditions prevailed, implies that stress had a major influence on most of the agronomic traits measured. The overall increased seedling vigor, reduced stalk and root lodging, and higher grain yields, together with the reduced pollen sensitivity to high temperature exposure, displayed by the selected progenies suggest an overlap of gametophytic and sporophytic heat stress tolerance. These results support the contention that gamete selection can be a means of shifting allelic frequencies in desirable directions such that large numbers of haploid genotypes can be screened during pollination resulting in a non-random population of progenies.

Literature Cited

Herraro, M.P.; Johnson, R.R. High temperature stress and pollen viability of maize. Crop Sci. 20:796-800; 1980.

Mulcahy, D.L.; Bergamini-Mulcahy, G.; Ottaviano, E., editors. Biotechnology and ecology of pollen. New York: Springer Verlag, 1986.

Mulcahy, G.B.; Mulcahy, D.L.; Pfahler, P.L. The effect of delayed pollination in Petunia hybrida. Acta. Bot. Neerl. 31:97-103; 1982.

Ottaviano, E.; Sari Gorla, M.; Pe, E. Male gametophyte selection in maize. Theor. Appl. Genet. 63:249-254; 1982.

Petolino, J.F.; Cowen, N.M.; Thompson, S.A; Mitchell, J.C. Gamete selection for heat-stress tolerance in maize. J. Plant Physiol. 136:219-224; 1990.

Pfahler, P.L.; Mulcahy, D.L.; Barnabas, B. The effect of forced-shedding on pollen traits, seed setting, and transmission of various maize (Zea mays L.) endosperm mutant loci. Acta Bot. Neerl. 35:195-200; 1986a.

Pfahler, P.L.; Mulcahy, D.L.; Barnabas, B. The effect of pre-pollination styler treatments on seed set and pollen transmission of various maize (Zea mays L.) mutant loci. Acta. Bot. Neerl. 35:201-207.

Sari Gorla, M.; Frova, C.; Benelli, G.; Ottaviano, E. The extent of gametophytic-sporophytic gene expression in maize. Theor. Appl. Genet. 72:42-47; 1986.

Stadler, L.J. Gamete selection in corn breeding. Jour. Amer. Soc. Agron. 36:988-989; 1944.

Ter-Avanesian, D.V. The effect of varying the number of pollen grains used in fertilization. Theor. Appl. Genet. 52:77-79; 1978.

Zamir, D. Pollen gene expression and selection: applications in plant breeding. In: Tanksley, S.D.; Orton, T.J., eds., Isozymes in plant genetics and breeding. New York: Elsevier, 1983:p. 313-330.

Zamir, D.; Gaddish, I. Pollen selection for low temperature adaption in tomato. Theor. Appl. Genet. 74:545-548; 1987.

Zamir, D.; Tanksley, S.D.; Jones, R.A. Haploid selection for low temperature tolerance of tomato pollen. Genetics 101:129-137; 1982.

58
Production of Highly Cold Tolerant Maize Inbred Lines by Repeated Gametophytic Selection

Géza Kovács and Beáta Barnabás

Pollen biological investigations have verified lately that approximately 65% of the analysed structural genes in plant genomes are expressed both in sporophytic and gametophytic generations (Tanksley et al., 1981; Willing et al., 1986; Sari-Gorla et al., 1986). This extent of genetic overlap gives the possibility of in vitro and in vivo pollen genotype selection. Pollen genotype selection would be more effective than the sporophytic selection since large numbers of pollen grains can be exposed to selection pressure on the haploid level (without the confounding effect of dominance, epistasis, etc.) as was suggested by Mulcahy (1979) and Pfahler (1983).

There are several advantages to use haploid pollen to screen desirable genes. These studies have indicated that selection for resistance to certain environmental stress factors (e.g. low/high temperature, drought, salinity, agrochemicals, pathotoxins, heavy metals, etc.) have been effective in various species (Zamir et al., 1982; Sacher et al., 1983; Smith and Moser, 1984; Walden and Greyson, 1986).

In theory, one should be able to use pollen selection for any agronomic trait which is expressed during the gametophytic life cycle and suitable selection protocol can be devised. Unfortunately most of the experiments were carried out on detecting interspecific differences, and up to now only Ottaviano and his colleagues (1989) have reported on successful pollen genotype selection based on intraspecific variation of herbicide tolerance in maize.

Correlation between gametophytic and sporophytic cold tolerance

In our lab we started with the pollen selection experiments in 1985 on the basis of the pollen storage experiences (Barnabas and Kovacs, 1988). The results from long term preservation of pollen grains suggested that the applied method had a strong selective effect on the different pollen genotypes, and this could be detected in the next sporophytic generation, too. In vitro germination capacity of the pollen grains of various maize inbreds at 10oC differed considerably (Table 1) and the pollen germination characteristics highly correlated with the low temperature germination of the kernels. Both pollen grains and the kernels germinated in a relatively high percentage if the inbred was cold tolerant and there was no or very low germination in the case of sensitive ones.

Table 1. Pollen germination characteristics of cold tolerant (K1E) and cold sensitive (K1J) genotypes at low temperature

Genotype	Germination %	Pollen tube length (m)
K1E	88.6	1418
K1J	30.2***	249***

*** - significant at 0.001 probability level.

Germination was examined on the stigma (6 hours after pollination).

These results suggest that the genetic overlap exists in cold tolerance, and the intraspecific variation of cold tolerance can be detected on gametophytic level too. In our first genetic experiment we found that the pollen germination percentage was reflected for the germination percentage of the kernels, while the pollen tube growth correlated with the early seedling growth and the genetic determination of pollen germination and tube growth was independent as at sporophytic level. The intraspecific variation and the association of pollen germination characteristics with the related sporophytic characters gives the possibility to elaborate suitable selection techniques.

Pollen selection

Our first selection experiment was very successful (Kovacs, 1989), but later on we met with unexpected difficulties. In the first experiments the duration of cold treatment was 24 hours, and the results were very confounding. Sometimes we had successful results, sometimes had not. The application of the necessary longer cold treatment resulted in a very low seed set (Table 2).

Table 2. Effect of duration of cold treatment on pollen fertilizing ability of K1E x K1J F1 maize hybrid

Treatment	Moisture content (%)	Seed set %	Seed No./ear
Fresh pollen	54	91.0a	460.1a
4°C for 48 hours	52	65.2b	358.0ab
4°C for 72 hours	54	31.0c	246.0b
4°C for 1 week	54	14.1c	92.4c

Treatments followed by the same letter are not significant.

Most of the pollen grains lost their viability and fertilizing ability during this period.

For successful preservation of the viability and fertility of the pollen grains generally we have to dry them before the pollen storage experiments. The low moisture content is the key point of short term pollen preservation at 4oC too, and the combination of pollen drying with the cold treatment was also necessary in the selection experiments (Table 3).

Table 3. Effect of pollen drying on the fertilizing ability of cold treated pollen grains

Treatment	Moisture content (%)	Seed set (%)
Fresh pollen	56	90.6a
Dried pollen	19	81.7a
4°C for 48 hours	19	82.1a
4°C for 72 hours	19	70.5a,b
4°C for 1 week	19	58.2b

Treatments followed by the same letter are not significant.

As a selection pressure the short term storage of the preliminary dried pollen at 4oC for one week guarantees that only the cold tolerant pollen grains can survive and they can transmit their tolerance into the next offspring (Table 4).

Table 4. Low temperature germination percentage and seedling characteristics of the F2 kernels obtained from the fresh and delayed pollen treatments

Character	Pollination treatment	
	Fresh	Delayed
Germination %	46.00	84.00***
Shoot length (cm)	0.60	2.60**
Root length (cm)	3.50	5.40*
Fresh weight of shoots (g)	0.28	0.58**
Fresh weight of roots (g)	0.27	0.67**

*, **, *** - significant at 0.05, 0.01 and 0.001 probability level, respectively.

The segregation of low temperature germination of the F2 sporophytic generation indicated that the pollen selection was very effective, only cold tolerant seeds could be obtained (Figure 1a). Since the female gametophyte cannot be selected in the same way, the F3 offspring will segregate and in this population there will be some cold sensitive genotype again. With the combination of

Fig. 1. Distribution of low temperature germination.

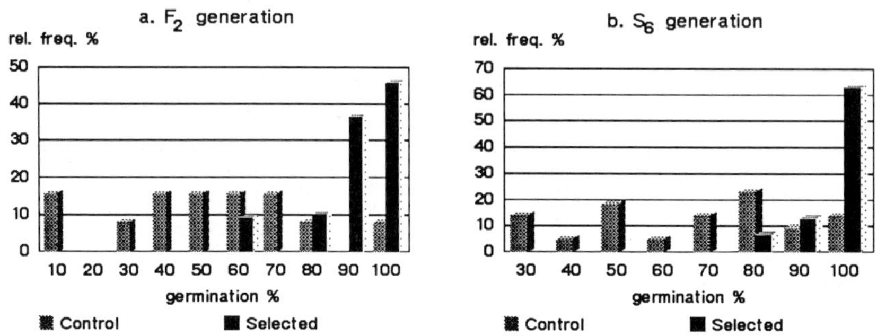

sporophytic and gametophytic selection or with repeated pollen selection can be guaranteed that during the inbreeding process the cold sensitive genotypes will be eliminated (Figure 1b). In the control (unselected) S6 generation all the different classes of cold tolerance might occur, in the selected populations only highly cold tolerant lines could be obtained. Most of these selected S6 lines have stronger cold tolerance than the original parents, the transgressive segregation is very common if we apply repeated pollen selection.

Table 5. Effect of repeated pollen selection on some agronomical characteristics of the open pollinated S6 generation

Agronomic traits	Unselected (control)	Selected
Emergence %	81.4	95.3*
Seed set %	67.6	81.9**
Yield/plant (g)	62.7	141.1***
Seed No./plant	483.9	447.1

*, **, *** - significant at 0.05, 0.01 and 0.001 probability level, respectively.

The field productivity of the selected lines gave very interesting results (Table 5). Most of the traits examined were better that of the unselected population. The inbreeding depression which generally exists in this case was not prominent. Presently this phenomena is under investigation.

In conclusion we suggest that pollen selection may represent a new important breeding tool and can be easily incorporated into many corn breeding programs.

References

1. Barnabas, B.; Kovacs, G. Perspectives of Pollen amd Male Gamete Selection in Cereals. In: Wilms, H.Z.; Keijzer, C.J., eds. Plant Sperm cells as Tools for Biotechnology. Pudoc, Wageningen, p. 137-147. 1989.

2. Kovacs, G. Selection for cold hardiness in pollen populations of maize. Novenytermeles, 38: 9-13. 1989.

3. Mulcahy, D.L. The rise of angiosperms: a genecological factor. Science, 26: 20-23. 1979.

4. Pfahler, P.L. Comparative Effectiveness of Pollen Genotype Selection in Higher Plants. In: Mulcahy, D.L.; Ottaviano, E., eds. Pollen: Biology and Implications for Plant Breeding. Elsevier Biomedical, New York, p. 366-367. 1983.

5. Sacher, R.F.; Mulcahy, D.L.; Staples, R.C. Developmental Selection during Self Pollination of Licopersicon x Solanum F1 for Salt Tolerance of F2. In: Mulcahy, D.L.; Ottaviano, E., eds. Pollen: Biology and Implications for Plant Breeding. Elsevier Biomedical, New York, p. 329-334. 1983.

6. Sari-Gorla, M.; Frova, C.; Binelli, G.; Ottaviano, E. The extent of gametophytic-sporophytic gene expression in maize. Theor. Appl. Genet. 72: 42-47. 1986.

7. Smith, G.A.; Moser, H.S. In vitro sporophytic screening and gametophytic verification of herbicide tolerance in sugarbeet. Agronomy Abstracts, p. 88. 1984.

8. Zamir, D.; Tanksley, S.D.; Jones, R.A. Haploid selection for low temperature tolerance of tomato pollen. Genetics, 101: 129-173. 1982.

9. Walden, D.B.; Grezson, R.I. Maize Pollen Research: Preliminary Report from Two Projects Investigating Gamete Selection. In: Mulcahy, D.L.; Ottaviano, E., eds. Biotechnology and Ecology of Pollen. Springer-Verlag, Berlin, p. 139-145. 1986.

10. Willing, R.P.; Bashe, D.; Mascarenhas, J.P. An analysis of the quantity and diversity of messenger RNAs from pollen and shoots of Zea mays. Theor. Appl. Genet. 75: 751-753. 1988.

59
Herbicide Tolerance in Maize – Genetics and Pollen Selection

M. Sari Gorla, S. Ferrario, L. Gianfranceschi, and M. Villa

Introduction

Herbicide use is a well established necessity in modern agriculture, but it has given rise to problems mainly related to environmental impact and weed resistance evolution (Schulz et al., 1990, Gressel, 1987). These could be solved by developing molecules ensuring environmental safety and endowed with low persistence in the soil, thus not selecting resistant weeds. However, frequently molecules with these desirable characteristics are not selective: they are not able to control all plant species, except for the crop of interest. For this reason, efforts are made to obtain crops resistant to specifically designed herbicides.

According to the mechanism of resistance and, consequently, to the genetic basis of its control, different approaches can be adopted. Single gene based resistances can be manipulated by gene transfer, while for traits controlled by a multifactorial system this technology is not feasible. A different possibility is provided by gametophytic selection: due to the peculiar characteristics of a segregating pollen population (large size, haploidy, wide genetic variability), this is expected to be much more effective than sporophytic selection, and its success, provided the genes conferring the tolerance are expressed both in pollen and in the plants, does not depend on the type of genetic control of the trait (Ottaviano and Mulcahy 1989).

EPTC is a thiocarbamate herbicide (S-ethyl dipropil thiocarbamate) which has many favourable characteristics, but it is not completely harmless for corn; in fact it is used in association with a safener, which is able to protect the crop, while maintaining the herbicide effect against the weeds. The primary herbicide effect is due to inhibition of fatty acid synthesis, causing membrane alteration and, as a final effect, the blocking of photosynthesis and respiration. The herbicidal compound is formed on metabolic sulfoxidation of EPTC by monoxygenase enzymes; then the EPTC-sulfoxide is detoxified through the formation of an EPTC-sulfoxide-glutathione conjugate, catalyzed by the glutathione transferase enzyme (Lay and Casida, 1976). The effect of safeners in protecting corn from EPTC injury is in fact based on induction of high glutathione (GSH) levels and enhancing glutathione transferase (GST) activity. The process does not occur in weed.

The purpose of this study was to explore the genetic variability in maize of tolerance to EPTC by means of pollen and sporophyte assay, in order to detect the genetic basis of tolerance and evaluate the efficiency of gametophytic selection for genetic manipulation of the trait.

Materials and Methods

A large collection of inbred lines (40) of different genetic origin were included in the sample used for the screening of the tolerant and susceptible genotypes at pollen and plant level.

Pollen assay

Fresh pollen was collected on three different days from plants grown in greenhouse and directly used for pollen assay. Each day's collections of pollen was split into two lots: one was grown on solid artificial medium (Cheng and Freeling, 1976), the other on medium supplemented with different EPTC doses: 25, 50, 75 mg/l a.i. After an incubation of 3 h at 27°C, the Petri dishes were fixed with Farmer's liquid, stained with cotton blue in lactic acid, and used for measuring pollen grain germination percent (on 500 grains) and pollen tube length (on 50 tubes). For both traits, pollen response to EPTC was expressed as ratio (%) between treated and control.

Sporophyte assay

Seeds soaked for 20 h in an EPTC solution (550mg/l) were sown in pots filled with dry sand and covered with cellophane bags, in order to prevent herbicide volatilization. During seedling growth water was added without removing the bag, and 15 days after sowing the response to the treatment was assessed by means of visual ratings, from 1 (normal) to 5 (completely malformed). For the genetical analysis and the data presentation, the seedlings were re-classified as normal if they fell into class 1 or 2 (little or no injury) as normal, or as injured, if they fell into classes 3, 4 or 5 (great injury, necrotic areas).

Genetic and statistical analysis

Two inbreds were chosen as tolerant (B73) and susceptible (H99) parents, from which F1's, F2's, F3's and BC's were produced. Due to the incomplete penetrance and variable expressivity of the trait, the observed phenotypic distributions in the parents and F1's were used to predict the F2 and F3 generation frequencies (Powers et al. 1950): according to the genetic model postulated, the frequencies of the different genotypes in the segregating generations were computed; P1 and P2 distributions were used to predict the homozygous genotype performance, F1 to predict heterozygous performance. The observed F2 and F3 data were compared with the expected frequencies thus obtained by goodness of fit X2 test.

Pollen selection

Pollen produced by F1 plants was submitted to selective pressure during the last maturation stage or during its function. In the first case male inflorescence of F1 plants of crossed tolerant and susceptible genotypes were enclosed in a plexiglass chamber containing EPTC vapors for different times (from 3 to 24 h) during the pollen shedding period; the pollen produced was used to pollinate female plants of recessive genotype. In the second type of treatment the female plants, after pollination by pollen from F1 plants, were enclosed in the chamber with EPTC vapors.

GST and GSH assay

GST assay on roots and pollen was carried out according to Komives et al (1985); GSH content in roots was quantitatively determined according to Lay and Casida (1976).

Results and Discussion

Though EPTC is known to be not completely selective for maize, the exploration of the genetic variability of plant response to the herbicide, measured as proportion of normal seedlings produced after treatment, revealed a wide range of variability, from a complete susceptibility (no normal seedlings produced), to a very good degree of tolerance (about 100% normal seedlings). One line of the first group (H99) and one of the second (B73) were chosen for the genetic analysis of the tolerance and the gametophytic selection experiment. These genotypes, analyzed by pollen assay, revealed a good concordance with the sporophytic performance for both germination percentage and tube growth rate (Table 1).

Genetic and biochemical analysis

Frequency distribution of injury degree of parental lines revealed an incomplete penetrance and variable expressivity of the tolerance: the tolerant and susceptible lines showed the highest frequency of normal and completely damaged plants, respectively, but the first genotype also produced some damaged plants and the second genotype some normal plants. The F1' phenotype was very similar to that of the susceptible parental line, thus suggesting that the trait is mainly due to recessive alleles. This is confirmed by the segregating generations; F2 and F3 distributions can be interpreted on the basis of segregation of few genes.

On the basis of the observed segregation and considering the EPTC metabolic pathway in maize, it is possible to hypothesize a genetic control of tolerance sustained by at least two genes. The mechanism involves a step where the toxic compound is produced by a monoxygenase, and at least a second detoxifying step, operated by the GST-GSH system. Thus it is reasonable to think that EPTC resistance can be achieved by absence of sulfoxide production, by inactivity of monoxygenase, or/and by a very efficient detoxification process (high levels of GSH and of GST). The first possibility is indicated by the observation (Ekler and Stephenson, 1987) that monoxygenase inhibitors have an antagonistic effect with EPTC in corn. The implication of the detoxifying system is suggested by the correlation between GST activity and plant tolerance (see Table 1): roots of the tolerant line revealed a greater GST activity in the presence of EPTC and of safener, while, in the susceptible line GST activity increase was induced only by the presence of the safener. GSH content did not seem to be a limiting factor, since it showed a similar trend of induction in both the susceptible and tolerant lines. The effect was dramatic in two lines particularly

Table 1. Tolerant and susceptible lines responses to EPTC. Gametophyte and sporophyte characterization.

	susceptible line		tolerant line	
Pollen response				
		SE		SE
Germination %	19.0	1.4	73.0	1.7
Tube growth rate	28.0	3.2	40.0	1.8
GST activity (U)	7.0	1.4	15.0	1.4
Sporophyte response				
Normal seedlings %	3.0	1.3	78.0	3.3
Roots GST activity (U):				
Control 12.1	1.4		12.7	
EPTC treatment	12.3		15.2	
Safener treatment	16.2		25.2	
Roots GSH content (mg/g):				
Control	247.1	35.9	310.8	
EPTC treatment	378.2		476.0	
Safener treatment	593.3		704.0	

susceptible to EPTC and revealing a null isozyme variant of GST both in roots and in pollen (data not reported).

Table 2 gives the results of the genetic analysis; F2 and F3 expected frequencies were computed on the basis of P1, P2 and F1 performances and assuming the segregation of two genes, the first conferring tolerance in the presence of the recessive allele, the second by the effect of a dominant allele. The results indicate that a model based on only two genes could be sufficient to explain the observed differences between the parental lines.

Gametophytic selection

Pollen produced by hybrids plants between tolerant and susceptible parents was submitted to selection. Due to the high volatility of the product, the treatment was performed by subjecting the tassel of the pollen producer plant and the silks of the female plants to EPTC vapors. The BC progeny produced was evaluated for herbicide tolerance (Table 3). Seedlings produced by treated pollen revealed greater tolerance than BC non-treated progeny: the relative proportion of normal and damaged seedlings in the first case was in favour of the normal, while the opposite relationship between damaged and normal plants was observed in the control BC population. The observed response to selection, applied at pollen level, on the

Table 2. Observed and expected frequencies of plant injury in parental inbreds, F1 crosses, F2 and F3 generations.

Generation	N	Injured	Normal		
H99	70	97.0%	3.0%		
B73	70	22.0%	78.0%		
F1	100	95.0%	5.0%		
				X2	P
F2 Expected frequencies		227.9	50.04		
Observed frequencies		227	51	0.02	0.9
F3 Expected frequencies		996.2	256.8		
Observed frequencies		968	255	0.80	0.4

resulting sporophyte, proved that the parallel feature of pollen and plant is due to the expression of the same genes in the gametophytic and in the sporophytic phases.

These results confirm the possibility of including pollen selection in breeding programs in order to increase the effectiveness of selection methods; as indicated by the results obtained for tolerance to temperature (Zamir et al., 1982), heavy metal (Searcy and Mulcahy, 1985), sulfonylureas (Sari Gorla et al., 1989), salt (Ottaviano et al., 1990), pollen selection seems particularly suitable for stress resistance selection: it allows very rare alleles, particularly if recessive, and complex genic combination to be selected. Its main limitation for plants producing trinucleate pollen, as many crops do, is that, since mature grain is metabolically silent, effective treatments have to be applied during microspore development or pollen function; hence the selection pressure that can be applied without injuring the whole plant is a limiting factor. New technologies are being developed to overcome this obstacle.

Table 3. Response to gametophytic selection in BC sporophytic progeny

	CONTROL: BC from non-treated F1 pollen	POLLEN SELECTION: BC from treated F1 pollen
Normal seedlings %	39.4	70.4
Injured seedlings %	60.6	29.6
N	150	170

X2 : 31.6; P : 0.00

Acknowledgements

Research supported by Ministero Agricoltura Foreste, Italy, Special Project "Tecnologie Avanzate in Agricoltura".

Literature Cited

Cheng, D.S.; Freeling, M. Methods of maize pollen germination in vitro, collection, storage and treatment with toxic chemicals; recovery of resistant mutants. MGC News Letter 50: 11-13; 1976.

Eekler, Z.; Stephenson, G.R. Synergism and antagonism of herbicides with monooxygenase inhibitors. British Crop Protection conference-Weeds 10B- 5: 1105-1112; 1987.

Gressel, J. Appearance of single and multi-group herbicide resistances and strategies for their prevention. British Crop Protection Conference- Weeds 5-3: 479-488; 1987.

Komives, A.V.; Komives, T.; Dudka, F. Effects of thiocarbamate herbicides on the activity of glutathione S-transferase in maize. Cereal Res. Commun. 13: 253-257; 1985.

Lay, M.M.; Casida, J.E. Dichloroacetamide antidotes enhance thiocarbamate sulfoxide detoxification by elevating corn root glutathione content and glutathione S-transferase activity. Pestic. Biochem. Physiol. 6:442-456; 1976.

Ottaviano, E.; Mulcahy, D.L. Genetics of Angiosperm pollen. Advances in Genetics 26: 1-64; 1989.

Ottaviano, E.; Sari Gorla, M.; Mulcahy, D.L. Pollen selection : efficiency and monitoring. In Z. Ogita; Market, C.L., eds. In : Isozymes: structure, function and use in Biology and Medicine. New York: Wiley-Liss, pp. 575-588; 1990.

Powers, L.; Locke, L.F.; Garret, J.C. Partitioning method of genetic analysis applied to quantitative characters of tomato crosses. Technical Bulletin, U.S. Dept. Agriculture 998: 1-56; 1950.

Sari Gorla, M.; Ottaviano, E.; Frascaroli, E.; Landi, P. P. Herbicide-tolerant corn by pollen selection. Sex Plant Reprod 2: 65-69; 1989.

Schulz, A.; Wengenmayer F.; Goodman, H.M. Genetic engineering of herbicide resistance in higher plants. Critical Rev. in Plant Sci. 9: 1-15.

Searcy, K.B.; Mulcahy, D.L. Pollen selection and the gametophytic expression of metal tolerance in Silene dioica (Caryophillaceae) and Mimulus guttatus (Scrophulariaceae). Amer.J.Bot. 72: 1700-1706; 1985.

Zamir, D.; Tanksley, S.D.; Jones, A.J. Haploid selection for low temperature tolerance of tomato pollen. Genetics 101: 129-137; 1981.

60

Gametophytic Selection for Early Maturity in Tomato (*Lycopersicon esculentum* Mill.)

Mark L. Crispi and Lincoln C. Peirce

The development of early-maturing tomato cultivars has been and continues to be a main objective of both processing and fresh market tomato breeding programs. Tomato earliness has been measured several ways, including length of time from seeding to fruit ripening (expressed in days and in growing degree-days), and number and/or weight of ripe fruit by a particular harvest date (Bernier and Ferguson, 1962; George and Peirce, 1969; Gibrel et al., 1982). To facilitate genetic studies of the inheritance of earliness, investigators have divided the time from seeding to date of first ripe fruit into components (Fogle and Currence, 1962), including number of days to first flower (anthesis), first flower to first fruit set, and first fruit set to first ripe fruit (Bernier and Ferguson, 1962; Fogle and Currence, 1950; Gibrel et al., 1982; Kerr, 1955; Powers and Lyon, 1941). Studies of earliness and its components in field-grown tomatoes have indicated that both are quantitatively inherited and greatly influenced by environmental conditions (Bernier and Ferguson, 1962; Cuarto and Cubero, 1982; Fogle and Currence, 1950; Gibrel et al., 1982; Kerr, 1955; Powers and Lyon, 1941) obscuring true genetic value in segregating populations. Selection of gametes would be one way to increase selection pressure, but if gametophytic selection were to be effective for earliness, shifts in the derived population for one or more earliness components must occur, with a consequent reduction in number of days from seeding to first ripe fruit.

One factor that contributes to the earliness of a tomato cultivar in northern areas is its capacity to set fruit early in the season when ambient temperatures are low (Daubeny, 1961; Kerr, 1955). Robinson and Echim (1988) found that in vitro pollen function at 10ſC is correlated with capacity to set fruit on the first inflorescence when field temperatures are low. They theorized that application of low temperature stress might be useful in selecting gametes that carry genes contributing to early maturity. Crispi (1991) observed temperature x cultivar interactions for in vitro pollen germination and tube growth that suggested temperature could be used to select pollen with early maturity genes. In this study, field tests of progeny derived from selective fertilizations under normal and low temperatures were evaluated for days from seeding to first flower and days from seeding to first ripe fruit. The predictive association of number of days from seeding to first flower and number of days from seeding to first ripe fruit was enhanced by transplants and black polyethylene mulch, presumably because environmental extremes had been eliminated (Peirce and Crispi, 1989).

Material and Methods

Plant Material

From a preliminary study of 21 tomato cultivars in 1987, four were selected (Table 1) based upon pollen germination and tube growth after 24 hours on artificial medium at three temperatures (8, 15, or 28°C). The preliminary data (not shown) suggested that low temperature might be applied to inhibit selectively pollen germination and/or subsequent tube growth of the four cultivars.

Table 1. Maturity class, 1987 number of days from seeding to first flower and days from seeding to first ripe fruit of four tomato cultivars used for gametophytic selection and pollen analysis.

		1987 Days from Seeding to:	
Cultivar	Maturity Class	First Flower	First Ripe
Gemstate	Early	46	87
Rocket	Early	46	85
Crimsonvee	Late	50	109
Harvestvee	Late	57	108

Selective Self-Fertilizations

The four cultivars were crossed within and between maturity classes. Seedlings of each F1 hybrid were grown in the greenhouse at 24°C day and 19°C night. Eight plants of each of two F1 hybrids were placed in two growth chambers, four plants per hybrid per growth chamber. One growth chamber provided 12 hour cycles of 25°C light and 20°C dark. Photosynthetically active radiation (PAR) was 200-400 µmol s^{-1} m^{-2}. The second chamber provided an identical time cycle and PAR but was maintained at 15°C light and 10°C dark. Relative humidity (approximately 50%) and temperature variation (+/- 1°C) were similar in both chambers. Pollen was collected from each plant to be pollinated, and these plants were also emasculated (as described by Rick,1980). Fresh self-pollen then was applied to obtain F_2 seeds. These hand-emasculations were made to insure against any self-pollination that may have occurred before the temperature treatment was applied. Seventy-two hours after pollination, the 16 plants were moved from the growth chambers into the greenhouse. Immediately after removing the plants from the growth chambers, pollinated styles were cut at midlength with a sterile razor blade to prevent late fertilizations under greenhouse conditions (24°C day and 19°C night). Another 16 plants (eight plants of each of two other F_1 hybrids) were then brought to the chambers to be pollinated. This two-hybrid cycling continued until all F_1 hybrid plants had been pollinated. Seeds were harvested from ripe fruit and stored.

Field Evaluation

In the spring, F_2 seeds were sown in the greenhouse (24°C day and 19°C night), hardened, and then hand-transplanted into a prepared field four weeks after seeding. The field was divided into four consecutive blocks, each of which contained 24 randomized plots. Each experimental plot consisted of staggered, double rows of 10 plants (20 plants per plot), spaced 0.61m apart on beds mulched with black plastic (1m wide and 38μm thick). Plots were 1.8m apart on center. Guard rows surrounded the entire experiment.

The dates of first flower (anthesis) and first ripe fruit (breaker stage) were recorded for each plant in each plot. Dates of first flower and first ripe fruit were converted to number of days from seeding. Summary statistics were calculated for each entry, and frequency distributions were constructed for these data. Contingency Chi-squares (Strickberger, 1976) were used to compare frequency distributions of populations derived from the two pollination temperature treatments.

Modifications

Three consecutive tests of selective self-fertilizations were grown in the field and are referred to as Experiments a, b, and c. The major differences between these tests were: time interval between pollination and style cutting was 72 hours for Experiments a and c and 96 hours for Experiment b; location of style cut was at style midlength for Experiments a and b and just below the stigmatic surface for Experiment c.

Results

When data from the three experiments were combined, low temperature self-pollination significantly reduced both percentage fruit set (number of comparisons = 33) and the number of seeds per fruit (number of comparisons = 29) (Wilcoxon's Signed Rank Test, P = 5%).

Numbers of days from seeding to first flower and of seeding to first ripe fruit of F_2 populations derived from normal-temperature, self-pollination treatments were compared with data from similar genetic populations derived from the low-temperature treatment. The mean number of days from seeding to first flower differed significantly (t-test, P = 5%) for one comparison each in Experiments a and b, and for four comparisons in Experiment c (Table 2). In four of these six comparisons, the mean number of days of the population derived from the low temperature self-pollination was less than that of the population derived from the normal-temperature self-pollination (Table 2). When frequency distributions of number of days from seeding to first flower of F2 populations derived from normal- and low-temperature self-pollinations were compared, two comparisons each of Experiments a and b, and four comparisons of Experiment c showed significant differences (Table 3).

The mean number of days from seeding to first ripe fruit differed significantly (t-test, P = 5%) for four comparisons in Experiment a, one comparison in Experiment b, and none of the comparisons in Experiment c (Table 4). In two of these five comparisons, the mean number of days of the population derived from the low temperature self-pollination was less than that of the population derived from normal-temperature self-pollinations (Table 4). Frequency distributions of days from seeding to first ripe of F_2 populations derived from the two self-pollination treatments differed significantly for four comparisons in Experiment a, none of the comparisons in Experiment b, and one comparison in Experiment c (Table 5).

Table 2. Significant comparisons of days from seeding to first flower data of F2 populations derived from normal- and low-temperature self-pollination treatments for Experiments a, b, and c.

		Days from seeding to first flower								
		Exp.a			Exp.b			Exp.c		
Entry[1]	Temp. (°C)	Mean	SE	n	Mean	SE	n	Mean	SE	n
G x H	25/20							45.1	0.4	74
"	15/10							44.0	0.3	80
R x G	25/20							43.7	0.5	68
"	15/10							42.0	0.4	65
R x C	25/20				45.4	0.3	73			
"	15/10				44.0	0.3	65			
R x H	25/20	48.3	0.3	80				44.5	0.4	79
"	15/10	46.8	0.2	79				45.6	0.4	72
C x G	25/20							43.3	0.4	73
"	15/10							45.8	0.4	63

1 Parental cultivars designated as follows: Gemstate (G); Rocket (R); Crimsonvee (C); Harvestvee (H).

Table 3. Significant contingency Chi-square analyses of Experiments a, b, and c, comparing frequency distributions of days from seeding to first flower of progeny derived from normal and low temperature self-pollinations of F1 hybrids.

	Experiment[1]					
	a		b		c	
Progeny group	Chi-square	df	Chi-square	df	Chi-square	df
R x G					26.35 *	15
R x C			27.71 **	11		
R x H	27.19 *	11			30.98 **	15
C x G					27.39 *	15
C x R			23.72 *	13		
H x G	28.51 **	13				
H x C					24.63 *	14

[1] Chi-square value significant at the 5% (*), or 1% (**) probability level.

Table 4. Significant comparisons of days from seeding to first ripe fruit data of F2 populations derived from normal- and low-temperature self-pollination treatments for Experiments a, b, and c.

		Days from seeding to first ripe								
		Exp. a			Exp. b			Exp. c		
Entry[1]	Temperature (°C)	Mean	SE	n	Mean	SE	n	Mean	SE	n
G x H	25/20	88.7	0.7	49						
"	15/10	91.8	0.9	41						
R x C	25/20				88.3	0.4	69			
"	15/10				86.5	0.3	64			
R x H	25/20	88.9	0.6	80						
"	15/10	86.6	0.4	79						
H x G	25/20	88.3	0.5	76						
"	15/10	91.4	1.3	13						
H x C	25/20	98.4	1.9	11						
"	15/10	101.3	0.5	58						

[1] Parental cultivars designated as follows: Gemstate (G); Rocket (R); Crimsonvee (C); Harvestvee (H).

Table 5. Significant contingency Chi-square analyses of Experiments a, b, and c, comparing frequency distributions of days from seeding to first ripe fruit of progeny derived from normal- and low-temperature self-pollinations of F1 hybrids.

	Experiment[1]					
	a		b		c	
Progeny group	Chi-square	df	Chi-square	df	Chi-square	df
G x C	27.80 *	16				
R x H	31.99 *	19				
H x G	32.19 *	17				
H x C	24.20 *	14			14.76 *	15

[1] Chi-square value significant at the 5% (*), or 1% (**) probability level.

Discussion

In this study, gametophytic selection was not successful in changing either the number of days from seeding to first flower or the number of days from seeding to first ripe fruit. There was no appreciable effect of the timing of the style cut after pollination or of the location of the style cut.

The cornerstone of any gametophytic selection scheme is that the genes of the desired sporophytic character also must be expressed in the gametophytic phase (Tanksley et al., 1981) and that an appropriate selection pressure must be employed (Zamir, 1983). In this study, gametophytic selection was utilized to try to select for characters that were not exclusively controlled by the sporophyte. A possible reason for the lack of response to gametophytic selection is that the genes controlling days from seeding to first flower and days from seeding to first ripe fruit are not expressed in the pollen phase. However, the temperature x cultivar interactions for in vitro pollen germination and tube growth observed by Crispi (1991) suggested that temperature could be used to favor and select pollen with early maturity genes. Other workers have used temperature stress during pollen development and pollen storage successfully in gametophytic selection (Mulinix and Iezzoni, 1988; Rodriquez-Garay and Barrow, 1988). However, Maisonneuve et al. (1986) reported no differential gametophytic response to temperature stress, and Maestro and Alvarez (1988) found that effects of temperature on muskmelon (Cucumis melo L.) in vitro pollen germination and pollen tube behavior cannot always be used to predict in vivo behavior.

Several aspects of this study could be modified to increase the likelihood of obtaining significant shifts in the mean of progeny generations, as for example, decreasing the time after pollination from 72 to perhaps 48 hours to increase selection pressure; increasing the size of the F_2 populations planted in the field to improve sampling of progeny derived from the self-pollination treatments; and, modifying stress temperatures to enhance selection pressure.

Choice of parents may be the most important factor. Parents must have large enough differences to achieve significant and useful population shifts (Maisonneuve et al., 1986). To date, the only successful gametophytic selection in tomato utilized interspecific variation (Sacher et al., 1983; Zamir et al., 1981; Zamir et al., 1982; Zamir and Vallejos, 1983; Zamir and Gadish, 1987). In our study, the differences between early and late cultivars for number of days from seeding to first flower and number of days from seeding to first ripe were consistent but not as great as would be expected from interspecific crosses.

Literature Cited

Bernier, C.C.; Ferguson, A.C. A study of the association between earliness, fruit size and early yield in determinate tomatoes. Can. J. Plant Sci. 4:635-641; 1962.

Crispi, M.L. Gametophytic selection for early maturity in tomato (Lycopersicon esculentum Mill.). Ph.D. Dissertation. Durham, NH: University of New Hampshire; 1991.

Cuartero, J.; Cubero, J.I. Genotype-environment interaction in tomato. Theor. Appl. Genet. 61: 273-277; 1982.

Daubeny, H.A. Earliness in tomato varieties with special reference to ability to set fruit at low temperatures. Proc. Amer. Soc. Hort. Sci. 78: 445-449; 1961.

Fogle, H.W.; Currence, T.M. Inheritance of fruit weight and earliness in a tomato cross. Genetics 35: 363-380; 1950.

George, B.F.; Peirce, L.C. The influence of population density and competition on phenotypic stability of tomato plants. J. Amer. Soc. Hort. Sci. 94: 65-67; 1969.
Gibrel, G.; Boe, A.A.; Simpson, W.R.; Everson, D.O. Evaluation of F1 hybrid tomato cultivars for earliness, fruit size, and yield using diallel analysis. J. Amer. Soc. Hort. Sci. 107: 243-247; 1982.
Kerr, E.A. Some factors affecting earliness in the tomato. Can. J. Agric. Sci. 35: 300-308; 1955.
Maestro, M.C.; Alvarez, J. The effects of temperature on pollination and pollen tube growth in muskmelon (Cucumis melo L.). Scientia Hort. 36: 173-181; 1988.
Maisonneuve, B.; Hogenboom, N.G.; den Nijs, A.P.M. Pollen selection in breeding tomato (Lycopersicon esculentum Mill.) for adaptation to low temperature. Euphytica 35: 983-992; 1986.
Mulinix, C.A.; Iezzoni, A.F. Microgametophytic selection in two alfalfa (Medicago sativa L.) clones. Theor. Appl. Genet. 75: 917-922; 1988.
Peirce, L.C.; Crispi, M.L. Relationship between flowering and ripening dates modified in tomatoes by polyethylene mulch and row covers. HortSci. 24: 781-782; 1989.
Powers, L.; Lyon, C.B. Inheritance studies on duration of developmental stages in crosses within the genus Lycopersicon. J. Agric. Res. 63: 129-148; 1941.
Robinson, R.W.; Echim, T. Genetic differences in pollen germination and tube development in relation to fruit set at low temperature. Rpt. Tom. Genet. Coop. 38: 40-41; 1988.
Rodriguez-Garay, B.; Barrow, J.R. Pollen selection for heat tolerance in cotton. CropSci. 28: 857-858; 1988.
Sacher, R.F.; Mulcahy, D.L.; Staples, R.C. Developmental selection during self-pollination of Lycopersicon x Solanum F1 for salt tolerance of F2. In: Mulcahy, D.L.; Ottaviano, E., eds. Pollen: biology and implications for plant breeding. New York: Elsevier Science Publishing Co., Inc.; 1983: p. 329-334.Strickberger, M.W. Genetics. 2nd ed. New York: MacMillan Publishing Co.; 1976.
Tanksley, S.D.; Zamir, D.; Rick, C.M. Evidence for extensive overlap of sporophytic and gametophytic gene expression in Lycopersicon esculentum. Science 213: 453-455; 1981.
Zamir, D. Pollen gene expression and selection: applications in plant breeding. In: Tanksley, S.D.; Orton, T.J., eds. Isozymes in plant genetics and breeding. Part A. Amsterdam: Elsevier Science Publishers B.V.; 1983: p. 313-330.
Zamir, D.; Gadish, I. Pollen selection for low temperature adaptation in tomato. Theor. Appl. Genet. 74: 545-548; 1987.
Zamir, D.; Vallejos, E.C. Temperature effects on haploid selection of tomato microspores and pollen grains. In: Mulcahy, D.L.; Ottaviano, E., eds. Pollen: biology and implications for plant breeding. New York: Elsevier Science Publishing Co., Inc.; 1983: p. 335-342.
Zamir, D.; Tanksley, S.D.; Jones, R.A. Low temperature effect on selective fertilization by pollen mixtures of wild and cultivated tomato species. Theor. Appl. Genet. 59: 235-238; 1981.
Zamir, D.; Tanksley, S.D.; Jones, R.A. Haploid selection for low temperature tolerance of tomato pollen. Genetics 101: 129-137; 1982.

61
Mechanisms in the Pistil that Regulate Gametophyte Population in Peach (*Prunus persica*)

Maria Herrero

During the gametophytic phase there is a luxurious production of individuals. However,from all of them, only a few achieve fertilization and pass into the next generation. While many are randomly lost, others are left behind during the time that goes from pollination to fertilization.

This process can be specially relevant if this reduction is implicated in gametophytic selection (Mulcahy,1979). While there is evidence that a situation of high competence during pollen tube growth decreases the genetic variability of the offspring (Ter-Avanesian, 1978; Schlichting et al.,1987) and that a correlation exists between pollen tube growth rate and the sporophytic vigour of the next generation(Mulcahy, 1974; Ottaviano et al., 1980) there is little evidence on how this competence is worked out and on which is the role that the pistil can play.

Gametophytic reduction is most apparent in the male gametophyte; but there is also a reduction in the number of female gametophytes and in many species far more ovules than seeds are produced (Lloyd, 1980; Stephenson; 1981). Different explanations have accounted for this fact (Charlesworth, 1989). However the information concerning which are the physiological events leading to it is scarce.

Since the processes that take place during the progamic phase (Linskens, 1986) appear closely regulated it is to be expected that the reduction in the number of gametophytes should follow a pattern. Here this reduction is monitored in peach (*Prunus persica* L. Batsch) and is related to the pistil structures the pollen tubes have to traverse.

Male gametophyte reduction

In peach a reduction occurs in the male gametophyte population from the number of pollen grains that land in one stigma to the number of pollen tubes that get to the base of the style. The biggest reduction occurs when the pollen tube penetrate the stigma and get into the transmitting tissue, since from all pollen grains present in one stigma only 53 % get into this tissue. A further reduction occurs along the stylar transmitting tissue and 40 % get 1/4 down, 26 % to half the style , 12 % to 3/4 down and finally only 2 % reach to the base of the style.

The reduction when the pollen tubes penetrate the transmitting tissue can be associated with the change that occurs in the pollen tubes from an autotrophous to an heterotrophous way of life (Herrero and Dickinson,1981; Mulcahy and Mulcahy, 1983). While growth at the stigma surface depends on the pollen grain own reserves and is very similar to pollen grain germination in an artificial culture media, when the pollen tubes get into the transmitting tissue pollen tubes rely on the reserves that are located in this tissue and probably a certain degree of adaptation to this new media is required.

Once in the transmitting tissue a reduction also occurs in the number of pollen tubes. A similar reduction in the number of pollen tubes with distances travelled down the style has been observed in other species (Sedgley, 1976; Cruzan, 1989). When comparing compatible versus incompatible matings, a bigger reduction is observed in incompatible matings (Herrero and Dickinson, 1980, Kahn and De Mason, 1986). This reduction has been interpreted as a manifestation of the incompatibility reaction. However, in all these species a reduction has also been recorded in compatible matings in the number of pollen tubes that travel down the style and no clear explanation has accounted for this fact.

In peach, if the stylar transmitting tissue is examined in cross section, it is apparent that while the external width of the style is very similar all along the stylar tract, a clear reduction can be seen in the width of the transmitting tissue. Thus the transmitting tissue has the shape of a funnel inside the stylar cylinder; and a kind of physical restriction is imposed on the space available for pollen tube growth that can help to reduce its number. A similar reduction in the area of tissue able to support pollen tube growth has been recorded in *Nicotiana glauca* (Cruzan, 1986). However, in peach, this is not the only cause and if the amount of carbohydrates along the stylar transmitting tissue is considered, this difference in shape is streaked since the amount of carbohydrates in this tissue decreases from the area located underneath the stigma to the lower style. Since pollen tube growth along the style is heterotrophic (Herrero and Dickinson, 1979) and reserves disappear as the pollen tubes pass by (Vasil, 1974; Sedgley, 1979); the fact that the provision of reserves is reduced along the stylar length could well play a role reducing gametophyte population through its trophic dependence on the pistil.

Female gametophyte reduction

In peach a reduction is also observed in the number of female gametophytes. From all the flowers present at flowering time only between 10 and 30 per cent become fruits. Likewise from the two ovules present in one flower usually only one becomes a seed while the other one degenerates. This situation is common to other *Prunus* species.

Ovule degeneration can be easily followed in peach because, when an ovule starts to degenerate, translocation is cut down and callose is layered down in the chalazal end (Arbeloa an Herrero, 1985). This layering of callose was first observed in other *Prunus* species by Anvari and Stösser (1978) and later studied in almonds (Pimienta and Polito, 1982).

Ovule degeneration has been followed along the time and has been related to the time when fertilization takes place. In peach it can be seen that 60 per cent of the ovules degenerate before fertilization takes place. This is not due to a premature degeneration but rather to a delayed fertilization. Fertilization occurs particularly late in this specie since it takes about three weeks from pollination to fertilization. This long time is not due to a slow growth of the pollen tubes along the style, for this process can take around one week but it is due to the fact that once the pollen tubes get to the base of the style they enter a lag phase that appears to be controlled by the changes that occur in the pistil tissues (Herrero and Arbeloa, 1989). The pollen tubes first stop when they meet the obturator, that is a placental protuberance that faces the micropyle. Despite its early description (Juel, 1918), and the fact that it has been described in different species as been a path for pollen tube growth (Tilton and Horner, 1980; Tilton et al.; 1984) not a clear function has yet been found to it. In peach it regulates access of the pollen tubes to the ovule, acting as a leaving bridge (Arbeloa and Herrero, 1987) for when the pollen tubes get to the ovary they stop and growth is not resumed until five days later, when starch disappears from the obturator cells and a secretion is produced. Once this secretion is produced callose is layered down at the obturator, thus isolating again the ovule. In *Zea mays* a different mechanism relating to the control of pollen tube entrance into the ovary has been described (Heslop-Harrison et al., 1985) in which the first pollen tubes entering the ovary wall induce a loss of turgidity in the cells at the base of the style that prevents other tubes from entering this region.

A second stop is produced when the pollen tubes face the micropyle. While more work needs to be done to understand this stop it appears clear that it is not due to lack of reserves since the pollen tubes seem to wonder around with no direction (Arbeloa and Herrero, 1991) and probably, like in other species, a kind of chemotropic stimulus can be required (Tilton and Lersten, 1981; Jensen et al., 1983; Lord and Kohorn, 1986). Once the pollen tubes enter the micropyle, growth is still a non straight forward process, and the pollen tubes seem to wonder around inside the micropyle or before entering the nucellus.

At all these wondering points some pollen tubes are lost for ever while the ovules degenerate. However degeneration before fertilization is not the only cause of reduced fertilization; since a proportion of the ovules remain with no signs of degeneration and still unfertilized. While some changes that take place in the ovule are related to early embryo sac nutrition and protection (Arbeloa and Herrero, 1991), information is still lacking on how male-female interaction at this later phase is mediated. A better understanding of the changes that take place in the ovule and how they relate to pollen tube directional growth will shed some light in the understanding of these mechanisms.

Pistil mediated reduction

The progamic phase appears as specially well devised for the control of male-female interaction (Heslop-Harrison and Heslop-Harrison,1985). While much work has been done on pollen pistil interaction in incompatible matings (Heslop-Harrison, 1983) that supports this view, evidence is still scarce on the study of this interaction in compatible matings. However this interaction appears as specially relevant if depending on this dialogue only a few male gametophytes are going to be selected and only a few ovules are going to become seeds. The evidence put forward points to the idea that, in peach, this interaction appears closely controlled and that the pistil is well prepared to mediate this reduction. While the stigma and the style appear as a domain specially well suited for male gametophyte reduction, in the ovary and in the ovule itself there appear to be mechanisms that can play a role controlling the number of female gametophytes that pass into the next generation. The fact that pollen tubes have a trophic dependence on the pistil (Herrero and Dickinson,1979) confers the pistil with a role controlling pollen tube kinetics (Herrero and Arbeloa, 1989), but also controlling pollen tube dynamics, thinning down the number of pollen tubes that get to the ovary. In peach this reduction is mediated by the fact that there is a physical restriction in the space available for pollen tube growth in the transmitting tissue and in the amount of carbohydrates available along this tissue.

Likewise in the ovary a number of changes take place that appear to be involved in controlling pollen tube access to the embryo sac. While much attention has been given to pollen tube growth along the style, the ovary has been a neglected area. However, it appears as a scenery specially well suited for male-female interaction (Herrero et al.,1988). In peach this interaction appears mediated through particular structures that are also present in other unrelated species. The fact that to these structures a clear role has still not been allocated opens the field to a wider frame if similar mechanisms can operate in other species.

References

Anvari, S. F.; Stösser, R. 1978. Eine neue fluoreszenzmikroskopische Methode zur Beurteilung der Befruchtungs-Fähigkeit der Samenanlagen bei *Prunus*. Z. Pflanzenzücht 81: 333-336.

Arbeloa, A.; Herrero, M. 1985. Valoración de la translocación al ovulo y de la esterilidad femenina en melocotonero. An. Estac. Exper. Aula Dei 17(3-4): 214-220.

Arbeloa, A.; Herrero, M. 1987. The significance of the obturator in the control of pollen tube entry into the ovary in peach (*Prunus persica*). Ann. Bot. 60: 681-685.

Arbeloa, A.; Herrero, M. 1991. Development of the ovular structures in peach (*Prunus persica*). New Phytol. In press.

Charlesworth, D. 1989. Why do the plants produce so many more ovules than seeds? Nature 338: 21-22.

Cruzan, M.B. 1986. Pollen tube distributions in *Nicotiana glauca*: Evidence for density dependent growth. Amer. J. Bot. 73(6): 902-907.

Cruzan, M.B. 1989. Pollen tube attrition in Eoythronium grandiflorum. Amer. J. Bot. 76(4): 562-570.

Herrero, M.; Arbeloa, A. 1989. Influence of the pistil on pollen tube kinetics in peach (*Prunus persica*). Amer. J. Bot. 76(10): 1441-1447.

Herrero, M.; Arbeloa, A.; Gascon, M. 1988. Pollen pistil interaction in the ovary in fruit trees. In: 'Sexual Reproduction in Higher Plants' (M. Cresti, P. Gori, E. Pacini Eds.) Springer Verlag: 297-302.

Herrero, M.; Dickinson, H.G. 1979. Pollen-pistil incompatibility in *Petunia hybrida*: changes in the pistil following compatible and incompatible intraspecific crosses. J. Cell Sci. 36: 1-18.

Herrero, M.; Dickinson, H.G. 1980. Pollen tube growth following compatible and incompatible intraspecific pollinations in *Petunia hybrida*. Planta 148: 217-221.

Herrero, M.; Dickinson, H.G. 1981. Pollen tube development in *Petunia hybrida* following compatible and incompatible intraspecific matings. J. Cell Sci. 47: 365-383.

Heslop-Harrison, J. 1983. Self-incompatibility: phenomenology and physiology. Proc. Roy. Soc. London, Ser. B, Biol. Sci. 218: 371-395.

Heslop-Harrison, J.; Heslop-Harrison, Y. 1985. Surfaces and secretions in the pollen-stigma interaction: a brief review. J. Cell Sci. Suppl. 2: 287-300.

Heslop-Harrison, J.; Heslop-Harrison, Y; Reger, B.J. 1985. The pollen- stigma interaction in the grasses. 7. Pollen tube guidance and the regulation of tube number in *Zea mays* L. Acta Botanica Neerlandica 34, 193-211.

Jensen, W. A.; Ashton, M.E.; Beasley, C.A. 1983. Pollen tube-embryo sac interaction in cotton. In: D.L. Mulcahy and E. Ottaviano (eds.), Pollen biology and implications for plant breeding, 67-72. Elsevier Biomedical, New York.

Juel, H.O. 1918. Beiträge zur Blütenanatomie und zur systematik der Rosaceen. Kunglica Svenska Vetenskapsakademiens Handlingar 58: 5.

Kahn, J.L.; De Mason, D.A. 1986. A quantitative and structural comparison of *Citrus* pollen tube development in cross-compatible and self-incompatible gynoecic. Can. J. Bot. 64: 2548-2555.

Linskens, H.F. 1986. Recognition during the progamic phase. In M. Cresti and R. Dallai (eds.), Biology of reproduction and cell motility in plants and animals, 21-32. University of Siena, Siena.

Lloyd, D.G. 1980. Sexual strategies in plants. 1.- An hypothesis of serial adjustment of maternal investment during one reproductive session. New Phytol. 86: 69-79.

Lord, E.M.; Kohorn, L.U. 1986. Gynoecial development, pollination and the path of pollen tube growth in the tepary bean *Phaseolus acutifolius*. Amer. J. Bot. 73: 70-78.

Mulcahy, D.L. 1974. Correlation between speed of pollen tube growth and seedling weight in *Zea mays*. Nature 249: 491-493.

Mulcahy, D.L. 1979. The rise of Angiosperms: a genecological factor. Science 206: 20-23.

Mulcahy, G.B.; Mulcahy, D.L. 1983. A comparison of pollen tube growth in bi- and trinucleate pollen. In D.L. Mulcahy and E. Ottaviano (Eds.), Pollen: biology and implications for plant breeding, 29-33. Elsevier Biomedical, New York.

Ottaviano, E.; Sari-Gorla, M.; Mulcahy, D.L. 1980. Pollen tube rates in *Zea mays*: implications for genetic improvement of crops. Science 210: 437-438.

Pimienta, E.; Polito, V.S. 1982. Ovule abortion in 'Nonpareil' almond (*Prunus dulcis* (Mill) D.A. Webb). Am. J. Bot. 69(6): 913-920.

Schlichting, C.D.; Stephenson, A.G.; David, L.D.; Winsor, J.A. 1987. Pollen competition and offspring variance. Evol. Trends. Plants., 1(1): 35-40.

Sedgley, M. 1976. Control by the embryo sac over pollen tube growth in the style of the avocado (*Persea americana* Mill). New Phytol. 77: 149-152.

Sedgley, M. 1979. Structural changes in the pollinated and unpollinated avocado stigma and style. J. Cell Sci. 38: 49-60.

Stephenson, A.G. 1981. Flower and fruit abortion:proximate causes and ultimate functions. Ann. Rev. Ecol. Syst. 12: 253-279.

Ter-Avanesian, D.V. 1978. The effect of varying the number of pollen grains used in fertilization. Theor. Appl. Genet. 52: 77-79.

Tilton, V.R.; Horner, H.T. 1980. Stigma, style and obturator of *Ornithogalum caudatum* (Liliaceae) and their function in the reproductive process. Am. J. Bot. 67: 1113-1131.

Tilton, V.R.; Lersten, N.R. 1981. Ovule development in *Ornithogalum caudatum* (Liliaceae) with a review of selected papers on angiosperm reproduction. III. Nucellus and megagametophyte. New Phytol. 88: 477-504.

Tilton, V.R.; Wilcox, L.W.; Palmer, R.G.; Albertsen, M.C. 1984. Stigma, style and obturator of soybean , *Glycine max* (L.) Herr. (Leguminoseae) and their function in the reproductive process. Am. J. Bot. 71: 676-686.

Vasil, I.K. 1974. The histology and physiology of pollen germination and pollen tube growth on the stigma and in the style. In H.F. Linskens (ed.), Fertilization in higher plants, 105-118. North-Holland, Amsterdam.

62
Gametophytic Selection in Rice

Yo-Ichiro Sato

Asian cultivated rice (Oryza sativa) and its wild relatives (O.rufipogon) are differentiated into sub-species (O. sativa indica and japonica), and ecogeographic races, respectively. Between those subspecies or races, and between those species and related taxa of AA genome, various reproductive barriers have been reported, a part of which are caused by gametophytic selection. Genetical and physiological studies on the gametophytic selection may thus provide a deep comprehension of its role for the reproductive isolation between and within those species of rice.

Conditioning gametophytic selections under various environments have been recently observed in varietal crosses of O. sativa, which are recognized to have played a role for the differentiation. In this chapter,new findings of gametophytic competition in rice, conditional gametophytic competitions in particular, are introduced in addition to a review of previous works.

1. Pre-mating factors
Genic sterility

Genic sterility, mainly arises in male gametes, is most common pre-mating mechanism for eliminating particular gametes in rice. Sterile pollen grains, which are usually not filled with starch, are easily distinguished from normal ones filled with starch. Sterility in rice arises by spontaneous and artificial mutation, gene-cytoplasm interaction, or gene-gene interaction (hybrid sterility). Occurrences by mutation and gene-cytoplasmic interaction were reviewed by Kaul (1987) and Virmani and Shinjo (l988).

Hybrid sterility is caused by gametophytic, sporophytic or sporogametophytic facts. In case of gametophytic sterility, abortion occurs only in gametes having a particular set of alleles of different loci (Oka,l953; l974). Sporophytic sterility arises in gametes produced on sporophytes having particular set of alleles of different loci (Oka and Doida, l962; Yokoo, l984). In this case, abortion occurs regardless of genotype of gametes. Sporogametophytic sterility occurs in gametes having a particular allele (Sx) which are produced by heterozygotes (SxSy) in indicax japonica (Kitamura, l961; Ikehashi and Araki, l986), O. sativa x O.rufupogon (Sano, l989), and O. sativa x O. glaberrima (Sano et al., l979;Sano, l983).

Certation and other selections

There are some occurrences of segregation distortion in F2 populations derived from varietal crosses of O. sativa, although no abortive pollen grains are observed.

Iwata et al (l964) reported a segregation distortion at the locus wx(responsible for glutinous endosperm and pollen grains). Pollen grains produced by F1 of glutinous cv.

x non-glutinous cv. were segregated into 1 glutinous vs. 1 non-glutinous ratio. Nevertheless, recessive homozygotes(wx wx) were less frequent than expected values. This phenomenon is called certation. There have been reported twelve loci responsible for events of certation up to the present (Sato et al., 1987)

Physiological mechanism causing certation still remains unknown, mainly because studies on gene expression in gametophytes are limited. Certation may be regarded as a functional sterility of pollen grains which are incapable of pollen germination. It is also probable that certation occurs by differential pollen tube elongation. It should be included in post-mating competition under this assumption.

Pham and Bougerol (1987) reported occurrences of segregation distortion at a number of isozyme loci in O. sativa x O. glaberrima crosses. Some of them could not be explained by neither sterility nor certation, although remains were considered to be caused by sporogametophytic sterility. Guiderdoni (1988) observe segregation distortion in F2 populations derived from indica x japonica crosses. The distortion was, however, disappeared in the populations raised through anther culture. Causing factors of the setwo cases are not fully studied.

Competition among egg-mother cells (EMCs)

A competition was found recently among egg-mother cells (EMCs) produced in F1 plants derived from glabrous japonica x non-glabrous (normal) indica. Glabrous kernel in rice is controlled by a recessive allele at gl locus. In F2 population derived from the F1, gl gl homozygotes were segretated in a range of 14-18% (Table 1), which are significantly lower than expected one assuming Mendelian segregation (25%). In BC1 F1 population in which F1 was used as male parent, glabrous vs. normal plants were

Table 1. Segregation distortion observed at gl locus

Population	No. of plants Normal	Glab. (glgl)	Total	% of glab. Plants	X2
Ac. 130(normal, indica) x Ac. 221 (glabrous, japonica)					
F2-Aa)	262	53	315	16.8	11.2**
-B	238	41	279	18.1	15.8**
-C	177	39	216	18.1	4.5*
Ac.221xF1	24	21	45	46.7	0.2n.s
F1xAc.221	41	21	62	33.9	6.5**
F2, observed (A+B+C)	677	133	810	16.4	
F2, expected (based on	675.0	135.0	810.0	0.04n.s	

Populations A, B and C were derived from F1 (Ac.221 x Ac.130) grown under Taichung (24on), Mishima (35oN) and short day plot in Mishima.

segregated in 1:1 ration. In the reciprocal BC1 F1s, when F1 was used as female parent, 41 normal and 20 glabrous plants (approx. 2:1) were segregated. Expected segregation rate of gl gl homozygotes in F2 is 1/6,when 1:1 and 2:1 segregations in male and female gametes, respectively, are assumed. This value gives a good fit to observed frequency (Table 1).Because seed fertility showed no difference between gl gl homozygotes and heterozygotes in BC1 population, the competition may occur among EMCs.

2. Post-mating factors
Partial self-incompatibility in O. longistaminata

In rice, self incompatibility could not be developed because of high selfing rate. Instead, a partial self incompatibility was reported in strains of O. longistaminata by Hayar (l967). In this species, out crossing rate exceeds 60%, suggesting a certain mechanism to prevent a complete selfing. Chu et al. (l969) observed differential pollen germination rate and pollen tube growth on stigmata of its own and of different strains, inspite that own pollen could partly germinate and penetrate pistil tissue.

Alien Pollen Primacy in indica x japonica cross

Female preference, or preferential fertilization is also observed in varietal crosses of O. sativa. Pollination of mixed pollen masses of an indica cultiva r IR36 and

Table 2. Alien pollen primacy found in an indica x japonica cross

Cross	Resulting sporophytes from fertilization by pollen of		
	T65(japonica)	IR36 (indica)	Total
T65 x (T65 + IR36)	57	100	157
IR36x (T65 + IR36)	67	51	118

a japonica cultivar T65 resulted more heterozygotes than homozygotes on both IR 36 and T65 stigmata (Sato, l988, alien pollen primacy, see Table 2).

Sato (unpublished) observed pollen tube growth at 4 and 8 minutes after pollen deposit in reciprocal crosses between IR 36 and T65. Pollen tube elongated rapidly on alien stigmata than on own stigmata, indicating that alien pollen is more preferable than its own one.

3. Conditional gametic selection
Conditional male sterility

Gametophytic competitions are often conditional. Wei et al. (l988)reported male sterility under a long day condition (PGMS). Pollen of a particular cultivar became all abortive when mother plant was grown under a long day condition (14.5 hrs.), while they were normal under a short day condition. A genetic study supported an assumption to be controlled by a single recessive gene.

Conditional male sterility due to high temperature was also found in a mutant lines derived from a Japanese rice cultivar Reimel (TGMS; Maruyama,in press). Pollen grains of this line became all sterile under a high temperature (31 C) during pollen formation. TGMS is also controlled by a single recessive gene.

Gametophytic competitions under stress conditions

Low temperature often gives damages for pollen development and pollen tube growth. However, resistance of gamete against a certain level of low temperature is heritable. Ishikawa (person. communication) carried out a mixing pollination experiment by using cold-resistant cultivar SL and non-resistant cultivar IR36. IR36 pollen contributed more frequently than SL pollen under a high temperature (30 C), while pollen from SL and IR36 were given the same chance of fertilization under low temperature (25 and 20C).

Segregation rate in F2s at a number of isozyme loci differed depend on environments under which respective F1 plants were fertilized (Sato et al.,1990; Sato, unpublished). In general, alleles derived from japonica tended to be increased in F2s of which F1s were grown under stress conditions,i.e., ozone treatment, cold and drought (Table 3).

Table 3. Gametophytic competition under stress conditions in F2s of indica x japonica crosses (Sato et al., 1990; Sato, unpublished)

Locus	Conditions where F1s were frown		
	+ozone 1)	low temp. 2)	drought 3)
Acp-1	i=j	i>j	i=j
Amp-2	=	<	=
Amp-3	<	=	>
Cat-1	=	=	<
Est-2	=	=	=
Pgi-1	<	<	=
Pgi-2	no segregation	<	=
Pox-2	=	<	=

> and < show that indica-derived and japonic-derived alleles were more than each other, respectively. = shows no significant difference.
1) An ozone (03) exposure (0.1 ppm) during flowering.
2) 25 C in day and 17 C in height time during flowering
3) Without standing water furring flowering

Conclusion

Mulcahy (l979) discussed a significance of gametophytic selection on the rise and rapid evolution of angiosperms. In higher plants, gametophytic competition has been playing an important role for acquirement of a rapid distribution under stress conditions. In rice, cultivars rapidly expanded its distribution to cool and dry zones within thousands of years, despite of a narrow ecological niche of wild progenitor. Various kinds of gametophytic competitions, particularly conditional competitions might have enabled cultivars rapidly expanding to stress environments.

References

Chu, Y.E.; Morishima, H.; Oka, H.I. Partial self-incompatibility found inOryza perennis subsp. bartii. Jpn. J. Genet. 44: 225-229; l969.

Guiderdoni, E. Comparative gametic selection in sexual and anther culture derivatives from five distant F1 hybrids of rice. Rice Genet. Newsl. 6:91-93; l989.

Ikehashi, H.; Araki, H. Genetics of F1 sterility in remote crosses of rice. In: IRRI; l986; p. ll9-130.

Iwata, N.; Nagamatsu, T.; Omura, T. Abnormal segregation of waxy and apiculus coloration by a gametophyte gene belonging to the first linkage group in rice. Japan. J. Breed. 14:33-39; l964.

Kaul, M.L.H. Male sterility in higher plants. Monographs on Theoretical and Applied Genetics. New York: Springer-Verlag; l987.

Kitamura, E. Genetic studies on sterility observed in hybrids between distantly related varieties of rice, Oryza sativa L. Bull. Chugoku Agr. Exp. Sta. Ser A8;141-205; l962.

Maruyama, K.; Araki, H.; Kato, H. Thermosensitive genetic male sterility induced by irradiation. In IRRI ed. Rice Genetics l990. Minila; IRRI; l991;in press.

Mulcahy, D.L. The rise of angiosperms: A genecological factor. Science 206:20-23; l979.

Nayer, N.E. Prevalence of self-incompatibility in Oryza barthii Cheval; its bearing on the evolution of rice and related taxa. Genetica 38:521-527:1967.

Oka, H.I. The mechanism of sterility in the intervarietal hybrid. Japan. J.Breed. 2(4):217-224; 1953

Oka, H.I. Analysis of genes controlling F1 sterility in rice by use of isogeneic lines. Genetics. 77:521-534; l974.

Oka, H.I.; Doida, Y. Analysis of the genic basis of hybrid breakdown in rice. Jpn. J. Genet. 37:24-35; l962.

Pham, J.L.; Bougerol, B. Abnormal segregation patterns in crosses between Oryza sativa and O. glaberrima. Rice Genet. Newsl. 6:80-82; l989.

Sano, Y. A new gene controlling sterility in F1 hybrids of two cultivated rice species. J. Hered. 74:435-439; l983.

Sano, Y. Gamete eliminator detected in the wild progenitor of Oryza sativa. Rice Genet. Newsl. 6:93-95; l990.

Sano, Y.; Chu, Y.E.; Oka, H.I. Genic analysis for the F1 sterility between Oryza sativa and O. glaberrima. Jpn. J. Genet. 54:121-132; l979.

Sato, Y.I. Alien pollen primacy detected in rice. Rice Genet. Newsl.5:80-81; l988.

Sato, Y.I.: Sano, Y,; Nakagahra, M. Gene symbols for gametic effect,sterility and weakness. Rice Genet. Newsl. 4:46-51; l987.

Sato, Y.I., Fujinuma, Y. and Aihara, A. Effect of ozone exposure on F1gametic series in rice hybrids. Rice genet. Newsl. 7:98-99; 1990.
Virmani, S.S.; Shinjo, C. Current status of analysis and symbols formale-sterile cytoplasms and fertility-restoring genes. Rice genet. Newsl.5:9-15; 1988.
Wei, M.L.; Cai, C.M.; Zhou, Y.C. A photoperiod-sensitive genic male-sterilerice newly discovered. Rice Genet. Newsl. 5:89-90; 1988.
Yokoo, M. Female sterility in an Indica-Japonica cross of rice. Japan. J.Breed. 34:219-227; 1984.

63
Germination as a Component of Pollen Competitive Ability in *Hibiscus moscheutos*

Allison A. Snow and Timothy P. Spira

Pollen tube competition is widely cited as a possible mechanism for nonrandom fertilization and gametophytic selection (see recent reviews by Snow, 1986; Mulcahy and Mulcahy, 1987; Walsh and Charlesworth, 1991). Two general approaches have been used to vary the intensity of pollen competition and study effects on progeny fitness. One involves varying the stylar distance which pollen tubes must traverse to reach the ovules. This technique has been extremely useful in studies of maize, for example, which has silks of vastly different lengths on the same ear (e.g., Sari-Gorla and coworkers, this volume). Competition among pollen tubes should be strongest in the longest silks.

The other commonly used approach is to vary the amount of pollen applied to stigmas, such that competition is most intense when the number of pollen grains on the stigma is many times the number of available ovules. With both approaches, the expectation is that weak competition will allow a random sample of pollen tubes to sire seeds, while strong competition will select for only the fastest growing tubes. When gametophytic and sporophytic growth rates are correlated (e.g., Mulcahy, 1979), gametophytic selection can result in more vigorous offspring. Most studies of gametophytic selection rely on improved sporophtyic vigor as evidence that pollen tube competition leads to selection for "superior" pollen genotypes.

Here we employ a different method for studying the consequences of pollen tube competition in natural populations. We use genetic markers and direct observations of pollen tubes to determine whether nonrandom fertilization occurs. Instead of using pollen from a single donor to screen for genetic variation in pollen performance <u>within individuals</u>, we are testing for variation <u>among individuals</u>. Natural pollinators often deposit mixtures of pollen from different plants (e.g., Marshall and Ellstrand, 1985). Furthermore, natural rates of pollen deposition can lead to intense competition among pollen tubes for ovules (Snow, 1986). Therefore, nonrandom fertilization based on pollen donor genotype could be a major consequence of pollen tube competition in wild populations. Whether or not this process has an impact on sporophytic growth rates, nonrandom fertilization has the potential to affect plant fitness by influencing the number of seeds sired on other plants in the population. Our research program is designed to investigate this intriguing possibility. In this chapter we focus on whether individuals differ in percent pollen germination.

Recent Work on <u>Hibiscus</u> <u>moscheutos</u>

Pollen tube competition can be intense in natural populations of <u>Hibiscus</u> <u>moscheutos</u>

(rose mallow), hereafter referred to as Hibiscus (Spira, 1989; Spira et al., 1991). This self-compatible species is an herbaceous perennial found in marshes of the eastern United States. The large one-day flowers provide both nectar and pollen for foraging bees. In previous papers, we described the pollination ecology of Hibiscus (Spira et al., 1991) and specific techniques for measuring relative pollen tube growth rates (Snow and Spira, 1991a). To summarize, we documented that co-occurring plants often differed in average pollen tube growth rates. An isozyme marker was used to confirm that differences in pollen tube growth rate were correlated with differences in seed set following the application of pollen mixtures. In other words, faster growing pollen tubes sired more seeds than did slower growing pollen tubes, as expected. Therefore, we believe that variation in pollen competitive ability provides a possible mechanism for nonrandom fertilization.

In order to quantify relative pollen tube growth rates, we performed crosses which took advantage of the five separate stigmas of Hibiscus styles. Pollen from a pair of pollen donors was applied to two different stigmas, and relative growth rates were determined about mid-way down the style (details in Snow and Spira, 1991a). Thus, paired t-tests could be used to detect differences between donors when their pollen tubes were growing in the same stylar environment. This experimental design alleviates the problem of variation among flowers in the conditions for pollen tube growth (Aizen et al., 1990), thus providing a sensitive and biologically relevant measure of pollen competitive ability.

One of our most surprising findings was that when a pair of pollen donors exhibited a strong difference in pollen tube growth rates, this difference was consistent across a random sample of recipient flowers (Snow and Spira, 1991b). Given the evidence for pollen-style interactions, and the potential for "female choice" (e.g., Waser et al., 1987), we had expected to find that interactions between donors and recipients might overwhelm any effects of donors on pollen tube growth rates.

The timing and extent of pollen germination is one component of pollen competitive ability that may differ among individuals (Bertin, 1988). Nonrandom fertilization could occur if any of the following varies among plants: 1) pollen germinability (final percent germination on stigmas), 2) germination rates (the time it takes for pollen grains to germinate on stigmas), and/or 3) pollen tube growth rates in the style. Pollen germinability needs to be considered in studies of pollen tube competition because it is theoretically possible, for example, that a plant with faster-growing pollen tubes would have poor pollen germination. In this scenario, the advantage of having faster-growing tubes would be countered by the disadvantage of producing fewer tubes per pollen grain.

Methods and Results

In Hibiscus, our protocol for assessing pollen competitive ability involved collecting styles 3 hrs after pollination (Snow and Spira, 1991a,b). Aborted pollen grains were extremely uncommon (A. Snow, personal observation), and most grains germinated within one hour of pollination (Snow and Spira, 1991a). Thus, any differences in percent germination after 3 hrs were probably due to germinability rather than to differences in the time required for germination. Since we did not measure final percent germination, however, variation in germination rates cannot be discounted. In this paper, we focus on donor and donor-by-recipient effects on pollen germination after 3 hrs. Our goal is to assess whether pollen germination is a major stage at which differences among donors are found.

Table 1. ANOVAs for effects of pollen donor and recipient on percent pollen germination. Pollen donors with significantly faster-growing pollen tubes are underlined; mean percent germination shown in parentheses (averaged across 3 -5 recipients; see Snow and Spira, 1991b). Significance levels are * P < 0.05, ** P < 0.01, *** P < 0.001.

DONOR PAIR	SOURCE OF VARIATION	DF	F
Donor 17 vs. 26	Donor	1	1.37
	Recipient	2	0.04
(63 vs. 57 %)	Donor x Recipient	2	0.01
	Flower	28	2.28 *
Donor 32 vs. 33	Donor	1	0.74
	Recipient	2	3.75
(69 vs. 66 %)	Donor x Recipient	2	0.67
	Flower	34	3.02 ***
Donor 37 vs. 41	Donor	1	6.05 *
	Recipient	4	0.69
(70 vs. 63 %)	Donor x Recipient	4	0.59
	Flower	74	1.33
Donor 45 vs. 63	Donor	1	3.54
	Recipient	3	2.20
(75 vs. 69 %)	Donor x Recipient	3	2.35
	Flower	47	3.27 ***
Donor 53 vs. 51	Donor	1	0.94
	Recipient	3	6.87 ***
(71 vs. 73 %)	Donor x Recipient	3	0.82
	Flower	52	1.47
Donor 54 vs. 60	Donor	1	0.43
	Recipient	4	3.60 **
(71 vs. 71 %)	Donor x Recipient	4	0.71
	Flower	68	1.72 *

In 1989 we used pollen from six pairs of donors to pollinate 8 - 20 flowers on each of 3 - 5 recipients (details in Snow and Spira, 1991b). The plants in this experiment came from a natural population in Maryland, and were cultivated outdoors in large pots. For each flower, an average of 53 $\pm$ 22 pollen grains (mean $\pm$ 1 SD, N = 293) were deposited on each of two labelled stigmas, using pollen from a different donor on each lobe. Three hrs later, the stigma lobes were excised for pollen counts with the use of a dissecting microscope, and the style was preserved for later measurements of pollen tube number and relative growth rates. The proportion of pollen grains that germinated was calculated by dividing the number of tubes in the stylar branches by the number of grains on the stigma lobe. All proportions were arcsin-transformed for statistical analyses.

In a previous paper (Snow and Spira, 1991b), we reported that these same pairs of donors rarely differed in percent pollen germination, as determined by paired t-tests. Of the 24 possible comparisons for the six donor pairs within recipients, only two revealed a significant difference between donors. However, sample sizes were relatively small, and analysis of variance is needed to test for main effects of donor, recipient, and their interaction on the proportion of pollen grains that germinated. Here we report on ANOVAs in which flowers were nested within recipients, and donor and recipent were treated as fixed variables. When pollen donor was significant, a more conservative mixed model ANOVA was also used: donor was considered to be a random effect so the interaction mean square was used as the F-ratio denominator.

The ANOVAs showed that variation in pollen germination was often explained by variation among flowers and among recipients (Table 1). Pollen donor was significant in one pair out of six, Donor 37 vs. Donor 41. Although individual t-tests within recipients were not significant for this donor pair, pollen donor was significant as a main effect. When the more conservative mixed model ANOVA was used, pollen donor was still significant at $P < 0.05$ ($F = 10.31$). Donor 37 had greater percent germination than did Donor 41, and it also had faster growing pollen tubes (Table 1; Snow and Spira, 1991b). Interactions between donors and recipients were not detected.

Discussion

In this experiment, strong differences in pollen tube growth rates were detected for four of the six pairs of pollen donors (Snow and Spira, 1991b; Table 1). For these four pairs of donors, 10 out of 17 t-tests for differences in pollen tube growth rate were significant at $P < 0.05$. Moreover, ANOVAs showed that pollen donor was significant as a main effect when donors were tested across recipients. In contrast, differences in pollen germination after three hours were uncommon (only two t-tests were significant). These results are similar to those from previous experiments (Snow and Spira, 1991a,b). ANOVAs showed that pollen donor was significant as a main effect for one pair of donors, even though t-tests within recipients were not. Thus, additional variation in pollen germination was detected using this type of analysis. It is possible that t-tests would also be significant if sample sizes had been larger.

Differences in percent germination after three hrs are likely due to differences in germinability rather than the time required for germination, although this was not confirmed (see above). For the two pollen donors that differed in percent germination (Donor 37 and Donor 41), a parallel difference was seen in relative pollen tube growth rates. The donor with better germination also had faster-growing pollen tubes. With few exceptions, this pattern was seen in other crosses reported elsewhere (Snow and Spira, 1991a,b).

We conclude that in Hibiscus populations, individuals are more likely to differ in pollen tube growth rates than in pollen germinability. Some of the variation that we saw in pollen tube growth rates may be due to differences in the timing of pollen germination, but this would not be evident if three hrs is enough time for most pollen to germinate. Because differences in pollen germination were uncommon, and because they were associated with parallel differences in pollen tube growth rates, we plan to focus on pollen tube growth rates in future studies (it is extremely time-consuming to count pollen grains for measurements of percent germination). Our long-term goals are to determine the relative importance of genetic and environmental effects on pollen competitive ability in this species.

Acknowledgements

This work was supported by the U.S. National Science Foundation (BSR-890667 and BSR-8906959) and the Smithsonian Institution.

REFERENCES

Aizen, M.A., K.B. Searcy, and D.L. Mulcahy. Among- and within-flower comparisons of pollen-tube growth following self- and cross-pollinations of Dianthus chinensis (Caryophyllaceae). Am. J. Bot. 77:671-676; 1990.

Bertin, R.I. Paternity in plants. In: Lovett Doust, J.; Lovett Doust, L. (eds.), Plant Reproductive Ecology: Patterns and Strategies. Oxford Univ. Press, New York; pp. 30-59; 1988.

Marshall, D.L., and N.C. Ellstrand. Proximal causes of multiple paternity in wild radish. Am. Nat. 126:596-605; 1985.

Mulcahy, D.L. The rise of the angiosperms: a genecological factor. Science 206:20-23; 1979.

Mulcahy, D.L., and G.B. Mulcahy. The effects of pollen competition. American Scientist 75:44-50; 1987.

Snow, A.A. Pollination dynamics in Epilobium canum (Onagraceae): consequences for gametophytic selection. Am. J. Bot. 73:139-151; 1986.

Snow, A.A., and T.P. Spira. Differential pollen-tube growth rates and nonrandom fertilization in Hibiscus moscheutos (Malvaceae). Am. J. Bot. in press; 1991a.

Snow, A.A., and T.P. Spira. Pollen vigor and the potential for sexual selection in plants. Nature in press; 1991b.

Spira, T.P. Reproductive biology of Hibiscus moscheutos (Malvaceae). In: Bock, J.H.; Linhart, Y.B. (eds.), The Evolutionary Ecology of Plants. Westview Press, Boulder, Colorado; pp. 247-256; 1989.

Spira, T.P., A.A. Snow, D.F. Whigham, and J. Leak. Flower visitation, pollen deposition, and pollen-tube competition in Hibiscus moscheutos (Malvaceae); unpublished manuscript (in review); 1991.

Walsh, N., and D.D. Charlesworth. The evolutionary role of variation in pollen tube growth rates. Quarterly Review of Biology in press; 1991.

Waser, N.M., M.V. Price, A.M. Montalvo, and R.N. Gray. Female mate choice in a perennial wildflower, Delphinium nelsonii. Evolutionary Trends in Plants 1:29-33; 1987.

Ecology and Evolution

64
Developmental Genetics of Sexual Propagation in *Oenothera*

Cornelia Harte

In Oenothera a remarkable number of genes involved in the process of sexual propagation is present. To understand the role of these genes and the system of their cooperation it is necessary to consider first the specialities of the genetic system of the genus Oenothera.

Oenothera was involved in the rediscovery of Mendelian rules of inheritance by De Vries, but the next step was the observation that Mendelian inheritance is the exception in this genus. Most species proved to be constant hybrids containing two different sets of alleles on completely linked loci, separated in the process of gamete formation as Renner-complexes but without segregation and recombination in progenies from self pollination.

This put two major problems: How does the genus manage to keep the sets of alleles together and how is the heterozygosity maintained? They seem closely related, but are of very different kind. The first is a problem of cytogenetics, and the second one of genetic regulation of developmental processes involved in sexual propagation.

Cytogenetics of Oenothera

The first problem was solved in two steps by Renner (1917) who detected the complex-heterozygosity and Cleland (Cleland and Blakeslee, 1930) who found its cytological equivalent in multiple reciprocal translocations of whole chromosome arms. In both cases the big step foreward was really not based on new observations but on new interpretations of facts known already for some time: the twin hybrids in species crosses and chromosome catenation in late prophase and metaphase of the first meiotic division.

The chromosomes of Oenothera consist of a heterochromatic middle region and euchromatic arms of approximately equal length. A reciprocal exchange of chromosome arms between non homologous chromosomes with the breakage points in the heterochromatic parts leads in meiosis to a cross-like pairing of four chromosomes in pachytene and a ring-of-four in diakinesis and metaphase I, followed by regular separation of the centromere in anaphase I. The consequence is that not chromosomes of pairs but sets of four chromosome arms, arranged in a different way in two chromosomes at each of the two poles, are separated. In evolution of the genus all seven chromosomes of the haploid genome became involved, leading to chromosome complexes of 14 chromosome arms arranged in seven chromosomes in different ways. Two such groups form large rings of 14 chromosomes in meiosis. By this mechanism the seven chromosomes of each set, which entered the zygote in one gamete, are kept together during meiosis. The two sets of chromosomes, and consequently the alleles located on them, are separated without any rearrangement of chromosomes or alleles coming from the two parents. The genetical result is interchromosomal linkage. The chromosome complexes are the cytological equivalent of the Renner-complexes. The hybrid species does not generate a multitude of different gametes by segregation and recombination of the alleles from different loci, but only two types, containing the Renner-complexes in the form as they entered the zygote. With random pairing of the gametes, however, in the progeny of self pollination a segregation of the two homozygotes and the heterozygote is expected but not observed. The hybrid species reproduces only itself.

From Spores to Embryos

This constancy of the hybrid species is the second problem. There must exist mechanisms by which either the homozygotes are eliminated or which prevent the formation of homozygotes at all. These mechanisms must be active at some time between the formation of the haploid cells at the end of meiosis and maturation of the seeds, that means in sporogenesis, gametogenesis and embryogenesis.

Genes that cause complete sterility in one sex are observed. As example are mentioned the genes fr and ster for male sterility by a malfunction of the tapetum, and br related to female sterility by disturbances in the development of style and ovules. As they are not an integrated part of the special Oenothera-system of sexual propagation, they are not discussed here.

The elimination of homozygotic embryos is observed in Oe.Lamarckiana and explained as the effect of balanced zygotic lethals. When the two Renner-complexes contain each a recessive lethal on different loci, the lethality of the homozygotes is ensured and only heterozygotes can survive. This method of discarding 50% of the embryos is effective in maintaining heterozygosity, but increase in efficiency is possible.

A better way is to prevent the formation of lethal homozygotes. One possibility seems to eliminate gametes of a certain genotype, but a different one in both sexes. This can be achieved by genetically determined influences on development of the haploid generation.

The Male Line

In most hybrid species, only one of the two expected genotypes is found in the active male gametes. Pollen lethals are observed, but they are not the cause for the regular elimination of a certain genotype in the pollen of heterozygotic species. The elimination is based on general inactivity or special inhibition of apparently living pollen grains. In some species the inactive pollen grains, which do not germinate, can be recognized by the round starch grains in contrast to the oblong starch grains in the others, indicating a difference of metabolism in the last stage of pollen development (Halac and Harte, 1990).In Oenothera several genetically different types of plastids are observed. Some of these are incompatible with certain Renner-complexes. One of the results is that pollen grains, containing an incompatible combination of genom and plastom, can not germinate. After pollination with such pollen, the fertilizing male gametes have all the same genotype, irrespective of the hybrid structure of the species. This happens in

Oe.biennis with the RENNER-complexes albicans.rubens. The RENNER-complex albicans is in the haploid state incompatible with all known plastoms and therefore never inherited through the pollen. (Stubbe, 1989).

Another method of elimination of one genotype from the pollen is based on gametophytic incompatibility (Steiner, 1964). In Oenothera the S-locus belongs to the normal genom of the genus as is proved by several homozygotic species with multiple incompatibility alleles. When in a heterozygotic species one Renner-complex contains a Si-allele, the partner instead a Sf-allele for self compatibility, then pollen grains with the Si-allele will be inhibited on styles of their own species. Following self pollination this Renner-complex will be eliminated from the population of fertilizing male gametes. In outcrossing to other species with a different Si-allele, however, these pollen grains are active and transmittance of both Renner-complexes will give twin hybrids.

Both methods, general inactivation by genom-plastom incompatibility and special inhibition of pollen grains by Si-alleles, are efficient in creating a homogenous population of male gametes from the pollen of a heterozygotic species. So generating one type of homozygotic lethal embryos is prevented. However, in relation to the whole progeny the effect is zero. Because the pollen complex is present in 50% of the egg cells, the result of fertilization will still be a seed population with appr. 50% homozygotic lethal embryos, now all of the same genotype. The addition of a system for elimination of one genotype from the active pollen to a system of balanced lethals does by itself not lead to an increase in efficiency in propagation of a constant hybrid. To achieve this goal, a system with similar effect must be introduced in the female line.

The Female Line

In most hybrid species of Oenothera the pollen-complex is only found in the egg cells with very low frequency. In most fertile ovules the partner complex is present. Following fertilization by genetically homogenous pollen, most embryos will be heterozygotes. Again lethal alleles active in the haploid stage, which could be suspected, are not observed. The basis for the creation of a nearly homogenous population of mature egg cells is not lethality of megaspores, embryo sacs or egg cells of a certain genotype, but competition between megaspores.

At the end of meiosis the megaspores are arranged in a linear tetrad. In homozygotic species immediately after meiosis the megaspore in the micropylar position starts enlarging to form a 4-celled embryo sac, while the three others degenerate. In heterozygotic species with preference of one Renner-complex in the egg cells, however, this polarity of the tetrad is not observed. The degeneration occurs in the middle of the tetrad while the megaspores in the micropylar and the chalazal position start growing (Halac and Harte, 1977). The two gametophytes are competing. Sometimes two complete embryo sacs are formed. In most cases, however, one will win while the other degenerates before maturity.

In the linear megaspore tetrad the spores at the ends will always have different genotypes. When, as usual, the position of the megaspore in the tetrad decides its fate, a random distribution of the two Renner-complexes in the mature egg cells will be observed. When a competition occurs and the decision, which one will become an embryo sac is made at random, this can not change the frequency of the two genotypes. Only when in competing gametophytes the genotype has an influence on the decision which one will win, this can lead to a considerable decrease of the frequency of the "weaker" Renner-complex in the egg cells.

Considering the sequence of events and the statistics of the consequences for the different types of development in the ovule, it becomes clear that polarity has a key position. The first sign of polarity in the ovule is the orientation of the meiocyte, elongating in the direction of the main axis and developing a callose layer which is thick at the chalazal end and very thin at the micropyle in homozygotic species. The meiotic spindles are orientated along this axis and later the polar pattern of degeneration is in accordance with the assumption that callose formation is an isolating mechanism causally related to megaspore degeneration at the chalazal end. The polarization of the meiocyte can be so strong that in the chalazal cell of the dyad the second meiotic division is delayed or even missing.

In heterozygotic species with megaspore competition, the pattern of callose deposits is changed. The callose layer around the meiocyte is missing or very thin, callose is mainly found in the cross walls of the tetrad. It seems to be a very convincing correlation between the two patterns of callose deposits and megaspore degeneration. When, however, the two phenomena are observed in the segregating progeny of a hybrid between a homozygotic and a heterozygotic species, flavens.haplo-Hookeri from the cross Oe.suaveolens x Hookeri with a small ring of 4 chromosomes and 5 bivalents, the nice hypothesis breaks down (Sniesko and Harte, 1984). In plants of the F1-generation development in the ovules follows a pattern intermediate between the parents. In individual plants of the segregating F2-progeny the callose pattern can be like one of the parents or somewhere in between. The frequency of the different types leads to the interpretation that two genes cooperating in the process of callose deposition in the ovule can explain the observed segregation of plants with the Hookeri-type of callose deposits and the flavens-type of missing callose.

Considering the degeneration pattern in the megaspore tetrad, the correlation with the callose pattern disappears. Plants with the flavens-pattern of depolarized callose deposits can exhibit heteropolarity of megaspore degeneration, only the megaspore at the micropylar end developing into an embryo sac, or extreme homopolarity with a high frequency of twin embryo sacs as well. The frequency of plants of the two extreme types and intermediate ones with a mixture of homopolar and heteropolar tetrads can be explained by the action of two genes with alleles for strong polarity of the tetrad in haplo-Hookeri and for missing polarity in flavens. By the time when embryo sac development starts, polarity in the ovule is completely restored. The orientation of the egg cell apparatus in the embryo sac is unchanged with the egg cell at the micropylar side and complete inhibition of development at the chalazal end.

The conclusion is, that the basis for competition phenomena on the female side is the genetical regulation of development of polarity in the ovule by a small number of genes, active in the sporophyte during a relatively short period between formation of the meiocyte and beginning of embryo sac development.

The logical sequence of events is not the one supposed by Renner (1921), that the competition between genetically different megaspores is the cause for development of the chalazal megaspore. On the contrary, the genetically regulated disturbance of polarity in the ovule makes the competition possible. The few cases in which two pollen complexes were combined in a hybrid gave proof of decreased ability for normal megaspore development. Many degenerating ovules were found and only a few fertile embryo sacs led to poor seed set (Renner, 1940). In a hybrid species with disturbance of polarity in the ovule, the partner of such a pollen complex has just to take the free positions to create a nearly homogenous egg cell population.

Fertilization

All these ways of maintaining heterozygosity are still lacking in efficiency. Instead of embryos the gametophytes of certain genotypes are discarded. Oenothera has found a way to improve this too. Selective fertilization, based on genetically regulated attraction between embryo sacs and pollen tubes in the ovary which prevents the meeting of gametes with similar genotypes, is the final solution (Schwemmle, 1952). In both sexes all gametes have the same chance of fertilization, but they select their partners and the result is a progeny of only heterozygotic embryos.

Summary

In the genus Oenothera an exceptional combination of genes influencing the development of the haploid generation is observed. Some act directly on the gametophytes, others indirectly by the sporophyte. All give proof that the development of the gametophytes is genetically regulated just as in the sporophyte, and that it will be worth to look for more details of these processes with modern methods of developmental genetics.

Conclusion

What has Oenothera done? It has a tendency to create chromosomal abnormalities, and a mutation rate high enough to make balanced lethals and heterozygosity on many loci. Then it created a flower structure, not discussed here, that enforces self pollination in species that need heterozygosity and live on hybrid vigour, developed a senseless variability of extrachromosomal genes in the plastids, incompatible with those in the genomes, a crazy incompatibility system where the Sf-allele has the most important role, a disturbance of a vital polarity system at a crucial point of the generation cycle, and touchy gametes which play a game of I-do-not-like-you. Then the genus combines this bag full of unfavourable characters to make it most successfull in rapid evolution and distribution to new areas.

References

Cleland, R.E.; Blakeslee, A.F. Circle formation in Datura and Oenothera. Proc.Natl.Acad.Sci.U.S.A. 16:177-183; 1930

Noher de Halac, I.; Cismondi, I.A.; Harte, C. Pollen ontogenesis in Oenothera: a comparison of genotypically normal anthers with the male sterile mutant sterilis. Sexual Plant Reproduction 3:41-53; 1990

Noher de Halac, I.; Harte, C. Different patterns of callose wall formation during megasporogenesis in two species of Oenothera (Onagraceae). Plant Systematics and Evolution 127:23-38; 1977

Renner, O. Versuche ƒber die gametische Konstitution der Oenotheren. Z.Verer-bungsl. 18:121-294; 1921

Renner, O. Heterogamie im weiblichen Geschlecht und Embryosackentwicklung bei den Oenotheren. Z.Bot. 13:608-621; 1921

Renner, O. åber die Beziehungen zwischen Heterogamie und Embryosackentwick-lung und ƒber diplarrhene Verbindungen. Flora (Jena) 34:145-158; 1940

Schwemmle, J. Selektive Befruchtung als Erklèrung unerwarteter Kreuzungser-gebnisse. Biol.Zbl. 71:353-384; 1952

Sniezko, R.; Harte, C. Callose pattern and polarization phenomena in the ovules in the F2-hybrids between Oenothera Hookeri and Oe.suaveolens. Plant Systematics and Evolution 147:79-90; 1984

Steiner, E. Incompatibility studies in Oenothera: The distribution of Si-alleles in biennis-I populations. Evolution 18:370-378; 1964

Stubbe, W. Oenothera - An ideal system for studying the interactions of genome and plastome. Plant Molecular Biology Reporter 7:245-257; 1989

65
The Effects on Genetic Load of Selection in the Gametophyte Stage

D. Charlesworth and B. Charlesworth

Abstract

We have studied a multi-locus selection model in which mutations to deleterious alleles occur that may affect not only the diploid stage, but also the haploid stage before zygote formation. We investigated the reduction in inbreeding depression (as measured in the diploid stage of the life cycle) caused by the lowering of mutant allele frequencies due to selection in the pollen. This is important for our understanding of the rle of inbreeding depression in the maintenance of outcrossing. We also studied the theoretically expected relationship generated in this situation between the pollen fitness of different donor genotypes and the fitness of the diploid progeny that they would sire. We show that the covariance between these fitnesses is small, and that experimentally subjecting pollen to strong selection is expected to increase sporophyte fitness only slightly, unless pollen selection is weak under natural conditions.

Introduction

There is evidence that expression of many genes occurs in the pollen (male gametophyte) stage of flowering plants, and that many of these genes are also expressed in the diploid, sporophytic stage (reviewed by Haldane, 1932 and Tanksley et al. 1981). In nature, there may often be more pollen grains on the stigma than are necessary to fertilize all of a flower's ovules, creating an opportunity for competition between pollen tubes (Haldane, 1932). In theoretical and empirical studies of genetic load in flowering plants (and also in plants with more extensive gametophytic stages in their life cycles) it is therefore important to include the possibility of selection in the gametophyte as well as the sporophyte stage.

Here, we study a multi-locus selection model in which mutations to deleterious alleles occur, and in which selection may operate not only in the diploid stage, but also on haploid genomes before zygote formation. We have previously shown that this model is capable of accounting for the magnitude of inbreeding depression observed in natural plant populations with various degrees of inbreeding (Charlesworth et al., 1990, 1991), and is also consistent with observations of the degree to which fitness is changed by a few generations of inbreeding (Barrett and Charlesworth, in review). We here investigate the reduction in inbreeding depression (as measured in the diploid stage of the life cycle) caused by the lowering of mutant allele frequencies due to selection in the pollen. This is important for our understanding of the rle of inbreeding depression in the maintenance of outcrossing. We also model the consequences of changing the degree to which selection operated in the haploid stage. This enables us to model the experimental situation in which plants from a population that has experienced weak pollen selection (so that even pollen of low fitness can achieve fertilizations) are pollinated with large pollen loads so that the pollen will compete.

Models

The model assumptions were similar to those of previous work on mutational load at many loci subject to mutations with the same selection coefficient (Charlesworth et al., 1990, 1991; Kondrashov, 1985), with the additional feature of selection during the pollen stage. We denote by k the fraction of loci expressed in both pollen and sporophyte. The parameter c denotes a factor by which the selection coefficient for the sporophyte stage is multiplied to obtain the selection coefficient in the pollen. Here we assume either that k was equal to 1, with c taking various possible values between zero and one (model 1), or else that c was equal to 1 and k varies (model 2).

The population is assumed to breed by partial self-fertilisation, with selfing rate S. Outcrossing is done by computing the composition of an ovule pool and a pollen pool (in which pollen selection takes place). The selection coefficients were assumed to be the same for all selected loci. For the diploid stage of the life cycle, fitness for the genotype with j homozygous and i heterozygous mutations was expressed as

$$w_{ij} = (1-s)^j(1-hs)^i \tag{1}$$

where s is the selection coefficient against mutant alleles in homozygotes, and h is the dominance coefficient. This assumes multiplicative fitness interactions between the selected loci. For model 1, the fitness of the haploid genotype with j mutations was expressed as.

$$w_j = (1-sc)^j \tag{2}$$

For model 2 we followed separately the genotype frequencies for the two sets of loci, those expressed only in the sporophyte stage, and those expressed in both sporophyte and pollen. All the results were generated using Fortran programs written for a Macintosh computer. We use the notation W_s and W^x for the mean fitness of progeny produced by selfing and by outcrossing, respectively. The inbreeding depression is defined by $d = 1- (w_S/w_x)$. To model the experimental situation in which the intensity of pollen selection is changed, we changed the value of the weighting parameters, k or c, defined above, starting with a population had reached equilibrium with a given regime of pollen selection. After one or a small number of generations, we examined the effects of the change in pollen selection on the average fitness of outcrossed progeny, wx, (the progeny type usually studied in such experiments).

Results

Effects on the genetic load and inbreeding depression

Figure 1 compares the results from the two different selection models, for outcrossing populations. The two models are equivalent when no pollen selection occurs (c or k=0), or when c and k are both equal to 1, (i.e. when all loci are expressed in pollen with the same selection coefficient as in the diploid stage). There is a strong effect of pollen selection on the mean fitness and inbreeding depression values. Unless k is very high, model 2, in which only a fraction, k, of loci are expressed in pollen, has less effect on the inbreeding depression than model 1 in which all loci are selected in pollen, but at lower intensity than in the sporophyte. This is expected, because in the model with all loci expressed in pollen, the fact that pollen selection acts on haploid genotypes should cause large differences in the numbers of mutant alleles per individual at equilibrium.

When there is no selection in the gametophyte stage, the mean fitness increases with the selfing rate, and the inbreeding depression decreases (Crow, 1970; Charlesworth, Morgan and Charlesworth, 1990). This lowering of the genetic load with increased selfing does not necessarily happen when a high proportion of the loci affecting fitness are strongly expressed in the pollen (Figure 2). Inbreeding still lowers the numbers of mutations per individual, and thus the

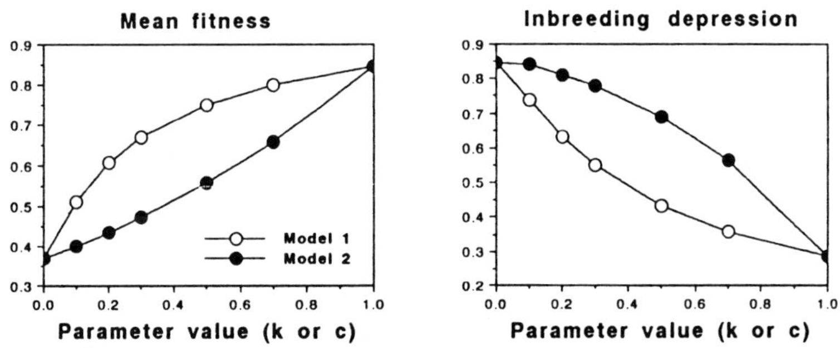

Figure 1. Comparison of the effects of different intensities of pollen selection under models 1 and 2 on the magnitudes of mean fitness of sporophytes, and inbreeding depression, in completely outcrossing populations at equilibrium under mutation at a rate of 1 per diploid genome per generation to detrimental alleles with selection coefficient s=0.2, and dominance coefficient h=0.1.

mean fitness values of both inbred (W^S) and outcrossed (W^X) progeny increase with S, but the mean fitness over the population as a whole may decrease because the mean fitness in highly selfing populations (W_S for $S=1$) does not increase sufficiently to become higher than the value for outbred populations (W_X for $S=0$). This is expected, because when there is expression in the haploid stage the selfing rate will have less influence on the frequencies of mutant alleles, compared with the situation when the loci influence on the frequencies of mutant alleles, compared with the situation when the loci are expressed only in the diploid stage. However, it should be noted that when this occurs the inbreeding depression will be low.

Effect of changing the intensity of selection in the pollen stage

If a population were at equilibrium under a regime of intense selection on pollen, a change to relaxed selection on pollen would be expected to produce a slight lowering of the meanfitness of the offspring, because the numbers of mutations per pollen grain should be small, due to the selection prior to the experiment. Only after enough generations have passed for new mutations to build up in the population under the new pollination regime would there be any substantial increase in the genetic load detectable in the diploid stage. In contrast, if the population used in an experiment had experienced low pollen selection prior to the experiment, and were then subjected to pollination with large amounts of pollen from unrelated plants, so that pollen grains were strongly

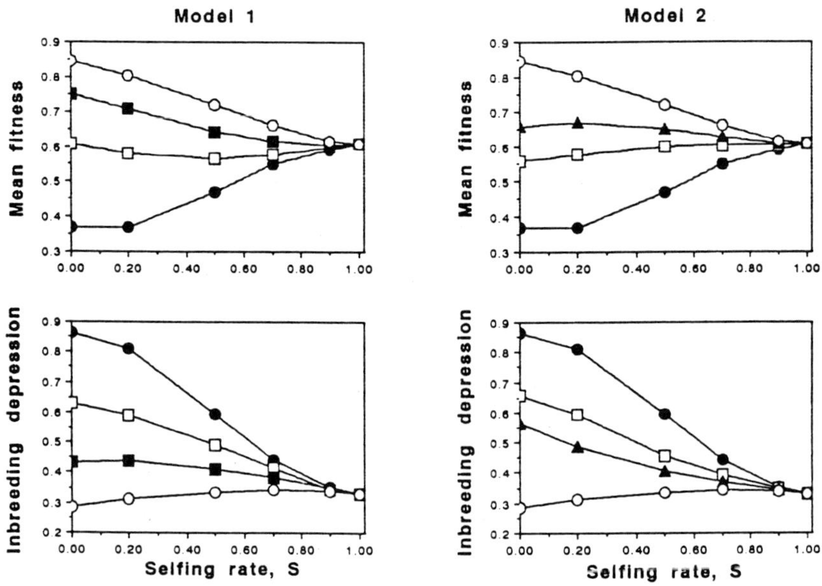

Figure 2. Mean sporophyte fitness and inbreeding depression values in populations undergoing partial self-fertilization. Models 1 and 2 are shown, and the parameters of the selection model are as for Figure 1. The symbols for the K or c values are as follows: closed circles 0, closed squares 0.2, open squares 05, triangles 0.7, open circles 1.0

competing, it is to be expected that the pollen genotypes with the fewest mutations would be the most successful, and the diploid progeny generation should show an increased mean fitness. It is of interest to find out the magnitude of this effect, since this is the maximum effect possible in such experiments.

The computer program can be used to calculate the mean progeny fitness after pollen competition (i.e. a high value of the k or c parameters) has been imposed on a population that was previously not undergoing selection in the pollen. Table 1 shows some of these results. For the parameter values examined, the effects appear to be of the order of a 10% increase in fitness at most. In agreement with these conclusions, the covariances between pollen fitness of donors and the average fitness of their progeny are very low. Furthermore, the between-donor genetic variances in progeny fitness were about 1/4 of the total genetic variances in fitness for outcrossing populations, and somewhat higherfor partially selfing populations. If the variance between donors in the fitness of the progeny they generate is not a high proportion of the total progeny generation variance in fitness, it is unlikely that the variation in sire quality could be detected experimentally.

Discussion

The model of mutational load which we have studied yields results suggesting that the magnitude of the effects of pollen selection on progeny fitness will not be very large, and that about 1/4 of the variation in zygote fitness is between progeny of individual pollen donors. Are these results due to special assumptions of the models studied, or to the particular values of the parameters that we chose for the calculations? As far as possible, we have used assumptions that should generate the strongest effects of pollen selection. For example, the biologically realistic case involves competition between the pollen of a finite (usually small) number of pollen donors, i.e. between pollen genotypes that might be produced from a limited set of donor plants. In our models, however, we assumed that non-selfed ovules are fertilized by pollen from a pollen pool composed of the pollen genotypes produced by the entire population. This pollen pool contains a wider range of pollen genotypes than would be found in a real set of donor plants. Thus there could be greater effects of pollen selection in our models than would be expected in real plant populations. We have also focused on the case of outbreeding populations, which show the greatest total variance in fitness and the greatest effects of pollen competition. Finally, calculations with other parameter values, including large selection coefficients for the mutant alleles, and different dominance coefficients, did not yield substantially larger effects. The only situations which generated large effects of increased pollen competition were when very strong selection ($s=0.9$) was assumed, together with a high mutation rate to such mutant alleles, or when the selection coefficient in pollen was assumed to be much higher than in the diploid stage. These are not biologically plausible assumptions (Crow, 1970).

These results correspond with the intuitive expectation that if pollen genotypes with fast pollen tube growth rates tend to produce progeny with high fitness (Mulcahy, 1979), this would lead to fixation of alleles for the best- competing phenotype, and thus to loss of genetic variation. Even with mutation, the magnitude of the variance maintained by mutation is unlikely to be great (Charlesworth, 1987), though it would certainly be true that fast pollen- tube growth will be an indication of overall high fitness in all life stages, as assumed by Mulcahy (1979), to the extent that loci expressed in pollen are expressed at other life stages.

We have not examined other possibilities for maintenance of genetic variance in fitness, such as the hypothesis that antagonistic pleiotropic effects of alleles for faster pollen tube growth may exist. This could maintain additive genetic variance in components of fitness, even when fitness itself has no additive genetic variance (Rose, 1982; Charlesworth, 1987). In this case, the expectation would be that genotypes with fast pollen-tube growth would have lower fitness at some other life stage, the opposite of Mulcahy's (1979) view.

Table 1. Calculated values of the expected increase in fitness after one generation of pollen selection.

Selfing rate (S)	Initial values (when c is zero) W_x	Mean fitness	% increase in C	W_x value
Parameter values: U=1.0, moderate selection, s=0.2; h=0.1				
0.0	0.3709	0.3674	0.5	5.28
"	"	"	1.0	10.10
0.2	0.4404	0.3690	0.5	4.50
"	"	"	1.0	8.67
0.5	0.6682	0.4697	1.0	4.34
0.9	0.8658	0.5918	1.0	1.52
Parameter values: U=0.1, strong selection, s=0.9; h=0.1				
0.01	0.9087	0.9071	0.5	3.36
"	"	"	1.0	4.44
0.2	0.9473	0.9278	0.5	1.88
"	"	"	1.0	2.49

Many recent studies of pollen-tube competition have found positive relationships between pollination intensity and progeny fitness (see Snow, 1990a, b). However, attempts to document the heritability of these differences have thus far not found significant heritabilities of these traits, indicating that nongenetic causes contribute (Mazer, 1987; Snow & Mazer, 1988; Schlichting et al., 1990; Snow, 1990a, b). Furthermore, differences in progeny quality have been found even in inbred strains, where one would expect genetic differences should be slight (Mulcahy, 1971; Ottaviano et al., 1980; Stephenson et al., 1986; Davis et al; 1987; see discussion in Charlesworth et al., 1987; Snow, 1990a, b). There is thus no clear evidence as yet that these effects are chiefly due to genetic differences. If the increased vigor of offspring proves to be heritable, the present results suggest that it may not be due to mutational load. It will thus be important to find out whether the additive genetic correlation between pollen tube growth rate and offspring vigor is positive or negative.

Acknowledgements

This work was supported by NSF grant BSR 8817976.

REFERENCES

Charlesworth, B. 1987. The heritability of fitness. In J.W. Bradbury and M.B. Andersson (eds.), Sexual Selection: Testing the Alternatives , pp. 21-40. John Wiley & Sons, New York.

Charlesworth, D., D.W. Schemske and V.L.Sork. 1987. The evolution of plant reproductive characters: sexual versus natural selection. In S.Stearns (ed.) The Evolution of Sex and its Consequences , pp. 317-335. Birkhauser-Verlag, Basel.

Crow, J. F. 1970 Genetic loads and the cost of natural selection. pp. 128-177. In K.-I. Kojima (ed.) .Mathematical Models in Population Genetics . Springer-Verlag, Berlin.

Davis, L.E., A.G. Stephenson and J.A. Winsor. 1987. Pollen competition improves performance and reproductive output of the common zucchini squash under field conditions. J. Amer. Soc. Hort. Sci. 112: 712-716.

Haldane, J.B.S. 1932. The causes of Evolution. Harper, New York.

Mazer, S.J. 1987. Parental effects on seed development and seed yield in Raphanus raphanistrum: implications for natural and sexual selection. Evolution 41: 355-371.

Mulcahy, D.L. 1971. A correlation between gametophytic and sporophytic characteristics in Zea mays L. Science 171: 1155-1156.

Mulcahy, D.L. 1979. The rise of the angiosperms: a genecological factor. Science 206: 20-23.

Rose, M. R. 1982. Antagonistic pleiotropy, dominance and genetic variation. Heredity 48: 63-78.

Schlichting, C. D., Stephenson, A. G., Small, L. A., and Winsor, J. A. 1990. Pollen loads and progeny vigor in Cucurbita pepo : the next generation. Evolution 44: 1358-1372.

Snow, A.A. 1990a. Effects of pollen load size and number of donors on sporophyte fitness in wild radish (Raphanus raphinistrum). Amer. Nat. in press.

Snow, A.A. 1990b. Effects of pollen load size on sporophyte competitive ability in two Epilobium species. In preparation. Snow, A.A. and S.J. Mazer. 1988. Gametophytic selection in Raphanus raphanistrum: a test for heritable variation in pollen competitive ability. Evolution 42: 1065-1075.

Stephenson, A.G., J.A. Winsor and L.E. Davis. 1986. Effects of pollen load size on fruit maturation and sporophyte quality in zucchini. In D.L. Mulcahy, G.B. Mulcahy, and E. Ottaviano (eds.), Biotechnology and Ecology of Pollen , pp. 429-434. Springer-Verlag, N.Y.

Tanksley, S.D., D. and C.M. Rick. 1981. Evidence for extensive overlap of sporophytic and gametophytic gene expression in Lycopersicon esculatum. Science 213: 453-455.

66
Natural and Artificial Hybrids in the Grasses

J.S. Heslop-Harrison

Wide hybridization

The production of interspecific and intergeneric (wide) hybrids is of considerable scientific interest because it enables interactions between two different genomes to be analyzed. Normally, a pollen tube follows the pathway shown in table 1 after germination of the grain. Analysis of the behavior of "alien" pollen tubes (from a different species) indicates some of the points where there is communication between the female sporophyte and male gametophyte. In mature hybrid plants, we have been able to analyze the positions of parental genomes and gene expression in ways that would be impossible in a species (see Heslop-Harrison, 1990; Heslop-Harrison and Bennett, 1990).

The production of wide hybrids is economically important because genes can be transferred between species. Many crop plants are themselves allopolyploids including genomes of separate origin: bread wheat (Triticum aestivum), durum wheat (T. durum or T.turgidum ssp. durum), and Brassicas (Brassica napus, B.juncea and B.carinata) have originated from natural wide hybrids, and the crop triticale (X Triticosecale) is a man-made hybrid. Other hybrids have almost certainly enabled natural gene movement from one species to another, and hence the barriers to wide hybridization are important to characterize because of implications for evolution and horizontal gene flow.

The success of making a wide hybrid depends on many factors. The most significant is the closeness of the parental species in evolution. In general, but with many important exceptions, two closely related species have a greater chance of making a successful wide hybrid than two distantly related species. Indeed, the viability and fertility of a hybrid is often taken as evidence that two organisms, no matter how dissimilar in phenotype, are in the same species. However, low fertility, or even sterility, is found in many plants that reproduce vegetatively or apomictically, and are nevertheless clear species.

In cereals, where the seed is the harvested product, high fertility is an agronomically important feature. Within the cereals, though, there is a remarkable variation in breeding behavior, showing that no single strategy is best. Some cereals are outbreeding: rye (Secale cereale), has strong self-incompatibility barriers; corn or maize (Zea mays) shows spatial separation of the anthers and ovules; Pennisetum americanum (pearl millet) is strongly protogynous so there is a temporal separation of stigma receptivity and pollen dehiscence. Other important cereals are largely inbreeding: barley (Hordeum vulgare) is strongly cleistogamous, and wheat is normally self-pollinated.

Such variability between species emphasizes the different barriers to pollen capture, germination, tube growth and fertilization that exist between species. Within cereal tribes, the success of many interspecific hybridizations depends as much on characteristics of the parents including structural differences in stigma and pollen morphology, as on evolutionary relationships between species. Here, I review some of the barriers that prevent successful wide hybridization within the cereals.

Pollen capture, germination and tube growth

The stigma provides not only a suitable chemical environment for pollen function but also the appropriate physical conditions for attachment and water transfer (Heslop-Harrison, 1987). There is a close co-adaptation of size of pollen grains, with the number, spacing and size of the stigma trichomes. Alien pollen may not be captured as easily on a stigma as pollen from the same species, and may not adhere to the stigma. This barrier is easily overcome by careful mechanical pollination, but alien pollen frequently will not germinate because the hydration conditions on, or attachment to, the stigma are unsuitable. Incompatibility barriers may operate, but they are often limited to intraspecific pollinations. Within the Gramineae, the maintenance of high humidity regimes is normally sufficient to allow the relatively similar stigma surfaces to hydrate the grain. For example, Heslop-Harrison et al. (1984) showed that atmospheric humidity was the principal factor controlling germination and pollen tube penetration on the stigma of Z. mays; the greatest success was achieved with humidities above 70%. Similarly, in making wide hybrids within the Triticeae, covering the pollinated ear with a polythene (impermeable) bag for 24h after pollination increases the chance of successful fertilization greatly; a permeable cellophane bag is unsatisfactory at this stage although essential at other times to avoid fungal growth.

After germination, a pollen tube should follow the pathway from stigma surface to embryo-sac that is summarized in table 1.

Table 1. Stages of pollen tube growth.

1) Penetration of the stigma
2) Growth into and through the transmitting tract of the style
3) Entering of the ovary cavity
4) Location and growth through the micropyle
5) Growth through the intercellular spaces of the nucellus
6) Penetration of an embryo-sac synergid

After J. Heslop-Harrison 1987

All the phases including guidance and nutrition require close harmony between the pollen and stigma, and size differences in alien pollen tubes can lead to abnormal tube growth. Alien tubes frequently do not enter the pollen tube transmitting tract. Guidance of the tube often goes awry: the tube may grow away from the ovary, in, for example, Z.mays x P.americanum crosses. Iwanami (1959) showed that pollen tubes grew normally through inverted grafts inserted into a stigma of Lilium, implying that once a direction of growth is established, the tube continues to grow along the pollen tube transmitting tract in one direction. Thus the initial mechanical interactions between tube and stigma trichome and the style are likely to establish the direction of growth.

Even if the tube successfully reaches the transmitting tract, growth may not be as fast as that of the species own pollen, and alien tubes may die before they reach the ovule. In P. americanum, the barrier zone (see Y. Heslop-Harrison, this volume) may stop pollination by sorghum and Z.mays. Heslop-Harrison et al. (1985) suggested that the barrier zone could control the penetration of alien tubes and so contribute to species isolation.

Different authors have obtained widely different success rates in making the same hybrids within the Triticeae. Major genotypic differences in crossability exist between different varieties, but technical factors also account for much variation. For example, the use of many alien pollen grains on one stigma has been considered to increase the chances of successful fertilization. However, the stigma may not be able to support hydration of many grains, and competition between the tubes may prevent normal growth of any one. Hence limited pollination may be more successful. At least in the cereals, the use of freshly dehisced pollen on days of copious anther dehiscence greatly increases hybrid production frequencies.

Analysis of the behavior of pollen tubes when they reach the ovary highlights another barrier to alien tubes. The defined nature of the barrier was not suspected from examination of pollen tubes in intraspecific crosses, since the tubes grow through the zone without any check to growth. Sitch and Snape (1987) examined wheats with different genotypes at two mapped crossability loci, kr1 and kr2. When pollinated with H.bulbosum, pollen germination and tube growth as far as the base of the stigma was similar in all genotypes. However, the frequency of pollen tube penetration of the ovary wall was severely reduced when the dominant Kr alleles were present, showing that a genetically controlled barrier operates at this point.

We have investigated the ion distribution at the base of the stigma in Z.mays, and found very high levels of calcium in the zone between the ovule wall and nucellus where the pollen tube grows (Heslop-Harrison and Reger, 1986). Perhaps a tendency to grow in or towards high calcium concentrations, combined with mechanical effects, enables Z.mays pollen grains to follow the nucellar surface to the micropyle. Artificial pollination of a sectioned Z.mays ovary wall and nucellus with P. americanum shows that the pollen grains germinate and, after some random growth, can follow the normal pollen tube growth pathway for some distance. Further work will help establish whether the stimulus is mechanical, nutritional, chemotactic or chemotropic. No successful fertilizations have yet been achieved by this method, and the pollen tube usually bursts after several hours growth.

In vitro fertilization

Another method that may become usable for wide hybrid production is in vitro pollination. Several authors have applied pollen to gynoecia or other parts of the female structures implanted in semi-solid media. Such studies have enabled analysis and timing of events associated with pollination and fertilization to be examined in detail under controlled conditions (Dupuis and Dumas, 1989).

Among mammals, in vitro fertilization using isolated egg and sperm nuclei has enabled various new hybrids to be made. Dutrilleaux and Rumpler (1977) have made new Lemur hybrids that give viable animals, while Brandriff et al. (1991) have shown that hamster x human hybrids will survive for one or two cell cycles. Perhaps in vitro fertilization in plants will be as helpful in extending limits of hybridization (see also Kranz and Lorz, this volume).

We have found the egg apparatus from Z. mays and T. aestivum wheat can be isolated by digestion of the ovule in weak pectinase/cellulase enzyme mixtures. The embryo sac can be dissected out and placed onto agar, where it will live for several hours. Alteration of digestion conditions enables control of the number of surrounding cells retained with the egg, to include synergids, and various numbers of nucellar and micropylar cells. Living pollen can be placed around the egg apparatus, and pollen tubes have been observed to grow near the synergids, but again, no fertilizations have yet been observed.

Embryo and endosperm stability

Once the pollen tube reaches the egg apparatus, it should penetrate a synergid and release the gamete nuclei. The limitations on fertilization by alien gametes have not yet been explored in plants, largely because of the barriers discussed above which prevent the pollen tube reaching the egg apparatus. Following the new hybrids produced by Zenkteler and Nitzche (1984), Laurie and Bennett (1986) have shown the cytology of intertribal hybrid zygotes in early divisions in wheat x maize hybrids. Similar intertribal hybrids have been reported between maize and oats (Avena spp.; Rines & Dahleen, 1990).

Techniques for making wide hybrids in the small grained cereals have been improving throughout this century. The first reports of rye x wheat hybrids were in 1876 (Wilson), but techniques now used enable such hybrids to be made routinely (see Simpson et al., 1980). Fertilization occurs in many hybrids between species and genera of the Triticeae, but the embryo or the endosperm often aborts. Where an embryo is produced, it can often be transferred onto an artificial medium 5 to 7 days after pollination, when it will develop and germinate. Similar techniques are extending the range of intergeneric hybrids in the Cruciferae (e.g. Agnihotri et al., 1990).

After fertilization, the hybrid embryo may abort or be unstable. Some genotypes give stable, diploid embryos, while others eliminate one parental chromosome set, giving a haploid embryo (Simpson et al., 1980). In hybrids between H.vulgare and H.bulbosum, the genes involved may affect the association of centromeres with microtubules and lead to uniparental chromosome elimination shortly after fertilization; we are investigating the stability of chromosomes, and find that chromosomes from a genome with a tendency to be lost lag during anaphase and have small centromeric structures (Schwarzacher et al. submitted). It is unlikely that the genes involved are important only in hybrids; they probably have a role in species too.

Conclusions

Many ways have been devised to overcome the barriers to wide hybridization between agriculturally important crops and their relatives. The production of hybrids has had immense economic benefits: many cereals include genes introduced by breeders who have overcome barriers to hybridization. Direct gene transfer might conceivably make crossing technology obsolete, although such molecular gene transfer is only now being used to move genes into major crops. Many hybridization barriers have been overcome by empirical methods, but scientific understanding has followed. For example, bud pollination in the Brassicas was later shown to involve the incompatibility proteins being produced late in maturation. The genetics of crossing barriers between closely related species, particularly within the Triticeae, has enabled elucidation of some genes related to the breeding systems. Further understanding of the subject of this volume, pollen and ovules, will lead to the production of even wider hybrids and wider gene transfer through sexual methods, along with new insights into the genetics and physiology of the life of sporophytes, fertilization and embryology.

I thank Venture Research International, BP International and the U.K. Agricultural and Food Research Council for support of the work. I thank Annals of Botany for sponsoring my attendance at the meeting.

References

Agnihotri, A.; Gupta, V.; Lakshmikumaran, M.S.; Shivanna, K.R.; Prakash, S.; Jagannathan, V. Production of Eruca-Brassica hybrids by embryo rescue. Plant Breed. 104: 281-289. 1990.

Brandriff, B.F.; Gordon, L.A.; Segraves, R.; Pinkel, D. The male-derived genome after sperm-egg fusion: spatial distribution of chromosomal DNA and paternal-maternal genomic association. Chromosoma 100: 262-266. 1991.

Dupuis, I.; Dumas, C. In vitro pollination as a model for studying fertilization in maize (Zea mays L.). Sex. Plant Reprod. 2: 265-269. 1989.

Dutrilleaux, B.; Rumpler, Y. Chromosomal evolution in Malagasy lemurs. II Meiosis in intra- and interspecific hybrids in the genus Lemur. Cytogenet. Cell Genet. 18: 197-211. 1977.

Heslop-Harrison, J. Pollen germination and pollen-tube growth. Int Rev Cytol. 107: 1-79. 1987.

Heslop-Harrison, J.S. Gene expression and parental dominance in hybrid plants. Development Supplement 21-28. 1990.

Heslop-Harrison, J.S.; Bennett, M.D. Nuclear architecture in plants. Trends in Genet. 6, 401-405. 190.

Heslop-Harrison, J.S.; Reger, B.J. X-ray microprobe mapping of certain elements in the ovary of Zea mays L. Ann. Bot. 57: 819-822. 1986.

Heslop-Harrison, Y.; Heslop-Harrison, J.; Reger, B.J. The pollen-stigma interaction in the grasses. 7. Pollen-tube guidance and the regulation of tube number in Zea mays L. Acta Bot. Neerl. 34: 193-211. 1985.

Heslop-Harrison, Y.; Reger, B.J.; Heslop-Harrison, J. The pollen-stigma interaction in the grasses. 6. The stigma ("silk") of Zea mays L. as a host to the pollens of Sorghum bicolor (L.) Moench and Pennisetum americanum (L.) Leeke. Acta Bot. Neerl. 33: 205-227. 1984.

Iwanami, Y. Physiological studies of pollen. J. Yokohama Municip. Univ. 116: 1-137. 1959.

Laurie, D.L.; Bennett, M.D. Wheat x maize hybridization. Can. J. Genet. Cytol. 28: 313-316. 1986.

Rines, H.W.; Dahleen, L.S. Haploid oat plants produced by application of maize pollen to emasculated oat florets. Crop Sci. 30: 1073-1078. 1990.

Simpson, E.; Snape, J.W.; Finch, R.A. Variation between Hordeum bulbosum genotypes in their ability to produce haploids of barley, Hordeum vulgare. A. Pflzucht. 85: 205-211. 1980.

Sitch, L.A.; Snape, J.W. Factors affecting haploid production in wheat using the Hordeum bulbosum system. 1. Genotypic and environmental effects on pollen grain germination and the frequency of fertilization. Euphytica 36: 483-496; 1987.

Wilson, A.S. Trans. Bot. Soc. Edinb. 12: 286.
Zenkteler, M.; Nitzche, W. Wide hybridization experiments in cereals. Theor. Appl. Genet. 68: 311-315. 1984.

67

Pollen Competition as an Evolutionary Parameter Involved in the Domestication of Allogamous Cereal: A Two-locus Symmetrical Model of the Evolution on Sympatric Condition of Wild and Cultivated Populations of Pearl Millet

Thierry Robert, Françoise Lamy, Michel Sandmeier, and Aboubakry Sarr

The importance of the haploid phase of the life-cycle of Angiosperms in their evolution has already been discussed (Mulcahy, 1979; Mulcahy and Mulcahy, 1987; Snow, 1990) and mostly studied through theoretical modeling (Pfahler, 1983; Hartl, 1975; Harding, 1975). Until now, two levels of consequences of haploid selection on evolution of population have been considered.

First, the existence of a genetic overlapping between gametophyte and sporophyte has been shown both at molecular level (Willing et al., 1988) and on quantitative characters (Mulcahy and Mulcahy, 1975). This has been used in population breeding schemes. Indeed, Ottaviano et al. (1988) have shown, on a synthetic population of maize, that selection based on pollen competitive ability produces an increase in mean kernel weight. One aspect of this phenomenon is that haploid selection on pollen populations can be very useful since providing the opportunity to eliminate deleterious alleles while lowering the cost with regard to sporophyte selection. But this genetic overlapping can also have negative consequences in the case of haplo-diploid antagonistic selection (Pernès, 1984).

Second, transmission of genes expressed only in the sporophyte can be modified by selection on the haploid phase if they are in statistic disequilibrium with genes selected in the haploid phase. This could produce genetic segregation distorsions in natural populations as well as and in breeding populations. Evaluation of these phenoma is therefore essential for efficient Genetic Resources management, especially in dynamic conservation programmes.

However, numerous results obtained in maize and especially from studies on gametophytic factors, Ga genes, have shown that homogametic fertilizations can be favoured by pollen competition and pollen-pistil interactions (Bianchi and Lorenzoni, 1975). Ga genes are found in cultivated lines as well as in wild populations of teosinte and would limit hybridization between adjacent wild and cultivated populations of teosinte and maize (Kermicle and Allen, 1990). Pollen competition can therefore play a role in the evolution of these populations in sympatric situation. In pearl millet, a preferential outcrossing diploid cereal, pollen competition with prepotency of self-pollen has already been described in cultivated lines. Experiments with mixtures of pollen from wild and cultivated plants deposited on stigmas of both wild and cultivated receptors have shown, in some extent, preferential wild/wild and cultivated/cultivated gamete association (Robert et al., 1991).

The impact of these pollen competition on evolution of the pearl millet primary gene pool is studied through a theoretical model simulating sympatry between a cultivated field and a wild population.

Model description

This model simulates sympatry between a cultivated pearl millet field and a population of wild relatives, taking into account the opposite selective pressures due to farmers and the natural selection. We also took into account the traditional agricultural techniques for pearl millet. Five fundamental assumptions were then made :

1- The cultivated or wild phenotype alternatives were assumed to be under the control of a biallelic locus C/S, named the domestication locus. Plants CC had the cultivated phenotype and plants SS had the wild phenotype. Heterozygotes CS had an intermediate phenotype (N'douls).

2- Farmers harvest only plants with a cultivated phenotype. A fraction of the seeds collected is put aside for seeding the following year. It was assumed that farmers cannot recognize the different phenotypes before flowering. On the other hand, selection is assumed to occur against cultivated and intermediate phenotypes in the wild population. The results given in this paper were obtained assuming co-dominance at the "domestication locus", the selective coefficients against CC and CS genotypes in the wild population being 1 and 0,5 respectively. It was also assumed no difference in pollen production between cultivated, intermediate and wild phenotypes.

3- Each generation, reciprocal gene flow, through pollen migration, occurs between the field and the wild population. We also assumed that pollen distribution on each population was homogenous. Seed migration does not occur. Only plants sown by the farmer could be found in the field. We also assumed the populations to be of infinite size.

4- The mating system is assumed to be outcrossing.

5- Pollen competition occurs in the cultivated population as well as in the wild population. Pollen competitive ability was assumed to have mendelian inheritance with a biallelic locus (A and B alleles). Pollen-pistil interactions led to preferential homogametic fertilizations : in AA styles, A pollen was more competitive than B pollen. In BB styles, B pollen was more competitive than A pollen. In AB styles, the pollens A and B had the same competitive ability. This locus was named the "homogamy locus". The relative fitnesses of pollens A and B in AA styles were 1 and 1-s respectively, where s was the gametophytic selection coefficient. In BB styles, these relative fitnesses became 1-s and 1 respectively. Thus, gametophytic selection was symmetrical. We assumed the same s for both populations.

The gametophytic selection coefficients used in this model were estimated from experimental data on wild and cultivated pollen competition (Robert et al., 1991). These estimates ranged between 0,3 and 0,7.

The other parameters of this model were :

- r : genetic recombination rate between the "domestication" locus and the "homogamy" locus.
- ms : pollen migration rate from the wild population to the field.
- mc : pollen migration rate from the field to the wild population.

In this paper, we only focus on the effect of pollen competition on the evolution of adjacent cultivated and wild populations. For this reason, mc and ms took only the values 0,1 or 0,2.

Simulation runs were made on a micro-computer, by using a programme written in Basic. The field and the wild population were assumed to be made up only with cultivated and wild phenotypes, respectively, at generation 0.

Results and discussion

This model is a mass-action homogamy model since it assumes that homogamy operates on all individuals of the populations. This kind of assortative mating can give, in general, two different stable equilibria solutions for the "homogamy locus" according to initial frequencies, all parameters being the same (Moore, 1979). All simulation runs are not discussed in this paper. The results given here were obtained with initial A allele frequencies equal to 0,6 and 0,4 in the field and in the wild population, respectively, since this illustrates the two types of equilibrium.

Pollen competition can promote reproductive isolation in sympatry through assortative mating

Table 1. Amount of allelic frequency difference at equilibrium between the two populations at the "homogamy locus". These values have been obtained for mc=0,1.

s	ms=0,1 r=0	ms=0,1 r=0,5	ms=0,2 r=0	ms=0,2 r=0,5
0,1	0,89	0	0,82	0
0,2	0,90	0	0,85	0
0,3	0,92	0,79	0,86	0
0,4	0,93	0,86	0,89	0,77
0,5	0,95	0,90	0,91	0,85
0,6	0,95	0,94	0,92	0,90
0,7	0,97	0,96	0,95	0,93

Table 1 gives the amount of allelic frequency difference at the "homogamy locus" between the two populations at equilibrium. In most cases, the system evolves towards divergence between the two populations at the homogamy locus despite the gene flow between them. The higher s, the higher divergence. Nevertheless, it can be seen that weak values of s cannot ensure the divergence at equilibrium when r=0,5. It is clear that interaction between s and r contributes to determine the equilibrium status. One may wonder if divergence at "homogamy locus" has incidences on the phenotypic structure of both populations pertaining to "domesticated" phenotype. In fact, histogramme 1 shows there is a decrease in the CS hybrid percentage in both populations when pollen competition promotes divergence at the homogamy locus.

Reproductive isolation between the two forms is therefore set up through assortative mating, i.e. pollen from the cultivated population is more efficient on stigmas of this population than on stigmas of the wild population, and reciprocally. It is noteworthy that, for ms=0,2 , hybrid percentage in the field can be reduced from 31,9% for s=0 to only 11,9% and 8,3% when s=0,7 and r=0,5 and 0, respectively. It is also noteworthy that linkage between the two locus is not a prerequisite to this settlement, although advantagous, since when r=0,5 , reproductive isolation between the two forms can also been set up. For example, when ms=0,2 and r=0 , divergence at the "homogamy" locus and thus decrease in the hybrid percentage in the field, is observed whatever s, whereas, when r=0,5 , divergence at equilibrium is obtained only when s> =0,4.

Dynamic of the domestication process

Figure 1 shows the evolution of CS hybrid percentage in the field from generation 0 to equilibrium. It can be seen that the higher s, the shorter the time span to reach this equilibrium. However, the ocurrence of pollen competition delays the equilibrium state with regard to absence of pollen competition, but only in a small extent. Nevertheless, the equilibrium being reached in an asymptotic way, the final values of hybrid percentage are approached very quickly in all cases. Thus, according to this model, domestication of outcrossing cereals could have been initiated very quickly.

CONCLUSION

In this model we assume that only one locus controls the cultivated/wild phenotype. Genetic studies on pearl millet have shown that one gene controls seed shedding (Joly-Ichenhauser, 1984). It has been suggested that this character was the first to be selected during the process of cereal domestication (Harlan, 1975). According to Iltis (1983) a catastrophic mutation with a pleiotropic effect could explain the evolution of maize from teosinte. Thus, it is at least plausible that a single gene initially determined the difference between cultivated and wild phenotypes for pearl millet as well as for maize.

Pollen competition promotes homogametic fertilization in cultivated/wild confrontations of pearl millet populations. This homogamy can lead to the establishment of a reproductive isolation between wild and cultivated populations in sympatric situations. Thus, pollen competition favours the maintenance of the phenotypic integrity of the two forms despite abundant pollen migration between them. This could be one explanation of the success of pearl millet domestication.

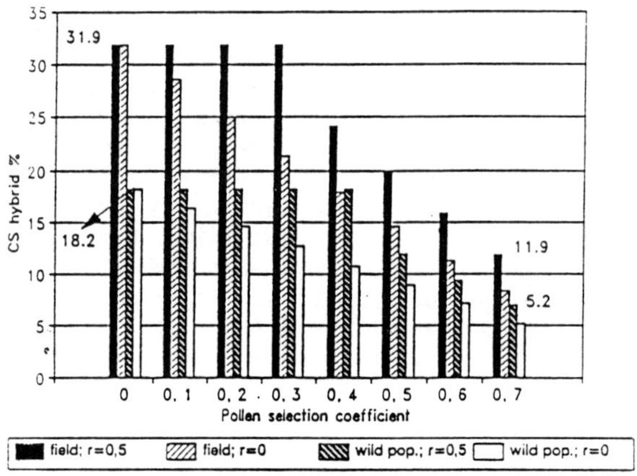

Histogramme 1. CS hybrid percentage at equilibrium in the field for m=0,2 and in the wild population for mc=0,2 as a function of s for mc=0,1 and for different r values.

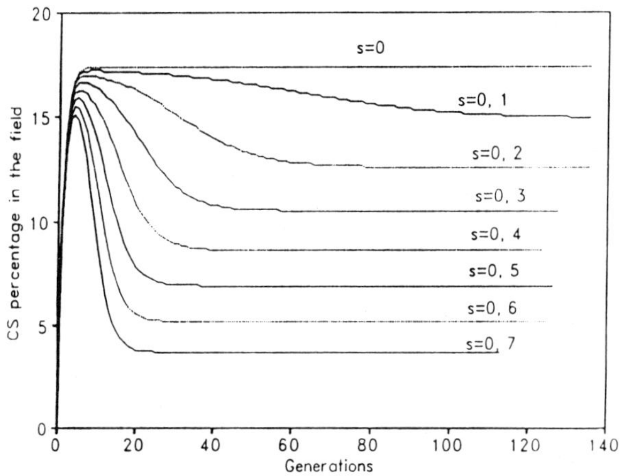

Figure 1. Evolution of CS hybrid percentage in the field for mc=0,1; ms=0,1 r=0 and t=1

It is now required to test the validity of these conclusions with regard to the real evolutionary situation in the center of origin of pearl millet. Indeed, gamete encounters is not only determined by the reproductive organization but also, in a first step, by mechanisms controlling pollen dispersion. Environmental parameters are also known to have an influence on gametophytic competition scenario (Sarr et al., 1988; Zamir and Gadish, 1987). We therefore plan to carry on experimental simulation of wild/cultivated confrontation of pearl millet populations in natural conditions. Such experiments are necessary in order to assess mechanisms controlling the recombination, sensu lato, and thus to improve Genetic Resources management strategies.

REFERENCES

Bianchi, A.; Lorenzoni, C. In: Mulcahy, D. L., ed. Gamete competition in Plants and Animals. North Holland Publishing Company; 1975:257-263.

Harding, J. In: Mulcahy, D. L., ed. Gamete competition in Plants and Animals. North-Holland Publishing Company. Amsterdam, the Netherlands; 1975:243-255.

Harlan, J. R. American society of agronomy, Inc. Crop Science Society of America Inc. Madison; 1975.

Hartl, D.L. D.L. Mulcahy ed. North-Holland Publishing Company. Amsterdam, the Netherlands, 1975:233-242.

Iltis, H. H. Science 222:886-894; 1983.

Joly-Ichenhauser, H. Hérédité Thèse de troisième cycle. Univ. Paris XI-Orsay, France, 1984.

Kermicle, J. L.; Allen, J. O. Maydica 35:399-408; 1990.

Moore, W. S. Heredity 42:173-186; 1979.

Mulcahy, d. I. Science 206:20-23; 1979.

Mulcahy, d. I.; Mulcahy, G. B. Theor. Appl. Genet. 46:277-280; 1975.

Mulcahy, d. I.; Mulcahy, G. B. American Scientist.75:44-50; 1987.

Ottaviano, E.; Sari Gorla, M.; Villa, M. Theor. Appl. Genet. 76:601-608; 1988.

Pernès, J. Tome II. Manuel. Agence de coopération culturelle et technique, ed., Paris, 1984.

Pfahler, P. L. : Mulcahy D. L.; Ottaviano, E. eds. Elsevier Science Publishing Co., 1983:361-366.

Robert, T.; Lespinasse, R.; Pernès, J.; Sarr, A. Génome 34:82-88; 1991.

Sarr, a.; sandmeier, m.; pernès, j. Génome 30:924-929; 1988.

Snow, A. A. Am. Nat. 136:742-758; 1990.

Willing, R. P.; bashe, D.; Mascarenhas, J. P. Theor. Appl. Genet. 75:751-753; 1988.

Zamir, D.; Gadish, I. Theor. Appl. Genet. 74:545-548; 1987.

68
Evolution of Pollen Morphology: The Hare and Tortoise Game

Isabelle Dajoz, Irene Till-Bottraud, and Pierre-Henri Gouyon

Introduction

It is widely acknowledged that the evolution of pollen grains has followed three main trends.

First of all, pollen tube growth rates increase from "primitive" to "evolved"species (Mulcahy & Mulcahy 1983). According to Hoekstra (1983) tube growth rate is multiplied by 2000 from Gymnosperms to families of Angiosperms such as Asteraceae or Poaceae. Increases in pollen tube growth rate are interpreted as ways to enhance gametophytic competition in the stigma and style (Mulcahy,1974; Mulcahy & Mulcahy, 1987). Pollen tubes that grow the most quickly will be the ones that will fertilize the ovules, and transmit their genes to the next sporophytic generation. Since there is important overlapping between the genes that are expressed in the pollen and the sporophyte, there is a correlation between sporophytic and gametophytic fitness (Mulcahy & Mulcahy,1975; Snow, 1988).

The life-expectancy of pollen grains has decreased from "primitive"(few-apertured, slow-germinating) to "evolved" ones, which usually only live for a few hours after being separated from the mother plant, but germinate very quickly (Brewbaker, 1967; Hoekstra & Bruinsma, 1975; Hoekstra, 1983).

Lastly, pollen grain morphological evolution has resulted, among other things,in an increase in aperture number (Walker & Doyle, 1975; Van Campo, 1976). Apertures are the only points in the pollen wall where the pollen tube can initiate, consequently, it seems logical that the more apertures a pollen grain has, the more quickly it germinates. Yet, no experimental testing of this hypothesis has been performed. The only data available deal with fossil pollen grains: they show that aperture increases from the first Angiosperm pollen grains to more recent ones (Van Campo, 1976).

In this chapter, our aim is to test whether an increase in aperture number leads to a higher fitness of the pollen grain. In modern Angiosperms, pollen aperture polymorphism is widely spread: about 1 out of 3 families contain genera or species where aperture number is variable. In Viola diversifolia (Violaceae), an endemic of the alpine vegetation zone in the Pyrenees (France),all plants from all populations studies produce pollen grains with 3 to 6 apertures (Dajoz, 1990). But 5- and 6-apertured grains are relatively rare,and only 3- and 4-apertured ones were taken into account. This makes it possible to compare the fitness of pollen grains with different aperture numbers but with a similar genetic background, since Ottaviano et al. (1980)have shown that the genotype of a pollen grain plays an important role on its chances to fertilize.

Experiments were carried out on 4 populations of V. diversifolia, called A, B,C and D. In population A, there is a flower color polymorphism, linked to statistically different proportions of the different pollen morphs between the different flower colors (Dajoz et al, 1991), and population A was divided into 3 subpopulations: A1 (white flowers), A2 (intermediate flowers) and A3 (violet flowers). In 1989, 10 plants from each of the 6 populations and subpopulations were transplanted from their natural habitat (2500-3000 meter elevation) to afield station 1600 meters elevation. In 1990, 20 plants from populations A1, Band D were collected in the field and planted in a greenhouse, but because of high mortality experiments could be conducted on much less individuals.

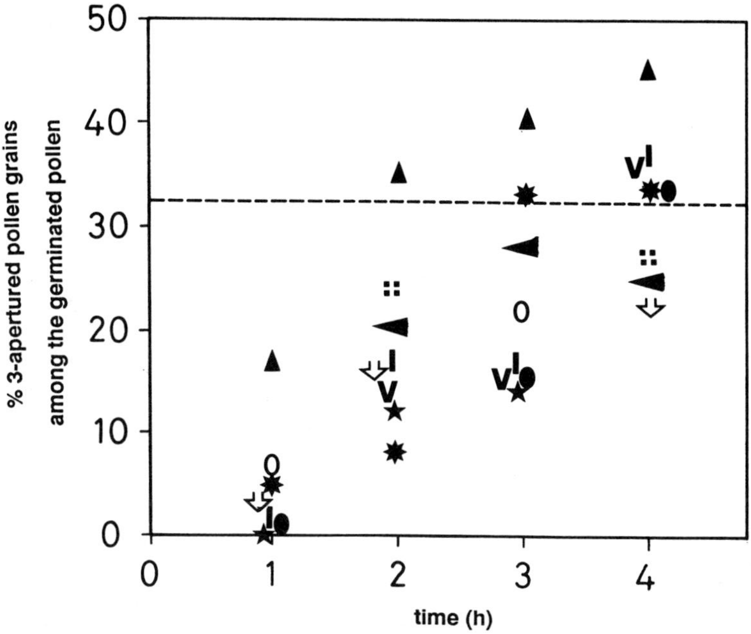

Figure 1: observed frequencies of 3-apertured germinated pollen grains at each sampling time, for the 10 plants studied in 1990. Each plant is represented by a different symbol. The dotted line represents the mean percentage of 3-apertured pollen grains in the anthers.

Pollen grains were germinated on an artificial medium (Hoekstra & Bruinsma,1975) complemented with 20% sucrose and 1% agarose. Each experiment was carried out with a set of 4 slides, and germination was stopped (with Alexander's stain; Alexander, 1969) after 1, 2, 3 and 4 hours. In 1989, each set of 4 slides contained pollen grains from various individuals belonging to the same population. In 1990, each set of slides contained pollen belonging to one individual. In both years the global proportions of 3- and 4-apertured pollen grains were determined in each slide, and the numbers of 3- and 4-apertured germinated grains were counted. We could thus compare (with a x2 test) the germination ability of 3- and 4-apertured pollen grains. Whenever possible, 50 pollen tubes were then measured in each slide, with a micrometer fitted into one of the oculars

Table 1: Results of ANOVAS on mean pollen tube length after different germination times.

1989: 3-way ANOVA model; effect of number of apertures(NAP), Population (POP) and experiment date (DATE) on mean pollen tube length.

1990: 3-way partially hierarchical ANOVA model; effect of NAP, POP, Individual(Individual being nested within Population, IND(POP) and experiment date on mean pollen tube length. The F values for DATE are not given in the table since they are always highly significant. Mean pollen tube length (in millimeters) is given in 1989 and 1990 for 3- and 4-apertured pollen grains (3 ap. and 4 ap.).

*:p<0.05;**:p<0.01;***:p<0.001.

Source	F Values			
	1 hour(df)	2 hours(df)	3 hours(df)	4 hours(df)
NAP				
1989	0.60(1)	7.26(1)**	12.66(1)***	14.57(1)***
1990	2.16(1)	4.88(1)*	11.89(1)***	21.41(1)***
POP				
1989	0.45(2)	3.39(5)**	14.94(5)***	16.65(4)***
1990	4.47(2)*	6.03(2)**	3.76(2)*	10.82(2)***
IND				
1990	2.53(5)*	2.96(6)**	2.92(6)*	5.41(5)***
Mean tube length(mm)				
1989				
3 ap	1.18	3.37	7.16	8.19
4 ap	1.29	3.05	6.07	7.34
1990				
3 ap	1.81	7.59	10.94	12.79
4 ap	1.57	6.33	9.86	11.03

of a Zeiss microscope. Analyses of variance were performed to compare the length of tubes from 3- and 4-apertured pollen grains, after different germination times and between the different populations, experiment dates (and individuals only in 1990).

The life expectancies of 3- and 4-apertured pollen grains were determined on the plants grown in the greenhouse in 1990. Flowers were picked and left to age in Petri dishes. Immediately after picking the flower, and after 1, 3, 5, 18, 24, 31 and 48 hours, pollen was put to germinate on the artificial medium for 2 hours. For each separation time and each population, the germination of 3- and 4-apertured pollen grains was compared with a x2 test.

Results and Discussion

After 1 hour of germination, most of the germinated pollen grains were 4-apertured. Figure 1 represents the proportion of 3-apertured pollen grains among all germinated grains after a given amount of time, for all the individuals studied in 1990. After one hour of germination, very few 3-apertured pollen grains germinated. This proportion increased as time passed by, until it reached the mean proportion of 3-apertured grains in the anthers,represented by the dotted line. The same results were obtained in 1989.

Consequently, after one hour on the germination medium, there were significantly less 3-apertured than 4-apertured germinated pollen grains. When germination time increased, no significant differences were observed,because 3-apertured pollen grains had caught up with 4-apertured ones.

Table 2: Results of x2 tests comparing the effectives of 3- and 4-apertured germinated pollen grains

after increasing separation times from the plant (n) is the number of plants studied in each population and for each separation time. In all cases where the x2 test is significant, 3-apertured grains are significantly more numerous than 4-apertured germinated ones. *:p<0.05; **:p<0.01;***:p<0.001;/:insufficient effectives for x2 test to be conducted.

Separation			Population			
time (hours)	(n)	A1	(n)	C	(n)	D
0	7	0.3	2	/	2	0
1	2	1.3	1	6.3*	1	0.5
2	5	0.1	3	0.2	2	0.4
3	2	0.2	1	/	1	0.1
5	6	14.2***	2	7.1*		
18	5	7.2**	2	0	2	4.7*
21		2	0.7			
24	8	12.8***	2	0.2	2	6.0*
31	1	6.3*	2	0		
48			2	7.6**		

-The longevity of 3-apertured grains is greater than that of 4-apertured grains

After 1 hour on the germination medium, there were no significant differences in tube length, certainly because of the differences in germination speed between 3- and 4-apertured pollen grains. But after 2, 3 and 4 hours on the germination medium, mean pollen tube length was significantly higher in 3-apertured pollen grains (table 1). Consequently, pollen tubes from 3-apertured grains grow more quickly than tubes from 4-apertured ones. Mean pollen tube length was slightly higher in 1990 than in 1989; in 1989, germination experiments were conducted in a field station 1600 meters elevation. Mean room temperature was colder than lab temperature where germination experiments took place in 1990. Temperature affects pollen tube growth rates (Hoekstra &Bruinsma, 1975; Williams & Rouse, 1990), this is why pollen tubes grown in cold conditions (1989) are shorter than those obtained in 1990. Consequently, the"date" effect was always very highly significant, since experiments were not conducted in controlled conditions. There were almost always significant differences in mean pollen tube length between the different populations and individuals sampled (except after 1 hour of germination in 1989).

The longer pollen grains are left to age, the higher the proportion of 3-apertured pollen grains. For example, immediately after picking the flowers,25.6% of all germinated pollen grains were 3-apertured, but when pollen grains are 48 hours old, this proportion turned to 48.2%! Table 2 shows that the effectives of 3-and 4-apertured germinated pollen grains began to different significantly after 5 hours. Consequently, 4-apertured grains seem to lose their ability to germinate more quickly than 3-apertured ones.

Conclusion

Three conclusions can be drawn from our study of in vitro germination of 3 and 4-apertured pollen grains:
4-apertured pollen grains germinate significantly more quickly than 3-apertured ones.

The differences in pollen tube growth rates observed between individuals and populations have already been widely described in the literature, together with the resulting gametophytic competition and its consequences on the fitness of the next sporophytic generation (Mulcahy, 1974; Galen et al., 1986; Mulcahy &Mulcahy, 1987; Ottaviano et al, 1988; Sarr et al., 1988; Marshall, 1989). But none of these works take into account the possible influence of pollen morphology on germination and pollen tube growth rates. Yet, the pollen morphs in V. diversifolia have different germination kinetics.

Results also show that the number of apertures of a pollen grain can strongly affect its life-expectancy. The more apertures a pollen grain has, the more quickly it loses it ability to germinate, and the shorter its life expectancy seems to be, all other things being equal. Many authors have studied the longevity of pollen grains in conjunction with their bi-or tricellular state at anther dehiscence (Brewbaker, 1967; Hoekstra & Bruinsma, 1975; Hoekstra, 1983). In V. diversifolia, all pollen grains are bi-cellular at anthesis (a species is always homogenous for this character, Brewbaker, 1967). The differences observed in the life expectancies of the 2 pollen morphs should rather be attributed to their numbers of apertures. Pollen wall is thinner in the apertures and the more apertures a grain has, the more susceptible to external stresses it is. Consequently, many-apertured pollen grains probably dry up much more quickly than few-apertured ones.

Shivanna & Heslop-Harrison (1981) have shown that when a pollen grain dries to a certain point, the plasmic membranes of its constituting cells deteriorate,and pollen can no longer rehydrate and become turgescent on the stigma, which is an absolute must for pollen tube initiation to take place.

From a schematic point of view, 3-apertured pollen grains (slow-germinating,but long-lived and with fast pollen tube growth) behave like the tortoise in the tale of the race between the hare and the tortoise. Four-apertured pollen grains (quick-germinating, but short-lived and with slowly growing pollen tubes) are "hare" strategists. The problem is to determine whether the situation observed (different pollen types in all flowers from all individuals)is the result of the action of selective forces (equilibrium) or whether it is just a fugitive state in the evolution of pollen morphology. Game theory models show that if we consider only the strategy of the gametophyte, no stable polymorphism can be maintained. However, when the strategy of the sporophyte is considered, different pollen types can be maintained within all individuals(Gouyon et al.; in prep.).

References

Alexander, M.P. 1969. Differential staining of aborted and non-aborted pollen. Stain Technology 44(3):117-122.

Brewbaker, J.L. 1967. The distribution and phylogenetic significance of binucleate and trinucleate pollen grains in the Angiosperms. Am. J. Bot.54(9):1069-1083.

Dajoz, I. 1990. Polymorphism chez le pollen et la fleur de Viola diversifolia(Gingins) Becker. Bull. soc. Bot. Fr. (1137), Actual. bot. (2):148-150.

Dajoz, I., Till-Bottraud, I. & Gouyon, P.H. 1991. Evolution of pollen morphology. Science, 252 (in press).
Galen, C.; Shykoff, J.A. & Plowright, R.C. 1986. Consequences of stigma receptivity schedules for sexual selection in higher plants. Am. Nat.127(4):462-476.
Gouyon, P.H.; Till-Bottraud, I; Venable, L & Dajoz, I. in prep.
Hoekstra, F.A. 1983. Physiological evolution in Angiosperm pollen: possible role of pollen vigor. in: Pollen: Biology and implications for plant breeding. D.L. Mulcahy & E. Ottaviano, eds., Elsevier Science Publishing Co., New York:29-33.
Hoekstra, F.A. & Bruinsma, J. 1975. Respiration and vitality of binucleate and trinucleate pollen. Physiol. Plant. 34:221-225.
Mulcahy, D.L. 1974. Correlation between speed of pollen tube growth and seedling height in Zea mays L. Nature 249:491-492.
Mulcahy, D.L. & Mulcahy, G. B. 1975. The influence of gametophytic competition on sporophytic quality in Dianthus chinesis. Theor. Appl. Genet. 46:277-280.
Mulcahy, G.B. & Mulcahy, D.L. 1983. A comparison of pollen tube growth in bi-and tri-nucleate pollen. in: Pollen: Biology and implications for plant breeding. D.L. Mulcahy & E. Ottaviano, eds. Elsevier Science Publishing, Co.,New York: 35-41.
Mulcahy, D.L. & Mulcahy, G.B. 1987. The effects of pollen competition.American Scientist 75: 44-50.
Ottaviano, E.; Sari-Gorla, M. & Mulcahy, D.L. 1980. Pollen tube growth rates in Zea mays: implications for the genetic improvement of crops. Science 210:437-438.
Ottaviano, E.; Sari-Gorla, M. & Villa, M. 1988b. Pollen competitive ability in maize: within population variability and response to selection. Theor. Appl. Genet. 76:601-608.
Sarr, A.; Sandmeier, M. & Pernes, J. 1988. Gametophytic competition in pearl millet, Pennisetum typhoides (Stapf et Hubb.). Genome 30:924-929.
Shivanna, K.R. & Heslop-Harrison, J. 1981. Membrane state and pollen viability. Ann. Bot. 47:759-770.
Snow, A.A. & Mazer, S.J. 1988. Gametophytic selection in Raphanusraphanistrum: a test for heritable variation in pollen competitive ability. Evolution 42(5):1065-1075.
Van Campo, M. 1976. Patterns of pollen morphological variation within taxa.in: The evolutionary significance of the exine. Eds., I.K. Ferguson & J.Muller. Academic Press, Inc., New York:163-183.
Walker, J.W. & Doyle, J.A. 1975. The bases of Angiosperm phylogeny: palynology.Ann. Mo. Bot. Gard. 62:664-725.
Williams, E.G. & Rouse, J.L. 1990. Relationships of pollen size, pistil length and pollen tube growth rates in Rhododendron and their influence on hybridization. Sex. Plant Reprod. 3:7-17.

69
Heterosis in Plants Starts Immediately after Fertilization

Minoru Yamada, Kazuhiro Suenga, and Kousuke Nakajima

The analysis of the mechanism of heterosis in plants is useful on strategy of hybrid breeding as a basis for the estimation of combining ability. For this purpose it is necessary to consider studies on heterosis from the viewpoint of the whole plant life cycle with two kinds of zygosity, diploidy and haploidy. At present two major theories on heterosis (ex.Yamada l987) have been proposed and the dominant gene linkage theory is generally accepted based on biometrical and physiological traits, which few data support the over-dominance theory.

Genetical analysis of physiological traits in plants has been initiated since Ashby's work (l930, l932). His hypothesis consists of four tenets and the most important and attractive one is as follows: hybrid vigour is nothing more than the maintenance of an initial advantage in embryo size for plant growth. After Ashby's proposal, more than 15 papers, some of them supporting and others rejecting his hypothesis, were published. All the reports we reviewed (Yamada, l985), however, were based on incomplete pollination schemes where the maternal effect was not eliminated and/or where on experimental design with evaluation of the error term was lacking.

Yamada (l985) and Yamada et al. (l984) confirmed the main tenet of Ashby's hypothesis, i.e. the presence of larger embryo in F1 compared with the embryos of the parental lines. In these two papers they conducted pollination experiments where the

Table 1. Differences of seed weight and of embryo weight between F1 seed and selfed seed in the ear of C164 resulting from themixed pollination experiments. (cited from YAMADA, l985).

		As male parent ear		
	P 51B	A257	B14	0-143
	(4)(5)	(4)(5)	(4)(5)	(4)(5)
1	-4.1**4.5**	6.6**2.3**	30.8**4.1**	26.6**5.3**
2	2.7**5.9**	7.3**2.5**	40.8**4.8**	29.0**5.9**
3	-5.8**4.8**	3.5**0.9**	25.9**3.4**	17.4**3.2**
4	-9.5**2.1**	17.4**4.0**	17.9**3.9**	16.2**4.1**
5	9.8**1.1**	19.9**3.9**	32.5**3.5**	25.4**6.8**
6	---------	2.5**2.5**	16.1**4.8**	17.2**3.7**
7	---------	21.6**1.9**	34.4**6.4**	27.1**5.4**
8	---------	22.2**7.7**	36.6**7.0**	20.5**6.3**
9	---------	---------	19.7**1.3**	23.1**6.4**
Mean	-0.4*3.7**	12.6**3.2**	28.3**4.4**	22.5**5.2**

Note: (4) - Seed weight (mg) and (5) - Embryo weight (mg)

maternal effect on ear pollinated was reduced, and weighed single embryos in mg unit using an electric balance. Part of the results are shown in Table 1. Although marker gene(s) of the materials used in the experiment involved different kernel genotypes, that is colored/colorless and sugary/non-sugary, the results confirmed that the embryo of the F1 was always larger than those of both parental inbred lines, independently of the genotypes. It is essential, indeed, to observe the embryo genesis immediately after fertilization under well-controlled experimental conditions, since heterosis in embryo size results from fertilization between different genotypes as an end-product (Yamada 1985).

Table 2. Embryo development in dry weight of selfed C164 and B14, and their reciprocal hybrids.

Days after pollination	Cross combination					
	C164 (as seed parent)			B14 (as seed parent)		
	selfed	crossed	(S/C)	selfed	crossed	(S/C)
9	0.01*	0.03	4.63	0.03	0.03	1.00
12	0.21*	0.48	2.65	0.37*	0.43	1.14
15	1.27*	2.17	1.73	1.25*	1.44	1.16
20	6.31	8.15	1.28	5.78*	6.40	1.11
30	14.9*	19.20	1.29	17.34*	20.73	1.20
40	20.94	24.39	1.19	24.81*	27.94	1.13

*significant at 5% level between the selfed and the crossed

Materials and Methods

The materials consisted of maize inbred lines of the denttype, Cl64 with colorless endosperm genotype, as well as B14 andA257 with a colored one, which were also used in the experiment of Yamada (l985).

Table 3. Embryo development in dry weight of selfed C164 and A257, and their reciprocal hybrids

Days after pollination	Cross combination					
	C164 (as seed parent)			A257 (as seed parent)		
	selfed	crossed	(S/C)	selfed	crossed	(S/C)
9	0.01	0.01	1.00	0.01*	0.02	3.65
12	0.15	0.27	1.87	0.04*	0.43	9.94
15	1.12*	1.75	1.56	0.45**	1.60	3.53
20	4.67*	6.96	1.49	3.91*	7.70	2.02
30	--	--	--	17.91*	25.96	1.45
40	--	--	--	27.35*	37.42	1.36

*, ** significant at 5% and 1% level between the selfed and the crossed,respectively.

Pollination technique adopted in this experiment consisted of split-ear pollination (Fairchild 1952) on the dividing plate of the ear pf C164. Pollination and fertilization at maturity were identified based on the kernel color by xenia, estimated at the immature stage and in alternative pollination. Sampling was performed 6, 9, 12, 15, 20, 30, and 40 (maturity) days. After pollination the number of pollinated ears exceeded three and more than 24 embryos were measured for every year.

Results and Discussion

The results obtained in C164 and B14, and in C164 andA257 including the reciprocal pollination, are shown in Table 2 and Table 3, respectively. The data revealed a significant difference between the selfed embryo and crossed one (F1), which was the larger F1 embryo. The difference was evident nine days after pollination in the combination of C164 with B14 (Table 2). In the combination of C164 and B14 the size of proembryo 6 days after pollination is shown in Table 4. Since it was impossible to

Table 4. Size of proembryo on the 6th day after pollination

Cross combination	Length Mean ± SD X 10 mm	Width Mean ± SD X 10 mm
C164 selfed	16.2 ± 0.4	8.7 ± 0.2
C164 x B14	21.7 ± 2.2	13.0 ± 0.5
B14 x C164	24.4 ± 1.3	12.6 ± 0.6
B14 selfed	18.8 ± 0.9	12.0 ± 1.0

measure the dry weight of the proembryo, the length and width were measured under a microscope. Significant and larger increments of the length and width in F1 including the reciprocal were confirmed, suggesting that the heterosis effect was already manifested 120 hours after fertilization.

Our data suggest that 1) heterosis starts immediately after fertilization and 2) the heterotic effect on biochemical traits controlled by mono- and/or dimorphic gene(s) should be detected and evaluated. It must be emphasized that the microspore and macrospore at the haploid phase just before fertilization, and diploid zygote just after fertilization, should be studied.

LITERATURE

1) Ashby, E. 1930. Studies in the inheritance of physiologicalcharacters. I A physiological investigation of the nature ofhybrid vigour in maize. Ann. Bot. 44: 457-469.

2) . 1932. . II. Further experiments upon the basis ofhybrid vigour and upon the inheritance of efficiency index andrespiration rate in maize. Ibid. 46:1007-1032.

3) Fairchild, R.S. 1952. Comparative development of the embryosof inbred and hybrid maize. Iowa State Co. J. Sci. 27:381-405.

4) Yamada, M. I985. Heterosis in embryo of maize. Zea mays L.Bull. Natl. Inst. Agrobiol. Resour. 1:85-98.

5) . 1987. (Hybrid breeding in crops and theory ofheterosis). (Proc. 7th Symp. Breed. 1986). 3-12 (in Japanese).

6) Yamada, M., T. Ishige and Y. Ohkawa 1985. Reappraisal ofAshby's hypothesis on heterosis of physiological traits in maize. Euphytica 34:593-598.

70

Evidence for Non-random Seed Setting of Ovules in the Pod of the Cruciferous Plants *Brassica juncea* and *Raphanus sativus*

Hyoji Namai and Ryo Ohsawa

Darwin (1876) had advocated that the process of sexual reproduction is the most important process in the life cycle of plants. Based on Darwin's theory of evolution, Michurin revealed the positive effects of "mixed pollination" for making distant hybrids (from "I.V.Michurin Selected Works", 1950). So far, several studies have been done on gametophytic selection and its efficacy for producing evolutionary change in plants (Mulcahy et al., 1975, Schlichting et al., 1987, Knox et al., 1990).

Over the last 20 years, several critical data on gametophytic selection have been reported, namely, (1) expression of the pollen genome during pollen germination and pollen tube growth (Heslop-Harrison, 1980, Willing and Mascarenhas, 1984), (2) pollen genotypic determination of both the speed of germination and pollen tube growth rates (Mulcahy, 1971, Ottaviano et al., 1975), (3) high similarity between gene expression of the pollen grain and sporophyte (Mulcahy, 1971, Willing and Mascarenhas, 1984), (4) non-random fertilization under conditions of pollen competition (Pfahler, 1967, Ockendon and Currah, 1978, Marshall and Ellstrand, 1986, Stephenson et al. ,1988), and (5) female choice which could operate at any of several times as follows: (a) selective pollination, (b) selective fertilization, and (c) zygotic female choice (seed and fruit abortion) (Mulcahy, 1979, Stephenson and Bertin, 1983).

Recently, we proposed a new concept of reproductive success rate pertaining to the pollen grains deposited onto a stigma (RSR-P) and that of the ovules in the pollinated flower (RSR-O) (Namai and Ohsawa, 1986). Our studies on the RSR-P and RSR-O in B. juncea (leaf mustard), Oryza sativa (rice) and Fagopyrum esculentum (buckwheat) clearly revealed a route for expanding genetic variation in offsprings derived from conditions of little or no pollen competition by limited pollination (Namai and Ohsawa, 1988, Namai, 1990).

In this chapter, we report some interesting results of our preliminary studies in Brassica juncea (leaf mustard) which is incompletely self-compatible and Raphanus sativus (radish) which is incompletely self-incompatible in order to clarify the evidence for non-random seed setting of ovules in pod. In the case of B. juncea, slight and moderate limited pollination with 31-200 pollen grains per stigma led to non-random seed setting in such a way that % seed set in the middle position of the ovary was lower than that in the other regions. On the other hand, in the case of R. sativus, both intensive limited pollination and unlimited pollination led to non-random seed setting such that the maximum % seed set ranged from the lower position (peduncular end) to the top (stylar end) of the ovary, a especially remarkable phenomenon in intensive limited pollination.

Recent study in B. juncea, leaf mustard

In B.juncea, both limited and unlimited self-pollination were artificially conducted by bud pollination onto 10 potted plants of a cultivar Hakarashina in a greenhouse. A special pollination manipulator made of a hairy thin wire was used for limited pollination at various intensity (refer to Namai and Ohsawa, 1986). After maturation, the pods were harvested and the number of ovules and mature seeds in each region of each pod were counted. In general, each pistil of B. juncea contains 18-24 ovules in B. juncea.

Table 1 summarizes the results from the study of limited and unlimited self-pollination in B. juncea.

Table 1. % seed set of each ovarian region of a pod in B. juncea

Ovarian region[1)]	Limited (<=30)	pollination (No. of pollen grains) (31-50)	(51-100)	(101-200)	Unlimited (>=300)
A	27.5a[2)]	41.5a	59.3a	76.9a	78.3a
B	28.0a	41.5a	55.6a	68.7ab	80.0a
C	23.8a	32.7ab	46.7ab	56.6 bc	83.3a
D	23.1a	24.9 b	37.8 b	52.0 c	84.2a
E	31.2a	32.6ab	47.4ab	65.3ab	85.7a
F	25.3a	32.6ab	49.9ab	65.2ab	83.6a
mean	26.5 c	34.3 c	49.5 bc	64.1ab	82.5a

1) Ovules of 18-24 in each ovary were divided into six parts: A = stylar end (top region), and F = peduncular end.

2) Means followed by the same letter in each column of A to F and in the mean numbers are not significantly different at 5% level of the probability according to Duncan's multiple range test.

Intensive limited pollination with 30 or less pollen grains per stigma as well as unlimited pollination with 300 or more pollen grains led to no difference of % seed set in each pod region, whereas slight and moderate limited pollination with 31-200 pollen grains led to non-random seed setting. In the case of slight limited pollination with 31-100 pollen grains, % seed set in the middle position of the ovary "D" was lower than that in the stylar region (top end), and in the case of moderate limited pollination with 101-200 pollen grains it was lower in the middle position than that in the other regions.

Recent study in R. sativus, radish

In R. sativus, both unlimited (A) and limited (B) pollination were also conducted in 16 cultivars as shown in Table 2. For unlimited pollination, four potted plants in a greenhouse were artificially self-pollinated by bud pollination at two to four days before flowering. For limited pollination, every nine plants of each cultivar were cultivated in separate netted isolation cages (W2.0m x L2.0m xH1.8m) and natural automatic self-pollination were conducted without any insect pollinator.

Then mature pods were harvested and the number of ovules as well as mature and immature seeds in each pod were counted. Each ovary had five to six ovules in almost all R. sativus cultivars. Mean and standard error of % pod set and % seed set of the 16 cultivars in limited pollination were 14.4+/-2.7% and 3.0+/-0.8%, respectively.

Table 2. Entries of R. sativus in the present study

Name of cultivars (Entries)	No. of ovules per pistil	% seed set[1] unlimited (A)	limited (B)
1.Nerima-Ohnaga-Takuan-Daikon	5.5	41.8	1.2
2.Hayafutori-Nerima-Daikon	4.5	42.6	2.2
3.Miura-Daikon	6.2	32.8	0.7
4.Ohkura-Daikon	5.4	58.8	3.2
5.Aokubi-Sofutori-Miyashige-Daikon	6.2	49.0	2.6
6.Aokubi-Miyashige-Nagafutori-Daikon	6.1	42.4	0.8
7.Hohryo-Daikon	5.9	21.5	6.7
8.Shogoin-Daikon	6.6	36.2	0.1
9.Minowase-Daikon	5.4	47.2	1.4
10.Benisuji-Daikon	5.8	25.8	4.5
11.Bansei-Sakurajima-Daikon	5.9	35.4	7.1
12.Kameido-Daikon	5.6	30.7	5.7
13.Akamaru-Hatsukadaiko	5.0	16.0	0.7
14.Kohhaku-Hatsukadaikon	4.0	20.9	0.0
15.Sakuranbo-Hatsukadaikon	3.4	11.9	10.7
16.Benimusume-Hatsukadaikon	8.0	37.5	0.2
Mean(+/-)s.e.	5.6+/-0.3	34.4+/-3.2	3.0+/-0.8

[1] (A): % seed set of the buds at two to four days before flowering.
(B): % seed set of the flowers in the conditions of intensive limited pollination in the isolation cages without any insect pollinator.

Table 3 summarizes the results from the study of limited and unlimited self-pollination in R.sativus. The % seed set was calculated from the data of flowers which set pods.

Table 3. % fertilized ovules (% f. o.) and % seed set (% s. s.) of each ovule in *R. sativus* (mean+-s.e.)

Ovarian region[1] (No.of cultivars)	Limited pollination (L) (% s. s.)	(% f.o.)[2]	Unlimited (%s.s.)	L/U [3]
A (14)	1.3+-0.9	2.0+-1.8	20.4+-4.7	0.06(0.10)
B (15)	3.0+-1.0	4.2+-1.3	24.4+-3.3	0.12(0.17)
C (16)	6.7+-1.7	9.9+-2.2	35.9+-3.2	0.19(0.28)
D (16)	14.1+-3.2	19.8+-4.2	54.8+-4.0	0.26(0.36)
E (16)	25.2+-5.5	40.1+-5.4	73.4+-2.4	0.34(0.55)
F (16)	68.6+-6.5	98.9+-0.6	97.7+-0.8	0.70(1.01)
Mean +- s.e.	19.8+-10.4	29.2+-15.1	36.7+-5.1	0.27(0.41)

[1] A = stylar end (one to three ovules at top region), and F = peduncular end (one ovule at bottom region).
[2] Total number of mature and immature seeds were divided by number of ovules per flower which set pod.
[3] The figures in parentheses were calculated from dividing % fertilized ovules (% f.o.) in limited pollination by % seed set in unlimited pollination.

Both limited pollination and unlimited pollination led to non-random fertilization and seed setting. The maximum % fertilization and % seed set ranged from lower position (peduncular end) to the top (stylar end) of the ovary and this phenomenon was especially remarkable in limited pollination.

Discussion and conclusion

Gametophytic selection as soft selection via pollen competition should result in increasing vigor and decreasing genetic variation among offspring (Mulcahy et al. 1975, Ter-Avenesian 1978). Schlichting et al. (1987) observed that variation of the progenies which were produced under conditions of intensive pollen competition decreased significantly than the progenies which were produced under conditions of little or no pollen competition by limited pollination in both birdsfoot trefoil, Lotus corniculatus, and zucchini, Cucurbita pepo. Stephenson et al. (1988) also showed evidence for non-random fertilization in zucchini. They observed that the first pollen tubes that penetrated into the ovary fertilized the ovules in a specific region of the ovary and the seeds produced by this region were more vigorous than those in the other regions of the ovary which were fertilized later.

In cruciferous plants, Hill and Lord (1986) found that significant variation in the order of fertilization of the linearly arranged ovules occurred within wild radish, Raphanus raphanistrum, and this variation suggested that prezygotic mechanisms of gametic selection operated to sort pollen tubes non-randomly to different ovule positions in the ovary. According to their study, the ovules in the stylar or middle positions were the fastest to be fertilized, followed by the ovules in the lower position which were physically the farthest from the stigma. It was, therefore, suggested that the quality of the pollen fertilized may also differ depending on the position of the ovules in the ovary. Hill and Load (1987) also expected that the differences in the fast fertilized ovule positions among cross combinations could be attributed to the structure of the transmitting tissues in the ovary, which allows pollen tubes to pass the ovules while inside the septum and exit this tissue lower in the ovary to fertilize ovules in that region. Marshall and Ellstrand (1986) also reported that competition between male gamete growing through the style and the ovary was generally expressed as variation in pollen tube growth rate in wild radish, Raphanus sativus.

On the other hand, based on the present studies, it was clearly revealed that non-random fertilization and seed setting generally occurred in cruciferous plants, especially slight or moderate intensity of pollen competition led to frequently non-random fertilization and seed setting in both Brassica juncea and Raphanus sativus. Non-random seed setting in Brassica juncea, however, was completely different from that in Raphanus sativus. In Brassica juncea subjected to slight and moderate limited self-pollination with 31 to 200 pollen grains per stigma, % seed set in the middle position of the ovary was lower than that in the others, and in Raphanus sativus subjected to unlimited self-pollination, the maximum % fertilization and seed setting ranged from the lower position (peduncular end) to the top (stylar end) of ovary.

In the case of hermaphrodite flowers like cruciferous plants, every flower must be visited successively by several insect pollinators. Furthermore, successive slight or moderate limited pollination with mixed pollen grains from various individuals and a total of 500 or more pollen grains per stigma, can be generally expected under ordinary growth conditions (Ohsawa and Namai 1988). Hill and Lord (1986) proposed that the variation in fertilization patterns could be attributed to variation in pollen tube growth rates in the central septum of the wild radish gynoecium. Therefore, it can be reasonably expected that at least under conditions of slight and moderate pollen competition, the degree of % fertilization and % seed set for ovules in each position of the ovary should vary in plant species, in conditions of a population and plant individuals as well as in quality and quantity of insect pollinators.

Consequently, it might be concluded that under conditions of slight or moderate pollen competition caused by artificially limited pollination as well as under conditions of intensive pollen competition caused by natural mixed pollinations, the order of fertilization and the degree of % seed set for ovules in each position of the ovary appear to be variable. Moreover, the ratio of some special traits of the seedlings obtained from each position of the ovary should vary critically.

References

1. Darwin, C. The Effects of Cross and Self Fertilization in the Vegetable Kingdom. London:John Murray; 1876.

2. Heslop-Harrison, J. The forgotten generation: some thoughts on the genetics and physiology of angiosperm gametophytes. In: Davis, D. R. and Hopwood, D. A., eds. The Plant Genome. Norwich:John Innes Charity; 1980:p. 1-14.

3. Hill, J. P. and Lord, E. M. Dynamics of pollen tube growth in the wild radish, Raphanus raphanistrum (Brassicacea). I. Order of fertilization. Evolution 40:1328-1333;1986.

4. Hill, J. P. and Lord, E. M. Dynamics of pollen tube growth in the wild radish, Raphanus raphanistrum (Brassicacea). II. Morphology, cytochemistry, and ultrastructure of the transmitting tissue, and the path of pollen tube growth. Amer. J. Bot. 74:988-997;1987.

5. Knox, R. B., Gaget, M. and Dumas, C. Mentor pollen techniques. Intl. Rev. Cytol. 107:315-332;1987.

6. Marshall, D. L. and Ellstrand, N. C. Sexual selection in Raphanus sativus: experimental data on nonrandom fertilization, maternal choice, and consequences of multiple paternity. Am. Nat. 127:446-461;1986.

7. Mulcahy, D. L. A correlation between gametophytic and sporophytic characteristics in Zea mays L. Science 171:1155-1156;1971.

8. Mulcahy, D. L. and Mulcahy, G. B. The influence of gametophytic competition on sporophyte quality in Dianthus chinensis. Theor. Appl. Genet .46:277-280;1975.

9. Mulcahy, D. L., Mulcahy, G. B. and Ottaviano, E. Sporophytic expression of gametophytic competition in Petunia hybrida. In:Mulcahy, D. L. ed. Gamete competition in plants and animals. Amsterdam:North-Holland; 1975:p227-232.

10. Mulcahy, D. L. The rise of the angiosperms: a genecological factor. Science 206:20-23;1979.

11. Namai, H. and Ohsawa, R. Variation of reproductive success rates of ovule and pollen deposited upon stigmas according to the different number of pollen on a stigma in angiosperm. In:Mulcahy, D. L., Mulcahy, G. B. and Ottaviano, E. eds. Biotechnology and Ecology of Pollen. New York:Springer-Verlag, 1986:p423-428.

12. Namai, H. and Ohsawa, R. Possibility of expanding genetic variation by limited pollination based on the reproductive success rate(RSR) of pollen grains deposited on stigma. In: Cresti, M., Gori, P. and Pacini, E. eds. Sexual Reproduction in Higher Plants. Berlin:Springer-Verlag, 1988: p63-68.

13. Namai, H. Pollination biology and reproductive ecology for improving genetics and breeding of common buckwheat, Fagopyrum esculentum (1). Fagopyrum 10:23-46;1990.
14. Ockendon, D. J. and Currah, L. Time of cross- and self-pollination affects the amount of self-seed set by partially self-incompatible plants of Brassica oleracea. Theor. Appl. Genet. 52:223-237; 1978.
15. Ohsawa, R. and Namai, H. Cross-pollination efficiency of insect pollinators (Shimahanaabu, Eristalis cerealis) in rapeseed, Brassica napus L. Japan. J. Breed. 38:91-102;1988.
16. Ottaviano, E., Sari-Gorla, M. and Mulcahy, D. L. Genetic and intergametophytic influences on pollen tube growth. In:Mulcahy, D. L. ed. Gamete Competition in Plants and Animals. Amsterdam:North-Holland, 1975: p125-130.
17. Pfahler, P. L. Fertilizition ability of maize pollen grains. II. Pollen genotype, female sporophyte and pollen storage interactions. Genetics 57:513-521; 1967.
18. Schlichting, C. D., Stephenson, A. G., Davis, L. E. and Winsor, J. A. Pollen competition and offspring variance. Evolutionary Trends in Plants 1:35-39;1987.
19. Stephenson, A. G. and Bertin, R. L. Male competition, female choice, and sexual selection in plants. In:Real, L. ed. Pollination Biology. New York: Academic Press, 1983:p109-149.
20. Stephenson, A. G., Winser, J. A., Schlichting, C. D. Evidence for non-random fertilization in the common zucchini, Cucurbita pepo. In: Cresti, M., Gori, P. and Pacini, E. eds. Sexual Reproduction in Higher Plants. Berlin:Springer-Verlag, 1988, p333-338.
21. Ter-Avenesian, D. V. The effect of varying the number of pollen grains used in fertilization. Theor Appl Genet 52:77-79; 1978.
22. Willing, R. P. and Mascarenhas, J. P. Analysis of the complexity and diversity of mRNAs from pollen and shoots of Tradescantia. Plant Physiol 75:865-868; 1984.

71
The Potential for Interference among the Pollen Grains of Wild Radish

Diane L. Marshall and Michael W. Folsom

The potential of pollen competition to sort among growing pollen tubes is widely recognized (Mulcahy and Mulcahy, 1987). Differential access to ovules may allow fertilization by the most vigorous pollen tubes. And, because gene expression in seedlings overlaps substantially with that in pollen tubes (Ottaviano and Mulcahy, 1990), pollen competition may result in production of superior progeny. In fact, increases in pollen competition have been shown to improve progeny growth in a variety of species (e.g., Mulcahy and Mulcahy, 1987; Winsor et al., 1987; Schlichting et al., 1990).

In these studies, pollen competition has been viewed primarily as exploitation competition. That is, the most vigorous pollen tubes preempt the resource—ovules. However, there may also be the potential for direct interference among the pollen grains on stigmas and among the pollen tubes in styles.

If only exploitation competition occurs, it can be predicted that, as more pollen and a greater diversity of pollen arrives on stigmas, ever faster genotypes might be observed. Selection could act primarily to increase the rate of pollen germination and pollen tube growth. If interference competition occurs, a decrease in germination and pollen tube growth rate might be seen at high densities or diversities of pollen. Selection could still act on pollen tube growth rate, but it might also act on mechanisms that produce negative effects on other pollen grains or that allow avoidance of interference. Thus, selection on pollen grains and tubes might act on a wider array of characters than we have supposed. However, selection for interference competitive ability should be influenced by the relatedness of pollen grains on the stigma and hence by whether pollen loads typically contain pollen from from multiple or single donors.

Interference among pollen grains might be affected by either the density or the diversity of pollen available. Here we concentrate on pollen diversity since there is little evidence in literature to test the importance of pollen diversity. Two previous studies altered the diversity of intraspecific pollen loads and found no effect on pollen germination (Schemske and Fenster, 1983; Murphy and Aarssen, 1990). However, several studies of heterospecific pollinations have suggested that some species have allelopathic pollen and can exhibit interspecific interference competition (Kanchan and Jayachandra, 1980; Thomson et al. 1981; Jimenez et al., 1983; Murphy and Aarssen, 1989).

As part of a larger study on the mechanisms that produce non-random mating, we compared the germination of pollen grains after mixed and single pollinations. These data directly address whether there is increased interference as pollen diversity increases.

Methods

The study was performed on wild radish, *Raphanus sativus*. This species is a self-incompatible annual that was grown in the greenhouse for these studies. We used 16 maternal plants and four pollen donors. The sixteen maternal plants included four full siblings from each of four lineages. The pollen donors were from four additional lineages. The eight lineages used were known to have entirely different S-alleles due to the crossing history of their seed and pollen parents. Two maternal plants from each lineage were grown under control conditions and watered frequently. The other two plants from each maternal lineage were given one half the control amount of water. These plants wilted daily.

The four pollen donors (A-D) were used to perform four kinds of single pollinations (A, B, C, and D) and six kinds of pairwise mixed pollinations (AB, AC, AD, BC, BD, and CD). In all cases pollen was collected in small petri dishes, mixed, and applied to stigmas with tissue wrapped forceps. This technique resulted in the application of large numbers of pollen grains in both mixed and single pollinations.

Each cross was repeated four times to measure pollen germination. Pollen was applied to emasculated, full anthesis flowers which were clear of self pollen. After pollination the style was lightly tapped to remove excess pollen from the stigmatic surface. Thirty minutes later the stigma was removed from the flower and placed in a drop of stain containing 1 mg/ml of both water soluble aniline blue and sodium azide in 0.033 M K_3PO_4. The sodium azide was added to the stain to inhibit further pollen germination. Pollen grains were scored as germinated if a pollen tube could be seen at a 100x total magnification.

Additional crosses were collected at longer time intervals to score fertiization of ovules and to determine the paternity of seeds after mixed pollinations (Marshall and Folsom, in press). Seed paternity was scored using electrophoretic markers (Marshall and Ellstrand, 1986).

Patterns of pollen germination were analyzed with ANOVA. Proportion of pollen germinated was arcsin square root transformed prior to analysis.

Table 1. Analysis of variance in the proportion of pollen germinated 30 minutes after pollination.

Independent variable	R^2	P
Maternal plant	0.246	0.0001
Pollen donor	0.014	0.0024
Maternal plant x Pollen donor	0.042	0.52

Results

Pollen germination differed among both maternal plants and pollen donors, ranging from 2.8% to 20.4% among maternal plants and 12.2% to 9.4% among pollen donors. The effect of the maternal plant was strongest and there were no maternal plant by pollen donor interaction effects (Table 1).

Pollen germination also differed among single and mixed pollinations (R^2= 0.01, $P < 0.0005$). Thirty minutes after single pollination, 10.6% of the pollen had germinated, while only 8.7% of the pollen had germinated 30 minutes after mixed pollinations. Relatively, 22% more pollen had germinated after single pollinations. The trend in pollen germination was the same in all four maternal families and in 12 of the 16 maternal plants.

Pollen germination was higher (10.7%) in plants given reduced water than in control plants (8.2%) ($P < 0.001$ in an ANOVA with maternal family, cross, replicate, and treatment as the main effects); however, there was no significant treatment by pollination interaction effect, so relative performance of pollen donors was not altered. The effect of single and mixed pollination on pollen germination was in the same direction on both stressed and control plants, but it was much stronger for control plants (Table 2).

Table 2. Proportion pollen germinated 30 minutes after single and mixed pollination on control and stressed plants. * — No significant difference in proportion germinating after single and mixed pollinations. ** — The proportions of pollen germinating after single and mixed pollination are signficantly different ($\underline{P}$ < 0.01).

	Control plants		Stressed plants	
Pollination:	Single	Mixed	Single	Mixed
	0.095	0.074**	0.11	0.10*

Discussion

While microgametophyte competition has been considered primarily as a race for access to fertilizable ovules, lower pollen germination after mixed than after single pollination suggests that direct interference among pollen grains may also occur. However, our experiment was not designed to determine the mechanism by which intereference was produced; therefore, alternative explanations are possible.

Interference among pollen grains might be a result of chemical inhibitors produced by the pollen grains. Chemical inhibition of pollen germination in interspecific mixed pollinations is known for a few species (Kanchan and Jayachandra, 1980; Thomson et al. 1981; Jimenez et al., 1983; Murphy and Aarssen, 1989), but is not known in intraspecific mixed pollinations. In fact, some specific attempts to test for chemical inhibition within species have produced negative results (Murphy and Aarssen, 1991). If chemical inhibition explains our results we would expect extracts of one donor's pollen grains to inhibit germination of pollen from other plants. If this can occur, then it becomes intriguing to ask how inhibition of pollen grains is limited in single donor crosses.

It is also possible that lowered pollen germination after mixed pollination involves the stigma as well as the two kinds of pollen. In wild radish this might occur because increasing pollen diversity in a mix increases the number of S-alleles present. Even in compatible crosses, it is conceivable that a great diversity of S-alleles could trigger a stigmatic response. While we cannot address this issue directly, the greater difference in mixed and single pollinations on control than on stressed plants suggests a potential for involvement of the stigma. Changing the condition of maternal plants, but not pollen donors, reduced the amount of inhibition.

Interference among pollen grains in mixed pollinations could provide a second level of pollen competition, in addition to exploitation competition, such that the overall competitive regime is more stringent after single than after mixed pollinations. This is an interesting possibility for wild radish, because it may provide a mechanism to explain a previously puzzling result. In several experiments, wild radish has been shown to fill multiply sired fruits to greater weights than singly sired fruits (Marshall and Ellstrand, 1986; Marshall, 1988; Marshall, 1990). The result is repeatable and consistent, but puzzling because its mechanism is not understood. Interference competition in mixed pollinations might explain the result.

If competition is more severe in more diverse groups of pollen on stigmas, then only very vigorous pollen tubes may ever enter the transmitting tissue in these crosses. This interpretation is supported by two results. First, while pollen germination is slower after mixed pollination, fertilization of ovules is not substantially slower after mixed pollination (Marshall and Folsom, in press). Perhaps this occurs because the winners in interference competition are sufficiently vigorous to catch up to pollen tubes resulting from single pollination. Second, in this as in other experiments, multiply sired fruits are filled to a greater weight (41.3 mg) than are singly sired fruits (32.8 mg) (Marshall and Folsom, unpublished data).

The presence of mixed pollen loads provides an opportunity for sorting among mates and may thus affect the paternity of seeds. Our results suggest that mixed pollen loads also create an opportunity for pollen interaction that may affect the quality of seeds and fruits. Confirmation of these interpretations requires duplication of these results in other species and investigation of the mechanisms of pollen interference.

Acknowledgements

This work was funded by National Science Foundation grants BSR-8818552 and BSR-8958233. We thank Anne Rice for formatting the text and Lynette Michelson for entering the data.

References

Jimenez, J.J.; Schultz, K.; Anaya, A.L.; Hernandez, J.; Espejo, O. Allelopathic potential of corn pollen. J. Chem. Ecol. 9:1011-1025; 1983.

Kanchan, S.; Jayachandra. Pollen allelopathy—a new phenomenon. New Phytol. 84:739-746; 1980.

Marshall, D.L. Post pollination effects on seed paternity: mechanisms other than microgametophyte competition operate in wild radish. Evolution 42: 1256-1266; 1988.

Marshall, D.L. Non-random mating in a wild radish, *Raphanus sativus*. Plant Species Biol. 5:143-156; 1990.

Marshall, D.L.; Ellstrand, N.C. Sexual selection in *Raphanus sativus*: experimental data on non-random fertilization, maternal choice, and consequences of multiple paternity. Am. Nat. 127: 446-461; 1986.

Marshall, D.L.; Folsom, M. W. Mechanisms of non-random mating in wild radish. In: Wyatt, R. ed. New Approaches to Plant Reproductive Ecology, New York: Chapman and Hall; in press.

Mulcahy, D.L.; Mulcahy, G.B. The effects of pollen competition. Amer. Sci. 75:44-50; 1987.

Murphy, S.D.; Aarssen, L.W. Pollen allelopathy among sympatric grassland species: *in vitro* evidence in *Phleum pratense* L. New Phytol. 112:295-305; 1989.

Murphy, S.D.; Aarssen, L.W. Occurrence and consequences of pollen allelopathy. Bull. Ecol. Soc. Am. 71 (2):266; 1990.

Ottaviano, E.; Mulcahy D.L. Genetics of angiosperm pollen. Adv. Genet. 26:1-64; 1990.

Schemske, D.W.; Fenster, C. Pollen-grain interactions in a neotropical *Costus*: effects of clump size and competitors. In: Mulcahy, D.L.; Ottaviano, E. eds. Pollen: biology and implications for plant breeding. New York: Elesevier Science Publishing, Inc.; 1983;

Schlichting, C.D.; Stephenson, A.G.; Small, L.E. Pollen loads and progeny vigor in *Cucurbita pepo*: the next generation. Evolution 44:1358-1372; 1990.

Thomson, J.D.; Andrews, B.J.; Plowright, R.C. The effect of a foreign pollen on ovule development in *Diervilla lonicera* (Caprifoliaceae). New Phytol. 90:777-783; 1981.

Winsor, J.A.; Davis, L.E.; Stephenson, A.G. The relationship between pollen load and fruit maturation and the effect of pollen load on offspring vigor in *Cucurbita pepo*. Am. Nat. 129:643-656; 1987.

72

Patterns of Bee Visitation to Flowers of Almond Cultivars in an Orchard as Determined by Molecular Genetic Marker Analysis of Pollen in the "Pollen Basket"

J.F. Jackson and G.R. Clarke

Introduction

Self-incompatibility of almond cultivars used in commercial orchards (Kester and Griggs 1959; Baker and Gathercole1977) means that gene flow by pollen between the different cultivars are essential to obtain nut set. A study made in a south Australian orchard showed that gene flow by pollen resulting in nut set is quite restricted, taking place essentially only between neighboring trees (Jackson and Clarke 1991). This is surprising, considering the comparatively large distances that the bees must travel within the orchard (at least twenty rows and more). To explain these results and the fact that only a small proportion of lowers gives rise to nuts, Jackson and Clarke (l991) suggest that the honeybee predominantly visits only one cultivar, either visiting single trees in times of pollen plenty or along rows of one cultivar when pollen is scarcer, and that cross-pollination results from accidental or rare visits involving two or more compatible cultivars. Here we examined the pollen brought back to the hive by individual honeybees, so as to trace patterns of flower visitations within the almond orchard.

Materials and Methods

The main almond cultivars present in the orchard investigated were Chellaston, Johnston, Ne Plus Ultra, Nonpareiland Mission (the latter three predominating), with a "2 to 1" arrangement of Non-Pareil rows to pollenizer rows. Honeybees were introduced during the flowering of the earliest cultivars (Chellaston), at approximately 4 hives per hectare. One of the hives had a "pollen trap" fitted (Stanley & Linskens 1974) a device which "scraped" off pollen in the form of pellets from the hind legs of bees as they crawled through into the hive after a pollen collecting flight.

3. Results and Discussion

In order to address the question as to why gene flow is so restricted in these commercial orchards, we examined the pollen pellets collected in the "pollen trap" placed on one of the stronger hives of bees. The pellets were extracted and gel electrophoresis and isozyme staining carried out as described previously for hand collected pollen (Jackson and Clarke, 1991;Jackson 1992). In all samples it was found that pollen from the bee pollen basket gave isozyme patterns similar to that of hand collected pollen. To our surprise, however, in most cases each pellet behaved as if it was a pure sample of

Table 1. Number of pollen pellets recovered from the pollen trap containing pollen from one almond cultivar only (>99%) for each of the 35 days of collection.

DATE	CHELLASTON	NE PLUS	JOHNSTON	NON-PAREIL	MISSION
2/8	70	16	10		
3/8	14	76	2		
4/8	22	66	2	10	
5/8	26	54	4	6	
6/8	8	54	2	26	
7/8	10	36		50	
8/8	2	46		48	
9/8	2	26	2	60	
10/8		12	4	78	
11/8		10		82	
12/8		4		88	
13/8		4		92	
14/8		8		92	
15/8		8		84	4
16/8				96	
17/8				76	18
18/8				86	10
19/8				62	16
20/8				44	32
21/8				40	38
22/8				8	76
23/8				30	40
24/8				32	52
25/8				52	42
26/8				42	30
27/8				28	56
28/8				40	56
29/8				34	50
30/8				38	44
31/8				30	48
1/9				44	38
2/9				38	54
3/9				36	40
4/9				6	68
5/9				6	16

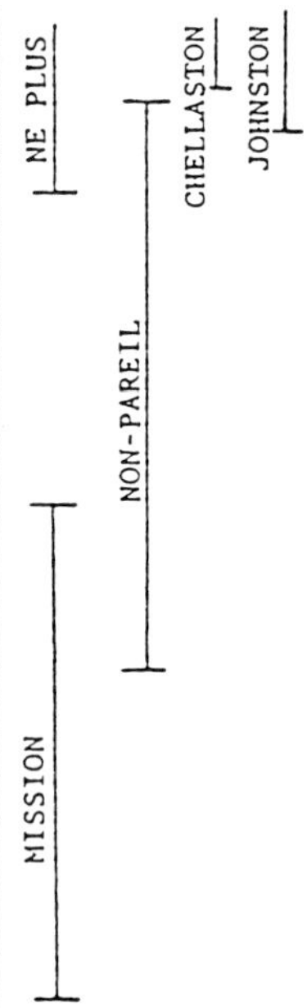

pollen, collected from only one almond cultivar (e.g. see Jackson and Clarke,l991). We mixed hand collected pollen in various proportions and showed that we could detect 1% or more of pollen from one cultivar and mixed with another. In what follows then, for simplicity, we will define mixed pollen pellet as having 1% or more of one cultivar mixed with another. In practice we believe we can detect less than 1% of one cultivar mixed with another in most cases, so that we refer to as "pure" pollen pellets are better than 99% from one particular almond cultivar. The results from analysis of the daily

The cultivar flowering periods (1% to 100%) are indicated at the right side of the table. This table should be used in conjunction with Table2.collection of pollen pellets are shown in Tables 1 and 2. As expected, Chellaston and Johnston, the early flowering varieties, made up a large proportion of the pollen pellets over the early period, while it was noticeable that some bees in the hive brought in Nonpareil pollen right from the start of flowering of that cultivar. Most of the pollen collected during this early period was predominantly from one cultivar or another (Table 1) and not mixed. In other words in most cases each has collected pollen from only one almond cultivar in any one collection flight.

Table 2. Number of pollen pellets recovered from the pollen trap containing pollen from more than one cultivar (and therefore "mixed"). This table should be used in conjuction with Table 1.

CHELL/JOHNSON	NE PLUS/NON-P	NON-P/MISSION	NOT ALMOND
2			
	6		
	6		
	2		
2	8		
	4		
	4		
	8		
	4		
	4		
	2	2	
	6		
		4	
	4	18	
	2	22	
		22	
		16	
		30	
		16	
		6	
	2	26	
	8	8	
		8	
		14	2
		16	2
		14	6
		8	10
		6	2
		14	
		2	18
		4	74

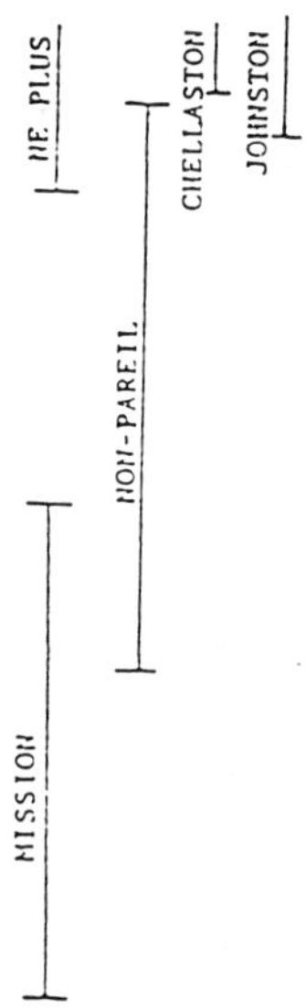

As the flowering period progressed, mixed pollen loads began to appear more often, particularly Nonpareil/ Mission mixes (Table 2). However, at no time did mixed loads exceed 30% of the pellets collected, and more often, even late in the flowering season, these mixed pollen pellets comprised less than 15% of the pollen pellets brought into the hive. After the late-flowering cultivar Mission had achieved 100% flowers opened, the bees in the hive began to collect pollen from other plant species (Table 2), until on the 5th of September, the majority of pollen loads was from sources other than almond, and the experiment was terminated.

A feature of the present investigation is that, except for some deviations late in the almond flowering season, individual bees in the hive under observation brought in pollen loads which were predominantly from only one almond cultivar, even though all almond cultivars in flower at the time were represented in the pollen pellets collected by different bees in the hive. These results are in accord with the hypothesis put forward previously that the honeybee predominantly visits only one cultivar, either visiting single trees or along rows of one cultivar, and that cross pollination results from accidental or rarer visits involving two cultivars. The present work goes further and suggests that late in the season, in the present orchard with its mix of cultivars, more bees visit and bring back significantly larger numbersof mixed loads of pollen after Nonpareil has achieved 100% flowers opened. These late mixed loads comprise no more than 30% of pollen coming in and are Nonpareil/Missionmixes.

References

Backer BT, Gathercole F (l977) Commerical almond growing. South Australian Department of Agriculture Bulletin no. 9/77.

Jackson JR (l992) Genotype determination in almond nuts for paternity analysis. Modern Methods of Plant Analysis, in press.

Jackson JR, Clarke GR (l991) Gene flow in an almond orchard. Theor.Appl.Genet. 82:169-173.

Kester DE, Griggs WH (l959) Fruit setting in the almond; the effect of cross-pollinating various percentages of flowers. Proc.Am.Soc.Hort. Sci. 74:206-213.

Stanley, RG, Linskens, HF (l974) Pollen. Biology, biochemistry, management. Springer, Berlin, Heidleberg, New York, pp. 87-115.

73
Environmental Effects on Pollen Characters and Paternity

Helen J. Young

Evidence of environmental effects on female reproductive success in plants has been well documented (reviewed in Stephenson, 1981; Roach and Wulff, 1987; Lee, 1988), with most studies finding that plants grown under conditions of adequate water, light, and nutrients have a greater rate of fruit initiation, greater seed number per fruit, and/or greater seed mass than plants grown in sub-optimal conditions. Through maternal effects, the male success of plants may also be influenced by the environmental conditions of the maternal plants: the fate of the seeds they sire may be influenced by the environmental conditions of the plant where their pollen is delivered.

In addition, environmental conditions can affect pollen characters that may, in turn, affect paternal success. For example, environmental conditions during pollen development have been shown to influence pollen production (Herpen, 1986; Vasek et al., 1987; McKone, 1989), pollen size (Bell, 1959), pollen germination (Zamir et al., 1981; Schlicting, 1986; Schoper et al., 1986), pollen viability (Sharma et al., 1985), and pollen tube growth rates (Miller and Schonhorst, 1968; Herpen and Linskens, 1981). However, the effect of such variation on male reproductive success remains largely unknown. Kuo et al. (1981) and Young and Stanton (1990) have found environmental effects during pollen development on the number of seeds sired in Brassica and Raphanus, respectively. This finding of paternal environmental effects on paternity forces a re-evaluation of the studies assuming purely genetic paternal effects (i.e. gametophytic selection experiments and studies partitioning variance into various genetic components). In this chapter I report on recent experiments on the genetic and environmental effects on pollen traits in Silene alba (=S. latifolia), Caryophyllaceae.

Methods

Ten male and two female individuals of S. alba were transplanted from various field sites in Virginia to 10 cm pots in the greenhouse. During the next 12 months, each individual was repeatedly clonally replicated. When the plants were still rosettes and before floral development had begun (early Feb. 1990), males of each genotype were placed under one of three growing regimes: one of each genotype was seldom watered (only when wilting had occurred) but received nutrient fertilizer ('Miracle-Gro') when watered ('water-limited'), one received water daily but no nutrient fertilizer ('nutrient-limited'), and the third received sufficient water and twice-weekly nutrient fertilizer ('control'). All female plants were grown under the control conditions.

As males began flowering (mid-March), three mature floral buds were collected and the anthers removed into microcentrifuge tubes for analysis of pollen number and pollen size. This was repeated three times during their flowering period (March 19-24, April 13-16, and April 30-May 1; collection dates 1, 2, and 3, respectively). Anthers dehisced at room temperature and pollen number and size were determined using an electronic particle analyzer (Elzone 280 PC). Non-viable pollen grains are about half the size of viable pollen, so the percent of pollen aborted can be measured from the pollen size distributions as percent of total pollen number that falls in the smaller size class.

Hand-pollinations were performed during April. On each of two clonal replicates of each female plant, single-donor pollinations were performed on open flowers. Pollen from each male plant of each treatment was used to pollinate a flower on each of the four female plants. Freshly opened male flowers were picked and the anthers were rubbed against the 5-lobed stigma of the female flower. As fruits matured 2-4 weeks later they were collected and seed number determined. Silene seeds are tiny (weighing about .0004-.0008 g. each) so entire seed sets were weighed; mean seed weight per flower was determined by dividing total seed weight by the number of seeds.

Results & Discussion

The experimental treatments had a marked effect on gross morphology of the plants, in particular, the water-limited plants were very small and produced smaller flowers than the control and nutrient-limited plants (Table 1).

There were no significant differences among treatments in pollen production per flower or pollen size (Table 1), though there were significant differences among male plants (Figure 1) and among the collection dates (Figure 2). Pollen production declined throughout the flowering period and pollen size declined (although only slightly) after the first collection date. The percent of aborted pollen varied significantly with treatment (Table 1). For collection date 2, flowers produced on the 'control' and 'nutrient-limited' plants had a lower percent aborted pollen than the 'water-limited' plants; on collection date 3, control plants had less aborted pollen than either of the other two treatments (Figure 2).

Seed production following single-donor pollinations and mean seed weight did not vary with treatment of the male plant (Table 2), though both traits varied among males and between the female recipients. Even within each recipient female, there were no treatment effects on either trait.

It appears that genotype plays a larger role than environmental variation in determining pollen traits in Silene alba. Variation in pollen production and pollen size were correlated with male genotype and not associated with the environmental conditions under which those males grew. But, experiments with wild radish also showed no environmental effects on these same two pollen traits, yet there were strong environmental effects on the ability of pollen to sire seeds under conditions of pollen competition (Young and Stanton, 1990). It will be important to examine other pollen traits in Raphanus, such as germination rates and pollen tube growth rates, to understand the mechanism underlying the differences in pollen performance. The lack of environmental effects on the number of seeds sired from single-donor crosses in Silene permits no predictions on the results of mixed pollinations because of the different results from the two pollination types in wild radish. However, the fact that there are environmental effects on the proportion of pollen that is viable in Silene does lead to the expectation of differences in overall male performance.

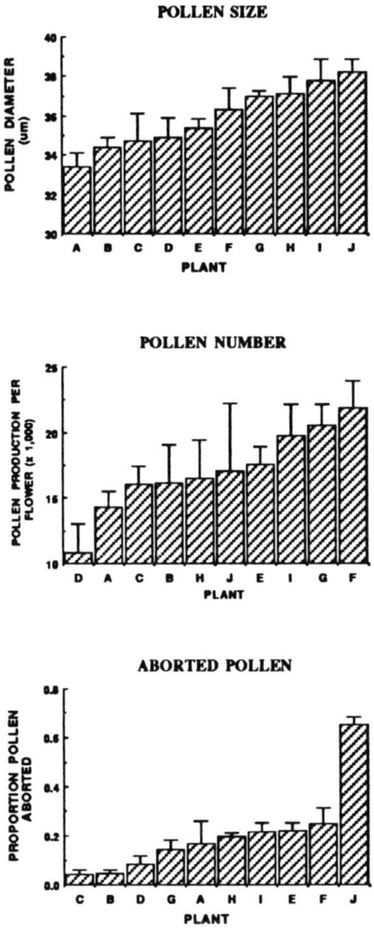

Figure 1. Variation in pollen traits among 10 males grown under 'control' conditions (see text for description of treatments). ANOVA results indicate that this variation is significant. Height of bar indicates mean values; 1 S.E.is plotted with vertical line.

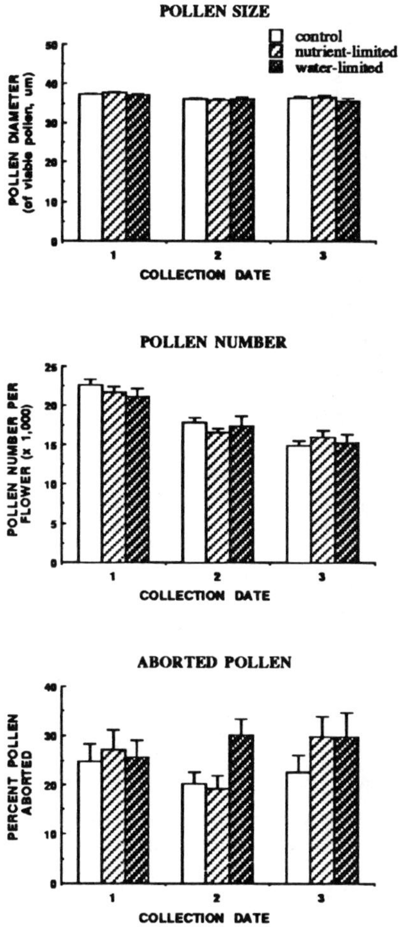

Figure 2. Effects of collection date and growth conditions of male plants on pollen traits. Height of bar indicates mean values; 1 S.E. is plotted with vertical line.

A non-genetic explanation for differential paternity and progeny performance among several pollen donors involves a series of developmental and ecological steps that are independent of the genotype of the gamete or zygote. Pollen traits that affect paternity are influenced by the environment. Pollen produced on plants in 'good' environments may be larger or possess more nutrients than pollen developed under poor conditions [protein content of pollen of *Petunia hybrida*

Table 1. Effect of environmental conditions during floral development on flower size and pollen traits in *Silene alba*.

Mean values (+S.D.) are presented. Mean values with different letters are significantly different at $p<0.05$ (Tukey's test). The p-values presented are those associated with F-values from ANOVA testing for effects of plant, collection date, and treatment. Flower size was measured on one date only.

	TREATMENT			
	CONTROL	WATER-LIMITED	NUTRIENT-LIMITED	ANOVA results
Mean corolla diameter (mm)	19.29^{b} (2.59)	17.22^{c} (2.56)	20.54^{a} (2.37)	plant $p<0.001$ treatment $p<0.001$
Mean size of viable pollen (um)	36.6^{a} (1.84)	36.3^{a} (2.59)	36.7^{a} (1.89)	plant $p<0.001$ date $p<0.001$
treatment $p=0.3$				
Mean pollen production (per flower)	18,607^{a} (5052.9)	18,259^{a} (6,189.3)	18,203^{a} (4,422.6)	plant $p<0.001$ date $p<0.001$
treatment $p=0.7$				
Mean proportion of pollen that is aborted	0.22^{b} (0.190)	0.28^{a} (0.185)	0.25^{b} (0.195)	plant $p<0.001$ date $p<0.001$ treatment $p=0.05$

Table 2. Mean seed production per flower (+ S.D.) and mean seed weight (total weight of all seeds/number of seeds) resulting from single-donor pollinations using male plants of Silene alba grown under three environmental conditions.

ANOVAs were performed to test for effects of female (recipient) genotype, replicate females within genotype, pollen donor genotype, and treatment (growth environment of pollen donor).

	TREATMENT			
	CONTROL	WATER-LIMITED	NUTRIENT-LIMITED	ANOVA results
Seed number	138.4 (103.31)	152.1 (86.71)	142.0 (98.34)	recip. genotype $p<0.001$ female w/i recip. $p=0.005$
donor genotype $p<0.001$ treatment $p=0.59$				
Seed weight (grams)	.000584 (.000138)	.000579 (.000151)	.000575 (.000164)	recip. genotype $p<0.001$ female w/i recip. $p=0.005$ donor genotype $p<0.001$ treatment $p=0.59$

plants was influenced by the temperature at which the plants were grown (Herpen,

1981)]. The 'good' pollen grows more rapidly down the style, fertilizing ovules before the 'poor' pollen does. Ovules fertilized first become large resource sinks (Stanton, 1984a, Mazer et al., 1986), accumulating biomass and/or nutrients more rapidly than ovules fertilized later by 'poor' pollen. The resulting seeds sired by 'good' pollen are either larger or possess more nutrients and thereby produce larger, more competitive seedlings solely because they have more resources available to them (Stanton, 1984b). The presence of environmental effects on pollen characters acts to reduce the amount of variation available for selection to act on. If pollen characters that influence paternity and/or offspring characteristics are affected by the environmental conditions of the paternal plant, then genetic models that assume no paternal environmental effects ignore an important source of variation. In quantitative genetic studies, differences between paternal half-sibs are considered to result solely from additive genetic variation. Attempts are made to control maternal effects by carefully controlling the conditions of plant development (Roach 1984) and using plants of similar size as maternal parents; the paternal environment is not controlled, with the assumption that pollen quality is genetically determined and not influenced by the environment. Given the evidence of environmental effects on the ability of radish pollen to sire seeds, ignoring this source of variation can lead to an oversimplification of the mechanics governing pollen traits and paternity.

REFERENCES

Bell, C.R. Mineral nutrition and flower to flower pollen size variation. Am. J. Bot.46:621-624;1959.

Herpen, M.M.A. van. Effect of season, age, and temperature on the protein pattern of pollen and styles in Petunia hybrida. Acta. Bot. Neerl. 30:277-287; 1981.

------. Biochemical alterations in the sexual partners resulting from environmental conditions before pollination regulate processes after pollination. In D.L. Mulcahy, G.B.Mulcahy, and E. Ottaviano (Eds.) Biotechnology and Ecology of Pollen. Springer-Verlag, NY.; 1986: pp. 131-133.

----- and H.F. Linskens. Effect of season, plant age, and temperature during plant growth on compatible and incompatible pollen tube growth in Petunia hybrida. Acta Bot. Neerl. 30:209-218; 1981.

Kuo, C.G., J.S. Peng, and J.S. Tsay. Effect of high temperature on pollen grain germination, pollen tube growth, and seed yield of Chinese cabbage. HortScience 16:67-68; 1981.

Lee, T.D. Patterns of fruit and seed production. In J. Lovett Doust and L. Lovett Doust (Eds.) Plant Reproductive Ecology: Patterns and Strategies. Oxford University Press, New York.; 1988: pp. 179-202.

Mazer, S.J., A.A. Snow, and M.L. Stanton. Fertilization dynamics and parental effects upon fruit development in Raphanus raphanistrum: consequences for seed size variation. Am. J. Bot 73:500-511; 1986.

McKone, M.J. Intraspecific variation in pollen yield in bromegrass (Poaceae: Bromus). Am. J. Bot. 76:231-237; 1989.

Miller, M.K. and M.H. Schonhorst. Pollen growth of alfalfa in vitro as influenced by grouping of grains on the medium and greenhouse versus field sources. Crop Science 8:525-526; 1968.

Roach, D.A. Ecological genetics of life-history characteristics in Geranium carolinianum. PhD thesis, Duke University, Durham, NC; 1984.

----- and R.D. Wulff. Maternal effects on plants. Ann. Rev. Ecol. Syst. 18:209-235; 1987.

Schlichting, C.D. Environmental stress reduces pollen quality in Phlox: compounding the fitness deficit. In D.L. Mulcahy, G.B. Mulcahy, and E. Ottaviano (Eds.) Biotechnology and Ecology of Pollen. Springer-Verlag, NY.; 1986: pp. 483-488.
Schoper, J.B., R.J. Lambert, and B.L. Vasilas. Maize pollen viability and ear receptivity under water and high temperature stress. Crop Science 36:1029-1033; 1986.
Sharma, S., P.C. Jindal. and K.S. Chauhan. A note on pollen sterility in grapes under salt stress. In T.M. Varghese (Ed.) Recent Advances in Pollen Research. Allied Publishers Private Ltd.; New Delhi. ; 1985.: pp. 201-202.
Stanton, M.L. Developmental and genetic sources of seed weight variation in Raphanus raphanistrum L. (Brassicaceae). Am. J. Bot. 71:1090-1098; 1984a.
-----. Seed variation in wild radish: effect of seed size on components of seedling and adult fitness. Ecology 65:1105-1112; 1984b.
Stephenson, A.G. Flower and fruit abortion: proximate causes and ultimate functions. Ann. Rev. Ecol. Syst. 12:253-279; 1981.
Vasek, F.C., V. Weng, R.J. Beaver, and C.K. Huszar. Effects of mineral nutrition on component of reproduction in Clarkia unguiculata. El Aliso 11:599-618; 1987.
Young, H.J. and M.L. Stanton. Environmental stress influences pollen quality in wild radish. Science 248:1631-1633; 1990.
Zamir, D., S.D. Tanksley, R.A. Jones. Low temperature effect on selective fertilization by pollen mixtures of wild and cultivated tomato species. Theor. and Applied Genetics 59:235-238; 1981.

74

Pollen and Ovules in Evolutionary Studies

I. Till-Bottraud, M. Morgante, R.A. Curiel, I. Dajoz, R. Giannini, M.L. Gonzales, P.H. Gouyon, A.M. Olivieri, V.D.E. Solorzano, V.E. Solorzano, and G.G. Vendramini

The posters illustrate three different aspects of the studies of pollen and ovules in Evolutionary biology. One is the study of the evolution of pollen morphology per se (see article in this volume) and the modelling of the selective forces that act on the maintenance of pollen aperture polymorphism. Another, in a genetic resources approach, points out to the variability of pollen size and fertility between landraces in a Vicia faba collection. The third one uses peculiarities of the reproductive biology of Gymnosperms to estimate outcrossing rates and selection in early life-cycle phases.

Evolution of pollen morphology : game theory models

Following the results obtained by Dajoz et al. (this volume) on Viola diversifolia (i.e. that all stamens in all plants produce pollen with 3 and 4 -sometimes 5 and 6- apertures and that the 4-aperturate pollen germinates faster but grows shorter tubes and survives less than 3-aperturate pollen), game theory models are proposed to test wether these selective forces can maintain variability in aperture number within plants.
Additional hypotheses of the models are that (i) the differences between plants have a genetic basis (ii) pollen from two plants compete for the ovules of a third, (iii) pollen is not a limiting factor for fecundation and (iv) pollen from each of the two plants can be collected on the day of anthesis (all pollen are alive) or the day after (some of the many aperturate pollen is dead).

In the first model, all the pollen grains from a plant have the same aperture number. The analytical solution shows that no stable polymorphism can be maintained in a population: at equilibrium all the plants have the same type of pollen. This may explain the stability or the increase (or decrease) of aperture number in a taxon. The second model, in which all plants have pollen grains with different aperture numbers, shows that, for certain values of germination advantage and survival of the 4-aperturate pollen and of the time of pollen collection (figure 1), there can be, at equilibrium, a stable frequency of the different types within each plant and that all the plants should be similar for this frequency. This explains the existence of intra-individual "polymorphism" for aperture number. This polymorphism has indeed been frequently reported in Dicotyledons and is not surprising because the morphology of the pollen grain has a sporophytic determinism.

The models thus confirm the idea that differences in aperture number lead to a difference in fitness of pollen grains and that the "best" strategy can in some cases be a "bet hedging" one.

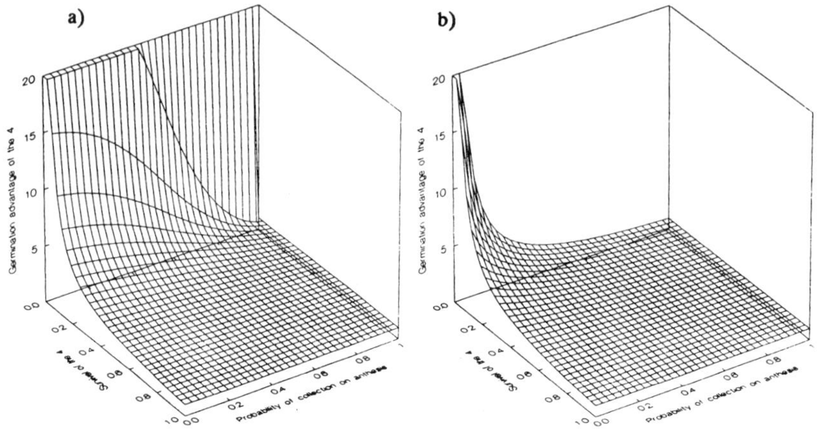

Figure 1 : Upper (a) and lower (b) boundaries of the parameter space between which a mixed strategy can be maintained. Under the lower boundary, all the pollen is 3-aperturate; over the upper one, all the pollen is 4-aperturate.

Variability of fertility and size of broad bean (Vicia faba L.) pollen grains

Eighty-four landraces were examined for pollen dimensions (length, width, length/width ratio) by using a magnifying camera or the photographic method and fertility as measured by Alexander's coloration (5 flowers on 3 plants per landrace).

Comparison of the two measurements techniques showed that the photographic method was easier to use, but proved more expensive. Most landraces showed a high pollen fertility but some had only 58.7% of fertile pollen. The size variability is correlated to the botanical variety to which the landrace belongs (table 1).

Pollen size can therefore be used as a marker of diversity in variability assessment of genetic resources in Vicia faba. The high pollen fertility encourages the use of landraces in breeding programs introducing traditional and adapted material into high yielding but unadapted varieties.

Table 1 : Mean size of pollen grains from the 84 landraces as classified according to botanical varieties. L: length; W: width; L/W: length/width ratio. Means followed by different letters are significantly different ($p<0.05$).

Variety	L	W	L/W
Major	60.77 a	37.57 ab	1.62 a
Equina	59.48 ab	38.17 a	1.55 b
Minor	58.80 b	37.26 b	1.59ab

Different effects of selection on paternal and maternal gene pools during germination in Pinus leucodermis Ant. seeds

In Gymnosperm seeds the endosperm, defined also as the megagametophyte, derives from the germination of the macrospore and is therefore genetically identical to the ovule. Simultaneous assay of embryo and endosperm genotypes allows hence the ascertainment of ovule and pollen contribution to the diploid individual. Moreover, in comparing embryos from dormant and germinated seeds, selection in early life-cycle phases can be detected. The assays were made on progenies of approximately 50 offspring from 20 trees.

Table 2: Allele frequencies at the four gene loci analyzed for ovule-derived alleles (Ovule, maternal gene pool) and pollen-derived alleles (Pollen, paternal gene pool) in dormant and germinated embryos.

Locus	Allele	Dormant embryos		Germinated embryos	
		Ovule	Pollen	Ovule	Pollen
GOT-A	1	0.155	0.077	0.167	0.131*
	2	0.845	0.923	0.833	0.869
GOT-C	1	0.047	0.020	0.045	0.024
	2	0.953	0.980	0.955	0.976
PGM-A	1	0.179	0.139	0.197	0.197*
	2	0.821	0.861	0.803	0.803
SKDH-A	1	0.733	0.870	0.759	0.820**
	2	0.267	0.130	0.241	0.180

** Chi-square test for homogeneity of pollen-derived allele frequencies in dormant and germinated embryos significant at the 1% level

The male gene pool can easily give the outcrossing rate. In this respect, Pinus leucodermis is characterized by significant levels of self-fertilization (16 and 24% respectively in two populations). In one population, dormant and germinated seeds were tested and no difference in fixation index was observed between the two phases, indicating that no selection occurs in the early stages.

On another hand, differences were observed in allele frequencies between the two phases at 3 of the 4 isozyme gene loci assayed, thus indicating that directional selection may act during germination. When the maternal and paternal gene pools were considered separately, differences were highly significant in the paternal gene pool but not in the maternal one (table 2). However, computer simulations (table 3) show that this phenomenon could be a consequence of the sampling procedure adopted, which was the same as the one used for outcrossing rate estimation: open-pollinated progeny arrays from single mother-tree were kept at a constant number in both phases, thus erasing the possible differential germination effects on gene frequencies.

Analysis of dormant and germinated seeds of Pinus leucodermis hence revealed that delayed sampling of offspring did not influence the outcrossing rate estimation and that there was no selection against offspring derived from selfing. Directional selection seems, on the contrary, to have acted on at least two of the four analyzed loci, but no definitive conclusion can be drawn because the sampling procedure was inappropriate.

Table 3 : Numeric examples of variation in allele frequencies observed under the two sampling procedures:
a) equal number of dormant and germinated embryos for each mother tree and b) equal number of dormant seeds for each mother tree pooled into a single lot and tested after germination (and therefore after selection). A biallelic locus with alleles A and a undergoing directional selection is considered. p = Frequency of A at the dormant seed stage. s = selection coefficient. Hardy-Weinberg proportion of adult mother trees and complete outcrossing are assumed. Pollen = paternally derived alleles; Ovule = maternally derived alleles.

	Sampling scheme a)			Sampling scheme b)		
	Embryos	Ovule	Pollen	Embryos	Ovule	Pollen
p=0.9, s(aa)=0.1 s(Aa)=0.05	0.903	0.902	0.905	0.905	0.905	0.905
p=0.5, s(aa)=0.1 s(Aa)=0.05	0.510	0.507	0.513	0.513	0.513	0.513
p=0.1, s(aa)=0.1 s(Aa)=0.05	0.104	0.102	0.105	0.105	0.105	0.105
p=0.9, s(aa)=0.2 s(Aa)=0.1	0.907	0.905	0.909	0.909	0.909	0.909
p=0.5, s(aa)=0.2 s(Aa)=0.1	0.521	0.514	0.528	0.528	0.528	0.528
p=0.1, s(aa)=0.2 s(Aa)=0.1	0.108	0.105	0.111	0.111	0.111	0.111

Conclusion

Pollen and ovules are used in two ways in evolutionary studies: they can be treated as parts of the plant (another organ of the sporophyte) and analyzed for their variability, as is the case in the Vicia faba study, or they can be studied as individual organisms, the gametophytes, independent of the sporophyte for the pollen, and on which selection can act. This is illustrated by the difference in fitness of pollen grains with different number of apertures. In Gymnosperms, the fact that the endosperm has exactly the same genotype as the oosphere allows the distinction between the male and female contribution to the zygote and permits very valuable gene flow and early selection estimates.

Index